OXFORD
UNIVERSITY PRESS

Cambridge International AS & A Level

Complete
Chemistry

Third Edition

W0051197

Ted Lister

Janet Renshaw

Exam-style questions:
Dr Nicholas Taylor

Advisory team:
Samuel Mao Hua Lee
Muhammad Talha
Ellen Wong

OXFORD
UNIVERSITY PRESS

Great Clarendon Street, Oxford, OX2 6DP, United Kingdom

Oxford University Press is a department of the University of Oxford. It furthers the University's objective of excellence in research, scholarship, and education by publishing worldwide. Oxford is a registered trade mark of Oxford University Press in the UK and in certain other countries

© Oxford University Press 2020

Index © James Helling 2020

The moral rights of the authors have been asserted

First published in 2020

British Library Cataloguing in Publication Data
Data available

978-1-38-200531-9

10 9 8 7 6 5

Paper used in the production of this book is a natural, recyclable product made from wood grown in sustainable forests. The manufacturing process conforms to the environmental regulations of the country of origin.

Printed and bound by CPI Group (UK) Ltd, Croydon, CR0 4YY

Acknowledgements

The authors would like to thank the editorial team, Ben Rout and Sharon Jordan, for their vital roles in producing the book.

The publisher and authors would like to thank the following for permission to use photographs and other copyright material:

Cover: Karl Gaff/Science Photo Library

Artworks: Aptara

Photos: p2-p3: Maxx-Studio/Shutterstock; p5: D-VISIONS/Shutterstock; p10: Science Museum (Circulation Department); p20: Fineart1/Shutterstock; p22: Science Photo Library; p24: Parmna/Shutterstock; p25: Martyn F. Chillmaid/Science Photo Library; p36: Andrew Lambert Photography/Science Photo Library; p44: David Iliff/Shutterstock; p49: Science Photo Library; p69: Petr Bonek/Shutterstock; p72: Martyn F. Chillmaid/Science Photo Library; p79: Martyn F. Chillmaid/Science Photo Library; p89: Martyn F. Chillmaid/Science Photo Library; p91: Todor Dobrev/Shutterstock; p97: Science Photo Library; p100: Science Photo Library; p104: Yasni/Shutterstock; p109: Martien van Gaalen/Shutterstock; p110 (T): Sciencephotos/Alamy Stock Photo; p110 (B): Auscape/Universal Images Group/Getty Images; p111 (R): Science Photo Library; p111 (L): Andrew Lambert Photography/Science Photo Library; p134 (T): Science Photo Library; p134 (C): Andrew Lambert Photography/Science Photo Library; p134 (B): Charles D. Winters/Science Photo Library; p135: Charles D. Winters/Science Photo Library; p145: Hong xia/Shutterstock; p146: Sciencephotos/Alamy Stock Photo; p153: Andrew Lambert Photography/Science Photo Library; p154: oneSHUTTER oneMEMORY/Shutterstock; p157: Zoonar GmbH/Alamy Stock Photo; p158: Richard Ellis/Alamy Stock Photo; p177: maradon 333/Shutterstock; p184 (T): Convery flowers/Alamy Stock Photo; p184 (C): Martyn F. Chillmaid/Science Photo Library; p187: Terry Poche/Shutterstock; p188: Reproduced courtesy of the International Association of Oil and Gas Producers (IOGP); p191: Science Photo Library; p203: Martyn F. Chillmaid/Science Photo Library; p209: Jerry Mason/Science Photo Library; p214: Martyn F. Chillmaid/Alamy Stock Photo; p223 (CL): Martyn F. Chillmaid/Science Photo Library; p223 (CL): Martyn F. Chillmaid/Science Photo Library; p223 (BR): Science Photo Library; p227: Andrew Lambert Photography/Science Photo Library; p231 (BR): Olga Popova/Shutterstock; p231 (TL): pzAxe/Shutterstock; p231 (TR): D. Pimborough/Shutterstock; p231 (BL): Ioana Davies (Drutu)/Shutterstock; p236 (TL): Marques/Shutterstock; p236 (TR): Ficmajstr/Shutterstock; p236 (CR): Olga Popova/Shutterstock; p236 (BL): You Touch Pix of EuToch/Shutterstock; p236 (C): Pongsak Tawansaeng/123RF; p236 (CL): Eric Wong/Shutterstock; p236 (BR): Alexey Kabanov/Shutterstock; p239 (TL): Larina Marina/Shutterstock; p239 (CR): Bloomberg/Getty Images; p240: Martyn F. Chillmaid/Science Photo Library; p246: Peter Menzel/Science Photo Library; p251: Mikael Karlsson/Alamy Stock Photo; p256 (T): Marc Pagani Photography/Shutterstock; p256 (B): Colin Cuthbert/Newcastle University/Science Photo Library; p260-p261: Ttstudio/Shutterstock; p267 (T): © 2010-2019 The Regents of the University of California, Lawrence Berkeley National Laboratory; p267 (B): © 2010-2019 The Regents of the University of California, Lawrence Berkeley National Laboratory; p272: Andrew Fletcher/Shutterstock; p279: Georgios Kollidas/Shutterstock; p280: © Corel Corporation 1994; p283: Andrew Lambert Photography/Science Photo Library; p285: Torsten Lorenz/Shutterstock; p299: Ivaschenko Roman/Shutterstock; p301: Belmonte/BSIP/Science Photo Library; p302: monticello/Shutterstock; p309: sciencephotos/Alamy Stock Photo; p333: Martyn F. Chillmaid/Science Photo Library; p338: Andrew Lambert Photography/Science Photo Library; p338: Andrew Lambert Photography/Science Photo Library; p341: avarand/Shutterstock; p344 (R): Andrew Lambert Photography/Science Photo Library; p344 (L): Martyn F. Chillmaid/Science Photo Library; p345 (TR): Martyn F. Chillmaid/Science Photo Library; p345 (CL): SunChan/iStockphoto; p351: Mediscan/Alamy Stock Photo; p355 (T): Molekuul/Science Photo Library; p355 (B): Science Photo Library; p362: Martyn F. Chillmaid/Science Photo Library; p369: PJF Military Collection/Alamy Stock Photo; p379: S-F/Shutterstock; p386: mikeledray/Shutterstock; p390: Evemilla/iStockphoto; p394: Studio Light and Shade/Shutterstock; p396: Science Museum/Science & Society Picture Library; p404 (T): imtmphoto/Shutterstock; p404 (B): Laszlo Mates/Alamy Stock Photo; p406 (T): Charles D. Winters/Science Photo Library; p406 (B): Michael Stokes/Shutterstock; p454: FLPA/Alamy Stock Photo; p411: Avalon/Photoshot License/Alamy Stock Photo; p415: ER Degginger/Science Photo

This Student Book refers to the Cambridge International AS & A Level Chemistry (9701) Syllabus published by Cambridge Assessment International Education.

This work has been developed independently from and is not endorsed by or otherwise connected with Cambridge Assessment International Education.

Contents

AS Level

Physical chemistry

Inorganic chemistry

A Level

Physical chemistry

Inorganic chemistry

Organic chemistry

Analysis

Introduction

Complete Chemistry has been written specifically to meet the requirements of the Cambridge International AS & A Level Chemistry syllabus (9701). As ever, this new edition aims to make your study of chemistry successful and interesting. A comprehensive student book, it presents chemistry in its wider context and emphasises its relevance as a vital contribution to society, industry and civilisation, as well as supporting the development of problem-solving skills.

The book is divided into 37 chapters:

- Chapters 1–22 cover AS Level
- Chapters 23–37 cover A Level

Note that the contents list is your syllabus matching grid.

The text draws on experimental evidence to develop key ideas and establish laws and theories. New ideas are presented in the book in a careful step-by-step manner to allow you to develop a firm understanding of key concepts. The supporting resources in this edition enable learners to develop confidence in the application of their knowledge to explain concepts and solve problems, and build on their mathematical and practical skills.

Key concepts are essential principles, theories and ideas that help you to develop a deeper comprehension of chemistry and to make relevant links between different topics. They are, in effect, the foundations upon which the whole subject is based. An awareness of key concepts allows you to see chemistry as an interrelated, coherent, albeit complex, whole.

Once you have mastered the key concepts, you will be able to use them to describe and explain facts and processes in detail and with accuracy, to solve problems, link ideas, and tackle related material that is completely new to you. The key concepts in chemistry cover the following ideas:

- **Atoms and forces** determine the physical and chemical properties of matter, including bonding and reactivity.
- **Experiments and evidence** help chemists to build models and formulate theories that explain the structure and properties of materials.
- **Patterns in chemical behaviour and reactions** predict the properties of substances and suggest how substances undergoing chemical reactions can give rise to new materials and innovation in synthetic routes to improve yield and lessen environmental impact.
- An understanding of **chemical bonding** allows chemists to predict patterns of reactivity, designing substances with desired physical, chemical and biological properties.
- **Energy changes** during a chemical reaction can be used to predict the extent of that reaction.

The layout of the book is designed to cover information in a clear, accessible way. Here is a summary of its features:

- **Learning outcomes** make clear the key concepts and understanding covered within each chapter.
- **Key terms** that you will need to be able to define and understand are highlighted using **purple type**. Definitions of these key terms are in the glossary on page 444.
- **Extension features** support and direct further reading that will develop breadth, depth and application of knowledge and skills.
- **Exam tips** support you to evidence your knowledge and understanding, avoiding misconceptions and refining your exam skills to give clear and concise responses.
- **Step-by-step worked examples** help you apply theory into practice.
- **Summary tests** appear at frequent intervals throughout the book to encourage you to read more critically and maximise your understanding. They provide a quick check on how well you have learnt and understood the factual content of the section you just finished studying.
- **End-of-chapter exam-style questions** test the full range of skills expected at AS and A Level, including application of knowledge, understanding, analysis, synthesis and evaluation. The questions cover the material in the preceding chapters and give you practice of the style of questions you can expect in a formal examination.

In the Exam-style Questions sections at the end of the chapters, you will find the following icon:

In the Enhanced Online Student Book, this icon will launch additional digital resources to support your learning further. This content includes:

- **Worksheets** containing additional questions and practice
- **Interactive online quizzes**, including **multiple-choice** question practice
- Further support to improve your **mathematical skills**
- Examples of **practicals** that may be carried out in your chemistry classes

Visit **www.oxfordsecondary.com/bookshelf** to redeem your token code and access the Enhanced Online Student Book.

Answers to questions in this book and the syllabus matching grid are also available on the support website, which can be accessed via the URL or QR code below:
www.oxfordsecondary.com/caie-al-complete-science

Terbium
5.8638

97

Bk

(247)

162.50
Dysposium
5.9389

98

Cf

D

Holm
6.021

99

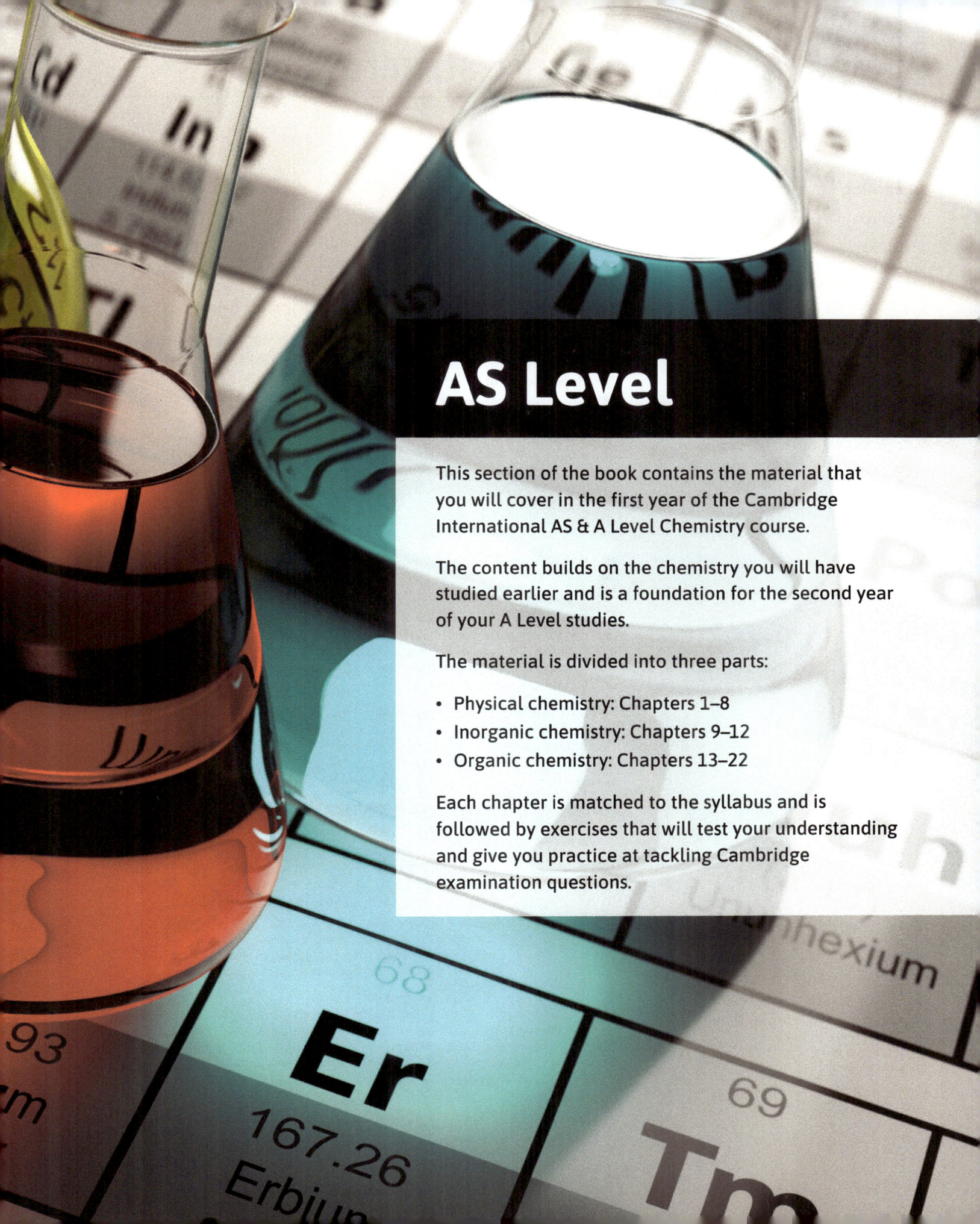

AS Level

This section of the book contains the material that you will cover in the first year of the Cambridge International AS & A Level Chemistry course.

The content builds on the chemistry you will have studied earlier and is a foundation for the second year of your A Level studies.

The material is divided into three parts:

- Physical chemistry: Chapters 1–8
- Inorganic chemistry: Chapters 9–12
- Organic chemistry: Chapters 13–22

Each chapter is matched to the syllabus and is followed by exercises that will test your understanding and give you practice at tackling Cambridge examination questions.

1 Atomic structure

1.1 Particles in the atom and atomic radius

Learning outcomes

On these pages you will learn to:

- describe the structure of the atom
- identify and describe the properties of subatomic particles
- explain the trends in atomic and ionic radius

Chemistry is all about atoms, and atoms are mostly empty space. Virtually all the mass of an atom is concentrated in its minute **nucleus**, which is surrounded by electrons a great distance away. The diameter of an atomic nucleus is of the order of 10^{-15} m, while the size of the atom is around 10^{-10} m. So the nucleus is $\frac{1}{100\,000}$ times smaller than the atom, similar to a football inside a football stadium.

Extension

Developing ideas of the atom

The Greek philosophers had a model in which matter was made up of a single continuous substance that produced the four elements—earth, fire, water, and air. The idea that matter was made of individual atoms was not taken seriously for another 2000 years. During this time alchemists built up a lot of evidence about how substances behave and combine. Their aim was to change other metals into gold. Here are a few of the steps that led to our present model.

In 1661, Robert Boyle proposed that there were some substances that could not be made simpler. These were the chemical elements, as we now know them.

In 1803, John Dalton suggested that elements were composed of indivisible atoms. All the atoms of a particular element had the same mass and atoms of different elements had different masses. Atoms could not be broken down.

In 1896, Henri Becquerel discovered radioactivity. This showed that particles could come from inside the atom. So the atom was not indivisible. The following year, J J Thomson discovered the electron. This was the first subatomic particle to be discovered. He showed that electrons were negatively charged, and electrons from all elements were the same.

As electrons had a negative charge, there had to be some source of positive charge inside the atom too. Also, as electrons were much lighter than whole atoms, there had to be something to account for the rest of the mass of the atom. Thomson suggested that the electrons were located within the atom in circular arrays, like plums in a pudding of positive charge (see Figure 1).

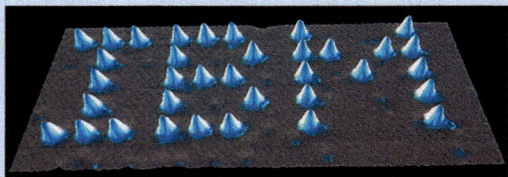

Figure 2 *Atoms can only be seen indirectly. This photograph of xenon atoms was taken by a scanning tunnelling electron microscope. Individual atoms are not visible to the eye even with the best optical microscopes*

In 1911, Ernest Rutherford and his team found that most of the mass and all the positive charge of the atom was in a tiny central nucleus. So, for many years, it has been known that atoms themselves are made up of smaller particles, called subatomic particles. The complete picture is still being built up in 'atom smashers' such as the one at CERN, near Geneva in Switzerland.

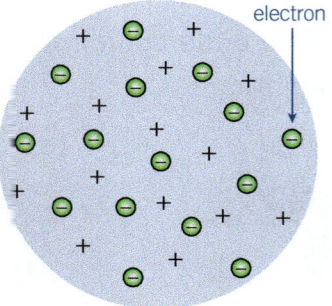

Figure 1 *The plum pudding model of the atom—electrons located in circular arrays within a sphere of positive charge*

The subatomic particles

Atoms are made of three fundamental particles—**protons**, **neutrons**, and **electrons**.

The protons and neutrons form the nucleus, in the centre of the atom.

- Protons and neutrons are sometimes called nucleons because they are found in the nucleus.
- The electrons surround the nucleus in shells of increasing radius.

The properties of the subatomic particles are shown in Table 1.

Table 1 *The properties of the subatomic particles*

Property	Proton p	Neutron n	Electron e
Mass / kg	1.673×10^{-27}	1.675×10^{-27}	0.911×10^{-30} (very nearly 0)
Charge / C	$+1.60 \times 10^{-19}$	0	-1.60×10^{-19}
Position	in the nucleus	in the nucleus	around the nucleus

These numbers are extremely small. In practice, *relative* values for mass and charge are used. The relative charge on a proton is taken to be +1, so the charge on an electron is −1. Neutrons have no charge (see Table 2).

Table 2 *The relative masses and charges of the subatomic particles*

	Proton p	Neutron n	Electron e
Relative mass	1	1	$\dfrac{1}{1840}$
Relative charge	+1	0	−1

In a neutral atom, the number of electrons must be the same as the number of protons because their charge is equal in size and opposite in sign.

The charges and masses mean that if a beam of fast-moving protons, neutrons and electrons travelling at the same velocity enters an electric field, the particles would be deflected as shown in Figure 4.

The arrangement of the subatomic particles

The subatomic particles (protons, neutrons, and electrons) are arranged in the atom as shown in Figure 5.

The protons and neutrons are in the centre of the atom (the nucleus), held together by a force called the **strong nuclear force**. This is much stronger than the **electrostatic force** of attraction that holds electrons and protons together in the atom, so it overcomes the repulsion between the protons in the nucleus. It acts only over the very short distances within the nucleus.

The nucleus is surrounded by electrons. Electrons are found in a series of energy levels, also referred to as orbits or shells, which get further and further away from the nucleus. The first shell can hold up to two electrons, the second, eight, and the third, 18. This is a simplified picture that will be developed in Section 1.3.

Figure 3 *A section of the Large Hadron Collider (LHC) at CERN is pictured here. Experiments here confirmed the existence of the Higgs Boson, a subatomic particle, in 2012*

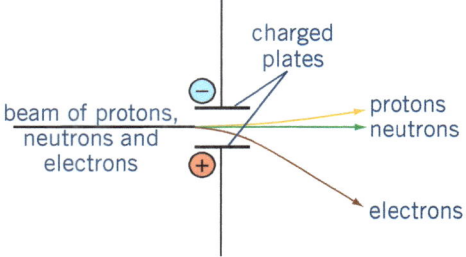

Figure 4 *A narrow beam containing protons, neutrons and electrons entering an electric field. Electrons are deflected more than the protons because they have less mass*

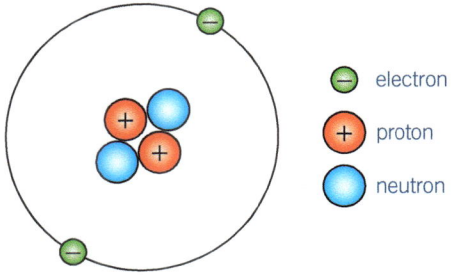

Figure 5 *The subatomic particles in a helium atom (not to scale)*

The nucleus, atomic number and mass number

Atoms consist of a tiny nucleus made up of protons and neutrons that is surrounded by electrons.

Proton number Z

The number of protons in the nucleus is called the **atomic number** or the **proton number, Z**.

The number of electrons in the atom is equal to the proton number, so atoms are electrically neutral. The number of electrons in the outer shell of an atom determines the chemical properties of an element (how it reacts) and what sort of element it is. The atomic number defines the chemical identity of an element.

atomic number (proton number) Z = number of protons

All atoms of the same element have the same atomic number. Atoms of different elements have different atomic numbers. It is the proton number that defines a particular element.

Mass number A

The total number of protons plus neutrons in the nucleus (the total number of nucleons) is called the **mass number, A**. This is also called the **nucleon number**, as both protons and neutrons are nucleons. It is the nucleons that are responsible for almost all of the mass of an atom because electrons weigh virtually nothing.

mass number A = number of protons Z + number of neutrons N

In a neutral atom the number of protons is equal to the number of electrons. In a positive ion, one or more electrons have been lost, so there are more protons than electrons. In a negative ion, one or more electrons have been gained, so there are more electrons than protons.

For example, in a sodium atom ($Z = 11$, $A = 23$) there are 11 protons, 11 electrons and 12 neutrons. In a Na^+ ion, there are 11 protons, 10 electrons and 12 neutrons.

In a fluorine atom ($Z = 9$, $A = 19$) there are 9 protons, 9 electrons and 10 neutrons. In an F^- ion, there are 9 protons, 10 electrons and 10 neutrons.

Atomic radii

Some key properties of atoms, such as size, are periodic. That means there are similar trends as you go across each period in the Periodic Table.

Atomic radii tell us about the sizes of atoms. As we shall see in Section 1.3, electrons can be thought of as clouds of negative charge rather than particles. You cannot measure the radius of an isolated atom because there is no clear point at which the electron cloud density around it drops to zero. Instead half the distance between the centres of a pair of atoms is used (see Figure 6).

The atomic radius of an element can differ, as it is a general term. It depends on the type of bond that it is forming—covalent, ionic, metallic and so on. The covalent radius is most commonly used as a measure of the size of the atom. Figure 7 shows a plot of covalent radius against atomic number.

The graph shows that:

- atomic radius is a periodic property because it *decreases across each period* and there is a jump when starting the next period
- atomic radius *increases* as you go *down the group*.

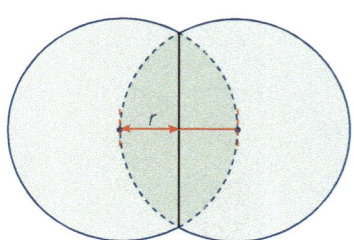

Figure 6 *Atomic radii are taken to be half the distance between the centres of a pair of atoms*

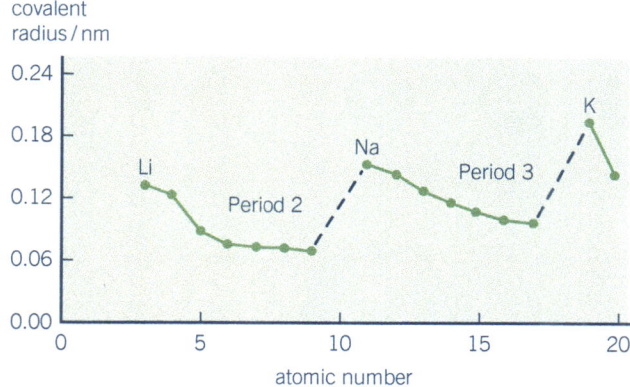

Figure 7 *The periodicity of covalent radii. The noble gases are not included because they do not form covalent bonds with one another*

Why the radii of atoms decrease across a period

You can explain this trend by looking at the electronic structures of the elements in a period, for example, sodium to chlorine in Period 3, as shown in Figure 8. The electrons in the outer shell are attracted by the positive charge on the nucleus, but this is partly cancelled out by the electrons in the inner shells. This is called **shielding**. So, for example, the electron in the outer shell of a sodium atom does not 'feel' the whole nuclear charge (11+) because of the shielding of the ten inner electrons.

Atom	Na	Mg	Al	Si	P	S	Cl
Size of atom							
	2,8,1	2,8,2	2,8,3	2,8,4	2,8,5	2,8,6	2,8,7
Atomic (covalent) radius / nm	0.156	0.136	0.125	0.117	0.110	0.104	0.099
Nuclear charge	11+	12+	13+	14+	15+	16+	17+

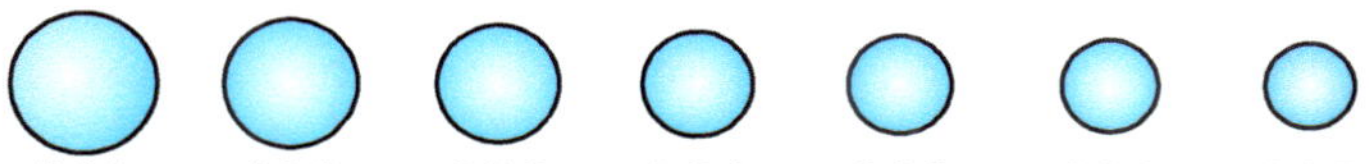

Figure 8 *The sizes and electronic structures of the elements sodium to chlorine*

As you move from sodium to chlorine you are adding protons to the nucleus and electrons to the outer main level, the third shell. The charge on the nucleus increases from +11 to +17. This increases the force of attraction between the nucleus and the outer shell electrons. There are no additional electron shells to shield the electrons from the nuclear charge. So the size of the atom *decreases* as you go across the period.

Why the radii of atoms increase down a group

Going down a group in the Periodic Table, the atoms of each element have one extra complete shell of electrons compared with the one before. So, for example, in Group 1 the outer electron in potassium is in shell 4, whereas in sodium it is in shell 3. So going down the group, the outer electron shell is further from the nucleus and the atomic radii increase.

Exam tip

It is a common mistake to think that atoms increase in size as you go across a period. While the nuclei have more protons (and neutrons), the radius of the atom depends on the distance of the outer electron shell from the nucleus.

Exam tip

Even metals can form covalent molecules, such as Na_2, in the gas phase. Since noble gases do not bond covalently with one another, they do not have covalent radii and so they are often left out of comparisons of atomic sizes.

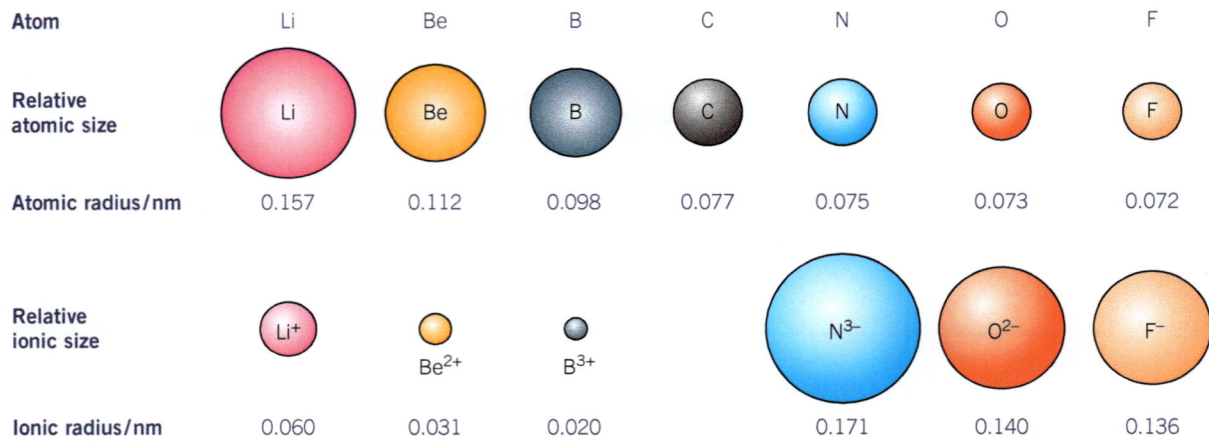

Atom	Li	Be	B	C	N	O	F
Relative atomic size	Li	Be	B	C	N	O	F
Atomic radius/nm	0.157	0.112	0.098	0.077	0.075	0.073	0.072
Relative ionic size	Li$^+$	Be^{2+}	B^{3+}		N^{3-}	O^{2-}	F$^-$
Ionic radius/nm	0.060	0.031	0.020		0.171	0.140	0.136

Figure 9 *Radii of the atoms and most stable ions of Li to F in Period 2*

Ionic radii

Many atoms form ions in order to gain full outer shells of electrons (see Section 3.2). They do this by loss or gain of electrons. Electron loss leads to positive ions and electron gain leads to negative ions. This has a large effect on the size of ions. Positive ions are smaller than their parent atoms. Negative ions are larger than their parent atoms (see Figure 9). As we go across a group, the positive ions get smaller as the increased nuclear charge draws the electrons in. Then there is a jump in size when negative ions start to form and the extra electrons in the outer shell cause extra repulsion.

Positive ions are smaller because they have lost the whole outer shell of electrons. Negative ions are larger because they have a full outer shell of electrons which repel each other.

Summary test 1.1

1 a Identify which of the following—protons, neutrons or electrons:
    i   are nucleons
    ii   have the same relative mass
    iii have opposite charges
    iv have no charge
    v   are found outside the nucleus
  b Explain why we assume that there are the same number of protons and electrons in an atom.

2 In the arrangement shown in Figure 4, explain why the protons are deflected less than the electrons.

3 State what happens to the size of atoms as you go from left to right across a period. Choose from increase, decrease, no change.

4 Suggest why a B^{3+} ion is smaller than a Li$^+$ ion.

5 Suggest why an F$^-$ ion is smaller than an N^{3-} ion.

6 State the number of protons, neutrons and electrons in the following:
  a $^{27}_{13}$Al
  b $^{27}_{13}$Al^{3+}
  c $^{19}_{9}$F
  d $^{19}_{9}$F$^-$

Isotopes

Every single atom of any particular element has the same number of protons in its nucleus and therefore the same number of electrons. But the number of neutrons may vary:

- Atoms with the same number of protons but different numbers of neutrons are called **isotopes**.
- Different isotopes of the same element react chemically in exactly the same way because they have the same electron configuration.
- Atoms of different isotopes of the same element vary in nucleon number because of the different number of neutrons in their nuclei.
- Different isotopes of the same element have different physical properties because they have different masses. This also means that they have different densities, because the sizes of the atoms are exactly the same.

All atoms of the element carbon, for example, have atomic number 6. That is what makes them carbon rather than any other element. However, carbon has three isotopes with mass numbers 12, 13, and 14 respectively (Table 3). All three isotopes will react in the same way, for example, burning in oxygen to form carbon dioxide.

Learning outcomes

On these pages you will learn to:

- distinguish between isotopes
- explain why isotopes of the same element have the same chemical properties
- explain why isotopes of the same element have different physical properties

Exam tip

Make sure you understand the terms atomic number (proton number) and mass number (nucleon number).

Table 3 *Isotopes of carbon*

Name of isotope	carbon-12	carbon-13	carbon-14
symbol	$^{12}_{6}C$	$^{13}_{6}C$	$^{14}_{6}C$
number of protons	6	6	6
number of neutrons	6	7	8
abundance	98.89%	1.11%	trace

Isotopes are often written like this: $^{13}_{6}C$. The superscript 13 is the mass number of the isotope and the subscript 6 is the atomic number.

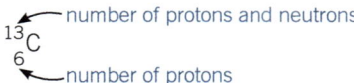

number of protons and neutrons

$^{13}_{6}C$

number of protons

Summary test 1.2

1 Isotopes are usually identified by the name of the element and the mass number of the isotope, as in carbon-13. However, isotopes of hydrogen have their own names. Hydrogen-2 is often called *deuterium* and hydrogen-3 is called *tritium*. Both of these isotopes behave chemically just like the most common isotope, hydrogen-1. State how many protons, neutrons, and electrons the atoms of the following have:
   a deuterium
   b tritium
2 Identify which of the following atoms (not their real symbols) are a pair of isotopes:
   $^{31}_{15}W$, $^{14}_{7}X$, $^{16}_{8}Y$, $^{15}_{7}Z$
3 For each element in question 2, state:
   a the number of protons
   b the mass number
   c the number of neutrons

Electrons, energy levels and atomic orbitals

In 1913, Niels Bohr put forward the idea that the atom consisted of a tiny positive nucleus orbited by negatively charged electrons, to form an atom like a tiny solar system. The electrons orbited in shells of fixed size. The movement of electrons from one shell to the next explained how atoms absorbed and gave out light. This was the beginning of what is called quantum theory.

In 1926, Erwin Schrödinger, a mathematical physicist, worked out an equation based on the idea that electrons had some of the properties of waves as well as those of particles. This led to a theory called quantum mechanics. It can be used to predict the behaviour of subatomic particles.

In 1932, James Chadwick discovered the neutron.

At the same time, chemists were developing their ideas about how electrons allowed atoms to bond together. One important contributor was the American, Gilbert Lewis. He put forward the ideas that:

- The inertness of the noble gases was related to their having full outer shells of electrons.
- Ions were formed by atoms losing or gaining electrons to obtain a noble gas electron arrangement.
- Atoms could also bond by sharing electrons to form full outer shells.

Lewis' theories are the basis of modern ideas of chemical bonding. They explain the formulae of many simple compounds, using the idea that atoms tend to gain the stable electronic structure of the nearest noble gas.

Evolving ideas of atomic structure

Early theories viewed the electron as a minute solid particle. Later theories suggest you can also think of electrons as smeared out clouds of charge, so you can never say exactly where an electron is at any moment. You can only state the probability that it can be found in a particular volume of space that has a particular shape. However, chemists still use different models of the atom for different purposes:

- Bohr's model can be used for a simple model of ionic and covalent bonding.
- The charge cloud idea is used for a more sophisticated explanation of bonding and the shapes of molecules.
- The simple model of electrons orbiting in shells is useful for many purposes, particularly for working out bonding between atoms.

You will be familiar with the electron diagrams in this section from your earlier studies. They lead on to the more sophisticated models of electron structure described later. However, they can still be useful, for example to predict and explain the formulae of simple compounds and the shapes of molecules.

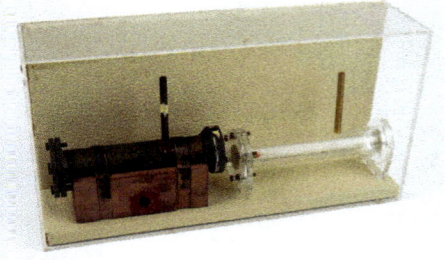

Figure 10 *Chadwick's apparatus for discovering the neutron. It looks remarkably crude compared with modern apparatus*

Electron shells

The first shell, which is closest to the nucleus, fills first, then the second, and so on. The number of electrons in each shell = $2n^2$, where n is the number of the shell, so:

- shell number 1 holds up to two electrons
- shell number 2 holds up to eight electrons
- shell number 3 holds up to 18 electrons.

The shell number is known as the **principal quantum number**.

Electron diagrams

If you know the number of protons in an atom, you also know the number of electrons it has. This is because the atom is neutral. You can therefore draw an electron diagram for any element. For example, carbon has six electrons. The four electrons in the outer shell are usually drawn spaced out around the atom (Figure 11).

Sulfur has 16 electrons. It has six electrons in its outer shell. It helps when drawing bonding diagrams to space out the first four (as in carbon), and then add the next two electrons to form pairs (Figure 12).

You can also draw electron diagrams of ions, as long as you know the number of electrons. For example, a sodium atom, Na, has 11 electrons, but its ion has 10, so it has a positive charge, Na+ (Figure 13).

An oxygen atom has eight electrons, but its ion has 10, so it has a negative charge, O^{2-} (Figure 14).

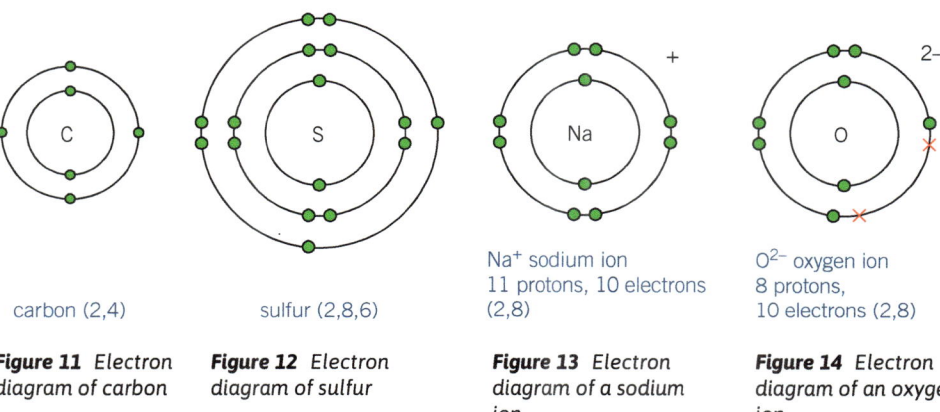

carbon (2,4)

sulfur (2,8,6)

Na+ sodium ion
11 protons, 10 electrons
(2,8)

O^{2-} oxygen ion
8 protons,
10 electrons (2,8)

Figure 11 *Electron diagram of carbon*

Figure 12 *Electron diagram of sulfur*

Figure 13 *Electron diagram of a sodium ion*

Figure 14 *Electron diagram of an oxygen ion*

You can write electron diagrams in shorthand:

- Write the number of electrons in each shell, starting with the inner shell and working outwards
- Separate each number by a comma.

For carbon, you write 2,4; for sulfur, 2,8,6; for Na+, 2,8.

Electron arrangements in atoms

In a simple model of the atom the electrons are thought of as being arranged in shells around the nucleus. The shells can hold increasing numbers of electrons as they get further from the nucleus—the pattern is 2, 8, 18, and so on.

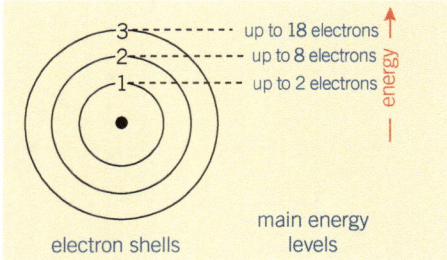

Figure 15 *Electron shells and energy levels*

Energy levels

Electrons in different shells have differing amounts of energy. They can therefore be represented on an energy level diagram. The shells are called main energy levels and they are labelled 1, 2, 3, and so on (Figure 15). As we have seen, each main energy level can hold up to a maximum number of electrons given by the formula $2n^2$, where n is the number of the shell. So, you can have two electrons in the first shell, eight in the next, 18 in the next, and so on.

Apart from the first level, which has only an s-sub-level, these main energy levels are divided into sub-levels, called s, p, d, and f, which have slightly different energies (Figure 16). Level 2 has an s-sub-level and a p-sub-level. Level 3 has an s-sub-level, a p-sub-level, and a d-sub-level.

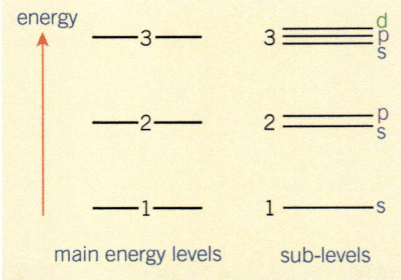

Figure 16 *Energy levels and sub-levels*

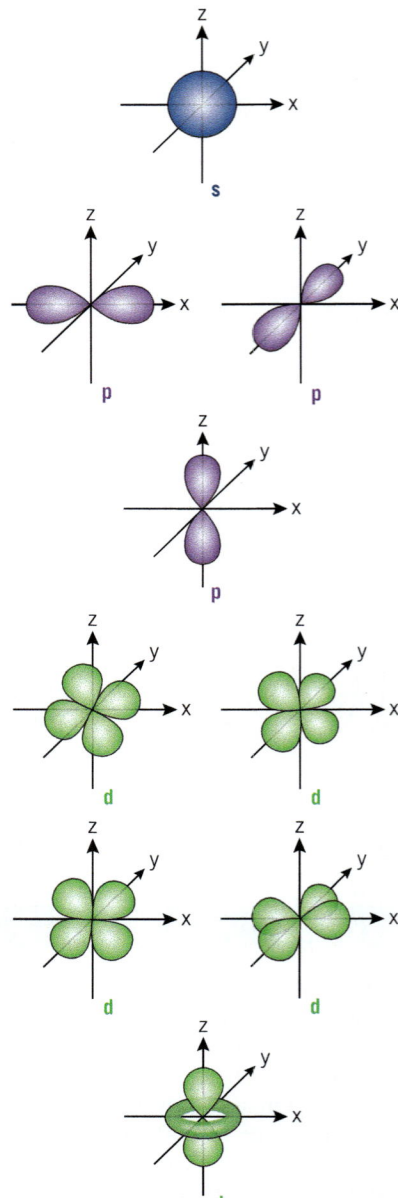

Figure 17 *The shapes of s-, p-, and d-orbitals*

Atomic orbitals

The electron is no longer considered to be a particle but a cloud of negative charge. An electron fills a volume in space called its **atomic orbital**. The concept of the shells and the sub-levels is then included in the following way.

- Different atomic orbitals have different energies. Each orbital has a number that tells us the main energy level that it corresponds to: 1, 2, 3, and so on.
- The atomic orbitals of each shell have different shapes, which in turn have slightly different energies. These are the sub-levels. They are described by the letters s, p, d, and f. The shapes of the s-, p-, and d-orbitals are shown in Figure 17. The shapes of f-orbitals are even more complicated.
- These shapes represent a volume of space in which there is a 95% probability of finding an electron and they influence the shapes of molecules.
- The first main energy level consists of a single s-orbital. The second shell has a single s-orbital and three p-orbitals of a slightly higher energy, the third shell has a single s-orbital, three p-orbitals of slightly higher energy, and five d-orbitals of slightly higher energy still, and so on (see Figure 17).
- Any single atomic orbital can hold a maximum of two electrons.
- s-orbitals can hold up to two electrons.
- p-orbitals can hold up to two electrons each, but always come in groups of three of the same energy, to give a total of up to six electrons in the p-sub-level.
- d-orbitals can hold up to two electrons each, but come in groups of five of the same energy to give a total of up to 10 electrons in the d-sub-level.

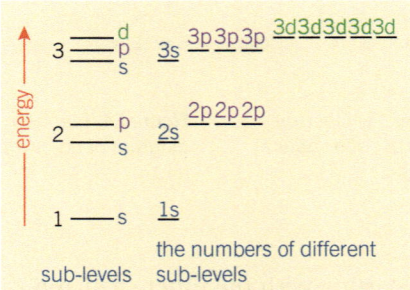

Figure 18 *The subdivisions of orbitals. Sets of orbitals with the same energy are described as 'degenerate'*

Table 4 summarises the number of electrons in the different levels and sub-levels.

Table 4 *The number of electrons in the different levels and sub-levels*

Main energy level (shell)	1	2		3			4			
Sub-level(s)	s	s	p	s	p	d	s	p	d	f
Number of orbitals in sub-level	1 (2e⁻)	1 (2e⁻)	3 (6e⁻)	1 (2e⁻)	3 (6e⁻)	5 (10e⁻)	1 (2e⁻)	3 (6e⁻)	5 (10e⁻)	7 (14e⁻)
Total number of electrons in main energy level	2	8		18			32			

Exam tip

You should know how many s-, p-, and d-orbitals there are in each main energy level. However, you only need to know the shapes of the s- and p-orbitals at this stage.

The energy level diagram in Figure 19 shows the energies of the orbitals for the first few elements of the Periodic Table. Notice that the first main energy level has only an s-orbital. The second shell has an s- and p-sub-level and the p-sub-level is composed of three p-orbitals of equal energy. The third main level has an s-, p-, and d-sub-level, and the d-sub-level is composed of five atomic orbitals of equal energy.

- Each 'box' in Figure 19 represents an orbital of the appropriate shape that can hold up to two electrons.
- Notice that 4s is actually of slightly lower energy than 3d for neutral atoms, although this can change when ions are formed.

Spin

Electrons also have a property called spin.

- Two electrons in the same orbital must have opposite spins.
- The electrons are usually represented by arrows pointing up or down to show the different directions of spin.

Putting electrons into atomic orbitals

Remember that the label of an atomic orbital tells us about the energy (and shape) of an electron cloud. For example, the atomic orbital 3s means the main energy level is 3 and the sub-level (and therefore the shape) is spherical.

There are three rules for allocating electrons to atomic orbitals.

1 Atomic orbitals of lower energy are filled first—so the lower shell is filled first and, within this level, sub-levels of lower energy are filled first.
2 Atomic orbitals of the same energy fill singly before pairing starts. This is because electrons repel each other.
3 No atomic orbital can hold more than two electrons.

The electron diagrams for the elements hydrogen to sodium are shown in Figure 20.

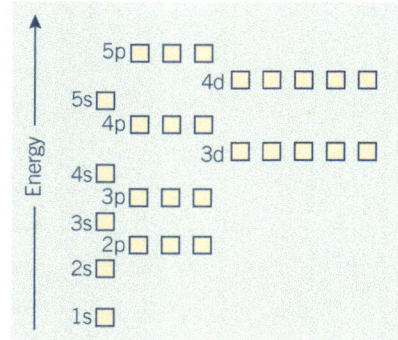

Figure 19 *The energy levels of the first few atomic orbitals. Although 4s has a lower energy than 3d it is further from the nucleus. 4s electrons tend to be lost before 3d electrons when ions are formed*

Exam tip

Although we use the term 'spin', the electrons are not actually spinning.

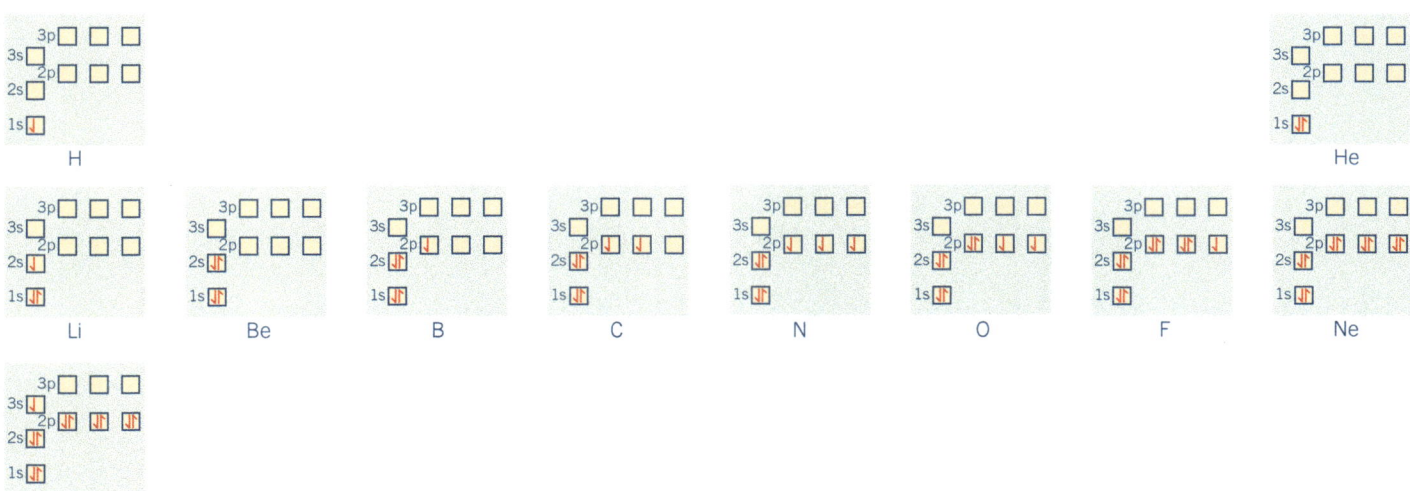

Figure 20 *The electron arrangements for the elements hydrogen to sodium – note how they obey the rules above*

The electron arrangements shown in Figure 20 are known as the **ground states** of these atoms because these are the lowest energy arrangements possible. It is possible for one or more electrons to occupy levels of higher energy.

If this results in an atom with one or more unpaired electrons, the resulting species is called a **free radical**.

Writing electronic structures

A shorthand way of writing electronic structures is shown here for sodium, which has 11 electrons:

$1s^2$ $2s^2$ $2p^6$ $3s^1$

2   8   1

Note how this matches the simpler 2,8,1 you used earlier.

Calcium, with 20 electrons would be:

$$1s^2\ 2s^2\ 2p^6\ 3s^2\ 3p^6\ 4s^2 \qquad \text{which matches 2,8,8,2}$$

Notice how the 4s orbital is filled before the 3d orbital because it is of lower energy.

After calcium, electrons begin to fill the 3d orbitals, so vanadium with 23 electrons is:

$$1s^2\ 2s^2\ 2p^6\ 3s^2\ 3p^6\ 3d^3\ 4s^2$$

Krypton with 36 electrons is: $1s^2\ 2s^2\ 2p^6\ 3s^2\ 3p^6\ 3d^{10}\ 4s^2\ 4p^6$

Sometimes it simplifies things to use the previous noble gas symbol. So the electron arrangement of calcium, Ca, could be written $[Ar]\ 4s^2$ as a shorthand for $[1s^2\ 2s^2\ 2p^6\ 3s^2\ 3p^6]\ 4s^2$, because $1s^2\ 2s^2\ 2p^6\ 3s^2\ 3p^6$ is the electron arrangement of argon.

You can use the same notation for ions. So a sodium ion, Na^+, would have the electron arrangement $1s^2\ 2s^2\ 2p^6$, one less than a sodium atom, $1s^2\ 2s^2\ 2p^6\ 3s^1$.

Exam tip

Practise working out the shorthand electronic structure of all the elements at least up to krypton (atomic number 36).

Summary test 1.3

1 Sketch the electron arrangement diagrams of atoms that have the following numbers of electrons:
   a 3
   b 9
   c 14

2 State, in shorthand, the electron arrangements of atoms with:
   a 4 electrons
   b 13 electrons
   c 18 electrons

3 Identify which of the following are atoms, positive ions, or negative ions. Give the size of the charge on each ion, including its sign. Use the Periodic Table to identify the elements A–E.

	Number of protons	Number of electrons
A	12	10
B	2	2
C	17	18
D	10	10
E	3	2

4 a Give the full electron arrangement for phosphorus.
   b Give the electron arrangement for phosphorus using an inert gas symbol as a shorthand.

5 a Give the full electron arrangements of:
      i Ca^{2+}
      ii F^-
   b Give the electron arrangements of **ai** and **aii** using an inert gas symbol as a shorthand.

Ionisation energy

Ionisation energy is the energy that has to be put in to form positive ions. Energy is given out when negative ions form. This is called **electron affinity**.

Ionisation energy is a consequence of the attraction between the nuclear charge and the electrons in the outer shell of an atom.

The patterns in first ionisation energies across a period provide evidence for electron energy sub-levels.

Ionisation energy

Electrons can be removed from atoms and the energy it takes to remove them can be measured. This is called ionisation energy, because as the electrons are removed the atoms become positive ions.

- Ionisation energy is the energy required to remove a mole of electrons from a mole of atoms in the gaseous state and is measured in kJ mol^{-1}.
- Ionisation energy has the abbreviation IE.

Removing the electrons one by one (successive ionisation energies)
You can measure the energies required to remove the electrons one by one from an atom, starting from the outer electrons and working inwards.

- The first electron needs the least energy to remove it because it is being removed from a neutral atom. This is the first IE.
- The second electron needs more energy than the first because it is being removed from a +1 ion. This is the second IE.
- The third electron needs even more energy to remove it because it is being removed from a +2 ion. This is the third IE.
- The fourth needs yet more energy, and so on.

These are called successive ionisation energies.

For example, for sodium:

$Na(g) \rightarrow Na^+(g) + e^-$ first IE $= +494$ kJ mol^{-1}

$Na^+(g) \rightarrow Na^{2+}(g) + e^-$ second IE $= +4560$ kJ mol^{-1}

$Na^{2+}(g) \rightarrow Na^{3+}(g) + e^-$ third IE $= +6940$ kJ mol^{-1}

and so on (see Table 5).

Table 5 *Successive ionisation energies of sodium*

Electron removed	1st	2nd	3rd	4th	5th	6th	7th	8th	9th	10th	11th
Ionisation energy/kJ mol^{-1}	494	4560	6940	9540	13352	16611	20115	25491	28934	141367	159079

Notice that the second IE is *not* the energy change for

$$Na(g) \rightarrow Na^{2+}(g) + 2e^-$$

The energy for this process would be (first IE + second IE).

If you plot a graph of the values shown in Table 5, you get Figure 21, shown at the top of the following page.

Notice that one electron is relatively easy to remove, then comes a group of eight electrons that are more difficult to remove, and finally two electrons that are very difficult to remove.

Learning outcomes

On these pages you will learn to:

- interpret ionisation energy data
- identify and explain the trends in ionisation energy
- use ionisation energy data to deduce the position of an element in the Periodic Table

Exam tip

The shape of the graph in Figure 21 has to be thought about carefully. The first electron removed is in the outer shell and the 10th and 11th electrons removed are in the innermost shell. Graphs like this give powerful evidence for the existence of electron shells.

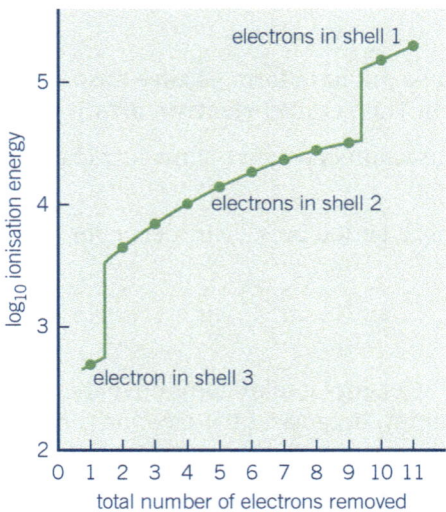

Figure 21 *The successive ionisation energies of sodium against number of electrons removed. Note that the log of the ionisation energy has been plotted in order to fit the large range of values on the scale*

This suggests that sodium has:

- one electron furthest away from the positive nucleus (easy to remove)
- eight electrons nearer in to the nucleus (harder to remove)
- two electrons very close to the nucleus (very difficult to remove because they are nearest to the positive charge of the nucleus).

This tells you about the number of electrons in each shell or orbit: 2,8,1. The eight electrons in shell 2 are in fact sub-divided into two further groups that correspond to the $2s^2$, $2p^6$ electrons in the second shell, but this is not visible on the scale of Figure 21. It is just visible in Figure 22.

You can find the number of electrons in each shell of any element by looking at the jumps in successive ionisation energies.

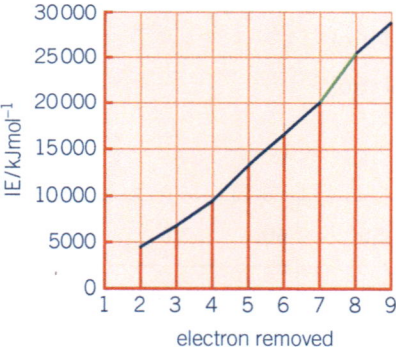

Figure 22 *The successive ionisation energies of the electrons in shell 2 in sodium. You can just see the jump between electron 7 and 8*

Trends in first ionisation energies across a period in the Periodic Table

The trends in first ionisation energies moving across a period in the Periodic Table can also give information about the energies of electrons in shells and sub-levels. Ionisation energies generally increase across a period because the nuclear charge is increasing and this makes it more difficult to remove an electron from the outer shell of the atom.

The data for Period 3 are shown in Table 6.

Table 6 *First ionisation energies of the elements in Period 3 in kJ mol⁻¹*

Na	Mg	Al	Si	P	S	Cl	Ar
494	736	577	786	1060	1000	1260	1520
nuclear charge increasing ⟶							

Plotting a graph of these values shows that the increase is not regular (Figure 23). In going from magnesium ($1s^2\ 2s^2\ 2p^6\ 3s^2$) to aluminium ($1s^2\ 2s^2\ 2p^6\ 3s^2\ 3p^1$), the ionisation energy actually goes down, despite the increase in nuclear charge. This is because the outer electron in aluminium is in a 3p orbital which is of a slightly higher energy than the 3s orbital. It therefore needs less energy to remove it (see Figure 24).

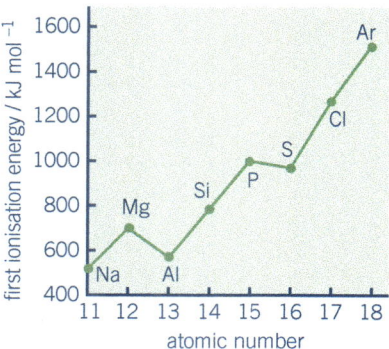

Figure 23 *Trends in first ionisation energies across Period 3*

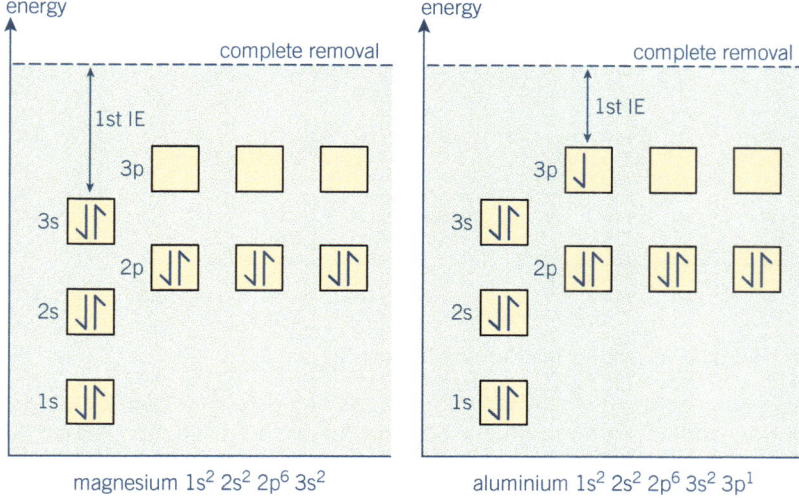

Figure 24 *The first ionisation energy of aluminium is less than that of magnesium*

> **Exam tip**
>
> As we go across a period the number of inner electrons shielding the electrons in the outer shell remains the same.

In Figure 23, notice the small drop between phosphorus ($1s^2\ 2s^2\ 2p^6\ 3s^2\ 3p^3$) and sulfur ($1s^2\ 2s^2\ 2p^6\ 3s^2\ 3p^4$). In phosphorus, each of the three 3p orbitals contains just one electron, while in sulfur one of the 3p orbitals must contain two electrons. The repulsion between these paired electrons makes it easier to remove one of them, despite the increase in nuclear charge (see Figure 25).

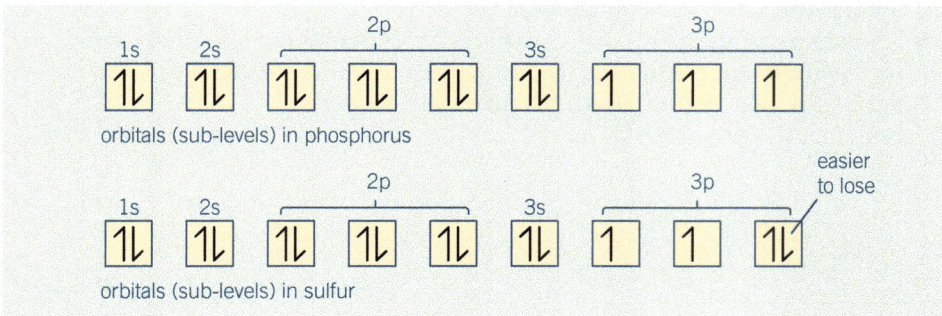

Figure 25 *Electron arrangements of phosphorus and sulfur*

Both these cases, which go against the expected trend, are evidence confirming the existence of s- and p-sub-levels.

Trends in ionisation energies down a group in the Periodic Table

Figure 26 shows that there is a general decrease in first ionisation energy going down Group 2 and the same pattern is seen in other groups. This is because the outer electron is in a shell that gets further from the nucleus in each case. So as the atomic radius increases, the ionisation energy decreases. The outer electrons in each atom in the same group feel the same nuclear charge, after taking into account the shielding of the inner electrons.

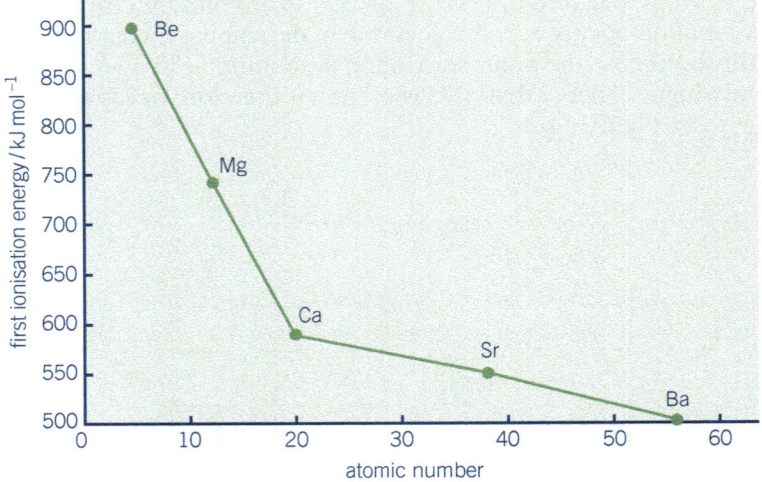

Figure 26 *The first ionisation energies of the elements of Group 2*

Going down a group, the nuclear charge increases. At first sight you might expect this to make it *more* difficult to remove an electron. However, the actual positive charge 'felt' by an electron in the outer shell is less than the full nuclear charge. This is because of the effect of the inner electrons shielding the nuclear charge.

Summary test 1.4

1 State why the second ionisation energy of any atom is larger than the first ionisation energy.
2 Sketch a graph similar to Figure 21 of the successive ionisation energies of aluminium (electron arrangement 2,8,3).
3 An element X has the following values (in kJ mol^{-1}) for successive ionisation energies: 1093, 2359, 4627, 6229, 37 838, 47 285.
   a Identify which group in the Periodic Table it is in.
   b Explain your answer to **a**.

1 a Define in terms of fundamental particles the term 'isotope'. *(2 marks)*

b Bromine exists as two isotopes of equal abundance, ^{79}Br and ^{81}Br.
  i Suggest a value for the relative atomic mass of bromine. *(1 mark)*
  ii Explain why atoms of ^{79}Br and ^{81}Br have the same chemical properties. *(1 mark)*
  iii Identify the overall charge on an atom of ^{81}Br, giving reasons for your answer. *(3 marks)*

2 Using the table below:

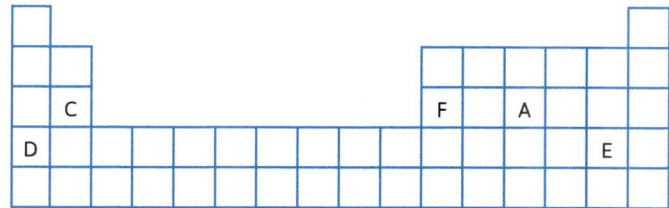

  a Identify the letter(s) corresponding to the element(s) in:
  i Group 2　**ii** Group 17　**iii** Period 3　**iv** s block
  v metals　**vi** p block *(6 marks)*

  b In terms of electronic configuration, identify one similarity and one difference between elements D and E. *(2 marks)*

3 a Define the term 'first ionisation energy'. *(3 marks)*

  b Explain the general trend in first ionisation energy:
  i across Period 3 *(4 marks)*
  ii descending Group 2. *(4 marks)*

  c Construct an equation to represent the second ionisation energy of magnesium. *(1 mark)*

d The table below shows the first ionisation energies of the elements of Period 3:

Element	Na	Mg	Al	Si	P	S	Cl	Ar
1st IE / kJ mol^{-1}	494	736	577	786	1060		1260	1520

  i Copy and complete the table to estimate the first ionisation energy of sulfur. *(1 mark)*
  ii Explain your answer to part **i** using your knowledge of atomic structure. *(2 marks)*

e State the element in Period 3 with the highest second ionisation energy. Explain your answer. *(2 marks)*

4 Give the electronic configuration of:
i Mg　**ii** Al^{3+}　**iii** S^{2-}　**iv** Fe　**v** Cr　**vi** Fe^{2+} *(6 marks)*

5 Element X is found in Period 3. Use the successive ionisation energy data in the table below to predict the identity of element X. Give reasons for your answer. *(2 marks)*

Ionisation	First	Second	Third	Fourth	Fifth	Sixth
IE / kJ mol^{-1}	786	1580	3230	4360	16091	19806

6 a Draw diagrams to show the shape of s and p orbitals. *(2 marks)*

  b State the maximum number of electrons that can be placed within a d orbital. *(1 mark)*

  c The boxes notation is used below to show the electronic configuration of phosphorus:

  [Ne] | ↑↓ |　　| ↑ | ↑ | ↑ |

  Give the electronic configuration for an atom of potassium using the boxes notation. *(1 mark)*

7 The 3$^-$ ion of Element Z has a noble gas configuration and the same number of neutrons as an atom of oxygen. Give the identity of Element Z. *(2 marks)*

2 Atoms, molecules and stoichiometry

2.1 Relative masses of atoms and molecules

Learning outcomes

On these pages you will learn to:

- define the atomic mass unit
- define and use relative atomic mass, relative isotopic mass, relative molecular mass and relative formula mass

Chemistry is all about atoms. The first person to suggest the idea of atoms was the Greek philosopher Democritus in around 400 BC.

Over 2 000 years later the British chemist John Dalton suggested that chemical elements were made of indivisible atoms. He proposed that all atoms of the same element had the same mass, and that atoms of different elements had different masses.

By the early 20th century it was realised that atoms of the same element could in fact differ in mass—this was the idea of isotopes (see Section 1.2). The masses of atoms are among their most important properties.

The unified atomic mass unit

The **unified atomic mass unit** is the standard unit for measuring atomic masses. It is defined as $\frac{1}{12}$ of the mass of an atom of carbon-12.

It is sometimes called the Dalton, Da, and its mass is 1.66×10^{-27} kg.

Relative atomic mass, A_r

Chemists usually use the idea of **relative atomic mass**, A_r. It is numerically the same as the unified atomic mass unit but has no units.

In the past, the relative atomic mass of hydrogen, the lightest element, was taken as 1. The average mass of an atom of oxygen is 16 times heavier, to the nearest whole number, so oxygen has a relative atomic mass of 16. Scientists now use the isotope carbon-12 as the baseline for relative atomic mass. This is because the mass spectrometer has allowed us to measure the masses of individual isotopes extremely accurately. The carbon-12 standard (defined below) is now accepted by all chemists throughout the world.

The relative atomic mass A_r is the weighted average mass of an atom of an element, taking into account its naturally occurring isotopes, relative to $\frac{1}{12}$ the relative atomic mass of an atom of carbon-12. It has no units.

$$\text{Relative atomic mass } A_r = \frac{\text{average mass of one atom of an element}}{\frac{1}{12} \text{ mass of one atom of } ^{12}\text{C}}$$

$$= \frac{\text{average mass of one atom of an element} \times 12}{\text{mass of one atom of } ^{12}\text{C}}$$

Relative isotopic mass is the relative mass of a particular isotope measured on the carbon-12 scale. In other words, it is the mass of an isotope of an element relative to $\frac{1}{12}$ of the relative atomic mass of an atom of carbon-12.

Figure 1 *Chemists can determine the masses of atoms in sample to a high degree of accuracy using a mass spectrometer*

Relative molecular mass M_r

Molecules can be handled in the same way, by comparing the mass of a molecule with that of an atom of carbon-12.

The **relative molecular mass**, M_r, of a molecule is the mass of that molecule compared to $\frac{1}{12}$ the relative atomic mass of an atom of carbon-12:

$$\text{Relative molecular mass } M_r = \frac{\text{average mass of one molecule}}{\frac{1}{12} \text{ mass of one atom of } ^{12}\text{C}}$$

$$= \frac{\text{average mass of one molecule} \times 12}{\text{mass of one atom of } ^{12}\text{C}}$$

You find the relative molecular mass by adding up the relative atomic masses of all the atoms present in the molecule and you find this from its formula.

Table 1 *Examples of relative molecular mass*

Molecule	Formula	A_r of atoms		M_r
water	H_2O	$(2 \times 1.0) + 16.0$	18.0	
carbon dioxide	CO_2	$12.0 + (2 \times 16.0)$	44.0	
methane	CH_4	$12.0 + (4 \times 1.0)$	16.0	

Relative formula mass

The term **relative formula mass** is used for ionic compounds because they do not exist as molecules. However, this has the same symbol, M_r, as relative molecular mass.

Table 2 *Some examples of the relative formula masses of ionic compounds*

Ionic compound	Formula	A_r of atoms	M_r
calcium fluoride	CaF_2	$40.1 + (2 \times 19.0)$	78.1
sodium sulfate	Na_2SO_4	$(2 \times 23.0) + 32.1 + (4 \times 16.0)$	142.1
magnesium nitrate	$Mg(NO_3)_2$	$24.3 + (2 \times (14.0 + (16.0 \times 3)))$	148.3

Exam tip

You need to learn the definitions for the unified atomic mass unit, relative atomic mass, relative isotopic mass, relative molecular mass and relative formula mass.

Summary test 2.1

1 Calculate the M_r for each of the following compounds.
   a CH_4
   b Na_2CO_3
   c $Mg(OH)_2$
   d $(NH_4)_2SO_4$
   Use these values for the relative atomic masses (A_r):
   C = 12.0, H =1.0, Na = 23.0,
   O = 16.0, Mg = 24.3, N = 14.0, S = 32.0

2 Imagine an atomic seesaw with an oxygen atom on one side. Deduce six combinations of other atoms that would make the seesaw balance. For example, one nitrogen atom and two hydrogen atoms would balance the seesaw. Use values of A_r to the nearest whole number.

The mole and the Avogadro constant

Figure 2 *Large numbers of coins or bank notes are counted by weighing them*

One atom of any element is too small to see with an optical microscope and impossible to weigh individually. So, to count atoms, chemists must weigh large numbers of them. This is how cashiers count money in a bank (Figure 2).

Working to the nearest whole number, a helium atom ($A_r = 4$) is four times heavier than an atom of hydrogen. A lithium atom ($A_r = 7$) is seven times heavier than an atom of hydrogen. To get the same number of atoms in a sample of helium or lithium as the number of atoms in 1 g of hydrogen, you must take 4 g of helium or 7 g of lithium.

In fact, if you weigh out the relative atomic mass of *any* element, this amount will also contain this same number of atoms.

The same logic applies to molecules. Water, H_2O, has a relative molecular mass M_r of 18. So, one molecule of water is 18 times heavier than one atom of hydrogen. Therefore, 18 g of water contain the same number of molecules as there are atoms in 1 g of hydrogen. A molecule of carbon dioxide is 44 times heavier than an atom of hydrogen, so 44 g of carbon dioxide contain this same number of molecules.

If you weigh out the relative molecular or formula mass M_r of a compound in grams you have this same number of entities.

The Avogadro constant, L

The actual number of atoms in 1 g of hydrogen atoms is unimaginably huge:

$$602\,000\,000\,000\,000\,000\,000\,000, \text{ usually written } 6.02 \times 10^{23.}$$

This number is the **Avogadro constant**, L (or Avogadro number). It is also the number of atoms in 12 g of carbon-12, and the number of molecules in 18 g of water.

The difference between the scale based on H = 1 and the scale based on ^{12}C that we use today is negligible.

The mole

The amount of substance that contains 6.02×10^{23} particles is called a **mole**.

The relative atomic mass of any element in grams contains one mole of atoms. The relative molecular mass (or relative formula mass) of a substance in grams contains one mole of entities. You can also have a mole of ions or electrons.

It is easy to confuse moles of atoms and moles of molecules. So you should always give the formula when working out the mass of a mole of entities. For example, 10 moles of hydrogen could mean 10 moles of hydrogen atoms or 10 moles of hydrogen molecules, H_2, which contain twice the number of atoms. Using the mole, you can compare the numbers of different particles that take part in chemical reactions.

Table 3 *Examples of moles*

Entities	Formula	Relative mass to nearest whole number	Mass of a mole / g (= molar mass)
oxygen atoms	O	16	16
oxygen molecules	O_2	32	32
sodium ions	Na^+	23	23
sodium fluoride	NaF	42	42

Number of moles

If you want to find out how many moles are present in a particular mass of a substance you need to know the substance's formula. From the formula you can then work out the mass of one mole of the substance.

You use:

$$\text{Number of moles } n = \frac{\text{mass } m \text{ (g)}}{\text{mass of 1 mole } M \text{ (g)}}$$

Exam tip

You can also use the term *molar mass*, which is the mass per mole of substance. It has units kg mol^{-1} or g mol^{-1}. The molar mass in g mol^{-1} is the same numerically as M_r.

Worked example

Finding the number of moles

How many moles are there in 0.53 g of sodium carbonate, Na_2CO_3?

Relative atomic masses (A_r): Na = 23.0, C = 12.0, O = 16.0,

so M_r of Na_2CO_3 = (23.0 × 2) + 12.0 + (16.0 × 3) = 106.0,

so 1 mole of sodium carbonate has a mass of 106.0 g.

Number of moles = $\frac{0.53}{106.0}$ = 0.0050 mol

Exam tip

The Avogadro constant is the same as the number of atoms in 1 g of hydrogen, H_2—not the number of molecules.

Worked example

Finding the number of atoms

You have 3.94 g of gold, Au, and 2.70 g of aluminium, Al. Which contains the greater number of atoms? (A_r Au = 197.0, A_r Al = 27.0)

Number of moles of gold atoms = $\frac{3.94}{197.0}$ = 0.020 mol

Number of moles of aluminium atoms = $\frac{2.70}{27.0}$ = 0.100 mol

There are more atoms of aluminium.

Summary test 2.2

1 Calculate the number of moles in the given masses of the following entities.
 a 32.0 g CH_4
 b 5.30 g Na_2CO_3
 c 5.83 g $Mg(OH)_2$
2 Identify which contains the fewest molecules:
 0.5 g of hydrogen H_2, 4.0 g of oxygen O_2 or 11.0 g of carbon dioxide CO_2.
3 Identify the entity in question 2 that contains the greatest number of atoms.
4 Calculate the number of entities in **1a–c** using the Avogadro constant.

Learning outcomes

On these pages you will learn to:

- write the formulae for ionic compounds
- construct and use balanced equations
- calculate empirical and molecular formulae
- describe anhydrous and hydrated compounds

Figure 3 *Limestone is largely made of calcium carbonate, CaCO₃*

Calculating empirical and molecular formulae

Chemical formulae tell us about the number of atoms combined together in a substance.

The empirical formula

The **empirical formula** (also called the simplest formula) is the formula that represents the simplest whole number ratio of the atoms of each element present in a compound. For example, the empirical formula of carbon dioxide, CO_2, tells us that for every carbon atom there are two oxygen atoms.

To find an empirical formula:

1 Find the masses of each of the elements present in a compound (by experiment).
2 Work out the number of moles of atoms of each element.

$$\text{Number of moles} = \frac{\text{mass of element}}{\text{mass of 1 mol of element}}$$

3 Convert the number of moles of each element into a whole number ratio.

Worked example

Finding the empirical formula of calcium carbonate

10.01 g of a white solid contains 4.01 g of calcium, 1.20 g of carbon, and 4.80 g of oxygen. What is its empirical formula?

(Relative atomic masses (A_r): Ca = 40.1, C = 12.0, O = 16.0)

Step 1: Find the masses of each element

Mass of calcium = 4.01 g

Mass of carbon = 1.20 g

Mass of oxygen = 4.80 g

Step 2: Find the number of moles of atoms of each element

A_r Ca = 40.1

Number of moles of calcium = $\frac{4.01}{40.1}$ = 0.10 mol

A_r C = 12.0

Number of moles of carbon = $\frac{1.2}{12.0}$ = 0.10 mol

A_r O = 16

Number of moles of oxygen = $\frac{4.8}{16.0}$ = 0.30 mol

Step 3: Find the simplest ratio

Ratio in moles of	calcium	:	carbon	:	oxygen
	0.10	:	0.10	:	0.30
So the simplest whole number ratio is:	1	:	1	:	3

The formula is therefore $CaCO_3$.

Worked example

Finding the empirical formula of copper oxide

0.795 g of black copper oxide is reduced to 0.635 g of copper when heated in a stream of hydrogen (Figure 4). What is the formula of copper oxide? A_r Cu = 63.5, A_r O = 16.0

Step 1: Find the masses of each element

Mass of copper = 0.635 g

We started with 0.795 g of copper oxide and 0.635 g of copper were left, so:

Mass of oxygen = 0.795 − 0.635 = 0.160 g

Step 2: Find the number of moles of atoms of each element

A_r Cu = 63.5; number of moles of copper = $\dfrac{0.635}{63.5}$ = 0.01

A_r O = 16.0; number of moles of oxygen = $\dfrac{0.16}{16.0}$ = 0.01

Step 3: Find the simplest ratio

The ratio of moles of copper to moles of oxygen is:

copper : oxygen

0.01 : 0.01   So the simplest whole number ratio is 1 : 1

The simplest formula of black copper oxide is therefore one Cu to one O, CuO. You may find it easier to make a table:

	Copper Cu	Oxygen O
mass of element	0.635 g	0.160 g
A_r of element	63.5	16.0
number of moles mass of element/A_r	$\dfrac{0.635}{63.5}$ = 0.01	$\dfrac{0.160}{16.0}$ = 0.01
ratio of elements	1	1

Finding the simplest ratio of elements

Sometimes you will end up with ratios of moles of atoms of elements that are not easy to convert to whole numbers. If you divide each number by the smallest number you will end up with whole numbers (or ratios you can recognise more easily). Here is an example.

Empirical formula

Compound X contains 50.2 g sulfur and 50.0 g oxygen. What is its empirical formula? A_r S = 32.1, A_r O = 16.0

Step 1: Find the number of moles of atoms of each element

A_r S = 32.1; number of moles of sulfur = $\dfrac{50.2}{32.1}$ = 1.564

A_r O = 16; number of moles of oxygen = $\dfrac{50.0}{16.0}$ = 3.125

Step 2: Find the simplest ratio

Ratio of sulfur : oxygen : 1.564 : 3.125

Now divide each of the numbers by the smaller number.

Ratio of sulfur : oxygen

$\dfrac{1.564}{1.564} : \dfrac{3.125}{1.564}$ = 1:2

The empirical formula is therefore SO_2. Sometimes you may end up with a ratio of moles of atoms such as 1:1.5. In these cases you must find a whole number ratio, in this case 2:3.

Finding the empirical formula of copper oxide

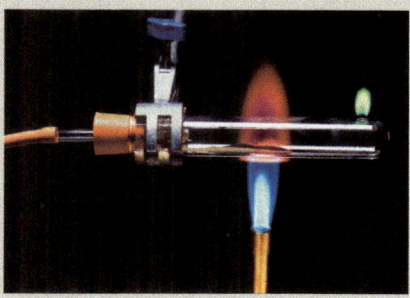

Figure 4 *Finding the empirical formula of copper oxide*

In this experiment, explain why:

1 There is a flame at the end of the tube.
2 This flame goes green.
3 Droplets of water form near the end of the tube.
4 The flame at the end of the tube is kept alight until the apparatus is cool.

Another oxide of copper

There is another oxide of copper, which is red. In a reduction experiment similar to that for finding the formula of black copper oxide, 1.43 g of red copper oxide was reduced with a stream of hydrogen and 1.27 g of copper were formed. Use the same steps as for black copper oxide to find the formula of the red oxide.

1 Find the masses of each element.
2 Find the number of moles of atoms of each element.
3 Find the simplest ratio.

- When calculating empirical formulae from percentages, check that all the percentages of the compositions by mass add up to 100%. (Don't forget any oxygen that may be present.)
- Remember to use relative atomic masses from the Periodic Table, *not* the atomic number.

Finding the molecular formula

The molecular formula gives the actual number of atoms of each element in one molecule of the compound. (It applies only to substances that exist as molecules.)

The empirical formula is not always the same as the molecular formula. There may be several units of the empirical formula in the molecular formula.

For example, ethane (molecular formula C_2H_6) would have an empirical formula of CH_3.

To find the number of units of the empirical formula in the molecular formula, divide the relative molecular mass by the relative mass of the empirical formula.

For example, ethene is found to have a relative molecular mass of 28.0 but its empirical formula, CH_2, has a relative mass of 14.0.

$$\frac{\text{Relative molecular mass of ethene}}{\text{Relative empirical formula mass of ethane}} = \frac{28.0}{14.0} = 2$$

So there must be two units of the empirical formula in the molecule of ethene. So ethene is $(CH_2)_2$ or C_2H_4.

Worked example

Molecular formula

An organic compound containing only carbon, hydrogen, and oxygen was found to have 52.17% carbon and 13.04% hydrogen. What is its molecular formula if $M_r = 46.0$?

100.00 g of this compound would contain 52.17 g carbon, 13.04 g hydrogen and (the rest) 34.79 g oxygen.

Step 1: Find the empirical formula

	Carbon	Hydrogen	Oxygen
Mass of element/g	52.17	13.04	34.79
A_r of element	12.0	1.0	16.0
Number of moles = mass of element/ A_r	$\frac{52.17}{12.0} = 4.348$	$\frac{13.04}{1.0} = 13.04$	$\frac{34.79}{16.0} = 2.174$
Divide through by the smallest	$\frac{4.348}{2.174} = 2$	$\frac{13.04}{2.174} = 6$	$\frac{2.174}{2.174} = 1$
Ratio of elements	2	6	1

So the empirical formula is C_2H_6O.

Step 2: Find M_r of the empirical formula

$$(2 \times 12.0) + (6 \times 1.0) + (1 \times 16.0) = 46.0$$

$$\quad C \qquad\quad H \qquad\qquad O$$

So, the molecular formula is the same as the empirical formula, C_2H_6O.

Extension

Combustion analysis

Organic compounds are based on carbon and hydrogen. One method of finding empirical formulae of new compounds is called **combustion analysis** (Figure 5). It is used routinely in the pharmaceutical industry. It involves burning the unknown compound in excess oxygen and measuring the amounts of water, carbon dioxide, and other oxides that are produced. The gases are carried through the instrument by a stream of helium.

The basic method measures carbon, hydrogen, sulfur, and nitrogen. It is assumed that oxygen makes up the difference after the other four elements have been measured. Once the sample has been weighed and placed in the instrument, the process is automatic and controlled by computer.

The sample is burnt completely in a stream of oxygen. The final combustion products are water, carbon dioxide, and sulfur dioxide. The instrument measures the amounts of these by infrared absorption. They are removed from the gas stream leaving the unreacted nitrogen which is measured by thermal conductivity. The measurements are used to calculate the masses of each gas present and hence the masses of hydrogen, sulfur, carbon, and nitrogen in the original sample. Oxygen is found by difference.

Traditionally, the amounts of water and carbon dioxide were measured by absorbing them in suitable chemicals and measuring the increase in mass of the absorbents. This is how the composition data for the worked example below were measured. The molecular formula can then be found, if the relative molecular mass has been found using a mass spectrometer.

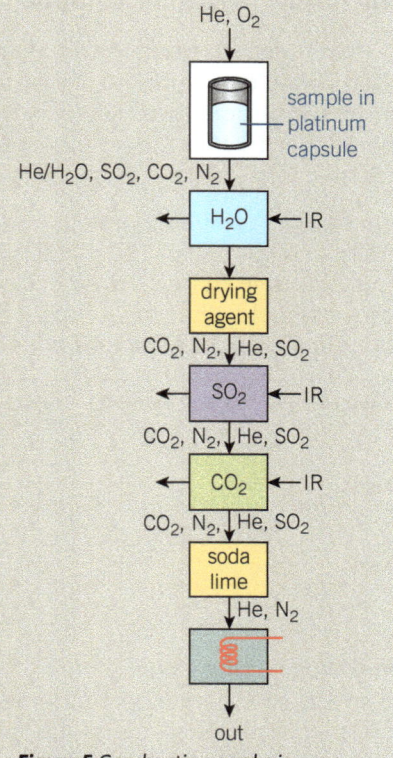

Figure 5 Combustion analysis

1 Soda lime is a mixture containing mostly calcium hydroxide, $Ca(OH)_2$. Construct a balanced symbol equation for the reaction of calcium hydroxide with carbon dioxide.

Answer: $Ca(OH)_2 + CO_2 \rightarrow CaCO_3 + H_2O$

Worked example

Molecular formula by combustion analysis

0.58 g of a compound X containing only carbon, hydrogen, and oxygen, gave 1.32 g of carbon dioxide and 0.54 g of water on complete combustion in oxygen. What is its empirical formula? What is its molecular formula if its relative molecular mass is 58.0?

To calculate the empirical formula:

carbon 1.32 g of CO_2 ($M_r = 44.0$) is $\dfrac{1.32}{44.0} = 0.03$ mol CO_2

As each mole of CO_2 has 1 mole of C, the sample contained 0.03 mol of C atoms.

hydrogen 0.54 g of H_2O ($M_r = 18.0$) is $= \dfrac{0.54}{18} = 0.03$ mol H_2O

As each mole of H_2O has 2 moles of H, the sample contained 0.06 mol of H atoms.

oxygen 0.03 mol of carbon atoms ($A_r = 12.0$) has a mass of 0.36 g

0.06 mol of hydrogen atoms ($A_r = 1.0$) has a mass of 0.06 g

Total mass of carbon and hydrogen is 0.42 g

The rest (0.58 − 0.42) must be oxygen, so the sample contained 0.16 g of oxygen.

0.16 g of oxygen ($A_r = 16.0$) is $\dfrac{0.16}{16.0} = 0.01$ mol oxygen atoms

So the sample contains 0.03 mol C, 0.06 mol H, and 0.01 mol O

Dividing by the smallest number 0.06 gives the ratio:

C	H	O	
3	6	1	so the empirical formula is C_3H_6O.

M_r of this unit is 58, so the molecular formula is also C_3H_6O.

Important ions

The charges on some other metal ions and some complex ions (those containing more than one element) are worth remembering.

Positive ions

Zn^{2+}, NH_4^+ (ammonium), Ag^+, Al^{3+}

Negative ions

NO_3^- (nitrate), CO_3^{2-} (carbonate), SO_4^{2-} (sulfate), OH^- (hydroxide), HCO_3^- (hydrogencarbonate), PO_4^{3-} (phosphate), N^{3-} (nitride)

Exam tip

Some elements can form ions that have different charges. These are identified by using the oxidation number (in Roman numerals) after the name of the ion. For example, copper forms two oxides: CuO in which the copper forms Cu^{2+} ions, and Cu_2O in which it forms Cu^+ ions. We call CuO copper(II) oxide and Cu_2O copper(I) oxide. Section 6.1 explains oxidation numbers in more detail.

Exam tip

Note the difference between nitrate and nitride. The ending –*ate* essentially means 'and oxygen as well'.

The formulae of ionic compounds

The formulae of compounds have to be found by experiment. However, it is often possible to work out the formulae of ionic compounds from the charges of the constituent ions. This is because compounds are electrically neutral, so the total charge on the positive ions must be equal to the total charge on the negative ions.

The charges on simple ions can be worked out from the position of the element in the Periodic Table. The metal elements in Group 1 always form ions with a single positive charge. Those in Group 2 form ions with two positive charges. Non-metal elements in Group 17 form ions with a single negative charge, and those in Group 16 form ions with two negative charges.

Group 1	Group 2	Group 13	Group 15	Group 16	Group 17
Li^+	Be^{2+}	Al^{3+}	N^{3-}	O^{2-}	F^-
Na^+	Mg^{2+}			S^{2-}	Cl^-
K^+	Ca^{2+}				Br^-
Rb^+	Sr^{2+}				I^-
Cs^+	Ba^{2+}				

Predicting formulae

Knowing the charges on common ions is useful as it allows you to predict the formulae of simple compounds. Compounds are neutral, so the number of positive charges must be the same as the number of negative charges. See the examples below:

Compound	Ions	Formula
Sodium bromide	$Na^+ + Br^-$	NaBr
Zinc sulfide	$Zn^{2+} + S^{2-}$	ZnS
Calcium nitrate	$Ca^{2+} + 2(NO_3^-)$	$Ca(NO_3)_2$
Magnesium phosphate	$3\,Mg^{2+} + 2(PO_4^{3-})$	$Mg_3(PO_4)_2$

Balanced equations

Equations represent what happens when chemical reactions take place. They are based on experimental evidence. The starting materials are reactants. After these have reacted you end up with products.

$$\text{reactants} \rightarrow \text{products}$$

Word equations only give the names of the reactants and products, for example:

$$\text{hydrogen} + \text{oxygen} \rightarrow \text{water}$$

Once the idea of atoms had been established, chemists realised that chemical entities react together in simple whole number ratios. For example, two hydrogen molecules react with one oxygen molecule to give two water molecules.

$$\text{2 hydrogen molecules} + \text{1 oxygen molecule} \rightarrow \text{2 water molecules}$$
$$2 \qquad : \qquad 1 \qquad : \qquad 2$$

The ratio in which the reactants react and the products are produced, in simple whole numbers, is called the **stoichiometry** of the reaction.

You can build up a stoichiometric relationship from experimental data by working out the number of moles that react together. This leads us to a balanced symbol equation.

Balanced symbol equations use the formulae of reactants and products. There are the same number of atoms of each element on both sides of the arrow. (This is because atoms are never created or destroyed in chemical reactions.) Balanced equations tell us about the amounts of substances that react together and are produced.

State symbols can also be added. These are letters, in brackets, which can be added to the formulae in equations to say what state the reactants and products are in: (s) means solid, (l) means liquid, (g) means gas, and (aq) means aqueous solution (dissolved in water).

Writing balanced equations

When aluminium burns in oxygen it forms solid aluminium oxide. You can build up a balanced symbol equation from this and the formulae of the reactants and product: Al, O_2, and Al_2O_3.

1 Write the word equation:

$$aluminium + oxygen \rightarrow aluminium\ oxide$$

2 Write in the correct formulae:

$$Al + O_2 \rightarrow Al_2O_3$$

This is not balanced because:
- There is one aluminium atom on the reactants side (left-hand side) but two on the products side (right-hand side).
- There are two oxygen atoms on the reactants side (left-hand side) but three on the products side (right-hand side).

3 To get two aluminium atoms on the left-hand side put a 2 in front of the Al:

$$2Al + O_2 \rightarrow Al_2O_3$$

Now the aluminium is correct but not the oxygen.

4 If you multiply the oxygen on the left-hand side by 3, and the aluminium oxide by 2, you have six O on each side:

$$2Al + 3O_2 \rightarrow 2Al_2O_3$$

5 Now you return to the aluminium. You need four Al on the left-hand side:

$$4Al + 3O_2 \rightarrow 2Al_2O_3$$

The equation is balanced because there are the same numbers of atoms of each element on both sides of the equation.

The numbers in front of the formulae (4, 3, and 2) are called coefficients.

6 You can add state symbols.

$$4Al(s) + 3O_2(g) \rightarrow 2Al_2O_3(s)$$

The equation tells you the numbers of moles of each of the substances that are involved. From this you can work out the masses that will react together: (using Al = 27.0, O = 16.0)

$4Al(s)$	$+ 3O_2(g)$	$\rightarrow 2Al_2O_3(s)$
4 moles	3 moles	2 moles
108.0 g	96.0 g	204.0 g

The total mass is the same on both sides of the equation. This is another good way of checking whether the equation is balanced.

Exam tip

Useful tips for balancing equations
- You must use the correct formulae—you *cannot* change them to make the equation balance.
- You can only change the numbers of atoms by putting a number, the coefficient, in front of formulae.
- The coefficient in front of the symbol tells you how many moles of that substance are reacting.
- It often takes more than one step to balance an equation, but too many steps suggests that you may have an incorrect formula.

Ionic equations

Ionic equations show the ions involved in the reaction. For example:

$$\text{sodium} + \text{water} \rightarrow \text{sodium hydroxide} + \text{hydrogen}$$

Sodium hydroxide is an ionic compound, so we can write it as $Na^+(aq) + OH^-(aq)$ to make this clear:

$$2Na(s) + 2H_2O(l) \rightarrow 2NaOH(aq) + H_2(g)$$

$$2Na(s) + 2H_2O(l) \rightarrow 2Na^+(aq) + 2OH^-(aq) + H_2(g)$$

Some ions take no part in the reaction and are called **spectator ions**, for example in the following reaction:

$$\text{silver nitrate} + \text{sodium chloride} \rightarrow \text{silver chloride} + \text{sodium nitrate}$$
$$AgNO_3(aq) + \quad NaCl(aq) \quad \rightarrow \quad AgCl(s) \quad + \quad NaNO_3(aq)$$

Ionically:

$$Ag^+(aq) + NO_3^-(aq) + Na^+(aq) + Cl^-(aq) \rightarrow AgCl(s) + Na^+(aq) + NO_3^-(aq)$$

Notice that $Na^+(aq)$ and $NO_3^-(aq)$ ions are unchanged. These are the spectator ions. So in fact the equation could be written without these, simply as:

$$Ag^+(aq) + Cl^-(aq) \rightarrow AgCl(s)$$

Hydrated compounds

You will be familiar with blue copper sulfate, with the formula $CuSO_4.5H_2O$. Blue copper sulfate is **hydrated**: there are five molecules of water associated with each $CuSO_4$. This water is part of the crystal structure and is called **water of crystallisation**. The water of crystallisation can be driven off by relatively gentle heating to leave behind white copper sulfate, which is described as **anhydrous** (dry).

$$CuSO_4.5H_2O(s) \rightarrow CuSO_4(s) + 5H_2O(g)$$

More equations with spectator ions

Here are some more examples of reactions that involve spectator ions.

1 $H_2SO_4(aq) + 2NaOH \rightarrow Na_2SO_4(aq) + 2H_2O(l)$

Ionic equation:
$2H^+(aq) + SO_4^{2-}(aq) + 2Na^+(aq) + 2OH^-(aq) \rightarrow 2Na^+(aq) + SO_4^{2-}(aq) + 2H_2O(l)$

Cancelling out the spectator ions gives: $2H^+(aq) + 2OH^-(aq) \rightarrow 2H_2O(l)$

2 $CuO(s) + Mg(s) \rightarrow MgO(s) + Cu(s)$

Ionic equation: $(Cu^{2+} O^{2-})(s) + Mg(s) \rightarrow (Mg^{2+} O^{2-})(s) + Cu(s)$

Cancel spectator ions: $Cu^{2+}(s) + Mg(s) \rightarrow Cu(s) + Mg^{2+}(s)$

3 $Na_2CO_3(aq) + 2HCl(aq) \rightarrow 2NaCl(aq) + CO_2(g) + H_2O(l)$

Ionic equation:
$2Na^+(aq) + CO_3^{2-}(aq) + 2H^+(aq) + 2Cl^-(aq) \rightarrow 2Na^+(aq) + 2Cl^-(aq) + CO_2(g) + H_2O(l)$

Cancel spectator ions: $CO_3^{2-}(aq) + 2H^+(aq) \rightarrow CO_2(g) + H_2O(l)$

4 $2NaI(aq) + Cl_2(aq) \rightarrow 2NaCl(aq) + I_2(aq)$

Ionic equation: $2Na^+(aq) + 2I^-(aq) + Cl_2(aq) \rightarrow 2Na^+(aq) + 2Cl^-(aq) + I_2(aq)$

Cancel spectator ions: $2I^-(aq) + Cl_2(aq) \rightarrow 2Cl^-(aq) + I_2(aq)$

5 $NaHCO_3(aq) + HCl(aq) \rightarrow NaCl(aq) + CO_2(g) + H_2O(l)$

Ionic equation:
$Na^+(aq) + HCO_3^-(aq) + H^+(aq) + Cl^-(aq) \rightarrow Na^+(aq) + Cl^-(aq) + CO_2(g) + H_2O(l)$

Cancel spectator ions: $HCO_3^-(aq) + H^+(aq) \rightarrow CO_2(g) + H_2O(l)$

Summary test 2.3

1 Calculate the empirical formula of each of the following compounds. (You could try to name them too.)
   a A liquid containing 2.0 g of hydrogen, 32.1 g sulfur, and 64.0 g oxygen.
   b A white solid containing 4.0 g calcium, 3.2 g oxygen, and 0.2 g hydrogen.
   c A white solid containing 0.243 g magnesium and 0.710 g chlorine.

2 3.888 g magnesium ribbon was burnt completely in air and 6.448 g of magnesium oxide was produced.
   a Calculate how many moles of magnesium and oxygen are present in 6.448 g of magnesium oxide.
   b Calculate the empirical formula of magnesium oxide.

3 Calculate the empirical formula of each of the following molecules.
   a cyclohexane, C_6H_{12}
   b dichloroethene, $C_2H_2Cl_2$
   c benzene, C_6H_6

4 M_r for ethane-1,2-diol is 62.0. It is composed of carbon, hydrogen, and oxygen in the ratio by moles of 1 : 3 : 1. Identify its molecular formula.

5 An organic compound containing only carbon, hydrogen, and oxygen was found to have 62.07% carbon and 10.33% hydrogen. Identify the molecular formula if $M_r = 58.0$.

6 A sample of benzene of mass 7.8 g contains 7.2 g of carbon and 0.6 g of hydrogen. If M_r is 78.0, determine:
   a the empirical formula
   b the molecular formula.

7 Deduce the formulae of the following ionic compounds using the charges on the ions: caesium nitrate; zinc chloride; ammonium bromide; calcium hydrogencarbonate.

8 Balance the following equations.
   a $Mg + O_2 \rightarrow MgO$
   b $Ca(OH)_2 + HCl \rightarrow CaCl_2 + H_2O$
   c $Na_2O + HNO_3 \rightarrow NaNO_3 + H_2O$

Reacting masses and volumes (of solutions and gases)

Learning outcomes

On these pages you will learn to:

- perform calculations using gas volumes
- perform calculations using solution concentrations

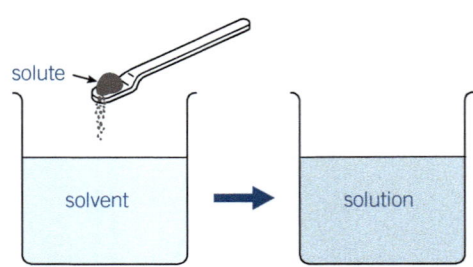

Figure 6 A solution contains a solute and a solvent

Knowing the quantities in which chemicals react is vital. It allows us to construct balanced equations for chemical reactions. We can measure reacting quantities by

- using known volumes of solutions of known concentration
- measuring the volumes of gases that react (under specified conditions of temperature and pressure, see Section 4.1)
- directly measuring masses of reactants and products.

Solutions

A solution consists of a solvent with a solute dissolved in it (Figure 6).

The units of concentration

The concentration of a solution tells us how much solute is present in a known volume of solution.

Concentrations of solutions are measured in mol dm^{-3}. 1 mol dm^{-3} means there is 1 mole of solute per cubic decimetre of solution; 2 mol dm^{-3} means there are 2 moles of solute per cubic decimetre of solution, and so on.

Worked example

Finding the concentration in mol dm^{-3}

1.17 g of sodium chloride was dissolved in water to make 500 cm^3 of solution. What is the concentration of the solution in mol dm^{-3}? A_r Na = 23.0, A_r Cl = 35.5

The mass of 1 mole of sodium chloride, NaCl, is 23.0 + 35.5 = 58.5 g.

$$\text{number of moles } n = \frac{\text{mass } m \text{ (g)}}{\text{mass of 1 mole } M \text{ (g)}}$$

So 1.17 g of NaCl contains $\frac{1.17}{58.5}$ = 0.020 mol

This is dissolved in 500 cm^3, so 1000 cm^3 (1 dm^3) would contain 0.040 mol of NaCl. This means that the concentration of the solution is 0.040 mol dm^{-3}.

The general way of finding a concentration is to remember the relationship:

$$\text{concentration } c \text{ (mol dm}^{-3}) = \frac{\text{number of moles } n}{\text{volume } V \text{ (dm}^3)}$$

Substituting into this gives:

$$\text{concentration} = \frac{0.020}{0.500} = 0.040 \text{ mol dm}^{-3}$$

The number of moles in a given volume of solution

You often have to work out how many moles are present in a particular volume of a solution of known concentration. The general formula for the number of moles in a solution of concentration c (mol dm^{-3}) and volume V (cm^3) is:

$$\text{Number of moles in solution, } n = \frac{\text{concentration } c \text{ (mol dm}^{-3}) \times \text{volume } V \text{ (cm}^3)}{1000}$$

Worked example

Moles in a solution

How many moles are present in $25.0\,cm^3$ of a solution of concentration $0.10\,mol\,dm^{-3}$?

From the definition,

$1000\,cm^3$ of a solution of $1.00\,mol\,dm^{-3}$ contains $1\,mol$

So $1000\,cm^3$ of a solution of $0.100\,mol\,dm^{-3}$ contains $0.100\,mol$

So $1.0\,cm^3$ of a solution of $0.100\,mol\,dm^{-3}$ contains $\frac{0.10}{1000} = 0.00010\,mol$

So $25.0\,cm^3$ of a solution of $0.10\,mol\,dm^{-3}$ contains $25.0 \times 0.00010 = 0.0025\,mol$

Using the formula gives the same answer:
$$n = \frac{c \times V}{1000} = \frac{0.10 \times 25.0}{1000} = 0.0025\,mol$$

Worked example

The reaction of calcium carbonate with acid
$$CaCO_3(s) + 2HCl(aq) \rightarrow CaCl_2(aq) + CO_2(g) + H_2O(l)$$

the reacting quantities are:
$$CaCO_3(s) + 2HCl(aq) \rightarrow CaCl_2(aq) + CO_2(g) + H_2O(l)$$
$$1\,mol \quad\quad 2\,mol \quad\quad 1\,mol \quad\quad 1\,mol \quad 1\,mol$$

1 How much $1\,mol\,dm^{-3}$ hydrochloric acid would just react with $1\,g$ of $CaCO_3$ to the nearest whole number?
M_r of $CaCO_3$ is $40 + 12 + (3 \times 16) = 100$

So $1\,g$ is $\frac{1}{100}\,mol\,CaCO_3$.

So we need $\frac{2}{100}\,mol$ of HCl, ie $0.02\,mol$.

$$\text{No of moles} = \frac{\text{concentration} \times \text{volume in cm}^3}{1000}$$

$$\text{Volume} = \frac{0.02 \times 1000}{1} = 20\,cm^3\,HCl \text{ to just react with all the } CaCO_3.$$

2 How much carbon dioxide would we expect to get from $1\,g$ of $CaCO_3$?
$1\,g\,CaCO_3$ is $\frac{1}{100}\,mol$ so we would get $\frac{1}{100}\,mol\,CO_2$.

Under room conditions, $1\,mol$ of any gas has a volume of $24000\,cm^3$.
So we would get $\frac{1}{100} \times 24000 = 240\,cm^3$ gas.

3 If we collected $200\,cm^3$ of gas, what would be the % yield of the reaction?
$$\text{Yield} = \frac{\text{actual volume of gas collected}}{\text{theoretical volume}} \times 100\% = \frac{200}{240} \times 100 = 83.3\%$$

4 What would be the mass of this gas?
M_r of CO_2 is $12 + (2 \times 16) = 44\,g$

We would get $\frac{1}{100}\,mol = 0.44\,g$

5 To be sure that the reaction goes to completion in a reasonable time, we decided to use an excess of hydrochloric acid. Is $25\,cm^3$ of $2\,mol\,dm^{-3}$ acid an excess?

$$\text{No of moles of HCl} = \frac{\text{concentration} \times \text{volume in cm}^3}{1000} = \frac{2 \times 25}{1000} = 0.05\,mol$$

We actually need $0.02\,mol$ of acid, so this is in fact an excess.

Exam tip

To get a solution with a concentration of $1\,mol\,dm^{-3}$ you have to add the solvent to the solute until you have $1\,dm^3$ of solution. You do *not* add $1\,mol$ of solute to $1\,dm^3$ of solvent. This would give more than $1\,dm^3$ of solution.

Exam tip

1 decimetre $= 10\,cm$, so one cubic decimetre, $1\,dm^3$, is:
$10\,cm \times 10\,cm \times 10\,cm = 1000\,cm^3$
This is the same as 1 litre ($1\,L$).

The small negative in $mol\,dm^{-3}$ means 'per', and is sometimes written as a slash, mol/dm^3.

📖 **Error in measurements**

Every measurement has some uncertainty (also known as error). In general, the uncertainty in a single measurement from an instrument is half the value of the smallest division. The uncertainty of a measurement may also be expressed by a ± sign at the end. For example the mass of an electron is given as $9.109\,382\,91 \times 10^{-31}$ kg $\pm\ 0.000\,000\,40 \times 10^{-31}$ kg, that is, it is between $9.109\,383\,31$ and $9.109\,382\,51 \times 10^{-31}$ kg.

For example, a 100 cm³ measuring cylinder has 1 cm³ as its smallest division so the measuring error can be taken as 0.5 cm³. So if you measure 50 cm³, the percentage error is $(\dfrac{0.5}{50}) \times 100\% = 1\%$

Calculate the percentage error if you use the measuring cylinder to measure:

a 10 cm³
b 100 cm³

Answers: a 5% b 0.5%

Reacting volumes of gases

Equal volumes of *all* gases contain the same number of particles (molecules or separate atoms). This is known as **Avogadro's law**. At room temperature and pressure (around 298 K or 25°C, and 100 kPa or 1 atmosphere) one mole of particles has a volume (called the molar volume) of 24 000 cm³. (More precisely, one mole of particles of any gas has a volume of 22 400 cm³ at standard temperature and pressure (273 K and 101 kPa)).

Worked example

Finding the equation for the reaction of zinc and hydrochloric acid

Find the balanced equation for the reaction of zinc and hydrochloric acid. The word equation is:

$$\text{zinc} + \text{hydrochloric acid} \rightarrow \text{zinc chloride} + \text{hydrogen}$$

(**Remember**: metal + acid → salt + hydrogen)

A_r of Zn = 65.4 g mol⁻¹

Using the apparatus shown in Figure 7, 0.215 g of zinc is added to excess dilute hydrochloric acid (HCl). Bubbles of hydrogen gas are produced and collected in the gas syringe. At the end of the reaction all the zinc is used up and 80 cm³ of hydrogen has been collected (at room temperature and pressure).

Exam tip

Excess hydrochloric acid means that there is more than enough acid to react with *all* the zinc. So the reaction stops when all of the zinc is used up. The zinc is the **limiting reagent**.

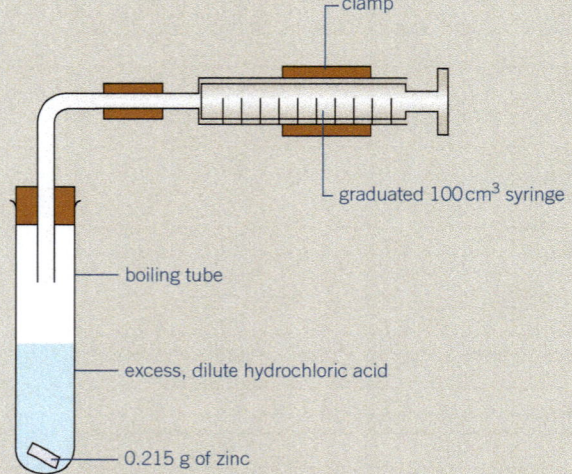

Figure 7 *Measuring the volume of hydrogen produced when zinc reacts with hydrochloric acid*

$0.215\,g$ of zinc is $\dfrac{0.215\,g}{65.4\,g\,mol^{-1}} = 0.0033\,mol\,Zn$

$80\,cm^3\,H_2$ is $\dfrac{80\,cm^3}{24\,000\,cm^3\,mol^{-1}} = 0.0033\,mol\,H_2$

The other product of the reaction is the salt zinc chloride, $ZnCl_2$.

We have found that 1 mol of Zn produces 1 mol H_2

So the unbalanced equation is:

$$Zn + HCl \rightarrow ZnCl_2 + H_2$$

In order to produce two atoms of hydrogen and two atoms of chlorine on the right-hand side of the equation, two molecules of HCl were used. So the balanced equation is:

$$Zn + 2HCl \rightarrow ZnCl_2 + H_2$$

Adding state symbols gives:

$$Zn(s) + 2HCl(aq) \rightarrow ZnCl_2(aq) + H_2(g)$$

Worked example

Combustion of gases

Propene, C_3H_8, is used as a fuel in camping gas stoves. It burns completely in a plentiful supply of oxygen to produce carbon dioxide and water.

a Write a balanced equation with state symbols for the complete combustion of propane.

b Determine the M_r of propane.

c 0.044 g of propane was burned in excess of oxygen. How many moles of propane is this?

d Calculate the volume this amount of propane occupies at room conditions. (Use the approximation that one mole of any gas occupies $24\,000\,cm^3$ under room conditions.)

e Calculate the volume of oxygen needed for the reaction to proceed.

f Calculate the volume of carbon dioxide produced.

g Explain why the volume of water produced is negligible under room conditions.

Solutions

a $C_3H_8(g) + 5O_2(g) \rightarrow 3CO_2(g) + 4H_2O(g)$

b M_r of propane $= (3 \times 12.0) + (8 \times 1.0) = 44.0\,g\,mol^{-1}$

c 0.044 g of propane is $\dfrac{0.044}{44} = 0.001\,mol$

d $24\,000 \times 0.001 = 24\,cm^3$

e 1 mol of propane reacts with 5 mol O_2, so the volume of O_2 required is $5 \times 24 = 120\,cm^3$

f 3 mol CO_2 is produced which has a volume of $3 \times 24\,cm^3 = 72\,cm^3$

g The water would be produced as a liquid rather than a gas.

Reacting masses

To calculate the reacting masses in a chemical reaction, we have to measure the masses of the starting material and product.

We can use a balanced equation to predict the mass of product formed. But this gives us the theoretical amount. In reality, we will obtain less than this because:

- The reaction may not go to completion.
- There may be other competing reactions.
- We will always lose some product, for example some product may be left in the filter paper, or droplets may be left in a flask.

The **yield** of a chemical reaction is:

$$\text{Yield} = \frac{\text{the mass of product actually obtained}}{\text{the theoretical mass of product predicted by the equation}}$$

This is usually expressed as a percentage.

Worked example

Finding the yield

We can make magnesium oxide by heating a known mass of magnesium ribbon in a crucible, until all the magnesium has reacted. The lid is lifted occasionally to allow air in (see Figure 8). The magnesium will then react with the oxygen in the air to form magnesium oxide.

Figure 8 *Making magnesium oxide by direct combination*

The balanced symbol equation for this reaction is:

$$2Mg(s) + O_2(g) \rightarrow 2MgO(s)$$

The reacting masses are (to the nearest whole number):

$$2Mg(s) + O_2(g) \rightarrow 2MgO(s)$$

$$24\,g \qquad 32\,g \qquad 56\,g$$

So if we started with 0.24 g of magnesium we would expect to produce 0.56 g of magnesium oxide.

However each time we lift the lid some magnesium oxide is lost as 'smoke'. In an experiment, 0.24 g magnesium produced only 0.50 g of magnesium oxide due to loss of smoke.

So the yield of this reaction was:

$$\frac{0.50\,g}{0.56\,g} \times 100\% = 89.3\%$$

Worked example

Formulae and equations

a What volume of a 2 mol dm⁻³ solution of sodium chloride contains 0.1 mol of sodium chloride?

Solution to a

1000 cm³ of the solution contains 2 mol NaCl

500 cm³ of the solution contains 1 mol NaCl

50 cm³ of the solution contains 0.1 mol NaCl

b Starting with a 2 mol dm⁻³ solution of sodium chloride, how would you make a 0.5 mol dm⁻³ solution?

Solution to b

1000 cm³ of the solution contains 2 mol NaCl

500 cm³ of the solution contains 1 mol NaCl

250 cm³ of the solution contains 0.5 mol NaCl

We need 0.5 mol in 1000 cm³ Of solution

So, we take 250 cm³ of the original solution and add distilled water to make it up to 1000 cm³

c 1.0 g of calcium carbonate is reacted with 100 cm³ of 2 mol dm⁻³ hydrochloric acid. Which reactant is in excess?

Solution to c

The equation for this reaction is:

$$CaCO_3(s) + 2HCl(aq) \rightarrow CaCl_2\ (aq) + CO_2(g) + H_2O(l)$$

M_r of $CaCO_3$ is 100 (to the nearest whole number), so 100 g is the mass of 1 mole of $CaCO_3$.

1.0 g of $CaCO_3$ is the mass of 0.01 mol, which will react with 0.02 mol HCl.

1000 cm³ of 2 mol dm⁻³ HCl contains 2 mol HCl

100 cm³ of 2 mol dm⁻³ HCl contains 0.2 mol HCl

Therefore the HCl was in excess and the reaction would stop when all the $CaCO_3$ was used up.

Summary test 2.4

1 Calculate the concentration in mol dm^{-3} of the following.
   a 0.500 mol acid in 500 cm^3 of solution
   b 0.250 mol acid in 2000 cm^3 of solution
   c 0.200 mol solute in 20 cm^3 of solution

2 Calculate how many moles of solute there are in the following.
   a 20.0 cm^3 of a 0.100 mol dm^{-3} solution
   b 50.0 cm^3 of a 0.500 mol dm^{-3} solution
   c 25.0 cm^3 of a 2.00 mol dm^{-3} solution

3 0.234 g of sodium chloride was dissolved in water to make 250 cm^3 of solution.
   a State the M_r for NaCl. A_r Na = 23.0, A_r Cl = 35.5
   b Calculate how many moles of NaCl is in 0.234 g.
   c Calculate the concentration in mol dm^{-3}.

4 In the reaction

$$Mg(s) + 2HCl(aq) \rightarrow MgCl_2(aq) + H_2$$

   2.60 g of magnesium was added to 100 cm^3 of 1.00 mol dm^{-3} hydrochloric acid.
   a State if there will be any magnesium left when the reaction finishes. Explain your answer.
   b Calculate the volume of hydrogen produced at 25°C and 100 kPa.

5 Determine the balanced equation for the reaction between sulfuric acid and sodium hydroxide.

6 In the worked example on page 34, 'Finding the equation for the reaction of zinc and hydrochloric acid':
   a Suggest how you might start the reaction without losing any of the hydrogen.
   b Explain why it is acceptable to measure the volume of hydrogen at room temperature and pressure.

7 Another student carried out the experiment above starting with 0.12 g of magnesium and obtained 0.20 g of magnesium oxide.
   a Calculate the theoretical mass of magnesium oxide. Use value of A_r to the nearest whole number.
      A_r Mg = 24
      A_r O = 16
   b Calculate the percentage yield.
   c Suggest two reasons why the yield was less than the yield in the experiment above.

(📖 **Launch additional digital resources for the chapter**)

1 a Define the term 'relative atomic mass'.   *(2 marks)*

  b Calculate the mass in kilograms, of a single atom of carbon.   *(2 marks)*

  c Give the formulae of the following compounds:
    i sodium carbonate   **ii** potassium nitrate
    iii calcium hydroxide   **iv** ammonium sulfate
    (4 marks)

2 Zinc sulfate can be prepared by the reaction between zinc and sulfuric acid according to the following equation:

$$Zn(s) + H_2SO_4(aq) \rightarrow ZnSO_4(aq) + H_2(g)$$

  a $25\,cm^3$ of $0.4\ mol\ dm^{-3}$ sulfuric acid was reacted with $0.7\,g$ of zinc.
    i Calculate which reactant is present in excess.
    (3 marks)
    ii Upon crystallisation, $1.36\,g$ of zinc sulfate crystals were isolated. Calculate the percentage yield.
    (3 marks)
    iii Suggest why the yield is not 100%.   *(1 mark)*

3 Dodecane, $C_{10}H_{22}$, is used as a component of jet fuel. At $298\,K$ it has a density of $0.75\,g\,dm^{-3}$.
(The gas constant $R = 8.31\,J\,K^{-1}\,mol^{-1}$)

  a Construct an equation for the complete combustion of dodecane.   *(2 marks)*

  b Calculate the number of moles of dodecane present in $0.2\,dm^3$ of dodecane at 298 K.   *(3 marks)*

  c Calculate the total volume of gas produced in m^3 from the complete combustion of $0.2\,dm^3$ of dodecane at 298 K.   *(2 marks)*

  d Use your answer to part **c** to calculate the volume of carbon dioxide produced.   *(1 mark)*

4 Anhydrous copper sulfate is a white solid that produces a blue solution. $4.79\,g$ of anhydrous copper(II) sulfate was added to water. Upon crystallisation, $7.49\,g$ of blue crystals of hydrated copper sulfate, $CuSO_4.XH_2O$ were isolated.

Calculate the value for X and hence give the formula for hydrated copper(II) sulfate.   *(6 marks)*

5 Barium chloride solution is used as a common test for sulfate ions SO_4^{2-} in inorganic solutions. When added to acidified sodium sulfate solution a white precipitate is observed.

  a Give a full chemical equation for this reaction. Include state symbols.   *(2 marks)*

  b Construct the simplest ionic equation for this reaction.   *(2 marks)*

  c Explain why it is necessary to acidify the sodium sulfate solution prior to the addition of barium chloride solution.   *(1 mark)*

  d Barium sulfate is used in medical imaging which requires it to be swallowed by the patient. Ba^{2+} ions are toxic. Explain why ingesting barium sulfate does not harm the patient.   *(2 marks)*

6 $25.0\,cm^3$ of sodium hydroxide solution of unknown concentration was titrated against $0.25\ mol\ dm^{-3}$ hydrochloric acid. This was repeated until two concordant results were obtained. The results are given in the table below:

	1	2	3	4	5
Titre / cm³	17.3	17.7	16.9	17.2	17.5

  a Identify the concordant values and calculate the mean titre volume to two decimal places.   *(2 marks)*

  b Use your answer to part **a** to calculate the concentration of sodium hydroxide solution. Give your answer to 2 decimal places.   *(3 marks)*

7 Antacid tablets contain calcium carbonate which neutralises excess stomach acid to relieve the symptoms of indigestion. An antacid tablet was crushed and reacted with $50\,cm^3$ of $0.5\ mol\ dm^{-3}$ hydrochloric acid. The resulting solution was diluted to $250\,cm^3$ using a volumetric flask.

$25.0\,cm^3$ of the resulting solution was titrated against $0.1\ mol\ dm^{-3}$ sodium hydroxide, requiring $18.3\,cm^3$ for complete neutralisation.

  a Calculate the concentration of hydrochloric acid in the $250\,cm^3$ volumetric flask.   *(4 marks)*

  b Use your answer to part **a** to deduce the number of moles of hydrochloric acid that reacted with the antacid tablet and hence calculate the number of moles of calcium carbonate present in a single antacid tablet.   *(3 marks)*

  c State the mass in mg, of calcium carbonate present in each antacid tablet.   *(2 marks)*

3 Chemical bonding

3.1 Electronegativity and bonding

The bonding between atoms is fundamental to the understanding of chemistry.

Why do chemical bonds form?

- The bonds between atoms always involve their outer electrons.
- Noble gases (Figure 1) are very unreactive.
- When atoms bond together they share, transfer or pool electrons to achieve a more stable electron arrangement, like those of the noble gases. Noble gases all have eight electrons in their outer shell, except helium, which has two.
- There are three types of strong chemical bonds—**ionic**, **covalent**, and **metallic**.

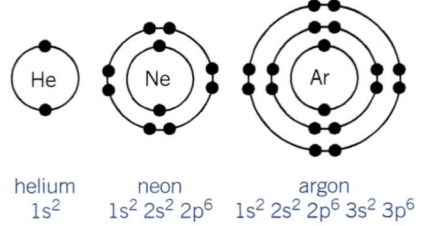

helium
$1s^2$

neon
$1s^2\ 2s^2\ 2p^6$

argon
$1s^2\ 2s^2\ 2p^6\ 3s^2\ 3p^6$

Figure 1 *Noble gases*

Electronegativity

Electronegativity is the power of an atom to attract electrons towards itself.

For example, fluorine is better at attracting electrons than hydrogen and so fluorine is said to be more electronegative than hydrogen.

Electronegativity depends on:

1 the nuclear charge – the higher the charge (proton number), the greater the electronegativity

2 the distance between the nucleus and the outer shell electrons – the greater the distance of the outer electron(s) from the nucleus, the smaller the electronegativity

3 the shielding of the nuclear charge by electrons in inner shells – the electrons in inner shells tend to cancel out the effect of the nuclear charge

Trends in electronegativity

Going down a group in the Periodic Table, electronegativity decreases (the atoms get larger) and there is more shielding by electrons in inner shells. So the attraction between the nucleus and the outer shell electrons is weaker.

Going across a period in the Periodic Table, the electronegativity increases. The nuclear charge increases, the number of inner shells shielding the nuclear charge remains the same and the atoms become smaller.

The Pauling scale is used as a measure of electronegativity. It runs from 0 to 4. The greater the number, the more electronegative the atom, see Table 1. The noble gases have no number because they do not, in general, form bonds.

So, the most electronegative atoms are found at the top right-hand corner of the Periodic Table (ignoring the noble gases which form few compounds), see Table 1. The most electronegative atoms are fluorine, oxygen, and nitrogen followed by chlorine. Atoms of low electronegativity are often described as **electropositive**.

Electronegativity and bonding type

The difference in electronegativity between a pair of atoms can be used to predict the type of bonding – covalent or ionic – that will occur between them. Two identical atoms (with the same electronegativity) will attract any outer electrons equally, so they will bond covalently. So two fluorine atoms, for example will share an outer electron each to form a covalent bond, F–F, which is non-polar. If two elements have very similar electronegativities, the bond between their atoms may also be non-polar (for example, a C–H bond).

In the case of two atoms with a large difference in electronegativity, the electrons will be strongly attracted to the more electronegative atom and one (or more) electrons will effectively be completely transferred from the less electronegative atom to the more electronegative one. This is **ionic bonding**. In sodium fluoride, for example, there is essentially complete transfer of an electron from sodium to fluorine and the bonding is $Na^+ F^-$.

In intermediate cases, the electrons may be shared between two atoms, but not equally. This is described as **polar** covalent bonding and the symbols $\delta+$ and $\delta-$ are used to indicate a partial transfer of charge. So hydrogen chloride may be written $H^{\delta+}–Cl^{\delta-}$.

Table 1 *Some values for Pauling electronegativity*

H 2.1							He
Li 1.0	Be 1.5	B 2.0	C 2.5	N 3.0	O 3.5	F 4.0	Ne
Na 0.9	Mg 1.2	Al 1.5	Si 1.8	P 2.1	S 2.5	Cl 3.0	Ar
						Br 2.8	Kr

Exam tip

When chemists consider the electrons as charge clouds, the term **electron density** is often used to describe the way the negative charge is distributed in a molecule.

Exam tip

If the difference in electronegativity between two atoms is greater than 2.1, the bonding between them is considered to be ionic.

Summary test 3.1

1 Identify which of the following are ionic compounds and explain why.
   a CO
   b KF
   c MgO
   d HF

2 Write $\delta+$ and $\delta-$ on the following formulae to show the polarity of the bonds.
   a H–Br
   b H–O–H
   c H–S–H

3.2 Ionic bonding

Learning outcomes

On these pages you will learn to:

- define ionic bonding
- use dot-and-cross diagrams to describe ionic bonding

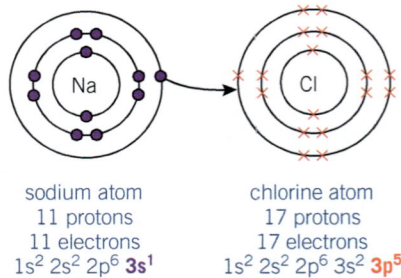

sodium atom
11 protons
11 electrons
$1s^2 2s^2 2p^6$ **3s^1**

chlorine atom
17 protons
17 electrons
$1s^2 2s^2 2p^6 3s^2$ **3p^5**

Figure 2 *A dot-and-cross diagram to show the transfer of the 3s^1 electron from the sodium atom to the 3p orbital on a chlorine atom. Remember that electrons are all identical whether shown by a dot or a cross*

Exam tip

When drawing dot-and-cross diagrams, it may help to draw rings to indicate full electron shells, but these are optional.

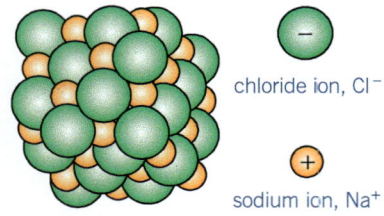

chloride ion, Cl$^-$

sodium ion, Na$^+$

Figure 4 *The sodium chloride structure. This is an example of a giant ionic lattice. The strong bonding extends throughout the compound and because of this it will be difficult to melt*

Metals have one, two, or three electrons in their outer shell and are electropositive. Therefore, the easiest way for them to attain the electron structure of a noble gas is to lose their outer electrons. Non-metals have spaces in their outer shells and are electronegative, so that the easiest way for them to attain the electron structure of a noble gas (page 40, Figure 1) is to gain electrons.

- Ionic bonding occurs between metals and non-metals.
- Electrons are transferred from metal atoms to non-metal atoms.
- Positive ions (cations) and negative ions (anions) are formed.

Sodium chloride (Figure 2) has ionic bonding.

- Sodium, Na, has 11 electrons (and 11 protons). The electron arrangement is $1s^2 2s^2 2p^6 3s^1$.
- Chlorine, Cl, has 17 electrons (and 17 protons). The electron arrangement is $1s^2 2s^2 2p^6 3s^2 3p^5$.
- An electron is transferred. The single outer electron of the sodium atom moves into the outer shell of the chlorine atom.
- Each outer shell is now full.

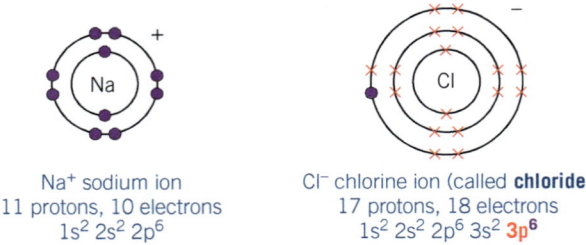

Na$^+$ sodium ion
11 protons, 10 electrons
$1s^2 2s^2 2p^6$

Cl$^-$ chlorine ion (called **chloride**)
17 protons, 18 electrons
$1s^2 2s^2 2p^6 3s^2$ **3p^6**

Figure 3 *The ions that result from electron transfer*

Both sodium and chlorine now have a noble gas electron arrangement. Sodium has the neon noble gas arrangement whereas chlorine has the argon noble gas arrangement (compare the ions in Figure 3 with the noble gas atoms on page 40, Figure 1).

The two charged particles that result from the transfer of an electron are called **ions**.

- The sodium ion is positively charged because it has lost a negative electron.
- The chloride ion is negatively charged because it has gained a negative electron.
- The two ions are attracted to each other and to other, oppositely charged, ions in the sodium chloride compound by electrostatic forces.

Therefore, ionic bonding is the result of electrostatic attraction between oppositely charged ions. The attraction extends throughout the compound. Every positive ion attracts every negative ion and vice versa. Ionic compounds always exist in a structure called a **lattice**. The strong bonding gives them a high melting temperature, and the presence of charged particles means they conduct electricity when molten or when dissolved in water. Figure 4 shows the three-dimensional lattice for sodium chloride with its singly charged ions.

The formula of sodium chloride is NaCl because for every one sodium ion there is one chloride ion.

Worked example

1 Magnesium oxide

Magnesium, Mg, has 12 electrons. The electron arrangement is $1s^2\ 2s^2\ 2p^6\ 3s^2$.

Oxygen, O, has eight electrons. The electron arrangement is $1s^2\ 2s^2\ 2p^4$.

This time, two electrons are transferred from the 3s orbitals on each magnesium atom. Each oxygen atom receives two electrons into its 2p orbital.

- The magnesium ion, Mg^{2+}, is positively charged because it has lost two negative electrons.
- The oxide ion, O^{2-}, is negatively charged because it has gained two negative electrons.
- The formula of magnesium oxide is MgO.

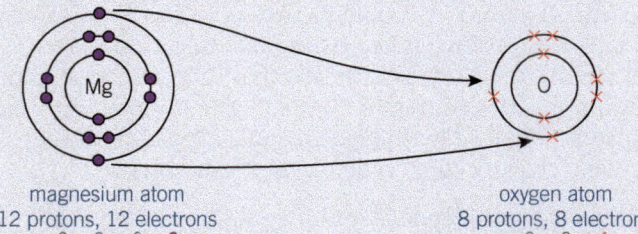 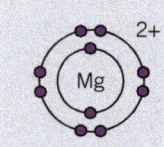 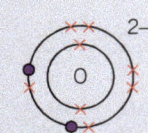

magnesium atom	oxygen atom	Mg^{2+} magnesium ion	O^{2-} oxygen ion (called **oxide**)
12 protons, 12 electrons	8 protons, 8 electrons	12 protons, 10 electrons	8 protons, 10 electrons
$1s^2\ 2s^2\ 2p^6\ \mathbf{3s^2}$	$1s^2\ 2s^2\ \mathbf{2p^4}$	$1s^2\ 2s^2\ 2p^6$	$1s^2\ 2s^2\ \mathbf{2p^6}$

Figure 5 *Ionic bonding in magnesium oxide, MgO*

2 Calcium fluoride

Calcium has 20 electrons ($1s^2\ 2s^2\ 2p^6\ 3s^2\ 3p^6\ 4s^2$)

Fluorine has nine electrons ($1s^2\ 2s^2\ 2p^5$)

The two outer electrons are transferred from the calcium atom, one to each of two fluorine atoms, forming one Mg^{2+} and two F^- ions, so the formula of calcium fluoride is CaF_2.

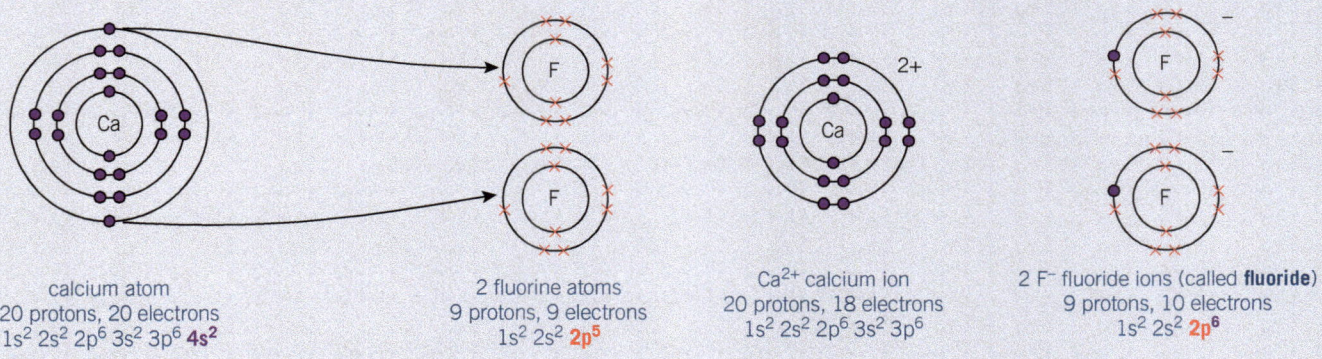

calcium atom	2 fluorine atoms	Ca^{2+} calcium ion	2 F^- fluoride ions (called **fluoride**)
20 protons, 20 electrons	9 protons, 9 electrons	20 protons, 18 electrons	9 protons, 10 electrons
$1s^2\ 2s^2\ 2p^6\ 3s^2\ 3p^6\ \mathbf{4s^2}$	$1s^2\ 2s^2\ \mathbf{2p^5}$	$1s^2\ 2s^2\ 2p^6\ 3s^2\ 3p^6$	$1s^2\ 2s^2\ \mathbf{2p^6}$

Figure 6 *Ionic bonding in calcium fluoride, CaF_2*

Summary test 3.2

1 Explain why ionic compounds have high melting temperatures.
2 Describe the conditions where ionic compounds conduct electricity.
3 Draw dot-and-cross diagrams to show the formation of the following ions. Include the electronic configuration of the atoms and ions involved.

   a the ions being formed when magnesium and fluorine react

   b the ions being formed when sodium and oxygen react.

4 Give the formulae of the compounds formed in question 3.
5 Look at the electron arrangements of the Mg^{2+} and O^{2-} ions. State the noble gas they correspond to.

3.3 Metallic bonding

Learning outcomes

On this page you will learn to:

- define and describe metallic bonding

Exam tip

The word **delocalised** is used to describe electron clouds that are spread over more than two atoms.

Exam tip

In Figure 7, the metal ions are shown spaced apart for clarity. In fact metal atoms are more closely packed, and so metals tend to have high densities.

Metals are shiny elements made up of atoms that can easily lose up to three outer electrons, leaving positive metal ions. For example, sodium, Na, 2,8,1 ($1s^2\ 2s^2\ 2p^6\ 3s^1$) loses its one outer electron, aluminium, Al, 2,8,3 ($1s^2\ 2s^2\ 2p^6\ 3s^2\ 3p^1$) loses its three outer electrons.

Bonding in metals

The atoms in a metal element cannot transfer electrons (as happens in ionic bonding) unless there is a non-metal atom present to receive them. In a metal element, the outer shells of the atoms merge. The outer electrons are no longer associated with any one particular atom. A simple picture of **metallic bonding** is that metals consist of a lattice of positive ions existing in a 'sea' of outer electrons. These electrons are **delocalised**. This means that they are not tied to a particular atom. Magnesium metal is shown in Figure 7. The positive ions tend to repel one another but this is balanced by the electrostatic attraction of these positive ions for the negatively charged 'sea' of delocalised electrons.

- The number of delocalised electrons depends on how many electrons have been lost by each metal atom.
- The metallic bonding spreads throughout, so metals have giant structures.

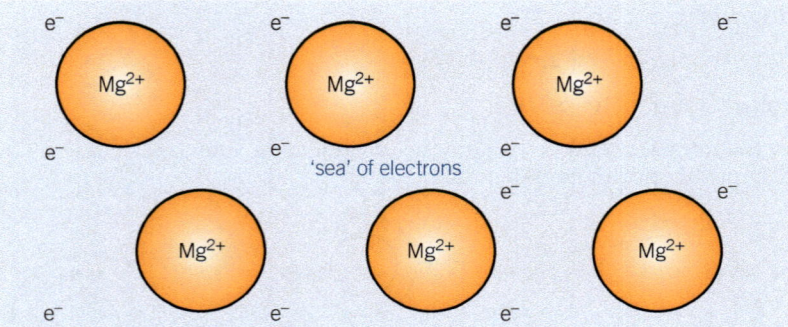

Figure 7 *The delocalised 'sea' of electrons in magnesium*

Summary test 3.3

1 Write the full electron arrangement of a calcium atom, Ca.

2 Which electrons will a calcium atom lose to gain a stable noble gas configuration?

3 State how many electrons each calcium atom will contribute to the delocalised sea of electrons that holds the metal atoms together.

Figure 8 *The delocalised 'sea' of electrons means that these metal power lines conduct electricity*

Covalent bonding and coordinate (dative covalent) bonding

Non-metal atoms need to receive electrons to fill the spaces in their outer shells.

Covalent bonding

- A **covalent bond** forms between a pair of non-metal atoms.
- A covalent bond has a shared pair of electrons.
- The atoms share some of their outer electrons so that each atom has a stable noble gas arrangement.

Forming molecules by covalent bonding

A small group of covalently bonded atoms is called a molecule. For example, chlorine exists as a gas that is made of molecules, Cl_2, see Figure 9.

Chlorine has 17 electrons and an electron arrangement $1s^2 2s^2 2p^6 3s^2 3p^5$. Two chlorine atoms make a chlorine molecule:

- The two atoms share one pair of electrons.
- Each atom now has a stable noble gas arrangement.
- The formula is Cl_2.
- Molecules are neutral because no electrons have been transferred from one atom to another.

You can represent one pair of shared electrons in a covalent bond by a dash, $Cl–Cl$.

Worked example

Methane

Methane gas is a covalently bonded compound of carbon and hydrogen. Carbon, C, has six electrons with electron arrangement $1s^2 2s^2 2p^2$ and hydrogen, H, has just one electron $1s^1$.

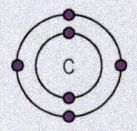

carbon
$1s^2 2s^2 2p^2$

hydrogen
$1s^1$

In order for carbon to attain a stable noble gas arrangement, there are four hydrogen atoms to every carbon atom.

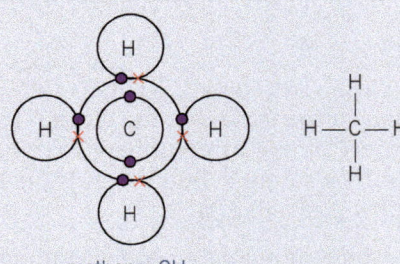

methane, CH_4

The formula of methane is CH_4. The four 2p electrons from carbon and the $1s^1$ electron from the four hydrogen atoms are shared.

chlorine atoms
$1s^2 2s^2 2p^6 3s^2 3p^5$

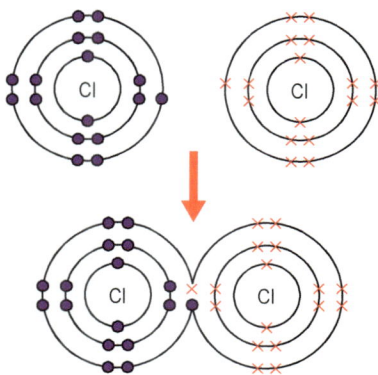

a chlorine molecule

Figure 9 *Formation of a chlorine molecule – the two atoms share a 3p electron from each atom*

Exam tip

The hydrogen atom has a filled outer shell with only two electrons ($1s^2$). It fills the first shell to get the structure of the noble gas helium. The carbon atoms have an electron arrangement $1s^2 2s^2 2p^6$.

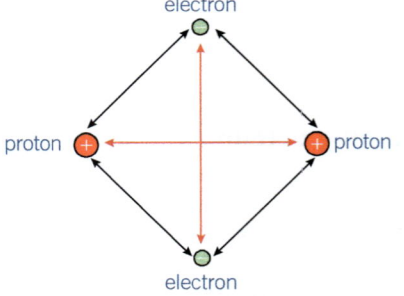

Figure 10 *The electrostatic forces within a hydrogen molecule*

> **Exam tip**
>
> Covalent bonding is defined as the electrostatic attraction between the nuclei of two atoms and a shared pair of electrons.

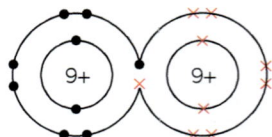

Figure 11 *Electron diagram of fluorine molecule*

Figure 12 *Electron cloud around fluorine molecule*

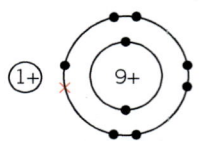

Figure 13 *Electron diagram of hydrogen fluoride molecule*

Figure 14 *Electron cloud around hydrogen fluoride molecule*

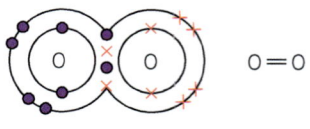

oxygen, O_2

Figure 15 *An oxygen molecule has a double bond which shares two 2p electrons from each atom*

How does sharing electrons hold atoms together?

Covalent bonds are held together by the electrostatic attraction between the nuclei and the shared electrons. This takes place within the molecule. The simplest example is hydrogen. The hydrogen molecule consists of two protons held together by a pair of electrons. The electrostatic forces are shown in Figure 10. The attractive forces are in black and the repulsive forces in red. These forces just balance when the nuclei are a particular distance apart.

Polarity of covalent bonds

Polarity is about the unequal sharing of the electrons between atoms that are bonded together covalently. It is a property of the bond.

Bond polarity measures the separation of charge resulting from the different electronegativities of the two atoms in a covalent bond. The electrons in the bond are attracted to, and displaced towards, the more electronegative atom.

Covalent bonds between two atoms that are the same

When both atoms are the same, for example, in fluorine, F_2, the electrons in the bond must be shared equally between the atoms (Figure 11) – both atoms have exactly the same electronegativity and the bond is completely non-polar.

If you think of the electrons as being in a cloud of charge, then the cloud is uniformly spread between the two atoms, as shown in Figure 12.

Covalent bonds between two atoms that are different

In a covalent bond between two atoms of different electronegativites, the electrons in the bond will not be shared equally between the atoms. For example, the molecule hydrogen fluoride, HF, shown in Figure 13.

Hydrogen has an electronegativity of 2.1 and fluorine of 4.0. This means that the electrons in the covalent bond will be attracted more by the fluorine than the hydrogen. The electron cloud is distorted towards the fluorine, as shown in Figure 14.

The fluorine end of the molecule is therefore relatively negative and the hydrogen end relatively positive, that is, electron deficient. You show this by adding partial charges to the formula:

$H^{\delta+}–F^{\delta-}$

Covalent bonds like this are said to be **polar**. The greater the difference in electronegativity, the more polar is the covalent bond.

You could say that although the H–F bond is covalent, it has some ionic character. It is going some way towards the separation of the atoms into charged ions. It is also possible to have ionic bonds with some covalent character

Multiple covalent bonds

In a double bond, four electrons are shared. The two atoms in an oxygen molecule share two pairs of electrons so that the oxygen atoms have a double bond between them (Figure 15). You can represent the two pairs of shared electrons in a covalent bond by a double line, O=O. Triple bonds may also be formed, for example in N≡N (see Table 2).

When you are drawing covalent bonding diagrams you may leave out the inner shells because the inner shells are not involved at all. Other examples of molecules with covalent bonds are shown in Table 2.

All the examples in Table 2 are neutral molecules. The atoms within the molecules are strongly bonded together with covalent bonds within the molecule. However, the molecules are not strongly attracted to each other.

Table 2 *Examples of covalent molecules. Only the outer shells are shown*

Formula	Name	Formula	Name	Formula	Name
H_2 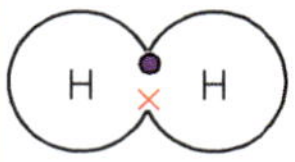	hydrogen Each hydrogen atom has a full outer main level with just two electrons	NH_3 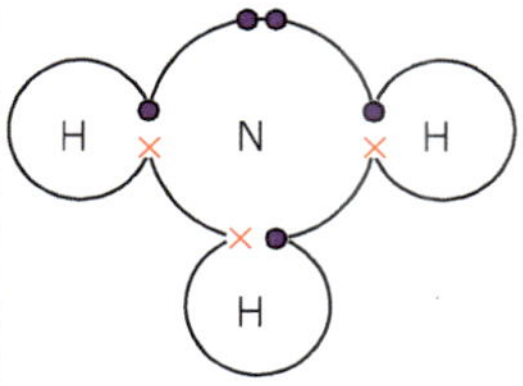	ammonia	N_2 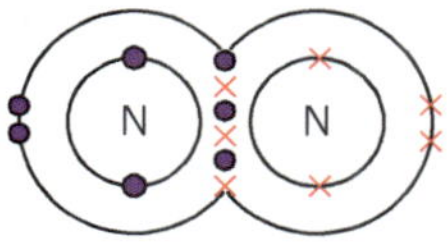	nitrogen There is a triple bond between the two nitrogen atoms
HCl 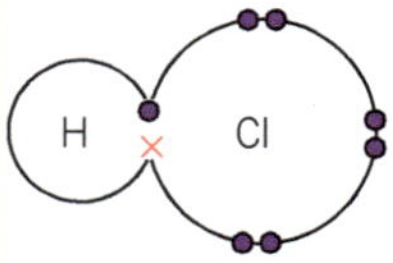	hydrogen chloride	C_2H_4 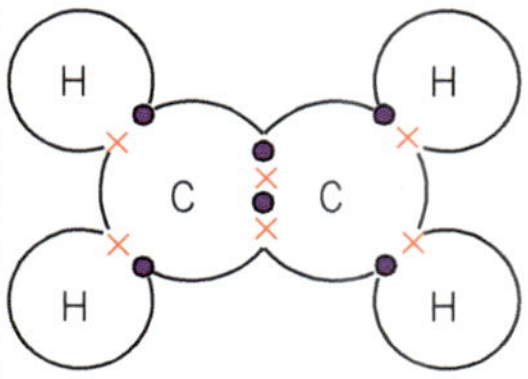	ethene There is a carbon–carbon double bond in this molecule	C_2H_6 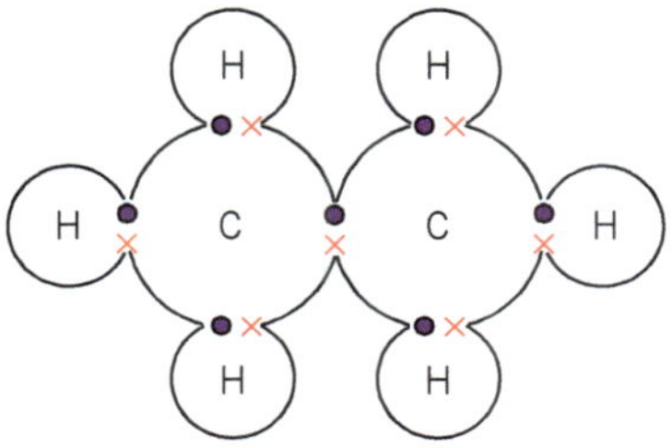	ethane There is one carbon-carbon bond and six carbon-hydrogen bonds
H_2O 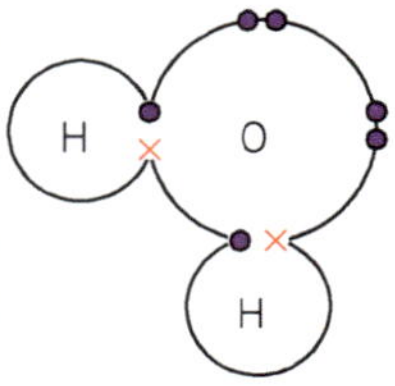	water	CO_2 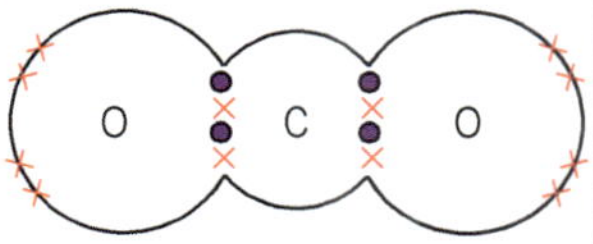	carbon dioxide There are two carbon–oxygen double bonds in this molecule	C_2H_2 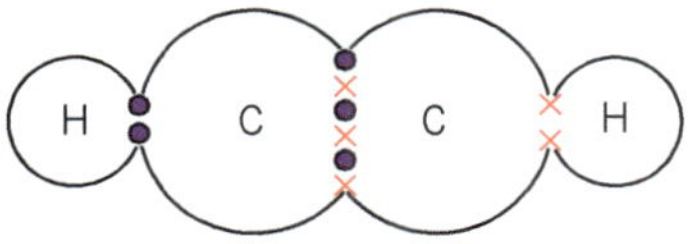	ethyne There is one carbon-carbon triple bond and two carbon-hydrogen bonds

Period 3 and beyond

Elements in Period 3 and beyond are able to hold more than eight electrons in their outer shell—this is called 'expanding their octet'. They can therefore form more than four single bonds (or their equivalent). Some examples are shown in Table 3, including two compounds of the noble gas xenon.

> **Exam tip**
>
> Another way of picturing covalent bonds is to think of electron orbitals on each atom merging to form a molecular orbital that holds the shared electrons.

Table 3 *Some covalent molecules with atoms from Period 3*

Formula	Name	Formula	Name
SO_2 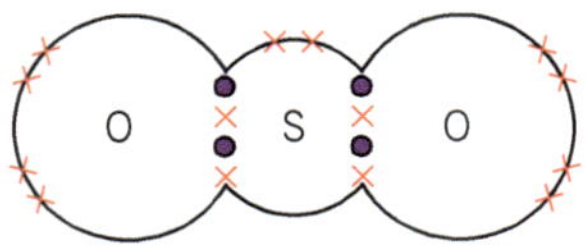	sulfur dioxide The sulfur has 10 electrons in its outer shell	XeF_2 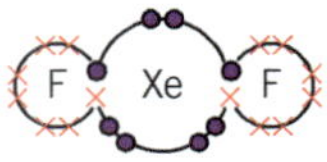	xenon difluoride The xenon has 10 electrons in its outer shell
PCl_5 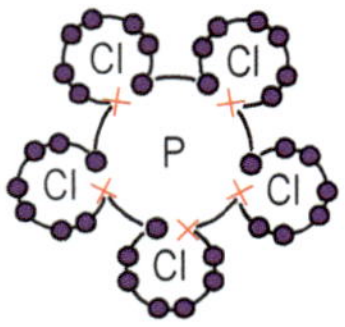	phosphorus pentachloride The phosphorus has 10 electrons in its outer shell	XeF_4 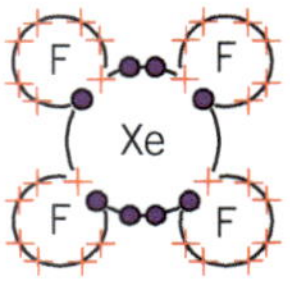	xenon tetrafluoride The xenon has 12 electrons in its outer shell
SF_6 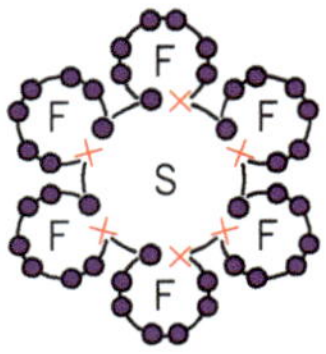	sulfur hexachloride The sulfur has 12 electrons in its outer shell	ClF_3 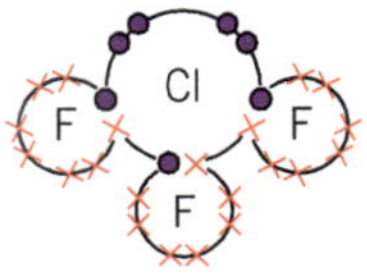	chlorine trifluoride The chlorine has 10 electrons in its outer shell

Coordinate bonding

A single covalent bond consists of a pair of electrons shared between two atoms. In most covalent bonds, each atom provides one of the electrons. But, in some bonds, one atom provides both the electrons. This is called **coordinate bonding**. It is also called **dative covalent bonding**.

In a coordinate or dative covalent bond:

- The atom that accepts the electron pair is an atom that does not have a filled outer main level of electrons—the atom is electron-deficient.
- The atom that is donating the electrons has a pair of electrons that is not being used in a bond, called a **lone pair**.

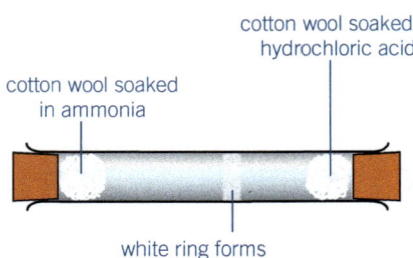

cotton wool soaked in ammonia

cotton wool soaked in hydrochloric acid

white ring forms

Figure 16 The ammonium ion is formed in the 'white ring' experiment by the reaction between hydrochloric acid and ammonia: $NH_3(g) + HCl(g) \rightarrow [NH_4^+Cl^-](s)$

Worked example

1 The ammonium ion

For example, ammonia, NH_3, has a lone pair of electrons. In the ammonium ion, NH_4^+, the nitrogen uses its lone pair of electrons to form a coordinate bond with an H^+ ion (a bare proton with no electrons at all and therefore electron-deficient).

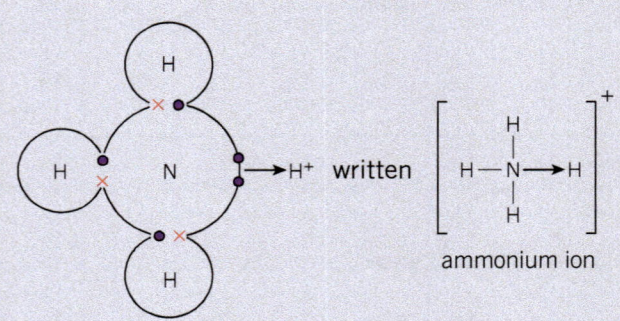

ammonium ion

Coordinate covalent bonds are represented by an arrow. The arrow points towards the atom that is accepting the electron pair. However, this is only to show how the bond was made. The ammonium ion is completely symmetrical and all the bonds have exactly the same strength and length.

- Coordinate bonds have exactly the same strength and length as ordinary covalent bonds between the same pair of atoms.

The ammonium ion has covalently bonded atoms but is a charged particle.

2 The Al_2Cl_6 molecule

Aluminium chloride forms molecules of formula $AlCl_3$ in the gas phase. The dot-and-cross diagram is as shown in Figure 17. Notice that the aluminium atom has only six electrons in its outer shell—it is termed electron deficient.

However, at lower temperatures, two coordinate bonds form between lone pairs of electrons on chlorine atoms and aluminium atoms as shown in Figure 18 to form a covalently bonded molecule of formula Al_2Cl_6. This 'double molecule' is called a **dimer**.

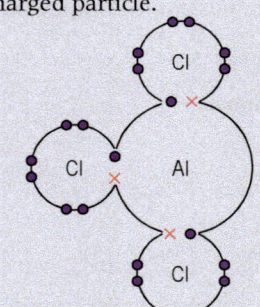

Figure 17 An electron 'dot-and-cross' diagram for $AlCl_3$

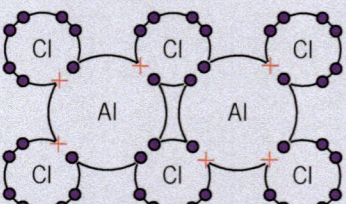

Figure 18 An electron 'dot-and-cross' diagram and a displayed structure to show the coordinate bonding in solid aluminium chloride, Al_2Cl_6.

Molecular orbitals

Dot-and-cross diagrams explain how atoms are held together in molecules and they can predict the formulae of simple molecules by considering electrons as charged particles. A more sophisticated picture of covalent bonding is given by considering the overlap of atomic orbitals, which are the volumes of space where an electron charge may be found.

Exam tip

Look back at Section 1.3 to remind yourself of the shapes of s- and p-orbitals.

Three examples are shown in Figure 19.

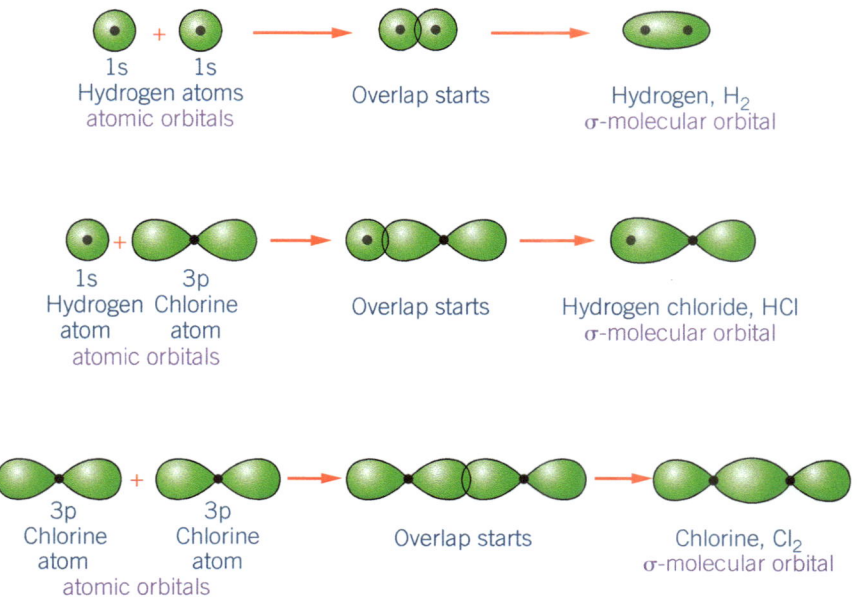

Figure 19 *The formation of sigma bonds in H_2, HCl and Cl_2*

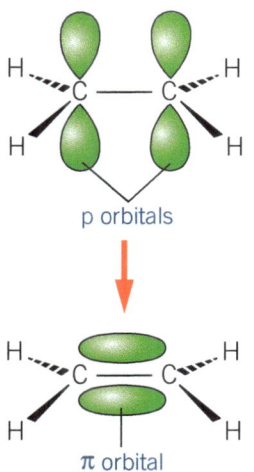

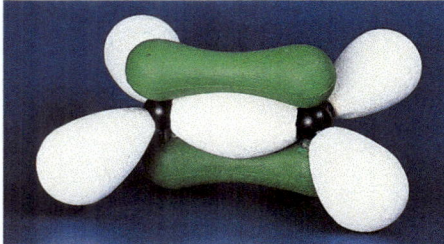

Figure 20 *The formation of a π bond in ethene*

The Cl_2 molecule shows overlap of two p-orbitals along a line joining the centres of the two chlorine atoms resulting in electron density between the two nuclei. The resulting molecular orbital is called a **sigma-orbital** (σ-orbital). p-orbitals can overlap in a different way to produce electron density above and below the line joining the nuclei. This helps to hold the resulting molecule together but is less effective than a σ-orbital. This type of molecular orbital is called a **π-orbital**. Figure 20 shows the formation of a π-orbital in ethene, C_2H_4.

Notice that when a π-orbital forms, a σ-orbital forms first, so that a double bond forms between the two carbon atoms as we have seen in Table 1.

Exam tip

A double covalent bond consists of a σ-bond plus a π-bond.

Hybridisation

A more sophisticated picture of the bonding uses the idea of hybrid molecular orbitals. The electron arrangement of carbon is $1s^2$, $2s^2$, $2p^2$. So at first sight it is hard to see how it can form four identical bonds. The answer to this is that the four orbitals in the outer shell of a carbon atom (one s-orbital and three p-orbitals) can mix together to form **hybrid orbitals**, a process called **hybridisation**. These hybrid orbitals can form bonds by overlapping with orbitals on other atoms.

In carbon, the orbitals involved in the outer shell can mix in three ways:

$s + 3 \times p \rightarrow 2 \times sp + 2 \times p$

$s + 3 \times p \rightarrow 3 \times sp^2 + 1 \times p$

$s + 3 \times p \rightarrow 4 \times sp^3 + 2 \times p$

The shapes are shown in Figure 21.

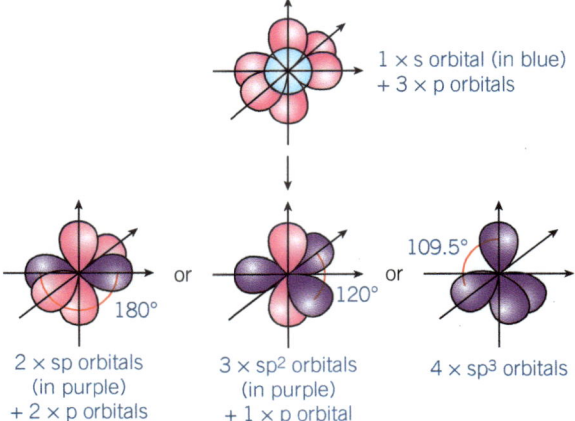

Figure 21 *Formation of hybrid orbitals—three possibilities*

Hybridisation may also be explained using the box model of atomic orbitals. The electron arrangement of carbon is shown this way below:

This arrangement has only two unpaired electrons and can therefore form only two covalent bonds.

However, the 2s orbital can mix with the three 2p orbitals, which have similar energy. The result is the formation of four sp^3 orbitals, each containing one electron. This allows the carbon atom to form four identical σ-bonds.

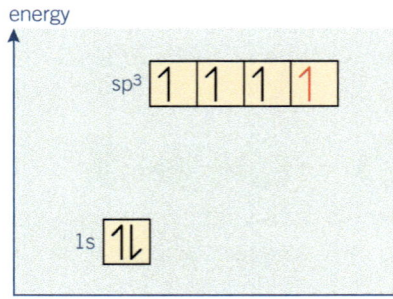

Notice that the four original orbitals always give rise to a total of four hybrids. The notation sp, sp^2 and sp^3 indicates the proportions of the original in the hybrid, so sp^3 is 25% s and 75% p, for example.

The shapes of the hybrids are as follows:

- sp-orbitals are at 180° to each other.
- sp^2-orbitals at 120° to each other.
- sp^3 are at 109.5°, i.e. pointing to the corners of a tetrahedron.

The bonding in ethane, C_2H_6

The ethane molecule is formed as shown in Figure 22.

The C–C bond is formed by overlap of two sp^3-orbitals to form a σ molecular orbital. Each C–H bond is formed by overlap of an sp^3-orbital on a carbon atom with an s-orbital on hydrogen to form a σ molecular orbital. The H–C–H and C–C–H angles are all 109.5°.

The bonding in ethene, C_2H_4

In ethene, each carbon is hybridised sp^2, leaving an unhybridised p-orbital.

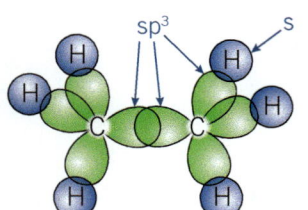

Figure 22 *The bonding in ethane*

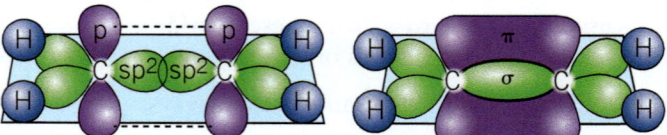

Figure 23 *The bonding in ethene*

Each of the four C–H bonds is formed by overlap of an sp^2-orbital on carbon with an s-orbital on hydrogen to form a σ-molecular orbital. The C–C bond is in two parts: a σ-orbital formed by overlap of two sp^2-orbitals and a π-bond formed by overlap of the p-orbitals on each carbon atom. The H–C–H and H–C–C angles are approximately 120°, but the H–C–H angles are in fact a little smaller because the four electrons in the double bond repel more than the two electrons in the single bonds giving H–C–H angles of about 118° and H–C–C angles of about 121°.

The bonding in hydrogen cyanide, H–C≡N

The carbon and nitrogen atoms are both hybridised sp. A σ-bond is formed between carbon and nitrogen by overlap of two sp-orbitals as well as two π-bonds formed by overlap of the p-orbitals to form a triple bond. The C–H bond is formed by overlap of the s-orbital on hydrogen with an sp-orbital on the carbon atom (see Figure 24). The bonding in nitrogen, N≡N, is similar (Figure 25).

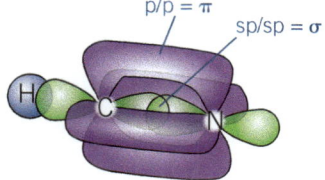

Figure 24 *The bonding in hydrogen cyanide*

Describing covalent bonds

Two of the properties of covalent bonds are bond length and bond energy.

Bond length is the distance between the centres of two atoms joined by a covalent bond. The shorter the bond between the same pair of atoms, the stronger it is. So the carbon–carbon double bond in ethene, C_2H_4 (length 0.134 nm) is stronger than the single bond in ethane, C_2H_6 (length 0.154 nm). 1 nm is 10^{-9} m.

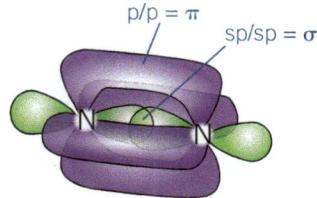

Figure 25 *The bonding in nitrogen*

Bond energy is a measure of the strength of a covalent bond. It is defined as the amount of energy that has to be *put in* to break one mole of a particular covalent bond in the gaseous state and separate the atoms completely. The greater the bond energy the stronger the bond. The C–C single bond in ethane has an energy of 350 kJ mol^{-1} and the C=C double bond in ethene has an energy of 610 kJ mol^{-1}.

The C=C bond in ethene consists of a σ-bond and a π-bond. So the energy of the π-bond is 610 − 350 = 260 kJ mol^{-1}. This means the π-bond can break leaving the σ-bond intact. This is what happens in most reactions of ethene.

Summary test 3.4

1 State what a covalent bond is.
2 Identify which of the following have covalent bonding and explain your answer.
  a Na_2O **b** CF_4 **c** $MgCl_2$ **d** C_2H_4
3 Draw a dot-and-cross diagram for hydrogen sulfide, a compound of hydrogen and sulfur.
4 State how many electrons there are in
  a sulfur in sulfur dioxide
  b phosphorus in phosphorus pentachloride.
5 Draw a dot-and-cross diagram to show a water molecule forming a coordinate bond with an H$^+$ ion.
6 Sketch and label the four C–H bonds in methane.

Learning outcomes

On these pages you will learn to:

- explain and predict the shapes of molecules

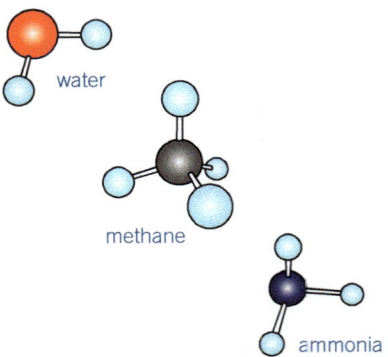

Figure 26 *The shapes of water, methane and ammonia molecules*

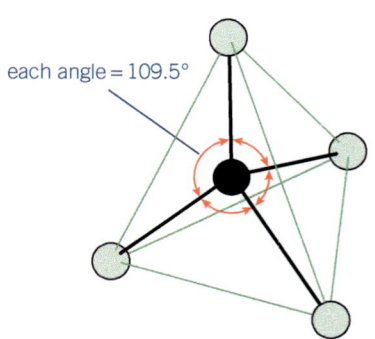

each angle = 109.5°

Figure 27 *A tetrahedron has four points and four faces*

Molecules are three-dimensional and they come in many different shapes (Figure 26).

Electron pair repulsion theory

You have seen that electrons in molecules exist in pairs in volumes of space called orbitals. You can predict the shape of a simple covalent molecule, for example, one consisting of a central atom surrounded by a number of other atoms, by using the ideas that:

- Each pair of electrons around an atom will repel all other electron pairs.
- The pairs of electrons will therefore take up positions as far apart as possible to minimise repulsion.

This is called the **electron pair repulsion theory**. Electron pairs may be a shared pair or a lone (non-bonding) pair.

The shape of a simple molecule depends on the number of pairs of electrons that surround the central atom. To work out the shape of any molecule you first need to draw a dot-and-cross diagram to find the number of pairs of electrons.

Two pairs of electrons

If there are two pairs of electrons around the atom, the molecule will be **linear**. The furthest away from each other the two pairs can get is 180° apart. Beryllium chloride, which is a covalently bonded molecule in the gas phase, despite being a metal–non-metal compound, is an example of this.

two groups of electrons

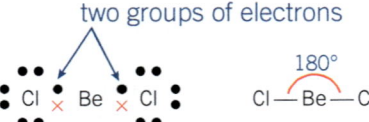

Carbon dioxide, O=C=O is also linear; here there are two groups of four electrons in each of the double bonds.

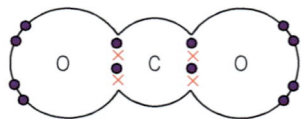

Three pairs of electrons

If there are three pairs of electrons around the central atom, they will be 120° apart. The molecule is planar (flat) and is described as **trigonal planar**. Boron trifluoride is an example of this.

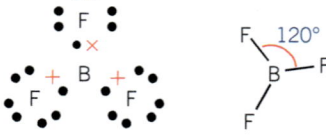

Four pairs of electrons

If there are four pairs of electrons, they are furthest apart when they are arranged so that they point to the four corners of a tetrahedron. This shape, with one atom positioned at the centre, is called **tetrahedral**, see Figure 27.

Methane, CH_4, is an example. The carbon atom is situated at the centre of the tetrahedron with the hydrogen atoms at the vertices. The angles here are 109.5°.

This is a three-dimensional, not planar, arrangement so the sum of the angles can be more than 360°.

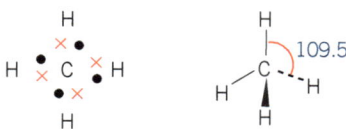

The ammonium ion is also tetrahedral. It has four groups of electrons surrounding the nitrogen atom. The fact that the ion has an overall charge does not affect the shape.

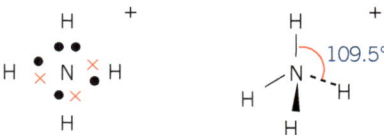

Five pairs of electrons

If there are five pairs of electrons, the shape usually adopted is that of a **trigonal bipyramid**. Phosphorus pentachloride, PCl_5, is an example.

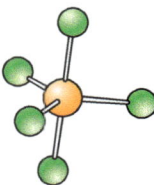

Figure 28 *The trigonal bipyramidal structure—the shape of phosphorus pentachloride*

Six pairs of electrons

If there are six pairs of electrons, the shape adopted is **octahedral**, with bond angles of 90°. The sulfur hexafluoride, SF_6, molecule is an example of this.

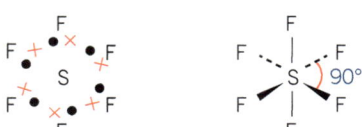

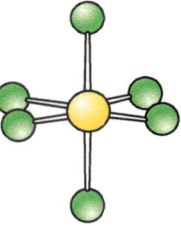

Figure 29 *The octahedral structure—the shape of sulfur hexafluoride*

Molecules with lone pairs of electrons

Some molecules have unshared (lone) pairs of electrons. These are electrons that are not part of a covalent bond. The lone pairs affect the shape of the molecule. Always watch out for the lone pairs in your dot-and-cross diagram because otherwise you might overlook their effect. Ammonia and water are good examples of molecules where lone pairs affect the shape.

Ammonia, NH_3

Ammonia has four pairs of electrons and one of the groups is a lone pair.

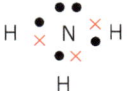

With its four pairs of electrons around the nitrogen atom, the ammonia molecule has a shape based on a tetrahedron. However, there are only three 'arms' so the shape is that of a **triangular pyramid**.

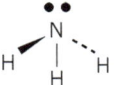

Another way of looking at this is that the electron pairs form a tetrahedron but the bonds form a triangular pyramid. (There is an atom at each vertex but, unlike the tetrahedral arrangement, no atom in the centre.)

Bonding pair–lone pair repulsion

The angles of a regular tetrahedron, see Figure 27, page 52, are all 109.5° but lone pairs affect these angles. In ammonia, for example, the bonding pairs of electrons are attracted towards the nitrogen nucleus and also the hydrogen nucleus. However, the lone pair is attracted only by the nitrogen nucleus and is therefore pulled closer to it than the shared pairs. So repulsion between a lone pair of electrons and a bonding pair of electrons is greater than that between two bonding pairs. This effect squeezes the hydrogen atoms together, reducing all the H–N–H angles. The approximate reduction of the angle is 2° per lone pair, so the bond angles in ammonia are approximately 107°:

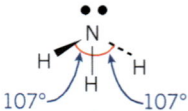

Water, H_2O

Look at the dot-and-cross diagram for water.

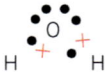

There are four pairs of electrons around the oxygen atom so, as with ammonia, the shape is based on a tetrahedron. However, two of the 'arms' of the tetrahedron are lone pairs that are not part of a bond. This results in a V-shaped or angular molecule. As in ammonia the electron pairs form a tetrahedron but the bonds form a V-shape. With two lone pairs, the H–O–H angle is reduced to 104.5°.

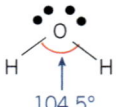

A summary of the repulsion between electron pairs

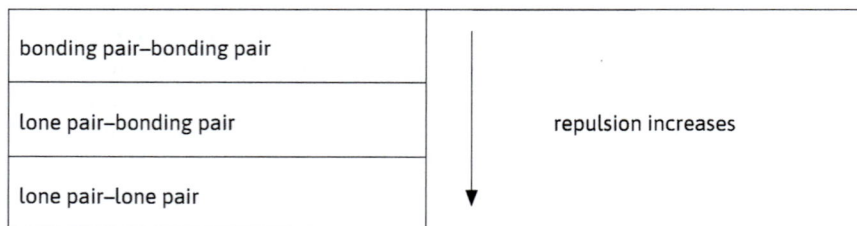

bonding pair–bonding pair	
lone pair–bonding pair	repulsion increases
lone pair–lone pair	

Summary test 3.5

1 Draw a dot-and-cross diagram for NF_3 and predict its shape.
2 Explain why NF_3 has a different shape from BF_3.
3 Draw a dot-and-cross diagram for the molecule silane, SiH_4, and describe its shape.
4 State the H–Si–H angle in the silane molecule.
5 Predict the shape of the H_2S molecule without drawing a dot-and-cross diagram.

Intermolecular forces, electronegativity and bond properties

Atoms in molecules are held together by strong covalent bonds *within* the molecules.

Forces acting *between* molecules

Molecules and separate atoms are attracted to one another by other, weaker forces called **intermolecular forces**. 'Inter' means between. If the intermolecular forces are strong enough, then molecules are held closely enough together to be liquids or even solids. These are also called van der Waals forces.

Intermolecular forces

There are three types of intermolecular (van der Waals) forces:

London dispersion forces act between all atoms and molecules		weakest
dipole–dipole forces act only between certain types of molecules		
hydrogen bonding acts only between certain types of molecules		strongest

Dipole–dipole forces and dipole moments

Polarity is the property of a particular bond, but molecules with polar bonds may have a **dipole moment**. This sums up the effect of the polarity of all the bonds in the molecule.

In molecules with more than one polar bond, the effects of each bond may cancel, leaving a molecule with no dipole moment. The effects may also add up and so reinforce each other. It depends on the shape of the molecule.

For example, carbon dioxide is a linear molecule and the dipoles cancel.

$$^{\delta-}O=C^{\delta+}=O^{\delta-}$$

Tetrachloromethane is tetrahedral and here too the dipoles cancel.

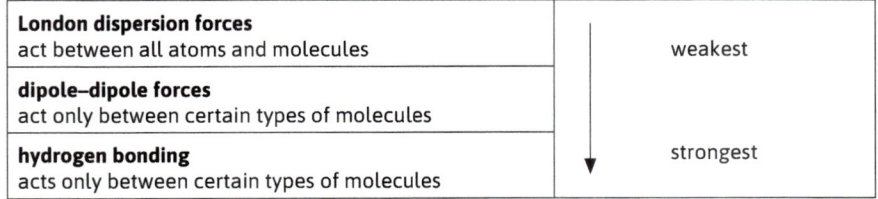

But in dichloromethane the dipoles do not cancel because of the shape of the molecule.

Consider the following molecules: CCl_4 (tetrahedral), BF_3 and $AlCl_3$ (trigonal), SF_6 (octahedral), and PCl_5 (trigonal bipyramidal). In each of these molecules, their symmetry means that the dipoles of the bonds cancel. So, the molecules have no overall dipole moment.

Learning outcomes

On these pages you will learn to:

- describe hydrogen bonding and van der Waals forces
- use the concept of hydrogen bonding to describe the properties of water
- use the concept of electronegativity to explain bond polarity and dipole moments

Exam tip

Do not confuse intermolecular forces with covalent, ionic, and metallic bonds, all of which are at least 10 times stronger.

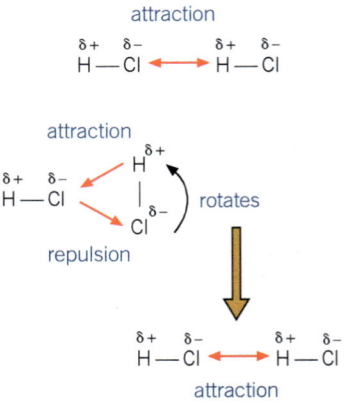

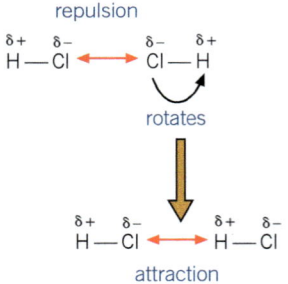

Figure 30 *Two polar molecules, such as hydrogen chloride, will always attract one another*

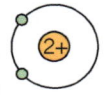 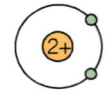 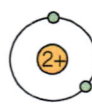

Figure 31 *These are just a few of the possible arrangements of the two electrons in helium. Remember, electrons are never in a fixed position*

Dipole–dipole forces act between molecules that have permanent dipoles. For example, in the hydrogen chloride molecule, chlorine is more electronegative than hydrogen. Therefore the electrons are pulled towards the chlorine atom rather than the hydrogen atom. The molecule therefore has a dipole and is written $H^{\delta+}$–$Cl^{\delta-}$.

Two molecules which both have dipoles will attract one another (see Figure 30).

Whatever their starting positions, the molecules with dipoles will 'flip' to give an arrangement where the two molecules attract.

London dispersion forces

All atoms and molecules are made up of positive and negative charges even though they are neutral overall. These charges produce very weak electrostatic attractions between all atoms and molecules. These are called **London dispersion forces**.

How do London dispersion forces work?

Imagine a helium atom. It has two positive charges on its nucleus and two negatively charged electrons. The atom as a whole is neutral but at any moment in time the electrons could be anywhere, see Figure 31. This means the distribution of charge is changing at every instant.

Any of the arrangements in Figure 31 mean the atom has a dipole at that moment. An instant later, the dipole may be in a different direction. But almost certainly the atom will have a dipole at any point in time, even though any particular dipole will be just for an instant. This is known as a **temporary dipole**. This dipole then affects the electron distribution in nearby atoms, so that they are attracted to the original helium atom for that instant. The original atom has induced dipoles in the nearby atoms, as shown in Figure 32, in which the electron distribution is shown as a cloud.

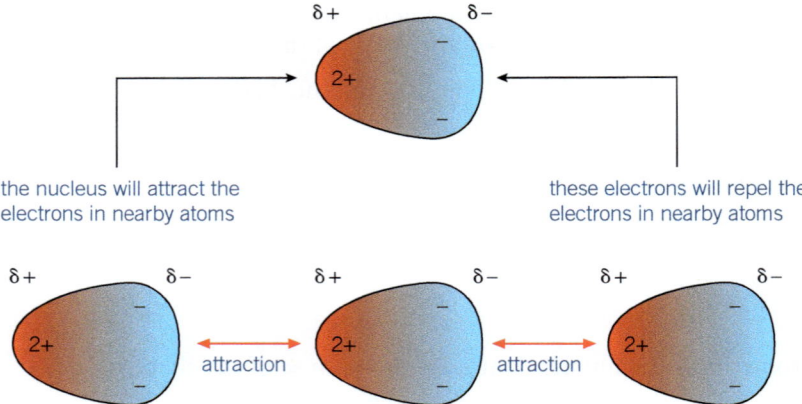

Figure 32 *Instantaneous dipoles induce dipoles in nearby atoms*

As the electron distribution of the original atom changes, it will induce new dipoles in the atoms around it, which will be attracted to the original one. These forces are sometimes called instantaneous dipole–induced dipole forces.

- London dispersion forces act between *all* atoms or molecules at *all* times.
- They are in addition to any other intermolecular forces.
- The dipole is caused by the changing position of the electron cloud, so the more electrons there are, the larger the instantaneous dipole will be.

Therefore the size of the London dispersion forces increases with the number of electrons present. This means that atoms or molecules with large atomic or molecular masses produce stronger London dispersion forces than atoms or molecules with small atomic or molecular masses.

This explains why:

- The boiling points of the noble gases increase as we go down the group.
- The boiling points of hydrocarbons increase with increased chain length.

Hydrogen bonding

Hydrogen bonding is a special type of intermolecular force with some characteristics of dipole–dipole attraction and some of a covalent bond. It consists of a hydrogen atom 'sandwiched' between two very electronegative atoms. There are conditions that have to be present for hydrogen bonding to occur. You need a very electronegative atom, with a lone pair of electrons, covalently bonded to a hydrogen atom. Water molecules fulfil these conditions. Oxygen is much more electronegative than hydrogen so water is polar (see Figure 33).

You would expect to find weak dipole–dipole attractions (as shown between hydrogen chloride in Figure 30) but in this case the intermolecular bonding is much stronger for two reasons:

1 The oxygen atoms in water have lone pairs of electrons.
2 In water, the hydrogen atoms are highly electron-deficient. This is because the oxygen is very electronegative and attracts the shared electrons in the bond towards it. The hydrogen atoms in water are positively charged and very small. These exposed protons have a very strong electric field because of their small size.

The lone pair of electrons on the oxygen atom of another water molecule is strongly attracted to the electron deficient hydrogen atom.

This strong intermolecular force is called a **hydrogen bond**. Hydrogen bonds are considerably stronger than dipole–dipole attractions, though much weaker than a covalent bond. They are usually represented by dashed lines, as in Figure 34.

When do hydrogen bonds form?

Water is not the only example of hydrogen bonding. In order to form a hydrogen bond there must be the following:

- a hydrogen atom that is bonded to a very electronegative atom. This will produce a strong partial positive charge on the hydrogen atom.
- a very electronegative atom with a lone pair of electrons. These will be attracted to the partially charged hydrogen atom in another molecule and form the bond.

The only atoms that are electronegative enough to form hydrogen bonds are oxygen, O, nitrogen, N, and fluorine, F. For example, ammonia molecules, NH_3, form hydrogen bonds with water molecules, see Figure 35.

The nitrogen–hydrogen–oxygen system is linear. This is because the pair of electrons in the N–H covalent bond repels those in the hydrogen bond between nitrogen and hydrogen. This linearity is always the case with hydrogen bonds.

 Shapes of molecules

Another factor that affects the strength of intermolecular forces is the shape of the molecule.

Long, thin molecules (such as unbranched hydrocarbon chains) can pack together closely with a large area of contact.

Small molecules have smaller areas of contact and therefore the attraction is smaller.

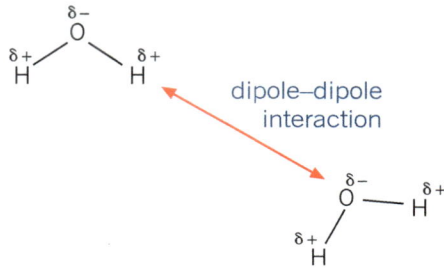

Figure 33 Dipole attraction between water molecules

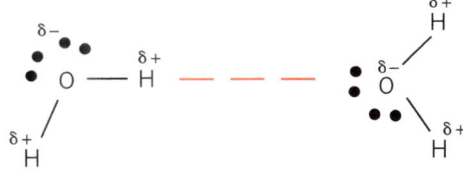

Figure 34 Hydrogen bond between water molecules

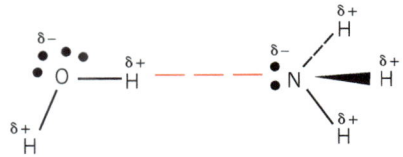

Figure 35 Hydrogen bond between a water molecule and an ammonia molecule

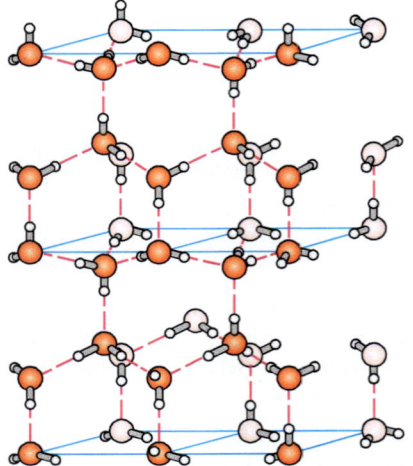

Figure 36 *The three-dimensional network of covalent bonds (grey) and hydrogen bonds (red) in ice. The blue lines are only construction lines*

The unusual properties of water

Water has unexpectedly high boiling and melting points (100°C and 0°C respectively) compared with methane (a molecule of comparable M_r) –164°C and –182°C. This is due to hydrogen bonding, which is not present in methane. Hydrogen bonding is also responsible for the high **surface tension** of water, the 'skin' effect that allows water boatman insects to skim the surface of ponds.

In water in its liquid state, the hydrogen bonds break and reform easily as the molecules are moving about. When water freezes, the water molecules are no longer free to move about and the hydrogen bonds hold the molecules in fixed positions. The resulting three-dimensional structure, shown in Figure 36, resembles the structure of diamond.

In order to fit into this structure, the molecules are slightly less closely packed than in liquid water. This means that ice is less dense than water and forms on top of ponds rather than at the bottom. This insulates the ponds and enables fish to survive through the winter. This must have helped life to continue, in the relative warmth of the water under the ice, during the Ice Ages.

Summary test 3.6

1 Explain why fluorine is more electronegative than chlorine.
2 Write δ+ and δ− signs to show the polarity of the bonds in a hydrogen chloride molecule.
3 Identify which of these covalent bonds is/are non-polar, and explain your answer.
   a H–H
   b F–F
   c H–F
4 Arrange the following covalent bonds in order of increasing polarity:
   a H–O, H–F, H–N
   b Explain your answer to part **a**.
5 Place the following elements in order of the strength of the van der Waals forces between the atoms (weakest first): Ar, He, Kr, Ne. Explain your answer.
6 Identify which one of the following molecules *cannot* have dipole–dipole forces acting between them: H_2O, HCl, H_2
7 Explain why hexane is a liquid at room temperature whereas butane is a gas.
8 Explain why covalent molecules are gases, liquids, or solids with low-melting temperature.
9 Draw two hydrogen bromide molecules to show how they would be attracted together by dipole–dipole forces.
10 Identify in which of the following does hydrogen bonding not occur between molecules: H_2O, NH_3, HBr, HF
11 Explain why hydrogen bonds do not form between:
   a methane molecules, CH_4
   b tetrachloromethane molecules, CCl_4.
12 Draw a dot-and-cross diagram for a molecule of water.
   a State how many lone pairs it has.
   b State how many hydrogen atoms it has.
   c Explain why water molecules form on average two hydrogen bonds per molecule, whereas the ammonia molecule, NH_3, forms only one.

⎕ Launch additional digital resources for the chapter

1 a Define the term 'electronegativity'. *(2 marks)*

b State and explain the trend in electronegativity.
i Descending Group 17 of the Periodic Table.
(4 marks)
ii Across Period 3 of the Periodic Table. *(4 marks)*

c Hydrogen reacts with halogens to form hydrogen halides HX, where X is a halide ion. Comment on the relative bond strength of the hydrogen halides H–F and H–Br, explain your answer. *(4 marks)*

2 a Sodium reacts violently with chlorine to produce a sodium chloride, a white crystalline solid with a high melting point.
i Identify the type of structure formed by sodium chloride. *(1 mark)*
ii Write a balanced equation for this reaction, including state symbols. *(2 marks)*
iii Explain why sodium chloride has a high melting point. *(3 marks)*
iv Predict the electrical conductivity of the product of this reaction, explain your reasoning. *(3 marks)*

b i Identify the type of bonding present in the element magnesium. *(1 mark)*
ii Use your knowledge of structure and bonding to describe, with the aid of a labelled diagram, the arrangement of particles in a crystal of magnesium. *(2 marks)*

3 Ammonia reacts in the presence of H^+ ions to form ammonium ions, NH_4^+ according to the following equation:

$$NH_3 + H^+ \rightarrow NH_4^+$$

a Draw the shape of the NH_3 molecule, including any lone pairs of electrons present that affect its structure. State the H–N–H bond angle. *(2 marks)*

b State the type of bond that is formed when H^+ reacts with a molecule of NH_3. Explain how this bond is formed. *(2 marks)*

c Use your answer to part **a** to describe how the H–N–H bond angle changes upon the formation of an ammonium ion. Explain this change. *(4 marks)*

4 The following molecules are covalent compounds:

HCl H_2 CO_2 H_2O CH_4 NH_3

a Group these molecules into polar and non-polar substances. *(3 marks)*

b Identify the strongest type of intermolecular force present for pure samples of each molecule. *(6 marks)*

5 Fluorine is a non-polar molecule which reacts with hydrogen to form the polar compound hydrogen fluoride, HF. The boiling point of fluorine is −188°C, whereas hydrogen fluoride has a boiling point of 19.5°C. Use your knowledge of bonding and intermolecular forces to explain these differences. *(6 marks)*

6 Ammonia, NH_3, is converted into ammonium nitrate solution by reaction with nitric acid according to the equation below. Ammonium nitrate is then re-crystallised for use as a fertiliser.

$$NH_3(g) + HNO_3(aq) \rightarrow NH_4NO_3(aq)$$

a Show, with the aid of a diagram, the hydrogen bonding that occurs between two molecules of ammonia. *(2 marks)*

b Ammonium nitrate is a white crystalline solid at room temperature. State the bonding present in a molecule of ammonium nitrate and explain why it has a melting point of 170°C. *(4 marks)*

7 Draw the shapes of the following molecules, stating the names of each type of structure and the bond angles present:

a BF_3 **b** SF_6 **c** H_2O **d** CO_2 *(8 marks)*

8 State and explain the trend in boiling points in the periodic table:

a descending Group 17 *(3 marks)*

b across Period 3 from Na to Al. *(3 marks)*

9 At 185°C Aluminium chloride, $AlCl_3$, is converted into its molecular form, a dimer with the formula Al_2Cl_6.

a Draw a diagram to show the shape of a molecule of $AlCl_3$, identify the bond angle and state the shape of the structure formed. *(3 marks)*

b Identify the type of bonding that occurs on formation of one molecule of the dimer Al_2Cl_6. Describe how this bonding occurs. *(3 marks)*

4 States of matter

The gaseous state: ideal and real gases and $pV = nRT$

Learning outcomes

On these pages you will learn to:

- explain the origin of pressure in a gas
- describe the assumptions made about ideal gases
- use the equation $pV = nRT$

Our understanding of how matter behaves and of the three states of matter—solid, liquid and gas—is based on the kinetic-particulate theory. This uses the idea that matter is made up of particles which are able to move and which attract one another.

The three states of matter

Matter is made of particles. Their movement and arrangement governs whether they are in the solid, liquid or gaseous state. This is summarised in Table 1.

Table 1 *The three states of matter*

	Solid	Liquid	Gas
Arrangement of particles	regular	random	random
Evidence	Crystal shapes have straight edges. Solids have definite shapes.	None direct but a liquid changes shape to fill the bottom of its container.	None direct but a gas will fill its container.
Spacing	close	close	far apart
Evidence	Solids are not easily compressed.	Liquids are not easily compressed.	Gases are easily compressed.
Movement	vibrating about a point	rapid 'jostling'	rapid
Evidence	Diffusion is very slow. Solids expand on heating.	Diffusion is slow. Liquids evaporate.	Diffusion is rapid. Gases exert pressure.
Models			

The particles in gases, are far apart and moving rapidly. The pressure of a gas is the result of collisions with the walls of its container. The volume of a given mass of any gas is not fixed. It changes with pressure and temperature. However, there are a number of simple relationships for a given mass of gas that connect the pressure, temperature, and volume. These lead to the ideal gas equation.

The gas laws

Experiments on the behaviour of gases have led to three approximate laws.

Boyle's law

The product of pressure and volume is a constant as long as the temperature remains constant:

$$\text{pressure } p \times \text{volume } V = \text{constant}$$

Charles' law

The volume is proportional to the temperature as long as the pressure remains constant:

$$\text{volume } V \propto \text{temperature } T \text{ and } \frac{\text{volume } V}{\text{temperature } T} = \text{constant}$$

Gay-Lussac's law (also called the constant volume law)

The pressure is proportional to the temperature as long as the volume remains constant:

$$\text{pressure } P \propto \text{temperature } T \text{ and } \frac{\text{pressure } P}{\text{temperature } T} = \text{constant}$$

Combining all these relationships gives us the equation:

$$\frac{\text{pressure } p \times \text{volume } V}{\text{temperature } T} = \text{constant for a fixed mass of gas}$$

The ideal gas equation

For n moles of gas, the ideal gas equation is:

$$\begin{array}{ccccccc}
\text{pressure} & \times & \text{volume} & = & \text{number of moles} & \times & \text{gas constant} & \times & \text{temperature} \\
p\,(\text{Pa}) & & V\,(\text{m}^2) & = & n & & R\,(\text{J K}^{-1}\text{mol}^{-1}) & & T\,(\text{K})
\end{array}$$

$$pV = nRT$$

Where R is a constant called the gas constant. The value of R is $8.31\ \text{J K}^{-1}\text{mol}^{-1}$.

This is the **ideal gas equation**. No gases obey it exactly, but at room temperature and pressure it holds quite well for many gases. It is often useful to imagine a gas which obeys the equation perfectly—an **ideal gas**.

For an ideal gas, we make the following assumptions:

1 Gases are made up of particles (atoms or molecules) in rapid random motion).
2 Gas pressure is caused by particles colliding with the walls of the container.
3 No energy is lost in the collisions of particles with the walls or with other particles.
4 The temperature of the gas is proportional to the average kinetic (moving) energy of the particles.
5 There are no forces of attraction between the particles.
6 We can ignore the volume of the particles themselves compared with the volume of the container.

Assumptions 1–4 are correct for real gases, but numbers 5 and 6 are not. There *are* attractive forces between the particles and these particles *do* have a volume.

 Notes on units

When using the ideal gas equation, consistent units must be used. If you want to calculate n, the number of moles:

- P must be in Pa (N m^{-2})
- T must be in K
- V must be in m^3
- R must be in $\text{J K}^{-1}\text{mol}^{-1}$

Extension

The units used here are part of the *Système Internationale* (SI) of units. This is a system of units for measurements used by scientists throughout the world. The basic units used by chemists are metre (m), second (s), kelvin (K), and kilogram (kg).

Real gases

The underlying assumptions of the ideal gas equation are:

- The particles in an ideal gas have zero volume.
- There are no forces of attraction between the particles.
- The particles do not gain or lose energy when they collide with each other or the walls of their container.

For **real gases** these assumptions are almost true at high temperatures and low pressures. Under these conditions the gas particles are far apart. The volume they take up is negligible compared with the volume of the container, and the forces of attraction between them are weak.

However, if we cool a real gas and/or increase its pressure, the equation begins to fail. The gas will eventually condense to form a liquid. In a liquid, the particles are touching, held together by the van der Waals forces of attraction between the particles. The stronger the forces of attraction, the greater the deviation from ideal gas behaviour.

Using the ideal gas equation

Using the ideal gas equation, you can calculate the volume of one mole of gas at any temperature and pressure. None of the terms in the equation refers to a particular gas, so this volume will be the same for any gas.

It is the space between the gas molecules that accounts for the volume of a gas. This may seem surprising. However, even the largest gas particle is extremely small compared with the space in between the particles.

Rearranging the ideal gas equation to find a volume gives:

$$V = \frac{nRT}{P}$$

The worked example tells you that the volume of a mole of *any* gas at room temperature and pressure is approximately 24 000 cm³ (24 dm³). For example, one mole of sulfur dioxide gas, SO_2 (mass 64.1 g), has the same volume as one mole of hydrogen gas, H_2 (mass 2.0 g).

In a similar way, pressure can be found using: $p = \dfrac{nRT}{V}$

Finding the number of moles n, of a gas

If you rearrange the equation $pV = nRT$ so that n is on the left-hand side, you get:

$$n = \frac{pV}{RT}$$

If T, P, and V are known, then you can find n.

Finding the relative molecular mass of a gas

If you know the number of moles present in a given mass of gas, you can find the mass of one mole of gas. This tells us the relative molecular mass.

The apparatus used to find the relative molecular mass of a gas in a pressurised canister is shown in Figure 1.

The canister was weighed.

1000 cm³ of gas was dispensed into the measuring cylinder, until the levels of the water inside and outside the measuring cylinder were the same. The pressure of the collected gas was now the same as atmospheric pressure.

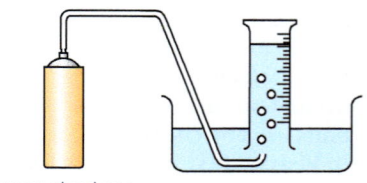

pressurised gas

Figure 1 *Measuring the relative molecular mass of a gas*

The canister was reweighed.

Atmospheric pressure and temperature were noted. ($R = 8.31\ \mathrm{J\,K\,mol^{-1}}$)

The results were as follows:

loss of mass of the can = 1.85 g
  temperature = 14°C = 287 K
 atmospheric pressure = 100 000 Pa
     volume of gas = 1000 cm^3 = 1000 × 10^{-6} m^3

$$n = \frac{pV}{RT}$$
$$= \frac{100\,000 \times 1000 \times 10^{-6}}{8.31 \times 287}$$
$$= 0.042\ \mathrm{mol}$$

0.042 mol has a mass of 1.85 g.

So, 1 mol has a mass of = 44 g

So, $M_r = 44$ (and the gas is propane)

Exam tip

Using 24 000 cm^3 as the volume of a mole of any gas is not precise and it is always necessary to apply the ideal gas equation in calculations.

Exam tip

To convert cm^3 into m^3, divide by 10^6 (or multiply by 10^{-6}).

Worked example

Finding the number of moles

How many moles of hydrogen molecules are present in a volume of 100 cm^3 at a temperature of 20.0 °C and a pressure of 100 kPa?

$R = 8.31\ \mathrm{J\,K^{-1}\,mol^{-1}}$

First, convert to the base units:

p must be in Pa, and 100 kPa = 100 000 Pa

V must be in m^3, and 100 cm^3 = 100 × 10^{-6} m^3

T must be in K, and 20°C = 293 K (add 273 to the temperature in °C)

Substituting into the ideal gas equation:

$$n = \frac{pV}{RT}$$

$$= \frac{100\,000 \times 100 \times 10^{-6}}{8.31 \times 293}$$

$$= 0.004\,11\ \mathrm{moles}$$

Summary test 4.1

1 a Calculate the volume of 2 moles of a gas if the temperature is 30°C, and the pressure is 100 000 Pa.
   b Calculate the pressure of 0.5 moles of a gas if the volume is 11 000 cm^3, and the temperature is 25°C.
2 Calculate how many moles of hydrogen molecules are present in a volume of 48 000 cm^3, at 100 000 Pa and 25°C.
3 State how many moles of carbon dioxide molecules would be present in question **2**? Explain your answer.

Bonding and structure

The properties of substances are governed by their structure and bonding.

Bonding describes the forces that hold the atoms together. It includes the three strongest types of bonding—ionic, metallic and covalent—as well as the weaker ones—London dispersion forces, dipole-dipole forces and hydrogen bonding.

Structure describes the geometrical arrangement of the atoms in space. There are two main types of structure. In a **giant structure** the atoms form an extended geometrical arrangement. A **molecular structure** consists of separate molecules.

Giant ionic structures

In an ionically bonded compound, the structure is held together by the electrostatic attraction of the positive and negative ions. The positive ions are surrounded by negative ions, and the negative ions are surrounded by positive ions.

Figure 2 shows the structure of sodium chloride in both space filling and exploded versions. Each Na^+ ion is surrounded by six Cl^- ions, and each Cl^- is surrounded by six Na^+ ions. The electrostatic attraction is strong and extends throughout the structure, giving it a high melting point (1074 K, 801°C). Magnesium oxide has the same structure, but because the ions have double the charge of those in sodium chloride the melting point is even higher (3125 K, 2852°C). A regular arrangement of ions which continues throughout the solid is called a lattice.

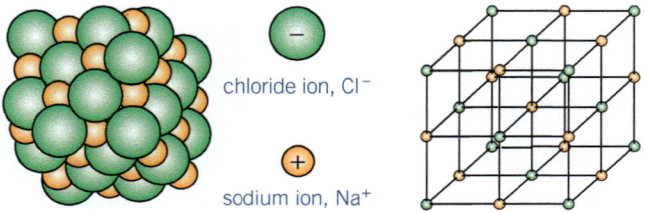

chloride ion, Cl^-

sodium ion, Na^+

Figure 2 *The giant structure of sodium chloride*

Properties of ionically bonded compounds

Ionic compounds are always solids at room temperature. They have giant structures and therefore high melting points. This is because in order to melt an ionic compound, energy must be supplied to break up the lattice of ions.

Ionic compounds conduct electricity when molten or dissolved in water (aqueous) but not when solid. This is because the ions that carry the current are free to move in the liquid state but are not free in the solid state (Figure 3).

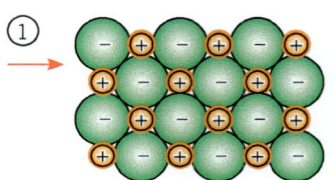

a small displacement causes contact between ions with the same charge...

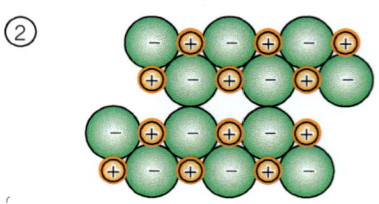

...and the structure shatters

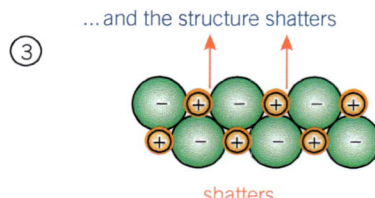

shatters

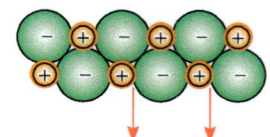

Figure 4 *The brittleness of ionic compounds*

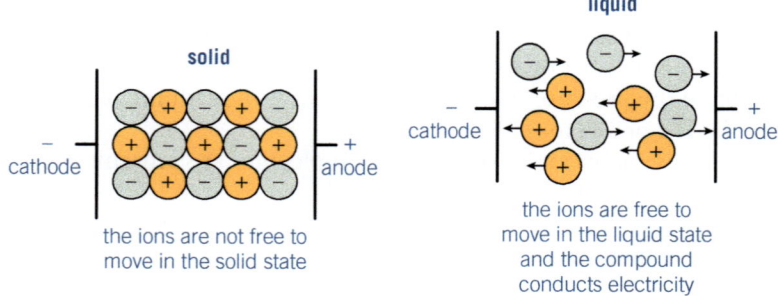

the ions are not free to move in the solid state

the ions are free to move in the liquid state and the compound conducts electricity

Figure 3 *Ionic liquids conduct electricity, ionic solids do not*

Ionic compounds are **brittle** and shatter easily when hit. This is because they form a lattice of alternating positive and negative ions (see Figure 4). A strike in the direction shown may move the ions and produce contact between ions with the same charge.

Simple molecular crystals

Molecular crystals consist of molecules held in a regular array by intermolecular forces. Covalent bonds *within* the molecules hold the atoms together, but they do not act *between* the molecules. Intermolecular forces are much weaker than covalent, ionic or metallic bonds, so molecular crystals have low melting points and low enthalpies of melting.

Iodine

Iodine (Figure 5) is an example of a molecular crystal. A strong covalent bond holds pairs of iodine atoms together to form I_2 molecules. Since iodine molecules have a large number of electrons, the van der Waals forces are strong enough to hold the molecules together as a solid. But van der Waals forces are much weaker than covalent bonds, giving iodine the following properties:

- crystals are soft and break easily
- low melting point (114°C, 387 K) and sublimes readily to form gaseous iodine molecules
- does not conduct electricity because there are no charged particles to carry current.

Ice

Ice is also a molecular crystal. In water in its liquid state, the hydrogen bonds break and reform easily as the molecules are moving about. When water freezes, the water molecules are no longer free to move about and the hydrogen bonds hold the molecules in fixed positions. This results in a three-dimensional structure, shown in Figure 36 in Section 3.6.

In order to fit into this structure, the molecules are slightly less closely packed than in liquid water. This means that ice is less dense than water and floats on top of liquid water on ponds.

Buckminsterfullerene - molecular footballs

More recently a number of new forms of pure carbon have been discovered. Chemists found the first one whilst they were looking for molecules in outer space. The structures of these new forms of carbon include closed cages of carbon atoms and also tubes called **nanotubes**. The most famous is, C_{60}, in which atoms are arranged in a football-like shape (Figure 6). Harry Kroto and colleagues received the Nobel Prize for the discovery. Now, scientists are investigating many uses for these new materials.

Giant molecular crystals

Covalent compounds are not always made up of small molecules. In some substances the covalent bonds extend throughout the compound and have the typical property of a giant structure held together with strong bonds—a high melting point. There are many examples of giant molecular crystals, including diamond and graphite.

Diamond and graphite

Diamond and graphite are both made of the element carbon only. They are **allotropes** of carbon. They are very different materials because their atoms are differently bonded and arranged.

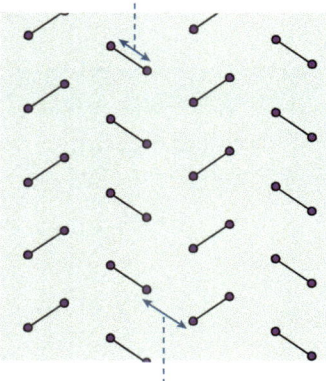

distance between a pair of covalently bonded iodine atoms = 0.267 nm

distance between a pair of iodine molecules (held by van der Waals forces) = 0.354 nm

Figure 5 *The arrangement of an iodine crystal*

The importance of hydrogen bonding

Although hydrogen bonds are only about 10% of the strength of covalent bonds, their effect can be significant—especially when there are a lot of them. The fact that they are weaker than covalent bonds, and can break or be made under conditions where covalent bonds are unaffected, is very important.

Exam tip

In liquid water, hydrogen bonding is responsible for the high surface tension of water—the 'skin' effect that allows a needle to 'float' (with care!) on the surface of water.

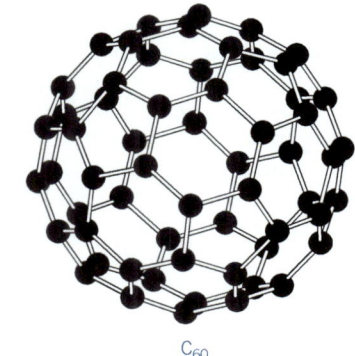

C_{60}

Figure 6 *Buckminsterfullerene—also called 'buckyballs'*

Diamond

Diamond consists of pure carbon with covalent bonding between every carbon atom. The bonds spread throughout the structure, which is why it is a giant structure.

A carbon atom has four electrons in its outer shell. In diamond, each carbon atom forms four single covalent bonds with other carbon atoms, as shown in Figure 7. These four electron pairs repel each other, following the rules of the electron pair repulsion theory. In three dimensions the bonds actually point to the corners of a tetrahedron (with bond angles of 109.5°).

Each carbon atom is in an identical position in the structure, surrounded by four other carbon atoms. Figure 8 shows this three-dimensional arrangement.

The atoms form a giant three-dimensional lattice of strong covalent bonds, which is why diamond has the following properties:

- very hard material (one of the hardest known)
- very high melting point, over 3700 K
- does not conduct electricity because there are no free charged particles to carry charge.

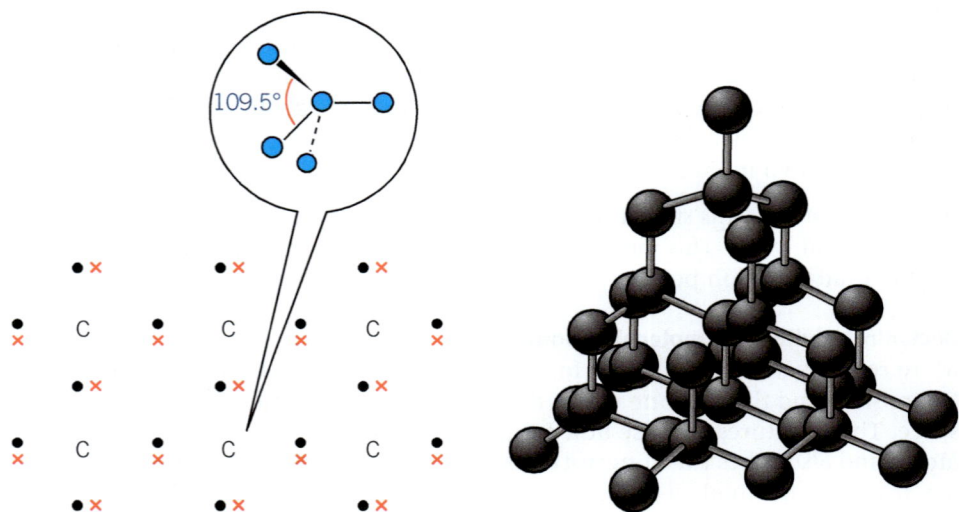

Figure 7 A dot-and-cross diagram showing the bonding in diamond

Figure 8 A three-dimensional diagram of diamond

Graphite

Graphite also consists of pure carbon but the atoms are bonded and arranged differently from diamond. Graphite has two sorts of bonding—strong covalent and the weaker van der Waals forces.

In graphite, each carbon atom forms three single covalent bonds to other carbon atoms. As predicted by electron pair repulsion theory, these form a flat trigonal arrangement, sometimes called trigonal planar, with a bond angle of 120° (Figure 9). This leaves each carbon atom with a 'spare' electron in a p-orbital that is not part of the three single covalent bonds.

This arrangement produces a two-dimensional layer of linked hexagons of carbon atoms, rather like a chicken-wire fence (Figure 10).

The p-orbitals with the 'spare' electron merge above and below the plane of the carbon atoms in each layer. These electrons can move anywhere within the layer. They are delocalised. This adds to the strength of the bonding and is rather like the delocalised sea of electrons in a metal, but in two dimensions only.

Exam tip

The element silicon has the same geometrical structure as diamond.

📖 **Allotropes of carbon**

Diamond is a three-dimensional giant structure. Graphite is a two-dimensional giant structure. Despite its impressive looking shape, buckminsterfullerene is a simple molecular structure.

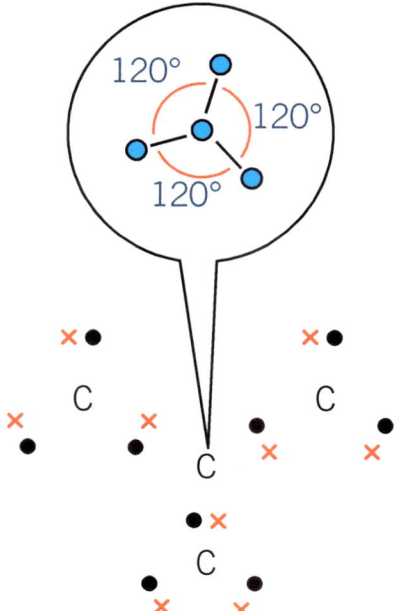

Figure 9 A dot-and-cross diagram showing the three covalent bonds in graphite

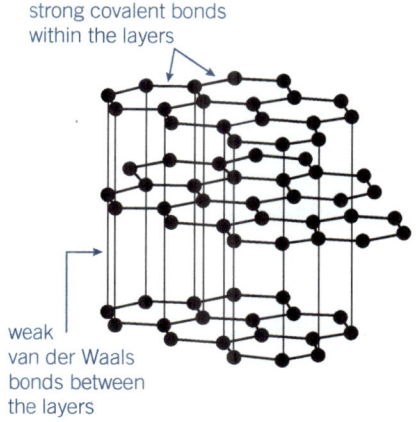

strong covalent bonds within the layers

weak van der Waals bonds between the layers

Figure 10 Van der Waals forces between the layers of carbon atoms in graphite

Extension

In 2010, Russian-born scientists Andre Geim and Konstantin Novoselov were awarded the Nobel Prize in Physics for their work on isolating single layers of graphite, called graphene. Geim and Novoselov were able to produce individual layers using a technique that involved removing sticky tape from graphite crystals.

These delocalised electrons are what make graphite conduct electricity (very rare for a non-metal). They can travel freely through the material, though graphite will only conduct along the hexagonal planes, not at right angles to them.

There is no covalent bonding between the layers of carbon atoms. They are held together by the much weaker van der Waals forces (see Figure 10). This weak intermolecular force of attraction means that the layers can slide across one another making graphite soft and flaky. It is used in pencils. The flakiness allows the graphite layers to transfer from the pencil to the paper.

- Graphite is a soft material.
- It has a very high melting point and in fact it breaks down before it melts. This is because of the strong network of covalent bonds, which make it a giant structure.
- It conducts electricity along the planes of the hexagons.

Silicon dioxide

Silicon dioxide (silica), the main constituent of sand, also has a giant molecular structure, Figure 11.

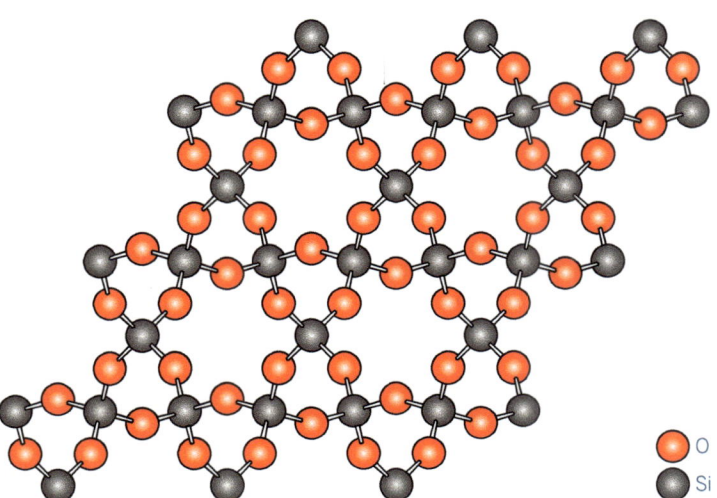

O
Si

Figure 11 The giant molecular structure of silicon dioxide

Each silicon atom is surrounded by four oxygen atoms covalently bonded to it. Each oxygen atom links two silicon atoms. The covalent bonding is strong and extends throughout the structure, giving it a high melting point (1983 K, 1710°C)

Properties of molecular crystals

- Covalently bonded compounds do not conduct electricity because there are no charged particles (ions or electrons) to carry the current. Graphite, with its unusual structure, is an exception.
- Giant molecular crystals have high melting points.
- Simple molecular crystals have low melting points.

Giant metallic structures and their properties

Metals also form giant structures, held together by the attraction of the positive metal ions to the 'sea' of electrons (see Section 3.3). Their properties have made them essential to the development of our civilisation.

Metals are good conductors of electricity and heat

The delocalised electrons can move throughout the structure of a metal. This explains why metals are such good conductors of electricity. An electron from the negative terminal of the supply joins the electron sea at one end of a metal wire; at the same time a different electron leaves the wire at the positive terminal, as shown in Figure 12.

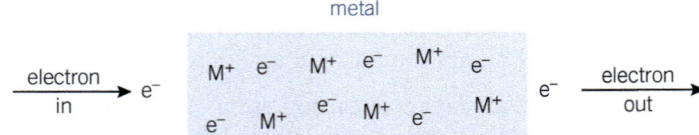

Figure 12 *The conduction of electricity by a metal*

Metals are also good conductors of heat—they have high thermal conductivities. The sea of electrons is partly responsible for this property. Energy is also spread by increasingly vigorous vibrations of the closely packed ions.

The good electrical conductivity of copper is why it is used for electric wiring. Its thermal conductivity is why it is sometimes used in cooking pans.

The strength of metals

In general, the strength of any metallic bond depends on the following:

- the charge on the ion—the greater the charge on the ion, the greater the number of delocalised electrons and the stronger the electrostatic attraction between the positive ions and the electrons.
- the size of ion—the smaller the ion, the closer the electrons are to the positive nucleus and the stronger the bond.

Metals tend to be strong. The delocalised electrons also explain this. These extend throughout the solid so there are no individual bonds to break.

Metals are malleable and ductile

Metals are **malleable** (they can be beaten into shape) and **ductile** (they can be pulled into thin wires). After a small distortion, each metal ion is still in exactly the same environment as before so the new shape is retained (see Figure 13).

Contrast this with the brittleness of ionic compounds described above.

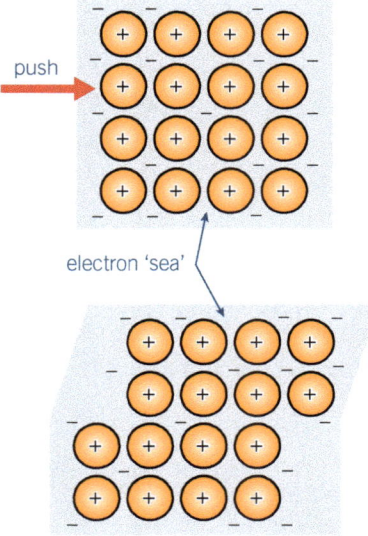

push

electron 'sea'

Figure 13 The malleability and ductility of metals

Figure 14 The malleability of this metal pipe allows a machine to bend it into shape without the pipe breaking

Metals have high melting points

Metals generally have high melting and boiling points because they have giant structures. There is strong attraction between metal ions and the delocalised sea of electrons. This makes the atoms difficult to separate.

Bonding, structure and properties

Table 2 summarises the properties of substances with ionic, covalent and metallic bonding.

Table 2 Summary of properties of substances with covalent, ionic, and metallic bonding

	Structure	Bond	Melting point, T_m	Solubility	Electrical conductivity		
					Solid	Liquid	Aqueous solution
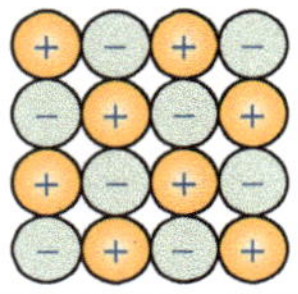	giant	ionic	high	may dissolve in water and polar solvents, not in non-polar solvents	no	yes	yes
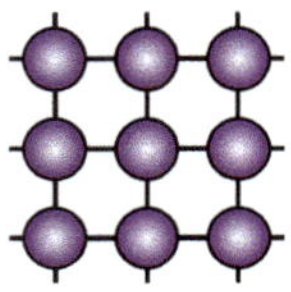	giant (macromolecular)	covalent	high	generally insoluble	no (except graphite and graphene)	no	no (usually insoluble)
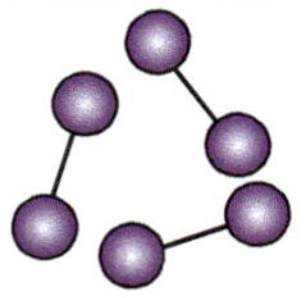	simple molecular	covalent	low	generally polar molecules dissolve in polar solvents (and non-polar molecules in non-polar solvents)	no	no	no (but may react to give ions)
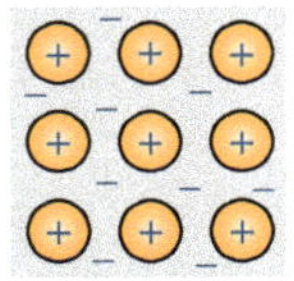	giant	metallic	high	insoluble but may react with the solvent	yes	yes	– does not dissolve but may react

The process of dissolving is complex and is discussed in more detail in Section 23.2.

Summary test 4.2

1 Give the formula of the ions present in the following compounds
   a sodium fluoride
   b magnesium oxide
   c magnesium fluoride

2 a Suggest how the melting points of these compounds compare.
   b Explain your answer.

3 Describe the difference between a giant molecular crystal and a simple molecular crystal in terms of the following:
   a bonding
   b properties

4 Phosphorus consists of P_4 molecules and has a melting point of 317 K while sulfur, S_8, has a melting point of 386 K. Explain this difference in terms of intermolecular forces.

5 Explain why graphite can be used as a lubricant.

6 Explain how graphite conducts electricity. How does it conduct differently from metals?

7 Explain why both diamond and graphite have high melting points.

8 Suggest why copper metal is used extensively for electrical wiring and plumbing.

9 The table below gives some information about four substances. Giving reasons for your answers:
   a Identify which substances have giant structures.
   b Identify which substance is a gas at room temperature.
   c Identify which substance is a metal.
   d Identify which substances are covalently bonded.
   e Identify which substance has ionic bonding.
   f Identify which substance is a giant molecule.

Substance	Melting point / K (°C)	Boiling point / K (°C)	Electrical conductivity	
			solid	liquid
A	1356 (1083)	2840 (2567)	good	good
B	91 (–182)	109 (–164)	poor	poor
C	1996 (1723)	2503 (2230)	poor	good
D	1266 (993)	1968 (1695)	poor	poor

Exam-style questions

$\boxed{\text{📄 Launch additional digital resources for the chapter}}$

1 A weather balloon was filled with helium at 25°C to a volume of $3.0\,m^3$. The balloon ascends to a height of 5000 metres where the temperature is –5°C. (The gas constant $R = 8.31\,J\,K^{-1}\,mol^{-1}$)

   a Assuming that atmospheric pressure remains constant, calculate the number of moles of helium used to fill the balloon. *(2 marks)*

   b Use your answer to part **a** to calculate the change in volume at a height of 5000 m. *(2 marks)*

   c Determine the mass of helium used. *(1 mark)*

2 $37.0\,cm^3$ of gas P was collected in a gas syringe. The initial mass of the gas syringe was 175.000 g, this increased to 175.057 g upon collection of gas P. Assume that the syringe remains at atmospheric pressure and 298 K.

   a Calculate the number of moles of gas P collected. *(3 marks)*

   b Gas P is a diatomic element. Use your answer to part **a** to calculate the molecular mass of gas P and hence state the identity of gas P. *(2 marks)*

3 A lighter contains a refillable gas container of volume $1 \times 10^{-4}\,m^3$. This is certified to a maximum operating pressure of 125 kPa at 298 K.

   a Calculate the maximum mass of butane (C_4H_{10}) that can used to refill the lighter. *(3 marks)*

   b State the pressure within the contained when half of the butane has been used. *(1 mark)*

   c The fuel was substituted for propane, C_3H_8. Calculate the maximum number of moles of propane required to fill the container and hence the maximum mass of propane that can be used. *(2 marks)*

4 Ammonia is synthesised using the Haber process according to the following equation:

$$2N_2(g) + 3H_2(g) \rightleftharpoons 2NH_3(g)$$

   a At 350°C in a container with a volume of $150\,m^3$ the equilibrium pressure was found to be $2.0 \times 10^7\,Pa$. Calculate the total number of moles of gas present at equilibrium. *(2 marks)*

   b To increase the equilibrium yield of ammonia, the volume of the container can be reduced to $100\,m^3$. Safety regulations stipulate that the maximum pressure should not exceed $2.8 \times 10^7\,Pa$. Calculate the pressure upon reducing the system volume to $100\,m^3$ and state whether this is safe. *(3 marks)*

5 A pure sample of a gas Z occupies a volume of $4.13 \times 10^{-3}\,m^3$ at a pressure of 150 000 Pa at 298 K. When gas Z was expelled from the container its mass was found to have decreased by 7 g.

   a Calculate the number of moles of Z present. *(2 marks)*

   b Calculate the molecular mass (M_r) of gas Z. *(1 mark)*

   c Gas Z was found to be a hydrocarbon. Use your answer to part **b** to suggest a molecular formula for gas Z. *(1 mark)*

6 Diamond and graphite are both allotropes of carbon with different properties and uses. Compare the bonding and physical properties of diamond and graphite, giving examples of how these differences are linked to their uses as materials. *(6 marks)*

7 The melting points of the chlorides of sodium and lithium are 1074 K and 1655 K respectively. Describe the bonding present in these substances and explain why lithium chloride has a greater melting point. *(3 marks)*

8 The hardness of the Group 1 metals decreases as you descend the group. Describe with the aid of a diagram, the bonding present in a crystal of sodium metal and explain the trend in hardness. *(4 marks)*

9 The buckminsterfullerene is an example of an allotrope of carbon with the molecular formula C_{60}. It is a black solid that does not conduct electricity and is commonly used in printing ink.

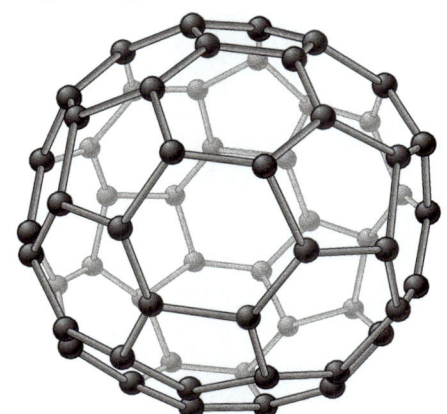

   a Calculate the relative molecular mass of a buckminsterfullerene molecule. *(1 mark)*

   b Identify the type of structure and state the bonding present in a buckminsterfullerene. *(2 marks)*

5 Chemical energetics

5.1 Enthalpy change, ΔH

Learning outcomes

On these pages you will learn to:

- explain that chemical reactions involve energy changes
- use the terms enthalpy change of reaction, standard conditions and bond energy
- calculate enthalpy changes from experimental results
- use average bond energies in calculations
- construct and interpret reaction pathway diagrams

Most chemical reactions give out or take in energy as they proceed. The energy involved may be in different forms—light, electrical, or, most usually, heat.

Thermochemistry

Thermochemistry is the study of heat changes during chemical reactions.

- When a chemical reaction takes place, chemical bonds break and new ones are formed.
- Energy must be put in to break bonds and energy is given out when bonds are formed, so most chemical reactions involve an energy change.
- The overall change may result in energy being given out or taken in.

Exothermic and endothermic reactions

Some reactions give out heat as they proceed. These are called **exothermic reactions**. Neutralising an acid with an alkali is an example of an exothermic reaction.

Some reactions take in heat from their surroundings to keep the reaction going. These are called **endothermic reactions**. The breakdown of limestone (calcium carbonate) to lime (calcium oxide) and carbon dioxide is an example of an endothermic reaction—it needs heat to proceed.

The amount of heat given out or taken in by a given reaction varies with the conditions—temperature, pressure, concentration of solutions, and so on. This means that you must state the conditions under which measurements are made.

When you measure a heat change at constant pressure, it is called an **enthalpy change**.

Enthalpy change, ΔH

Enthalpy has the symbol H, so enthalpy *changes* are given the symbol ΔH. The Greek letter Δ (delta) is used to indicate a change in any quantity.

- In an exothermic reaction the products end up with less energy than the starting materials, because they have lost heat energy when they heated up their surroundings. This means that ΔH is *negative* and has a negative sign.
- In an endothermic reaction the products end up with more energy than the starting materials, so ΔH is positive and has a *positive* sign.

It is *always* the case that a reaction that is endothermic in one direction is exothermic in the reverse direction.

For example, heating hydrated copper sulfate is an endothermic reaction. Blue copper sulfate crystals have the formula $CuSO_4.5H_2O$. The water molecules are bonded to the copper sulfate. In order to break these bonds and make white, anhydrous copper sulfate, heat energy must be supplied (Figure 1). This reaction takes in heat so it is endothermic:

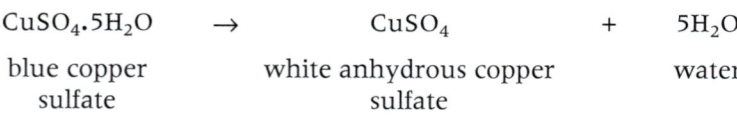

| $CuSO_4.5H_2O$ | $\rightarrow$ | $CuSO_4$ | + | $5H_2O$ |
| blue copper sulfate | | white anhydrous copper sulfate | | water |

ΔH is positive.

Figure 1 Heating copper sulfate

When you add water to anhydrous copper sulfate, the reaction gives out heat.

$$CuSO_4 \quad + \quad 5H_2O \quad \rightarrow \quad CuSO_4.5H_2O$$

| white anhydrous copper sulfate | water | blue copper sulfate |

In this direction the reaction is exothermic. **ΔH is negative**.

Standard conditions

There are the **standard conditions** for measuring enthalpy changes:

- pressure of 101 kPa (approximately normal atmospheric pressure)
- temperature of 298 K (around normal room temperature, 25°C).

(The standard state of an element is the state in which it exists at 298 K and 101 kPa.)

 Heat and temperature

Temperature is related to the average kinetic energy of the particles in a system. As the particles move faster, their average kinetic energy increases and the temperature goes up. But it doesn't matter how many particles there are—temperature is independent of the number present. Temperature is measured with a thermometer.

Heat is a measure of the total energy of all the particles present in a given amount of substance. It does depend on how much of the substance is present. The energy of every particle is included. So a bath of warm water has much more heat than a red hot nail because there are so many more particles in it. Heat always flows from high to low temperature, so heat will flow from the nail into the bath water, even though the water has much more heat than the nail.

When an enthalpy change is measured under standard conditions it is written as $\Delta H^{\ominus}_{298}$, although usually the 298 is left out and implied by the standard symbol $\ominus$.

It may seem strange to talk about measuring heat changes at a constant temperature because heat changes normally cause temperature changes. The way to think about this is to imagine the reactants at 298 K, see Figure 2. Mix the reactants and heat is produced (this is an exothermic reaction). This heat is given out to the surroundings.

A reaction is not thought of as being over until the products have cooled back to 298 K. The heat given out to the surroundings while the reaction mixture cools is the enthalpy change for the reaction, $\Delta H^{\ominus}$.

Some endothermic reactions that take place in aqueous solution absorb heat from the water and cool it down, for example, dissolving ammonium nitrate in water. Again you don't think of the reaction as being over until the products have warmed up to the temperature at which they started, taking in heat from the surroundings to do this.

Unless you remember this, it can seem strange that a reaction that is absorbing heat initially gets cold.

Pressure affects the amount of heat energy given out by reactions that involve gases. If a gas is given out, some energy is required to push away the atmosphere. The greater the atmospheric pressure, the more energy is used for this. This means that less energy remains to be given out as heat by the reaction. This is why it is important to have a standard of pressure for measuring energy changes.

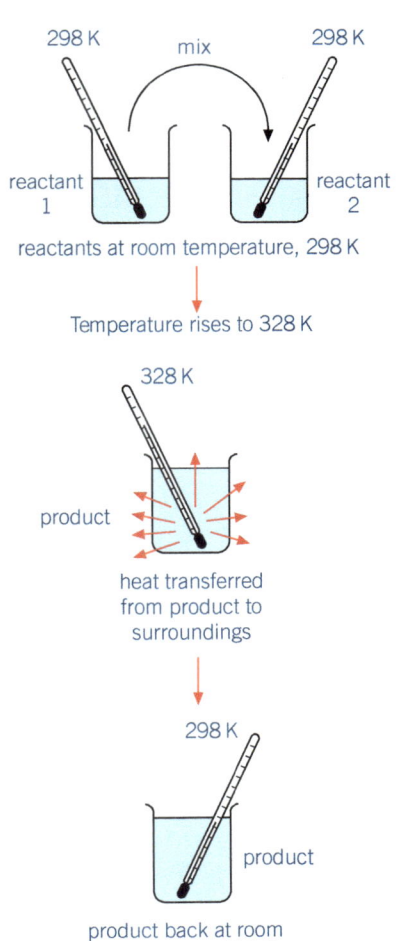

Figure 2 A reaction giving out heat at 298 K

The importance of the equation

The amount of heat given out or taken in during a chemical reaction depends on the quantity of reactants. This energy is usually measured in kilojoules per mole, $kJ\,mol^{-1}$. To avoid any confusion about quantities you need to give an equation.

The physical states (gas, liquid, or solid) of the reactants and products also affect the enthalpy change of a reaction. For example, heat must be put in to change liquid to gas, and is given out when a gas is changed to a liquid. This means that you must always include state symbols in your equations.

For example, in the combustion of methane, CH_4, one mole of methane reacts with two moles of oxygen:

$$CH_4(g) + 2O_2(g) \rightarrow CO_2(g) + 2H_2O(l) \qquad \Delta H = -890 \text{ kJ}\,mol^{-1}$$

890 kJ are given out when one mole of methane burns in two moles of oxygen.

Enthalpy level diagrams

Enthalpy level diagrams, sometimes called **energy level diagrams**, are used to represent enthalpy changes. They show the relative enthalpy levels of the **reactants** (starting materials) and the **products**. The vertical axis represents enthalpy, and the horizontal axis represents the extent of the reaction. We are usually only interested in the beginning of the reaction (100% reactants) and the end of the reaction (0% reactants and 100% products), so the horizontal axis is usually left without units.

Figure 3 shows a general enthalpy diagram for an exothermic reaction (the products have less enthalpy than the reactants) and Figure 4 shows an endothermic reaction (the products have more enthalpy than the reactants).

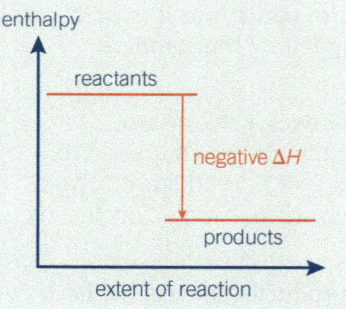

Figure 3 Enthalpy diagram for an exothermic reaction

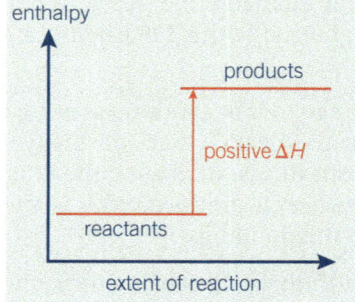

Figure 4 Enthalpy diagram for an endothermic reaction

Activation energy

For a reaction to take place, particles must collide. For a collision to result in a reaction, the molecules must have a certain minimum energy, enough to start breaking bonds. The minimum energy needed to start a reaction is called the **activation energy** and has the abbreviation E_A.

You can include the idea of activation energy on an enthalpy level diagram that shows the **reaction pathway**.

Exothermic reactions

Figure 5 shows the reaction profile for an exothermic reaction with a large activation energy. This reaction will take place extremely slowly at room temperature because very few collisions will have sufficient energy to bring about a reaction.

Figure 6 shows the reaction profile for an exothermic reaction with a small activation energy. This reaction will take place rapidly at room temperature because many collisions will have enough energy to bring about a reaction.

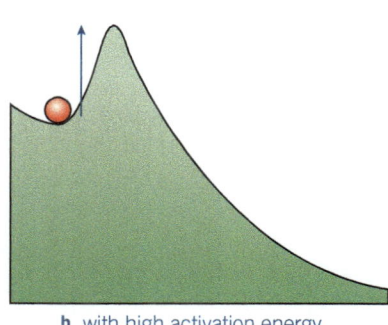

a with low activation energy

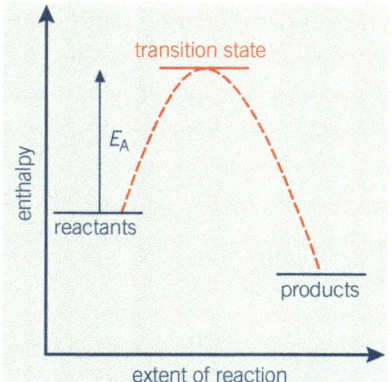

Figure 5 *An exothermic reaction with a large activation energy, E_A*

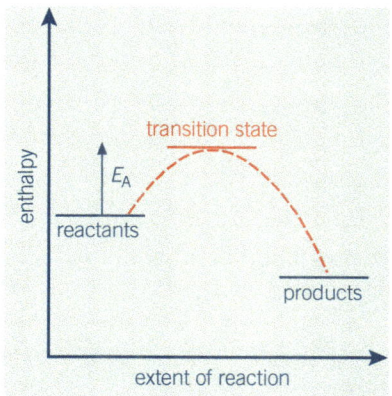

Figure 6 *An exothermic reaction with a small activation energy, E_A*

b with high activation energy

Figure 7 *Ball on a mountainside models*

The situation is a little like a ball on a hill (see Figure 7). A small amount of energy is needed in Figure 7a, to set the ball rolling, while a large amount of energy is needed in Figure 7b.

The species that exists at the top of the curve of an enthalpy level diagram is called a **transition state** or **activated complex**. Some bonds are in the process of being made and some bonds are in the process of being broken. Like the ball at the very top of the hill, it has extra energy and is unstable.

Endothermic reactions

Endothermic reactions are those in which the products have more energy than the reactants. An endothermic reaction, with activation energy E_A, is shown in Figure 8. The transition state has been labelled.

Notice that the activation energy is measured from the reactants to the top of the curve.

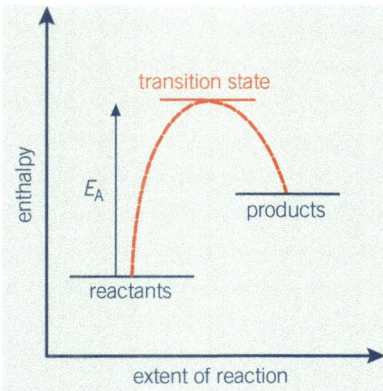

Figure 8 *An endothermic reaction with activation energy E_A*

Bond energies

When chemical change takes place:

- Bonds have to break and energy is taken in.
- New bonds form and energy is given out.

For covalently bonded compounds, we use the idea of **bond energy**. This is the amount of energy that must be *put in* to break a covalent bond so that the atoms move far away from each other. The same amount of energy is given out when this bond forms from separate atoms. When measured under standard conditions (298 K (25°C) and 101 kPa) the bond energy may be referred to as the **bond enthalpy**. The bond energy for a specific bond in a specific molecule may be known exactly. However, bond energies vary slightly for the same bond in different molecules. So we often use mean bond energies which are averages for the same bond in different molecules.

> **Exam tip**
>
> Exothermic reactions may also absorb energy from the surroundings to start the reaction. But overall, more energy is released.

Table 1 shows the bond energies (bond enthalpies) for some commonly encountered bonds.

Table 1 *Some bond enthalpies*

Bond	Bond enthalpy / kJ mol^{-1}
C—H	410
C—C	350
Cl—Cl	242
C—Cl	340
Cl—H	431
Br—Br	193
Br—H	366
C—Br	280

Note

The bond enthalpies for Cl–Cl and Br–Br are actual bond energies for the reactions Cl–Cl(g) → Cl· + Cl· and Br–Br(g) → Br· + Br·. The other values (eg for C–H) are mean bond energies for the same bond in different molecules.

You can use mean bond enthalpies to work out the enthalpy change of reactions, for example:

$$C_2H_6(g) \quad + \quad Cl_2(g) \quad \rightarrow \quad C_2H_5Cl(g) \quad + \quad HCl(g)$$
$$\text{ethane} \qquad \text{chlorine} \qquad \text{chloroethane} \quad \text{hydrogen chloride}$$

The mean bond enthalpies you will need for this example are given in Table 1.

The steps are as follows:

1 First draw out the molecules and show all the bonds. (Formulae drawn showing all the bonds are called **displayed formulae**.)

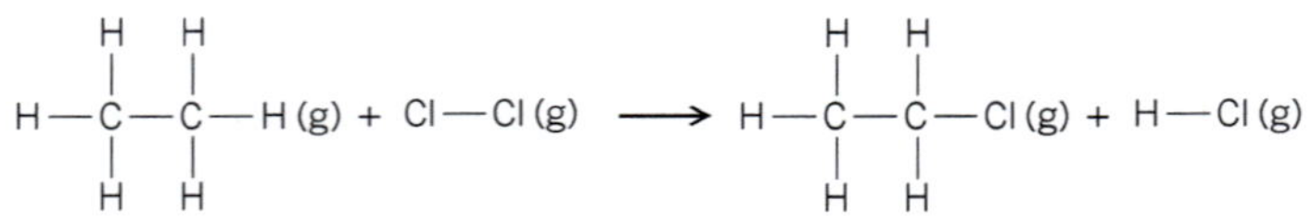

2 Now imagine that all the bonds in the reactants break leaving separate atoms. Look up the bond enthalpy for each bond and add them all up. This will give you the total energy that must be *put in* to break the bonds and form separate atoms.

You need to *break* these bonds:

6 × C–H	6 × 410 kJ mol^{-1}	= 2460 kJ mol^{-1}
1 × C–C	1 × 350 kJ mol^{-1}	= 350 kJ mol^{-1}
1 × Cl–Cl	1 × 242 kJ mol^{-1}	= 242 kJ mol^{-1}
		= 3052 kJ mol^{-1}

So 3052 kJ mol^{-1} must be put in to convert ethane and chlorine to separate hydrogen, chlorine, and carbon atoms.

3 Next imagine the separate atoms join together to give the products. Add up the bond enthalpies of the bonds that must form. This will give you the total enthalpy *given out* by the bonds forming.

You need to *make* these bonds:

5 × C–H	5 × 410 kJ mol^{-1}	= 2050 kJ mol^{-1}
1 × C–C	1 × 350 kJ mol^{-1}	= 350 kJ mol^{-1}
1 × C–Cl	1 × 340 kJ mol^{-1}	= 340 kJ mol^{-1}
1 × Cl–H	1 × 431 kJ mol^{-1}	= 431 kJ mol^{-1}
		= 3171 kJ mol^{-1}

So 3171 kJ mol^{-1} is given out when you convert the separate hydrogen, chlorine, and carbon atoms to chloroethane and hydrogen chloride.

Exam tip

Bond breaking processes where the electrons are shared equally between each atom are described as *homolytic*.

The difference between the energy put in to break the bonds and the energy given out to form bonds is the approximate enthalpy change of the reaction.

The difference is 3171 − 3052 = 119 kJ mol^{-1}

4 Finally work out the sign of the enthalpy change. If more energy was put in than was given out, the enthalpy change is *positive* (the reaction is endothermic). If more energy was given out than was put in, the enthalpy change is *negative* (the reaction is exothermic).

In this case, more enthalpy is given out than put in, so the reaction is exothermic and

$\Delta H = -119$ kJ mol^{-1}

Note that in practice it would be impossible for the reaction to happen like this. But Hess's Law (see Section 5.2) tells us that the enthalpy change will be the same whatever route is taken from starting material to products.

A shortcut
You can often shorten mean bond enthalpy calculations:

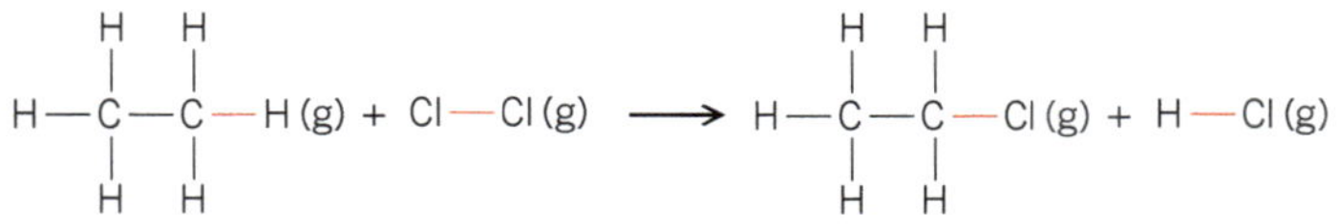

Only the bonds drawn in red make or break during the reaction, so you only need to break:

$$1 \times \text{C–H} = 410 \text{ kJ mol}^{-1}$$
$$1 \times \text{Cl–Cl} = 242 \text{ kJ mol}^{-1}$$
$$\text{Total energy put in} = \textbf{652 kJ mol}^{-1}$$
$$\text{You only need to make: } 1 \times \text{C–Cl} = 340 \text{ kJ mol}^{-1}$$
$$1 \times \text{H–Cl} = 431 \text{ kJ mol}^{-1}$$
$$\text{Total energy given out} = \textbf{771 kJ mol}^{-1}$$
$$\text{The difference is: } 771 - 652 = 122 \text{ kJ mol}^{-1}$$

More energy is given out than taken in so:

$$\Delta H = -119 \text{ kJ mol}^{-1} \text{ (as before)}$$

Measuring enthalpy changes

The general name for the enthalpy change for any reaction is the **standard molar enthalpy change of reaction, $\Delta H_r^\ominus$**. It is measured in kilojoules per mole, kJ mol^{-1} (molar means 'per mole'). You write a balanced symbol equation for the reaction and then find the heat change for the quantities in moles given by this equation.

For example, ΔH for:

$$2NaOH + H_2SO_4 \rightarrow Na_2SO_4 + 2H_2O$$

is the enthalpy change when two moles of NaOH react with one mole of H_2SO_4.

Standard enthalpies
Some commonly used enthalpy changes are given names, for example, the enthalpy change of formation, the enthalpy change of combustion and the enthalpy change of neutralisation. All of these quantities are useful when calculating enthalpy changes for reactions. In addition, enthalpies of combustion are relatively easy to measure for compounds that burn readily in oxygen. Their formal definitions are as follows:

The **standard molar enthalpy of formation, $\Delta H_f^{\ominus}$**, is the enthalpy change when one mole of substance is *formed* from its constituent elements under standard conditions, all reactants and products being in their standard states. For example, $\Delta H_f^{\ominus}$ for methane is the enthalpy change for $C(s,gr) + 2H_2(g) \rightarrow CH_4(g)$.

The **standard molar enthalpy of combustion, $\Delta H_c^{\ominus}$**, is the enthalpy change when one mole of substance is *completely burnt* in oxygen under standard conditions, all reactants and products being in their standard states. So $\Delta H_c^{\ominus}$ for methane is the enthalpy change for $CH_4(g) + 2O_2(g) \rightarrow CO_2(g) + 2H_2O(l)$

The **standard molar enthalpy of neutralisation, $\Delta H_{neut}^{\ominus}$**, is the enthalpy change when solutions of an acid and an alkali *react* together to produce one mole of water. $\Delta H_{neut}^{\ominus}$ of sodium hydroxide and hydrochloric acid is the enthalpy change for $NaOH(aq) + HCl(aq) \rightarrow NaCl(aq) + H_2O(l)$.

Measuring the enthalpy change of a reaction
There is no single instrument that measures the heat output of a reaction. We have to arrange for the heat produced by the reaction to go into a known mass of water and then measure the temperature change.
Then you need to know three things:
1 mass of the substance that is being heated up or cooled down, m
2 temperature change, ΔT
3 specific heat capacity of the substance, c

The **specific heat capacity, c**, is the amount of heat needed to raise the temperature of 1 g of the substance by 1 K. Its units are joules per gram per kelvin, or $J g^{-1} K^{-1}$. For example, the specific heat capacity of water is $4.18\ J g^{-1} K^{-1}$. This means that it takes 4.18 joules to raise the temperature of 1 gram of water by 1 kelvin. This is often rounded up to $4.2\ J g^{-1} K^{-1}$.

Then
$$q = mc\Delta T$$
where q is the heat change worked out from the experimental results.
A – sign is added because when there is a rise in temperature of the water it means that heat has been given out by the reaction (ie it is exothermic). So:
$$\Delta H = -mc\Delta T$$

The simple calorimeter
You can use the apparatus in Figure 9 to find the approximate enthalpy change when a fuel burns. The apparatus used is called a **calorimeter** (from the Latin *calor* meaning heat).

You burn the fuel to heat a known mass of water and then measure the temperature rise of the water. You assume that all the heat from the fuel goes into the water.

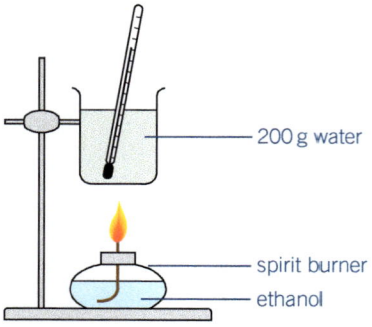

200 g water

spirit burner

ethanol

Figure 9 A simple calorimeter

Worked example

Working out the enthalpy change of combustion

The calorimeter in Figure 9 was used to measure the enthalpy change of combustion of methanol:

$$CH_3OH(l) + 1\tfrac{1}{2}O_2(g) \rightarrow CO_2(g) + 2H_2O(l)$$

0.32 g (0.01 mol) of methanol was burnt and the temperature of the 200.0 g of water rose by 4.0 K.

Heat change $= q = m \times c \times \Delta T$

$\qquad = 200.0 \times 4.2 \times 4.0 = 3360\ J$

0.01 mol gives 3360 J.

So 1 mol would give 336 000 J or 336 kJ.

$\Delta H_c = -340\ kJ\,mol^{-1}$ (negative because heat is given out)

The simple calorimeter can be used to compare the ΔH_c values of a series of similar compounds because the errors will be similar for every experiment. However, you can improve the results by cutting down the heat loss as shown in Figure 10.

Measuring enthalpy changes of reactions in solution
It is relatively easy to measure heat changes for reactions that take place in solution. The heat is generated in the solutions themselves and only has to be kept in the calorimeter. Expanded polystyrene beakers are often used for the calorimeters. These are good insulators (this reduces heat loss through their sides) and they have a low heat capacity so they absorb very little heat. The specific heat capacity of dilute solutions is usually taken to be the same as that of water, $4.2\,J\,g^{-1}\,K^{-1}$ (or more precisely $4.18\,J\,g^{-1}\,K^{-1}$).

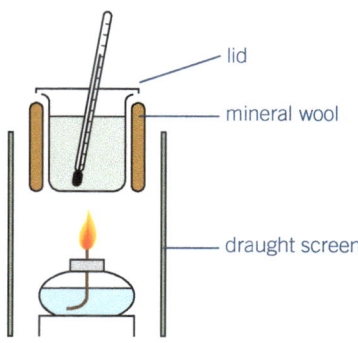

Figure 10 An improved calorimeter

Neutralisation reactions
Neutralisation reactions in solution are exothermic—they give out heat. When an acid is neutralised by an alkali the equation is:

$$acid + alkali \rightarrow salt + water$$

To find an enthalpy change for a reaction, you use the quantities in moles given by the balanced equation. For example, to find the molar enthalpy change of reaction for the neutralisation of hydrochloric acid by sodium hydroxide, the heat given out by the quantities in the equation needs to be found.

Worked example

Working out the enthalpy change for a neutralisation reaction

$HCl(aq)$	$+$	$NaOH(aq)$	$\rightarrow$	$NaCl(aq)$	$+$	$H_2O(l)$
hydrochloric acid		sodium hydroxide		sodium chloride		water
1 mol		1 mol		1 mol		1 mol

$50\,cm^3$ of $1.0\,mol\,dm^{-3}$ hydrochloric acid and $50\,cm^3$ of $1.0\,mol\,dm^{-3}$ sodium hydroxide solution were mixed in an expanded polystyrene beaker. The temperature rose by 6.6 K.

The total volume of the mixture is $100\,cm^3$. This has a mass of approximately $100\,g$ because the density of water and of dilute aqueous solutions is approximately $1\,g\,cm^{-3}$.

enthalpy change q	$=$	mass of water m	$\times$	specific heat capacity of solution c	$\times$	temperature change ΔT

$q = m \times c \times \Delta T$

$= 100 \times 4.2 \times 6.6 = 2772\,J$

Number of moles of acid (and also alkali)

$$n = \frac{\text{concentration } c\ (mol\,dm^{-3}) \times \text{volume } V\,(cm^3)}{100}$$

$= 1.0 \times \dfrac{50}{1000} = 0.05\,mol$

So 1 mol would give $\dfrac{2772}{0.05}\,J = 55\,440\,J = 55.44\,kJ$

$\Delta H = -55.44\,kJ\,mol^{-1}$

$\Delta H = -55\,kJ\,mol^{-1}$ (to 2 sf)

The sign of ΔH is negative because heat is given out.

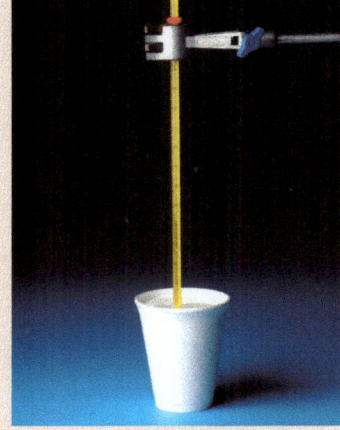

Figure 11 Polystyrene beakers make good calorimeters because they are good insulators and have low heat capacities

---

Exam tip

Remember to use the *total* volume of the mixture, $100\,cm^3$. A common mistake is to use $50\,cm^3$.

Extension

Allowing for heat loss

Although expanded polystyrene cups are good insulators, some heat will still be lost from the sides and top leading to low values for enthalpy changes measured by this method. This can be allowed for by plotting a cooling curve. As an example, the measurement of the heat of neutralisation of hydrochloric acid and sodium hydroxide is repeated using a cooling curve.

Before the experiment, all the apparatus and both solutions are left to stand in the laboratory for some time. This ensures that they all reach the same temperature, that of the laboratory itself.

Then proceed as follows:

1 Place 50 cm³ of 1.0 mol dm⁻³ hydrochloric acid in one polystyrene cup and 50 cm³ of 1.0 mol dm⁻³ sodium hydroxide solution in another.
2 Using a thermometer that reads to 0.1°C, take the temperature of each solution every 30 seconds for four minutes to confirm that both solutions remain at the same temperature, that of the laboratory. A line of 'best fit' is drawn through these points. It is likely there will be very small variations around the line of best fit, indicating random errors.
3 Now pour one solution into the other and stir, continuing to record the temperature every 30 seconds for a further six minutes.
4 The results are shown on the graph in Figure 12. The experiment can also be done using an electronic temperature sensor and data logging software to plot the graph directly.

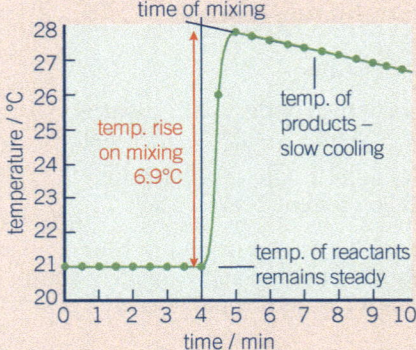

Figure 12 *Graph to show temperature as a neutralisation reaction proceeds*

On mixing, the temperature rises rapidly as the reaction gives out heat, and then drops slowly and regularly as heat is lost from the polystyrene cup. To find the best estimate of the temperature immediately after mixing, you draw the best straight line through the graph points after mixing and extrapolate back to the time of mixing. This gives a temperature rise of 6.9°C.

The calculation is as before.

$q = m \times c \times \Delta T = 100 \times 4.2 \times 6.9 = 2898 \text{ J}$

The number of moles of acid (and alkali) was 0.05 mol (as before).

So 1 mol would give $\dfrac{2898}{0.05}$ J $= 57\,960$ J $= 57.96$ kJ

$\Delta H_{\text{neut}} = -58 \text{ kJ mol}^{-1} \text{ (to 2 sf)}$

The sign of ΔH is negative because heat is given out.

Summary test 5.1

1 Consider this reaction:

$$C(s) + O_2(g) \rightarrow CO_2(g) \qquad \Delta H^\ominus_{298} = -394 \text{ kJ mol}^{-1}$$

    a State what the symbol Δ means.
    b State what the symbol H means.
    c State what the 298 indicates.
    d State what the minus sign indicates.
    e Explain whether the reaction is exothermic or endothermic.

2 Consider these reactions:

$$CH_4(g) + 2O_2(g) \rightarrow CO_2(g) + 2H_2O(l) \qquad \Delta H = -890 \text{ kJ mol}^{-1}$$

$C_6H_8O_7(aq)$	+	$3NaHCO_3(aq)$	$\rightarrow$	$Na_3C_6H_5O_7(aq)$	+	$3H_2O(l)$	+	$3CO_2(g)$
citric acid		sodium hydrogencarbonate		sodium citrate		water		carbon dioxide

$$\Delta H = +51 \text{ kJ mol}^{-1}$$

    a In each case explain whether the reaction is exothermic or endothermic.
    b In each case sketch an enthalpy level diagrams to show the reaction, including as much information as you can on your diagram
    c The first reaction above takes place in a gas burner. Explain why a spark or match is needed to start the reaction.

Note: questions 3–6 are about the reaction:

$$CH_3CH_3 + Br_2 \rightarrow CH_3CH_2Br + HBr$$

Sketch the displayed formulae of all the products and reactants so that all the bonds are shown.

3 **a** Identify the bonds that have to be broken to convert the reactants into separate atoms.
    b Calculate how much energy this takes.

4 **a** Identify the bonds that have to be made to convert separate atoms into the products.
    b Calculate how much energy this takes.

5 Describe what the difference is between the energy put in to break bonds and the energy given out when the new bonds are formed.

6 **a** State what ΔH is for the reaction (this requires a sign).
    b Identify whether the reaction is endothermic or exothermic.

7 0.74 g (0.010 mol) of propanoic acid was burnt in the simple calorimeter, shown in Figure 9. The temperature rose by 8.0 K. Calculate the value this gives for the enthalpy change of combustion of propanoic acid.

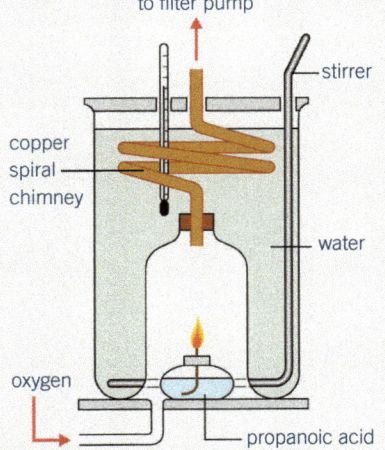

Figure 13 A flame calorimeter

8 The flame calorimeter above is an improved version of the simple calorimeters used for measuring enthalpy changes of combustion. State three improvements over the simple calorimeter.

9 50.0 cm³ of 2.00 mol dm⁻³ sodium hydroxide and 50.0 cm³ of 2.00 mol dm⁻³ hydrochloric acid were mixed in an expanded polystyrene beaker. The temperature rose by 11.0 K.
    a Calculate ΔH for the reaction.
    b Describe how this value will compare with the accepted value for this reaction.
    c Explain your answer to **b**.

5.2 Hess's Law

Learning outcomes

On these pages you will learn to:

- use Hess's Law to construct energy cycles
- calculate energy changes that cannot be found directly

The enthalpy changes for some reactions cannot be measured directly. To find these you use an indirect approach. Chemists use enthalpy changes that they can measure to work out enthalpy changes that they cannot measure. It is often easy to measure enthalpies of combustion. To do this, chemists use Hess's Law, first stated by Germain Hess, a Swiss-born Russian chemist, born in 1802.

Hess's Law

Hess's Law states that the enthalpy change for a chemical reaction is the same, whatever route is taken from reactants to products.

This is a consequence of a more general scientific law, the Law of conservation of energy, which states that *energy can neither be created nor destroyed*. So, provided the starting and finishing points of a process are the same, the energy change must be the same. If not, energy would have been created or destroyed.

Using Hess's Law

To see what Hess's Law means, look at the following example where ethyne, C_2H_2, is converted to ethane, C_2H_6, by two different routes. How can we find the enthalpy of reaction?

Route 1: The reaction takes place directly—ethyne reacts with two moles of hydrogen to give ethane:

$$\underset{\text{ethyne}}{C_2H_2(g)} + 2H_2(g) \rightarrow \underset{\text{ethane}}{C_2H_6(g)} \qquad \Delta H_1 = ?$$

Route 2: The reaction takes place in two stages.

a Ethyne, C_2H_2, reacts with one mole of hydrogen to give ethene, C_2H_4.
b Ethene, C_2H_4, then reacts with a second mole of hydrogen to give ethane, C_2H_6.

Hess's Law tells us that the total energy change is the same whichever route you take—direct or via ethene (or, in fact, by any other route). You can show this on a diagram called an **energy cycle**, also called a **thermochemical cycle**.

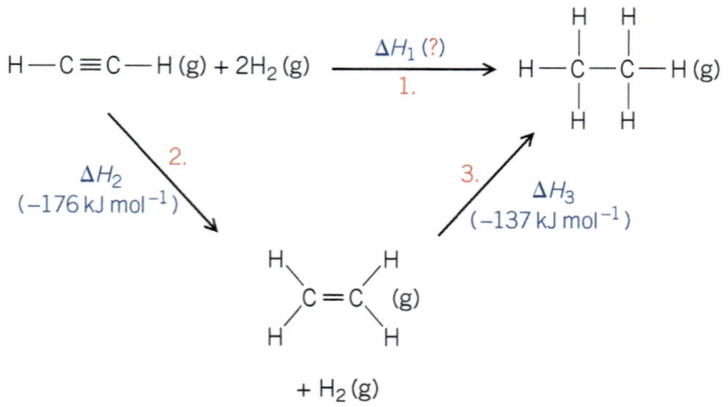

Hess's Law means that: $\Delta H_1 = \Delta H_2 + \Delta H_3$

The actual figures are: $\Delta H_2 = -176\ \text{kJ mol}^{-1}$

$\Delta H_3 = -137\ \text{kJ mol}^{-1}$

So $\Delta H_1 = (-176) + (-137) = -313\ \text{kJ mol}^{-1}$

This method of calculating ΔH_1 is fine if you know the enthalpy changes for the other two reactions. There are certain enthalpy changes that can be looked up for a large range of compounds. These include the enthalpy change of formation ΔH_f°, and enthalpy change of combustion, ΔH_c°. In practice, many ΔH_f° are calculated from ΔH_c° via energy cycles using Hess's Law.

Using the enthalpy changes of formation $\Delta H_f^\ominus$

The enthalpy of formation, $\Delta H_f^\ominus$, is the enthalpy change when one mole of compound is formed from its constituent elements under standard conditions, all reactants and products being in their standard states.

Another theoretical way to convert ethyne to ethane could be via the elements carbon and hydrogen.

- Ethyne is first converted to its elements, carbon and hydrogen. This is the reverse of formation and the enthalpy change is the *negative* of the enthalpy of formation. This is a general rule. The reverse of a reaction has the negative of its ΔH value. This is a consequence of Hess's Law.
- Then the carbon and hydrogen react to form ethane. This is the enthalpy of formation for ethane.

Hess's Law tells us that:

$$\Delta H_1 = \Delta H_4 + \Delta H_5$$

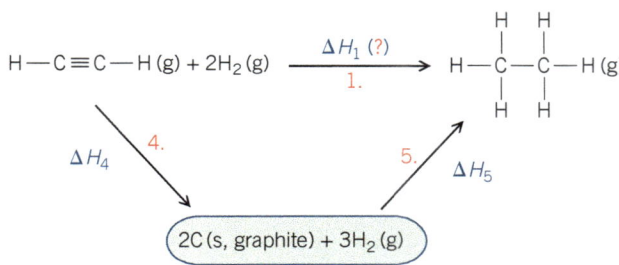

ΔH_5 is the enthalpy of formation, $\Delta H_f^\ominus$, of ethane whilst reaction 4 is the reverse of the formation of ethyne.

The values you need are: $\Delta H_f^\ominus(C_2H_2) = +228 \text{ kJ mol}^{-1}$

and $\Delta H_f^\ominus(C_2H_6) = -85 \text{ kJ mol}^{-1}$

So $\Delta H_4 = -228 \text{ kJ mol}^{-1}$
(remember to change the sign)

$\Delta H_5 = -85 \text{ kJ mol}^{-1}$

Therefore

$$\Delta H_1 = -228 + -85 = -313 \text{ kJ mol}^{-1}$$

This was the result you got from the previous method, as you should expect from Hess's Law.

Notice that in reaction 4 there are two moles of hydrogen 'spare' as only one of the three moles of hydrogen is involved. These two moles of hydrogen remain in their standard states and so no enthalpy change is involved.

$C_2H_2(g) \rightarrow 2C(s, \text{graphite}) + H_2(g)$ is the reaction you are considering, but you have:

$$C_2H_2(g) + 2H_2(g) \rightarrow 2C(s, \text{graphite}) + 3H_2(g)$$

However, this makes no difference. The 'extra' hydrogen is not involved in the reaction and it does not affect ΔH.

Exam tip

$\Delta H_f^\ominus$ for a compound may be exothermic (negative) or endothermic (positive).

Exam tip

$\Delta H_c^\ominus$ of an element is the same as $\Delta H_f^\ominus$ of its oxide.

Exam tip

Graphite is the most stable form of carbon (another form is diamond). It has a special state symbol (s, graphite).

Exam tip

For an element, $\Delta H_f^\ominus$ is zero by definition.

Thermochemical (energy) cycles using enthalpy changes of combustion

> **The enthalpy change of combustion, $\Delta H^{\ominus}_c$, is the enthalpy change when one mole of substance is completely burnt in oxygen under standard conditions.**

We will look again at the thermochemical cycle used to find $\Delta H^{\ominus}$ for the reaction between ethyne and hydrogen to form ethane.

$$C_2H_2(g) + 2H_2(g) \rightarrow C_2H_6(g)$$

This time we will use enthalpy changes of combustion.

Ethyne, hydrogen, and ethane all burn readily. This means their enthalpy changes of combustion can be easily measured. The combustion products of all three substances are carbon dioxide and water.

The thermochemical cycle is:

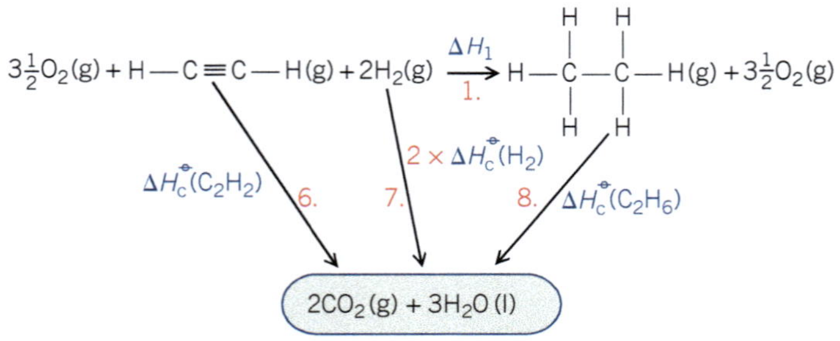

Putting in the values:

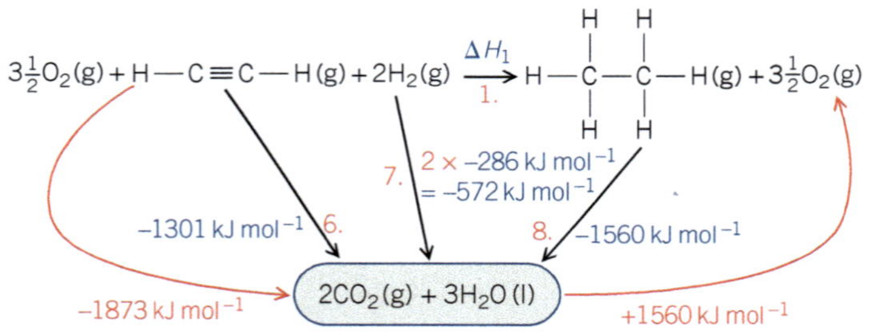

To get the enthalpy change for reaction 1 you must go round the cycle in the direction of the red arrows. This means reversing reaction 8 so you must change its sign.

So $\Delta H_1 = -1873 + 1560 \text{ kJ mol}^{-1}$

$\Delta H_1 = -313 \text{ kJ mol}^{-1}$ once again, the same answer as before

Notice that in reaction 1 there are $3\frac{1}{2}$ moles of oxygen on either side of the equation. They take no part in the reaction and do not affect the value of ΔH.

Finding $\Delta H^{\ominus}_f$ from $\Delta H^{\ominus}_c$

Enthalpy changes of formation of compounds are often difficult or impossible to measure directly. This is because the reactants often do not react directly to form the compound that you are interested in.

For example, the following equation represents the formation of ethanol from its elements.

$$2C(s, \text{graphite}) + 3H_2(g) + \frac{1}{2}O_2(g) \rightarrow C_2H_5OH(l)$$

Exam tip

Remember to multiply by the number of moles of reagents involved in each step.

This does not take place. However, all the species concerned will readily burn in oxygen so their enthalpy changes of combustion can be measured. The thermochemical cycle you need is:

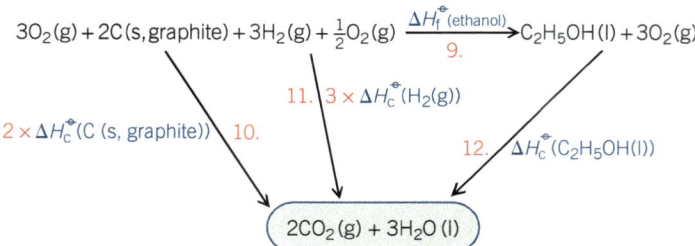

Putting in the values:

- $\Delta H_c^\ominus$ (C(s, graphite)) = −393.5 kJ mol^{-1}
- $\Delta H_c^\ominus$ (H$_2$(g)) = −286 kJ mol^{-1}
- $\Delta H_c^\ominus$ (C$_2$H$_5$OH(l)) = −1367.3 kJ mol^{-1}

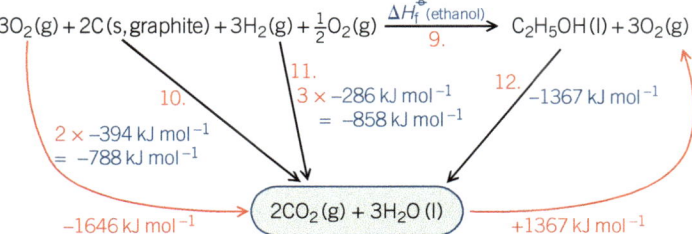

Note that in reaction 9 there are three moles of oxygen on either side of the equation that take no part in the reaction. This means that they do not affect the value of ΔH.

Note also that:

- $\Delta H_c^\ominus$ (C(s, graphite)) is the same as $\Delta H_f^\ominus$ (CO$_2$(g))
- $\Delta H_c^\ominus$ (H$_2$(g)) is the same as $\Delta H_f^\ominus$ (H$_2$O(l)).

To get the enthalpy change for reaction 9, you must go round the cycle in the direction of the red arrows. This means reversing reaction 12 so you must change its sign.

So, ΔH_9 = −1646 + 1367 kJ mol^{-1} = −279 kJ mol^{-1}

So, $\Delta H_f^\ominus$ (C$_2$H$_5$OH(l)) = −279 kJ mol^{-1}

There are exam-style questions to test your knowledge of the material in this chapter at the end of Chapter 6.

Summary test 5.2

1 Use the values of $\Delta H_f^\ominus$ in the table to calculate $\Delta H^\ominus$ for each of the reactions below using a thermochemical cycle.
   a $CH_3COCH_3(l) + H_2(g) \rightarrow CH_3CH(OH)CH_3(l)$
   b $C_2H_4(g) + Cl_2(g) \rightarrow C_2H_4Cl_2(l)$
   c $C_2H_4(g) + HCl(g) \rightarrow C_2H_5Cl(l)$
   d $Zn(s) + CuO(s) \rightarrow ZnO(s) + Cu(s)$
   e $Pb(NO_3)_2(s) \rightarrow PbO(s) + 2NO_2(g) + \frac{1}{2}O_2(g)$

Compound	$\Delta H_f^\ominus$ / kJ mol^{-1}
CH$_3$COCH$_3$(l)	−248
CH$_3$CH(OH)CH$_3$(l)	−318
C$_2$H$_4$(g)	+52
C$_2$H$_4$Cl$_2$(l)	−165
C$_2$H$_5$Cl(l)	−137
HCl(g)	−92
CuO(s)	−157
ZnO(s)	−348
Pb(NO$_3$)$_2$(s)	−452
PbO(s)	−217
NO$_2$(g)	+33

2 Calculate $\Delta H^\ominus$ for the reaction by thermochemical cycles:

$$H-\overset{\overset{\textstyle H}{|}}{\underset{\underset{\textstyle H}{|}}{C}}-C\overset{\displaystyle O}{\underset{\displaystyle H}{\diagup}} (l) + H_2(g) \longrightarrow H-\overset{\overset{\textstyle H}{|}}{\underset{\underset{\textstyle H}{|}}{C}}-\overset{\overset{\textstyle H}{|}}{\underset{\underset{\textstyle H}{|}}{C}}-O-H(l)$$

   a via $\Delta H_f^\ominus$ values
   b via $\Delta H_c^\ominus$ values

Compound	$\Delta H_f^\ominus$ / kJ mol^{-1}	$\Delta H_c^\ominus$ / kJ mol^{-1}
CH$_3$CHO(l)	−192	−1167
H$_2$(g)	−	−286
CH$_3$CH$_2$OH(l)	−277	−1367

6 Electrochemistry

6.1 Redox processes: electron transfer and changes in oxidation number (oxidation state)

Learning outcomes

On these pages you will learn to:

- calculate oxidation numbers
- balance chemical equations using oxidation numbers
- explain redox processes
- use the terms 'oxidising agent', 'reducing agent' and 'disproportionation'

Historically, **oxidation** was used for reactions in which oxygen was added. In this reaction copper has been oxidised to copper oxide. Oxygen is called an **oxidising agent**.

$$Cu(s) + \frac{1}{2}O_2(g) \rightarrow CuO(s)$$

Reduction described a reaction in which oxygen was removed. In this reaction copper oxide has been reduced and hydrogen is the **reducing agent**.

$$CuO(s) + H_2(g) \rightarrow Cu(s) + H_2O(l)$$

As hydrogen was often used to remove oxygen, the addition of hydrogen was also called reduction. In this reaction chlorine has been reduced because hydrogen has been added to it.

$$Cl_2(g) + H_2(g) \rightarrow 2HCl(g)$$

The reverse, where hydrogen was removed, was called oxidation.

The word **redox** is short for reduction–oxidation.

Redox reactions—electron transfer

By describing what happens to the electrons in the above reactions, you get a much more general picture.

When something is oxidised it loses electrons, and when something is reduced it gains electrons.

Redox reactions always involve the movement of electrons, so they are also called **electron transfer reactions**. You can see the transfer of electrons by separating a redox reaction into two half equations that show the gain and loss of electrons.

Worked example

1: Half equations

Look again at the reaction between copper and oxygen to form copper oxide:

$$Cu + \frac{1}{2}O_2 \rightarrow CuO$$

Copper oxide is an ionic compound. You can write the balanced symbol equation using $(Cu^{2+} + O^{2-})$ instead of CuO, to show the ions present in copper oxide:

$$Cu + \frac{1}{2}O_2 \rightarrow (Cu^{2+} + O^{2-})$$

Look at the copper. It has lost two electrons, so it has been oxidised.

$$Cu - 2e^- \rightarrow Cu^{2+} \text{ or } Cu \rightarrow Cu^{2+} + 2e^-$$

This is a **half equation**. It is usual to write half equations with plus electrons rather than minus electrons, that is:

$$Cu \rightarrow Cu^{2+} + 2e^- \text{ rather than } Cu - 2e^- \rightarrow Cu^{2+}$$

Next look at the oxygen. It has gained two electrons, so it has been reduced:

$$\frac{1}{2}O_2(g) + 2e^- \rightarrow O^{2-}$$

If you add the two half equations together, you end up with the original equation. Notice that the numbers of electrons cancel out.

$$Cu \rightarrow Cu^{2+} + \cancel{2e^-}$$

$$\frac{1}{2}O_2(g) + \cancel{2e^-} \rightarrow O^{2-}$$

$$Cu(s) + \frac{1}{2}O_2(g) \rightarrow (Cu^{2+} + O^{2-})(s)$$

Worked example

2: Half equations

When copper oxide reacts with magnesium, copper and magnesium oxide are produced:

$$CuO(s) + Mg(s) \rightarrow MgO(s) + Cu(s)$$

Write the equation with copper oxide as $(Cu^{2+} + O^{2-})$ and magnesium oxide as $(Mg^{2+} + O^{2-})$ to show the ions present.

$$(Cu^{2+} + O^{2-}) + Mg \rightarrow Cu + (Mg^{2+} + O^{2-})$$

Look at the copper. It has gained two electrons, so it has been reduced.

$$Cu^{2+} + 2e^- \rightarrow Cu$$

Look at the magnesium. It has lost electrons, so it has been oxidised.

$$Mg \rightarrow Mg^{2+} + 2e^-$$

Notice that the O^{2-} ion takes no part in the reaction. It is a **spectator** ion.

If you add these half equations you get:

$$Cu^{2+} + Mg \rightarrow Cu + Mg^{2+}$$

This is the **ionic equation** for the redox reaction.

The definition of oxidation and reduction now used is:

Oxidation Is Loss of electrons.
Reduction Is Gain of electrons.

By this definition, magnesium is oxidised by anything that removes electrons from it (not just oxygen) leaving a positive ion. For example, chlorine oxidises magnesium:

$$Mg(s) + Cl_2(g) \rightarrow (Mg^{2+} + 2Cl^-)(s)$$

Look at the magnesium. It has lost electrons and has therefore been oxidised:

$$Mg \rightarrow Mg^{2+} + 2e^-$$

Look at the chlorine. It has gained electrons and has therefore been reduced:

$$Cl_2 + 2e^- \rightarrow 2Cl^-$$

And adding the two half equations together, the electrons cancel out:

$$Mg(s) + Cl_2(g) \rightarrow (Mg^{2+} + 2Cl^-)(s)$$

You may find that adding arrows to the equation, which show the transfer of electrons, helps you keep track of them, as shown in Figure 1.

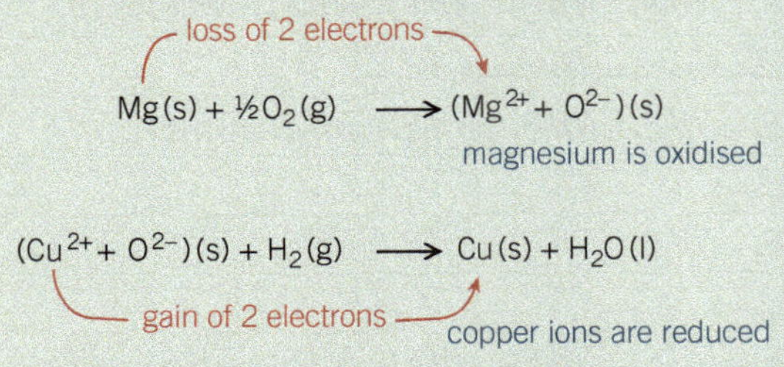

┌ loss of 2 electrons ┐
$Mg(s) + \frac{1}{2}O_2(g) \longrightarrow (Mg^{2+} + O^{2-})(s)$
magnesium is oxidised

$(Cu^{2+} + O^{2-})(s) + H_2(g) \longrightarrow Cu(s) + H_2O(l)$
└ gain of 2 electrons ┘ copper ions are reduced

Figure 1 *Writing the electrons that are transferred helps to keep track of them*

In a chemical reaction, if one species is oxidised (loses electrons), another must be reduced (gains them).

Oxidising and reducing agents

It follows from the above that:

- Reducing agents give away electrons—they are **electron donors**.
- Oxidising agents accept electrons—they are **electron acceptors**.

Oxidation numbers

Oxidation numbers are used to see what has been oxidised and what has been reduced in a redox reaction. Oxidation numbers are also called **oxidation states**.

The idea of oxidation numbers

Each element in a compound is given an oxidation number. In an ionic compound the oxidation number simply tells us how many electrons it has lost or gained, compared with the element in its uncombined state, so it is the charge on the ion. In a molecule, the oxidation number tells us about the distribution of electrons between elements of different electronegativity. The more electronegative element is given the negative oxidation number.

- Every element in its uncombined state has an oxidation number of zero.
- A positive number shows that the element has lost electrons and has therefore been oxidised. For example, Mg^{2+} has an oxidation number of +2.
- A negative number shows that the element has gained electrons and has therefore been reduced. For example Cl^- has an oxidation number of –1.
- The more positive the number, the more the element has been oxidised. The more negative the number, the more it has been reduced.
- The numbers always have a + or – sign unless they are zero.

> **📖 Rules for finding oxidation numbers**
>
> The following rules will allow you to work out oxidation numbers:
>
> 1 Uncombined elements have an oxidation number of 0.
> 2 Some elements always have the same oxidation number in all their compounds. Others usually have the same oxidation number. Table 1 gives the oxidation numbers of these elements.
> 3 The sum of all the oxidation numbers in a compound = 0, since all compounds are electrically neutral.
> 4 The sum of the oxidation numbers of a complex ion, such as NH_4^+ or SO_4^{2-}, equals the charge on the ion.
> 5 In a compound, the most electronegative element always has a negative oxidation number.

Exam tip

You should know the rules for finding oxidation numbers.

Table 1 The usual oxidation numbers of some elements in compounds

Element	Oxidation number in compound	Example
hydrogen, H	+1 (except in metal hydrides, e.g. NaH, where it is –1)	HCl
Group 1	always +1	NaCl
Group 2	always +2	$CaCl_2$
aluminium, Al	always +3	$AlCl_3$
oxygen, O	–2 (except in peroxides where it is –1, and the compound OF_2, where it is +2)	Na_2O
fluorine, F	Always –1	NaF
chlorine, Cl	–1 (except in compounds with F and O, where it has positive values)	NaCl

Working out oxidation numbers of elements in compounds

Start with the correct formula. Look for the elements whose oxidation numbers you know from the rules. Then deduce the oxidation numbers of any other element. Some examples are shown below.

Phosphorus pentachloride, PCl_5

Chlorine has an oxidation number of –1, so the phosphorus must be +5, to make the sum of the oxidation numbers zero.

Ammonia, NH_3

Hydrogen has an oxidation number of +1, so the nitrogen must be –3, to make the sum of the oxidation numbers zero. Also, nitrogen is more electronegative than hydrogen, so hydrogen must have a positive oxidation number.

Nitric acid, HNO_3

Each oxygen has an oxidation number of –2, making –6 in total.

Hydrogen has an oxidation number of +1.

So the nitrogen must be +5, to make the sum of the oxidation numbers zero.

Notice that nitrogen may have different oxidation numbers in different compounds. Here the nitrogen has a positive oxidation number because it is combined with a more electronegative element, oxygen.

Hydrogen sulfide, H_2S

Hydrogen has an oxidation number of +1, so the sulfur must be –2, to make the sum of the oxidation numbers zero.

Sulfate ion, SO_4^{2-}

Each oxygen has an oxidation number of –2, making –8 in total.

So the sulfur must be +6, to make the sum of the oxidation numbers equal to the charge on the ion.

Notice that sulfur may have different oxidation numbers in different compounds.

Black copper oxide, CuO

Oxygen has an oxidation number of –2, so the copper must be +2, to make the sum of the oxidation numbers zero.

Red copper oxide, Cu_2O

Oxygen has an oxidation number of –2, so each copper must be +1, to make the sum of the oxidation numbers zero.

Oxidation numbers are used in Roman numerals to distinguish between similar compounds where the metal has a different oxidation number. So, black copper oxide is copper(II) oxide and red copper oxide is copper(I) oxide. These compounds are shown in Figure 2.

Figure 2 The two oxides of copper—copper(II) oxide (left) and copper(I) oxide (right)

Redox equations: using oxidation numbers in redox equations

You have seen above that you can work out which element has been oxidised and which has been reduced in a redox reaction by considering electron transfer.

Another way of working is to remember that when an element is reduced it gains electrons and its oxidation number is reduced. For example, in $M^{3+} \rightarrow M^{2+}$ the number of plusses has been reduced, so M has been reduced. It follows that for $M^{2+} \rightarrow M^{3+}$ the number of plusses has been increased, so M has been oxidised.

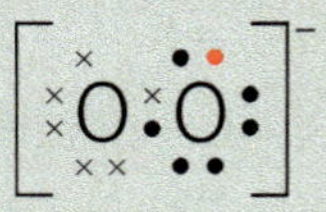

Remember that oxidation is loss of electrons (OIL) and reduction is gain of electrons (RIG).

You can also use oxidation numbers to help you to understand redox reactions.

Tracking redox processes

When an element is reduced, it gains electrons and its oxidation number goes down. In the reaction below, you can see that iron is reduced because its oxidation number has gone down from +3 to +2, whilst iodine is oxidised from −1 to 0:

$$\overset{+3}{Fe^{3+}} + \overset{-1}{I^-} \rightarrow \overset{+2}{Fe^{2+}} + \overset{0}{\tfrac{1}{2}I_2}$$

Even in complicated reactions, you can see which element has been oxidised and which has been reduced when you put in the oxidation numbers:

$$\overset{+5\ -2}{2IO_3^-} + \overset{+1\ +4\ -2}{5HSO_3^-} \rightarrow \overset{0}{I_2} + \overset{+6\ -2}{5SO_4^{2-}} + \overset{+1}{3H^+} + \overset{+1\ -2}{H_2O}$$

Iodine in IO_3^- is reduced (+5 to 0) and sulfur in HSO_3^- is oxidised (+4 to +6). The oxidation numbers of all the other atoms have not changed.

Balancing redox reactions

You can use the idea of oxidation numbers to help balance equations for redox reactions.

For an equation to be balanced:

- The numbers of atoms of each element on each side of the equation must be the same.
- The total charge on each side of the equation must be the same.

Worked example 1

The thermite reaction

This is a strongly exothermic reaction in which aluminium reacts with iron(III) oxide to produce molten iron. It can be used to weld railway lines.

The unbalanced equation is:

$$Fe_2O_3(s) + Al(s) \rightarrow Fe(l) + Al_2O_3(s)$$

Write the oxidation numbers above each element:

$$\overset{+3\ -2}{Fe_2O_3(s)} + \overset{0}{Al(s)} \rightarrow \overset{0}{Fe(l)} + \overset{+3\ -2}{Al_2O_3(s)}$$

If you look at the equation you can see that that only the iron and aluminium have changed their oxidation number. The oxygen is unchanged.

Each iron atom has been reduced by gaining three electrons so you can write the half equation:

$$Fe^{3+} + 3e^- \rightarrow Fe$$

Each aluminium atom has been oxidised by losing three electrons:

$$Al \rightarrow Al^{3+} + 3e^-$$

In the reaction, the number of electrons gained must equal the number of electrons lost. This means that there must be the same number of aluminium atoms as iron atoms. (The oxygen is a spectator ion.) You started with two iron atoms, so you must also have two aluminium atoms. The balanced equation is therefore:

$$Fe_2O_3(s) + 2Al(s) \rightarrow 2Fe(l) + Al_2O_3(s)$$

Figure 3 *The thermite reaction can be used to weld railway lines*

Exam tip

Practise the techniques used in balancing redox equations.

Worked example 2

Aqueous solutions

Sometimes in aqueous solutions, species take part in redox reactions but are neither oxidised nor reduced. You must balance them separately. These include water molecules, H^+ ions (in acid solution), and OH^- ions (in alkaline solution). Oxidation numbers only help us to balance the species that are oxidised or reduced.

Suppose you want to balance the following equation, where dark purple manganate(VII) ions react in acid solution with Fe^{2+} ions to produce pale pink Mn^{2+} ions and Fe^{3+} ions.

The unbalanced equation is:

$$MnO_4^- + Fe^{2+} + H^+ \rightarrow Mn^{2+} + Fe^{3+} + H_2O$$

1 Write the oxidation number above each element.

$$\overset{+7\,-2}{MnO_4^-} + \overset{+2}{Fe_2^+} + \overset{+1}{H^+} \rightarrow \overset{+2}{Mn^{2+}} + \overset{+3}{Fe^{3+}} + \overset{+1\,-2}{H_2O}$$

2 Identify the species that has been oxidised and the species that has been reduced.

$$\overset{+7}{MnO_4^-} \rightarrow \overset{+2}{Mn^{2+}}$$

Manganese has been reduced from +7 to +2, therefore five electrons must be gained.

$MnO_4^- + 5e^- \rightarrow Mn^{2+}$ (this equation is not chemically balanced)

$$\overset{+2}{Fe^{2+}} \rightarrow \overset{+3}{Fe^{3+}}$$

Fe has been oxidised from +2 to +3 so one electron must be lost.

$$Fe^{2+} \rightarrow Fe^{3+} + e^-$$

In order to balance the number of electrons that are transferred, this step must be multiplied by 5:

$$5Fe^{2+} \rightarrow 5Fe^{3+} + 5e^-$$

So, you know that there are $5Fe^{2+}$ ions to every MnO_4^- ion.

3 Include this information in the unbalanced equation, to balance the redox process.

$$MnO_4^- + 5Fe^{2+} + H^+ \rightarrow Mn^{2+} + 5Fe^{3+} + H_2O$$

(this equation is still not chemically balanced)

4 Balance the remaining atoms, those that are neither oxidised nor reduced. In order to 'use up' the four oxygen atoms on the left-hand side, you need $4H_2O$ on the right-hand side, which will in turn require $8H^+$ on the left-hand side.

$$MnO_4^- + 5Fe^{2+} + 8H^+ \rightarrow Mn^{2+} + 5Fe^{3+} + 4H_2O$$

Notice that this equation is balanced for both atoms and charge.

Disproportionation

In some chemical reactions, atoms of the same element can be both oxidised *and* reduced. For example, hydrogen peroxide decomposes to oxygen and water.

$$\overset{-1}{2H_2O_2} \rightarrow \overset{-2}{2H_2O} + \overset{0}{O_2}$$

Check that you can work out the oxidation number of each oxygen (shown in red) using the rules above.

Two of the oxygen atoms in the hydrogen peroxide have increased their oxidation number and two have reduced it.

Under different conditions, chlorine undergoes two disproportionation reactions with hydroxide ions:

$$\overset{0}{Cl_2(aq)} \quad + \quad 2OH^-(aq) \quad \xrightarrow{\text{cold}} \quad \overset{-1}{Cl^-(aq)} \quad + \quad \overset{+1}{ClO^-(aq)} \quad + \quad H_2O(l)$$

a   chlorine   dilute   chloride ion   chlorate(I) ion

and

$$\mathbf{b} \quad \overset{0}{3Cl_2(aq)} \quad + \quad 6OH^-(aq) \quad \xrightarrow{\text{hot}} \quad \overset{-1}{5Cl^-(aq)} \quad + \quad \overset{+5}{ClO_3^-(aq)} \quad + \quad 3H_2O(l)$$

    chlorine   conc   chloride ion   chlorate(V) ion

Summary test 6.1

1 The following questions are about the reaction:

$$Ca(s) + Br_2(l) \rightarrow (Ca^{2+} + 2Br^-)(s)$$

   a State which element has gained electrons.
   b State which element has lost electrons.
   c State which element has been oxidised.
   d State which element has been reduced.
   e Give the half equations for these redox reactions.
   f State the oxidising agent.
   g State the reducing agent.

2 Deduce the oxidation numbers of each element in the following compounds:
   a $PbCl_2$
   b CCl_4
   c $NaNO_3$

3 For the reaction: $CuO + Mg \rightarrow Cu + MgO$, give the oxidation numbers of oxygen before and after the reaction.

4 For the reaction: $2Cu + O_2 \rightarrow 2CuO$, give the oxidation numbers of oxygen before and after the reaction.

5 For the reaction: $FeCl_2 + \frac{1}{2}Cl_2 \rightarrow FeCl_3$, give the oxidation numbers of iron before and after the reaction.

6 State the oxidation number of the following:
   a P in PO_4^{3-}
   b N in NO_3^-
   c N in NH_4^+

7 The following questions are about the equation:

$$Fe^{2+} + \frac{1}{2}Cl_2 \rightarrow Fe^{3+} + Cl^-$$

   a State the oxidation numbers for each element.
   b Which element has been oxidised? Explain your answer.
   c Which element has been reduced? Explain your answer.
   d Give the half equations for the reaction.

8 a Use oxidation numbers to balance the following equations:
   i $Cl_2 + NaOH \rightarrow NaClO_3 + NaCl + H_2O$
   ii $Sn + HNO_3 \rightarrow SnO_2 + NO_2 + H_2O$
   b Give the half equations for i and ii.

9 This is a disproportionation reaction:

$$Cu_2O \rightarrow Cu + CuO$$

Deduce the oxidation number of each atom using the rules in Section 6.1. Identify which element disproportionates.

> 📖 Launch additional digital resources for the chapter

1 a Define in term 'enthalpy change of formation', ΔH_f°. *(2 marks)*

b Explain why the standard enthalpy of formation at 298 K of molecular oxygen (O_2) is zero. *(1 mark)*

2 Glucose is synthesised biologically by plants through photosynthesis according to the following reaction:

$$6CO_2 \ + \ 6H_2O \ \rightarrow \ C_6H_{12}O_6 \ + \ 6O_2$$

Construct a Hess' cycle using the data in the table below and calculate the value for the enthalpy of the photosynthesis reaction, ΔH_r. Give your answer to 3 significant figures.

Formula of substance	ΔH_f° / kJ mol^{-1}
CO_2	−393.5
H_2O	−285.8
$C_6H_{12}O_6$	−1273.3
O_2	0.0

(3 marks)

3 Carbon monoxide is used in the blast furnace to reduce iron(III) oxide. It is produced by the incomplete combustion of coke ($C(s)$) according to the following equation:

$$C(s) \ + \ \tfrac{1}{2}O_2(g) \ \rightarrow \ CO(g)$$

a Construct a Hess' cycle using the data in the table below and calculate the enthalpy of reaction, ΔH_r. for the above reaction.

Formula of substance	ΔH_c° / kJ mol^{-1}
CO_2	−393.5
H_2O	−285.8
$C_6H_{12}O_6$	−1273.3
O_2	0.0

(3 marks)

b Explain why it would not be possible to obtain an accurate value for ΔH_r experimentally. *(1 mark)*

4 When excess powdered zinc is added to $50.0 \, cm^3$ of $0.2 \, mol \, dm^{-3}$ copper(II) sulfate solution, the temperature rises by 9°C.

a Construct a balanced chemical equation for this reaction. *(2 marks)*

b Calculate the molar enthalpy change for this reaction. *(3 marks)*

5 Chloroalkanes are synthesised be reaction of an alkane in the presence of chlorine and UV light. The following equation shows the formation of chloromethane:

$$CH_4(g) \ + \ Cl_2(g) \ \rightarrow \ CH_3Cl(g) \ + \ HCl(g)$$

a Use the bond energy data in the table below to calculate the enthalpy change of reaction (ΔH_r) for the formation of chloroethane. *(4 marks)*

Bond	Bond energy / kJ mol^{-1}
C–H (average)	410
Cl–Cl	242
C–Cl (average)	340
H–Cl	431

b Use the enthalpy of formation data to calculate the enthalpy change (ΔH_r) for the above reaction. *(4 marks)*

Formula of substance	Enthalpy change of formation / kJ mol^{-1}
CH_4	−74.8
HCl	−92.3
CH_3Cl	−82.0

c Suggest why the values for parts **a** and **b** are different. *(2 marks)*

6 The enthalpy change of vaporisation of water is $+44 \, kJ \, mol^{-1}$. An electric kettle rated 3 kW ($1 \, kW = 1 \, kJ \, s^{-1}$) boils water at 100°C. If the kettle is left switched on, how many seconds will it take to evaporate 18 g of water? *(2 marks)*

7 a State the oxidation numbers for each element in the following species:
  i N_2 **ii** H_2O **iii** H_2O_2 **iv** $MgCl_2$ **v** H_2SO_4
  vi NO_3^- **vii** PO_4^{3-} **viii** Ca^{2+} **ix** Br^- **x** Ne
  (10 marks)

  b Fe^{2+} reacts with acidified $KMnO_4$ to produce Mn^{2+} and Fe^{3+} ions.
  i Construct a balanced redox equation for this reaction. *(1 mark)*
  ii Identify the oxidising agent in this reaction. *(1 mark)*
  iii Explain how the equation supports the need for an acid to be present for this reaction to occur. *(1 mark)*

8 Chlorine reacts with water according to the following equation:

$$Cl_2(g) + H_2O(l) \rightarrow HClO(aq) + HCl(aq)$$

  a State the oxidation numbers for chlorine in the following species:
  i Cl_2 **ii** $HClO$ **iii** HCl *(3 marks)*

  b Name the type of reaction that produces both products in this reaction. *(1 mark)*

9 When 0.1 mol of solid calcium oxide reacts with excess water to form calcium hydroxide solution, 107 kJ of heat is produced.

  a Write an equation for the reaction including state symbols. *(2 marks)*

  b Calculate the amount of heat energy produced in $kJ\ mol^{-1}$ for the reaction of calcium oxide with excess water. *(1 mark)*

  c Construct a reaction pathway diagram for the reaction between calcium oxide and excess water. Label the following:
  i *x*- and *y*-axis
  ii enthalpy change
  iii activation energy   *(3 marks)*

10 A student carried out an investigation to calculate the standard enthalpy of neutralisation for the reaction between nitric acid and sodium hydroxide according to the following method:

  1) 50 cm^3 of 0.5 mol dm^{-3} nitric acid was placed into a glass beaker and the temperature recorded at one-minute intervals for three minutes.

  2) At the fourth minute, 50 cm^3 of sodium hydroxide was added with stirring.

  3) The temperature was then recorded for a further seven minutes.

  a Plot a graph of temperature against time using the following experimental data:

Time / min	0	1	2	3	4	5	6	7	8	9	10	11
Temperature / °C	19.3	19.3	19.3	19.3	-	20.9	21.4	21.3	21.2	21.1	21.0	20.9

  (5 marks)

  b Deduce the temperature change, ΔT, at the fourth minute. Annotate the graph from part **a** to show all of your working out. *(3 marks)*

  c Using your answer to part **b** to calculate the standard molar enthalpy of neutralisation for this reaction. (Assume $c = 4.2\ J\ K^{-1}\ g^{-1}$) *(4 marks)*

  d Suggest how your answer to part **c** differs from the literature value, give a reason for your answer and give a change that could be made to improve the accuracy of your experimental result. *(2 marks)*

11 Write balanced equations and then construct individual redox half-equations for each of the following reactions:
  a Sodium with oxygen
  b Potassium with water
  c Magnesium with dilute sulfuric acid
  d Electrolysis of molten sodium chloride   *(12 marks)*

12 Deduce which of the following reactions are redox reactions:
  a $Cl_2 + 2OH^- \rightarrow Cl^- + ClO^- + H_2O$
  b $Cu^{2+} + 2OH^- \rightarrow Cu(OH)_2$
  c $H_2O + SO_3 \rightarrow H_2SO_4$
  d $2CrO_4^{2-} + 2H^+ \rightarrow Cr_2O_7^{2-} + H_2O$ *(4 marks)*

7.1 Chemical equilibria: reversible reactions, dynamic equilibrium

Chemists usually think of a reaction as starting with the reactants and ending with the products.

$$\text{reactants} \rightarrow \text{products}$$

However, some reactions are reversible. For example, as we saw in Section 5.1, when you heat blue hydrated copper sulfate it becomes white anhydrous copper sulfate as the water of crystallisation is driven off. The white copper sulfate returns to blue if you add water.

$$\begin{array}{ccc} CuSO_4.5H_2O & \rightleftharpoons & CuSO_4 + 5H_2O \\ \text{blue hydrated} & & \text{white anhydrous} \\ \text{copper sulfate} & & \text{copper sulfate} \end{array}$$

However, something different would happen if it were possible to do this reaction in a closed container. As soon as the products are formed they react together and form the reactants again, so that instead of reactants or products you get a mixture of both. Eventually you get a mixture in which the proportions of all three components remain constant. This mixture is called an **equilibrium mixture**.

Setting up an equilibrium

You can understand how an equilibrium mixture is set up by thinking about what happens in a physical process, like the evaporation of water. This is easier to picture than a chemical change.

First imagine a puddle of water out in the open. Some of the water molecules at the surface will move fast enough to escape from the liquid and evaporate. Evaporation will continue until all the water is gone.

But think about putting some water into a closed container. At first the water will begin to evaporate as before. The volume of the liquid will get smaller and the number of vapour molecules in the gas phase will go up. But as more molecules enter the vapour, some gas-phase molecules will start to re-enter the liquid (see Figure 1).

After a time, the rate of evaporation and the rate of condensation will become equal. The level of the liquid water will then stay exactly the same and so will the number of molecules in the vapour and in the liquid. The evaporation and condensation are still going on but at the same rate. This situation is called a **dynamic equilibrium** and is one of the key ideas of this topic.

In fact, you could have started by filling the empty container with the same mass of water vapour as you originally had liquid water. The vapour would begin to condense and, in time, would reach exactly the same equilibrium position.

The conditions for equilibrium

Although the system used here is very simple, you can pick out four conditions that apply to all equilibria:

a

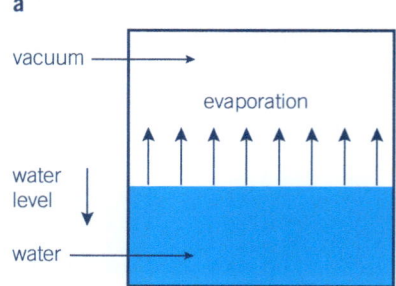

b

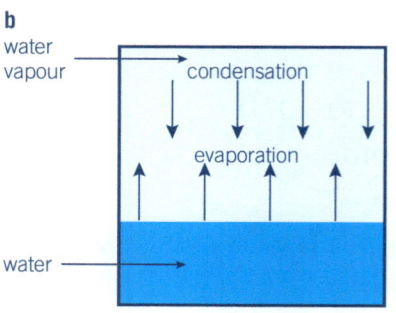

Figure 1 **a** Water will evaporate into an empty container. Eventually the rates of evaporation and condensation will be the same **b** Equilibrium is set up

- Equilibrium can only be reached in a **closed system** (one where the reactants and products can't escape). The system does not have to be sealed. For example, a beaker may be a closed system for a reaction that takes place in a solvent, as long as the reactants, products, and solvent do not evaporate.
- Equilibrium can be approached from either direction (in Figure 1, from liquid or from vapour) and the final equilibrium position will be the same (as long as conditions, such as temperature and pressure, stay the same).
- Equilibrium is a dynamic process. It is reached when the rates of two opposing processes, which are going on all the time (in Figure 1, evaporation and condensation), are the same.
- You know that equilibrium has been reached when the macroscopic properties of the system do not change with time. These are properties like density, concentration, colour, and pressure—properties that do not depend on the total quantity of matter.

A **reversible reaction** that can reach equilibrium is denoted by the symbol $\rightleftharpoons$, for example:

$$\text{liquid water} \rightleftharpoons \text{water vapour}$$

$$H_2O(l) \rightleftharpoons H_2O(g)$$

Chemical equilibria

The same principles that you have found for a physical change also apply to chemical equilibria such as:

$$\underset{\text{reactants}}{A + B} \quad \rightleftharpoons \quad \underset{\text{products}}{C + D}$$

- Imagine starting with A and B only. At the start of the reaction the forward rate is fast, because A and B are plentiful. There is no reverse reaction because there is no C and D.
- Then as the concentrations of C and D build up, the reverse reaction speeds up. At the same time the concentrations of A and B decrease so the forward reaction slows down.
- A point is reached where exactly the same number of particles are changing from A + B to C + D as are changing from C + D to A + B. Equilibrium has been reached.

One important point to remember is that an equilibrium mixture can have any proportions of reactants and products. It is not necessarily half reactants and half products, though it could be. The proportions may be changed depending on the conditions of the reaction, such as temperature, pressure, and concentration. But under constant conditions the proportions of reactants and products do not change.

Changing the conditions of an equilibrium reaction

Some industrial processes, such as the production of ammonia or sulfuric acid, have reversible reactions as a key step. In closed systems, these reactions would produce equilibrium mixtures containing both products and reactants. In principle, you would like to increase the proportion of products. For this reason, it is important to understand how to control equilibrium reactions.

The equilibrium mixture

It is possible to change the proportion of reactants to products in an equilibrium mixture. In this way you may be able to obtain a greater yield of the products. This is called changing the position of equilibrium.

> **Exam tip**
>
> Remember that at equilibrium, both forward and backward reactions occur at the same rate so the concentrations of all the reactants and products do not change.

Figure 2 Henri-Louis Le Châtelier was a French chemist who first put forward his 'Loi de stabilité déquilibre chimique' in 1884

- If the proportion of products in the equilibrium mixture increases, the equilibrium is said to have moved to the right, or in the forward direction.
- If the proportion of reactants in the equilibrium mixture increases, the equilibrium is said to have moved to the left, or in the backward direction.

You can often move the equilibrium position to the left or right by varying conditions like temperature, the concentration of species involved, or the pressure (in the case of reactions involving gases).

Le Chatelier's principle

Le Chatelier's principle is useful because it gives us a rule. It tells us whether the equilibrium moves to the right or to the left when the conditions of an equilibrium mixture are changed.

It states:

> **If a change is made to a system at dynamic equilibrium, the position of equilibrium moves to minimise this change.**

So if any factor is changed which affects the equilibrium mixture, the position of equilibrium will move to the right or left to oppose the change.

Le Chatelier's principle does not tell us how far the equilibrium moves so you cannot predict the quantities involved.

Changing concentrations

If you increase the concentration of one of the reactants, Le Chatelier's principle says that the equilibrium will shift in the direction that tends to reduce the concentration of this reactant.

Look at the reaction:

$$A + B \rightleftharpoons C + D$$

Suppose you add some extra A. This would increase the concentration of A. The only way that this system can reduce the concentration of A is by some of A reacting with B (to form more C and D). Adding more A uses up more B, and produces more C and D—this moves the equilibrium to the right. You end up with a greater proportion of products in the reaction mixture than before you added A. The same thing would happen if you added more B.

Or suppose you remove C as it is formed. The equilibrium would move to the right to produce more C (and D), using up A and B. The same thing would happen if you remove D as soon as it is formed.

Changing the overall pressure

Pressure changes only affect reactions involving gases. Changing the overall pressure will only change the position of equilibrium if there are a different number of molecules on either side of the equation.

An example of such a reaction is:

$N_2O_4(g)$	$\rightleftharpoons$	$2NO_2(g)$
dinitrogen tetraoxide		nitrogen dioxide
1 molecule		2 molecules
colourless		brown

Increasing the pressure of a gas means that there are more molecules of it in a given volume—it is equivalent to increasing the concentration of a solution.

If you increase the pressure on this system, Le Chatelier's principle tells us that the position of equilibrium will move to decrease the pressure. This means

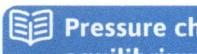
Pressure changes and equilibrium

Dinitrogen tetraoxide is a colourless gas and nitrogen dioxide is brown. You can investigate this in the laboratory, by setting up the equilibrium mixture in a syringe. If you decrease the pressure, by pulling out the syringe barrel, you can watch as the equilibrium moves to the right because the colour of the mixture gets browner, see Figure 3.

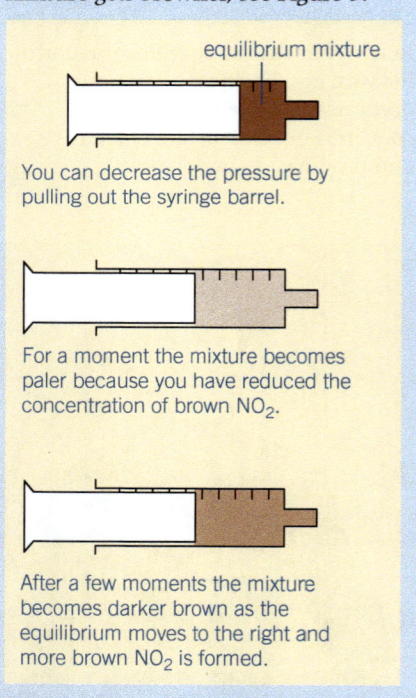

equilibrium mixture

You can decrease the pressure by pulling out the syringe barrel.

For a moment the mixture becomes paler because you have reduced the concentration of brown NO_2.

After a few moments the mixture becomes darker brown as the equilibrium moves to the right and more brown NO_2 is formed.

Figure 3 $N_2O_4(g) \rightleftharpoons 2NO_2(g)$ The equilibrium moves to the right as you decrease the pressure

Exam tip

The *rate* at which equilibrium is reached will be speeded up by increasing the pressure, as there will be more collisions in a given time.

that it will move to the left, because fewer molecules exert less pressure. If you decrease the pressure, the equilibrium will move to the right—molecules of N_2O_4 will decompose to form twice as many molecules of NO_2, thereby increasing the pressure.

Note that if there is the same number of molecules of gases on both sides of the equation, then pressure has no effect on the equilibrium position. For example:

$$H_2(g) \; + \; I_2(g) \;\; \rightleftharpoons \;\; 2HI(g)$$

2 moles   2 moles

The equilibrium position will not change in this reaction when the pressure is changed, so the proportions of the three gases will stay the same.

Changing temperature

Reversible reactions that are exothermic (give out heat) in one direction are endothermic (take in heat) in the other direction, see Section 5.1. The size of the enthalpy change is the same in both directions, but the sign is reversed.

Suppose you increase the temperature of an equilibrium mixture that is exothermic in the forward direction. An example is:

$$2SO_2(g) + O_2(g) \rightleftharpoons 2SO_3(g) \qquad \Delta H^\ominus = -197 \text{ kJ mol}^{-1}$$

The negative sign of ΔH means that heat is given out when sulfur dioxide and oxygen react to form sulfur trioxide in the forward direction. This means that heat is absorbed as the reaction goes in the reverse direction (to the left).

Le Chatelier's principle tells us that if you increase the temperature, the equilibrium moves in the direction that cools the system down. To do this it will move in the direction which absorbs heat (is endothermic)—to the left. The equilibrium mixture will then contain a greater proportion of sulfur dioxide and oxygen than before. In the same way, if we cool the mixture the equilibrium will move to the right and increase the proportion of sulfur trioxide.

Catalysts

Catalysts have no effect on the position of equilibrium so they do not alter the composition of the equilibrium mixture. They work by producing an alternative route for the reaction, which has a lower activation energy, see Section 5.1. This affects the forward and backward reactions equally.

Although catalysts have no effect on the position of equilibrium or the yield of the reaction, they do allow equilibrium to be reached more quickly. So they are important in industry.

The equilibrium constant K_c

In this section, we will look at equilibrium reactions mathematically. We will deal only with homogeneous systems—those where all the reactants and products are in the same phase (for example, all liquids).

Example: an esterification reaction

The reaction between ethanol, C_2H_5OH, and ethanoic acid, CH_3CO_2H, to produce ethyl ethanoate, $CH_3CO_2C_2H_5$, (an ester) and water is reversible and will eventually reach equilibrium.

If ethanol and ethanoic acid are mixed in a flask (stoppered to prevent evaporation) and left for several days with a strong acid catalyst, an equilibrium mixture is obtained in which all four substances are present. You can write:

 Temperature change and equilibrium

The effect of temperature on the dinitrogen tetraoxide/nitrogen dioxide equilibrium can be investigated using the same apparatus you used to investigate the effect of pressure. The reaction is endothermic as it proceeds from dinitrogen tetraoxide to nitrogen dioxide (the forward direction).

$$N_2O_4(g) \rightleftharpoons 2NO_2(g)$$
$$\Delta H^\ominus = +58 \text{ kJ mol}^{-1}$$

The gas mixture is contained in a syringe as before. The syringe is then immersed in warm water along with another syringe containing the same volume of air for comparison. The plunger of the syringe containing air will rise as the air expands. The plunger of the syringe containing the N_2O_4 / NO_2 mixture will also rise but by a greater amount. This indicates that more molecules of gas have been formed in this syringe. This is because the equilibrium has moved to the right; each molecule of N_2O_4 that disappears produces two molecules of NO_2. This is consistent with Le Chatelier's principle. When the mixture is warmed up, the equilibrium moves in the endothermic direction (to the right), and it absorbs heat which tends to cool the mixture down.

You should be able to predict the colour change that you would see during this experiment and also what would happen if the experiment were repeated in ice water.

$$C_2H_5OH(l) \;+\; CH_3CO_2H(l) \;\rightleftharpoons\; CH_3CO_2C_2H_5(l) \;+\; H_2O(l)$$

ethanol   ethanoic acid   ethyl ethanoate   water

The equilibrium mixture may be analysed by titrating the ethanoic acid with standard alkali (allowing for the amount of acid catalyst added). It is possible to do this without significantly disturbing the equilibrium mixture because the reversible reaction is much slower than the titration reaction.

The titration allows us to work out the number of moles of ethanoic acid in the equilibrium mixture. From this you can calculate the number of moles of the other components (and from this their concentrations if the total volume of the mixture is known).

If several experiments are done with different quantities of starting materials, it is always found that the ratio:

$$\frac{[CH_3CO_2C_2H_5(l)]_{eqm}\,[H_2O(l)]_{eqm}}{[CH_3CO_2H(l)]_{eqm}\,[C_2H_5OH(l)]_{eqm}}$$

has a constant value, provided the experiments are done at the same temperature. The subscript 'eqm' means that the concentrations have been measured when equilibrium has been reached.

Square brackets are used around a chemical formula to represent the concentration of that species in mol dm^{-3}.

For any reaction that reaches an equilibrium we can write the equation in the form:

$$aA + bB + cC \rightleftharpoons xX + yY + zZ$$

Then the expression

$$\frac{[X]^x{}_{eqm}[Y]^y{}_{eqm}[Z]^z{}_{eqm}}{[A]^a{}_{eqm}[B]^b{}_{eqm}[C]^c{}_{eqm}}$$

is constant provided the temperature is constant. We call this constant K_c. This expression can be applied to any reversible reaction. K_c is called the **equilibrium constant** and is different for different reactions. It changes with temperature. The units of K_c vary, and you must work them out for each reaction by cancelling out the units of each term, for example:

$$2A \rightleftharpoons 2C \qquad K_c = \frac{[C]^2}{[A]^2[B]}$$

Units are: $\dfrac{(\cancel{mol\,dm^{-3}})^2}{(\cancel{mol\,dm^{-3}})^2(mol\,dm^{-3})} = \dfrac{1}{mol\,dm^{-3}} = mol^{-1}\,dm^3$

The value of K_c is found by experiment for any particular reaction at a given temperature.

Figure 4 *Titrating the ethanoic acid to investigate the equilibrium position*

Worked example

Finding the value of K_c for the reaction between ethanol and ethanoic acid

0.10 mol of ethanol is mixed with 0.10 mol of ethanoic acid and allowed to reach equilibrium. The total volume of the system is made up to 20.0 cm³ (0.020 dm³) with water. By titration, it is found that 0.033 mol ethanoic acid is present once equilibrium is reached.

From this you can work out the number of moles of the other components present at equilibrium:

At start

C_2H_5OH (l)	+	CH_3CO_2H (l)	$\rightleftharpoons$	$CH_3CO_2C_2H_5(l)$	+	$H_2O(l)$
0.10 mol		0.10 mol		0 mol		0 mol

You know that there are 0.033 mol of CH_3CO_2H at equilibrium. This means that:

- there must also be 0.033 mol of C_2H_5OH at equilibrium. (The equation tells you that they react 1:1 and you know we started with the same number of moles of each.)

- $(0.10 - 0.033) = 0.067$ mol of CH_3CO_2H has been used up. The equation tells you that when 1 mol of CH_3CO_2H is used up, 1 mol each of $CH_3CO_2C_2H_5$ and H_2O are produced. So, there must be 0.067 mol of each of these.

At equilibrium

$$C_2H_5OH\,(l) \quad + \quad CH_3CO_2H\,(l) \quad \rightleftharpoons \quad CH_3CO_2C_2H_5\,(l) \quad + \quad H_2O\,(l)$$
$$\text{0.033 mol} \qquad\qquad \text{0.033 mol} \qquad\qquad \text{0.067 mol} \qquad\qquad \text{0.067 mol}$$

You need the concentrations of the components at equilibrium. As the volume of the system is 0.020 dm³ these are:

$$C_2H_5OH\,(l) \quad + \quad CH_3CO_2H\,(l) \quad \rightleftharpoons \quad CH_3CO_2C_2H_5\,(l) \quad + \quad H_2O\,(l)$$
$$\begin{matrix} 0.033/0.020 \\ \text{mol dm}^{-3} \end{matrix} \qquad \begin{matrix} 0.033/0.020 \\ \text{mol dm}^{-3} \end{matrix} \qquad \begin{matrix} 0.067/0.020 \\ \text{mol dm}^{-3} \end{matrix} \qquad \begin{matrix} 0.067/0.020 \\ \text{mol dm}^{-3} \end{matrix}$$

Enter the concentrations into the equilibrium equation:

$$K_c = \frac{[CH_3CO_2C_2H_5\,(l)]\,[H_2O\,(l)]}{[CH_3CO_2H\,(l)]\,[C_2H_5OH\,(l)]}$$

$$K_c = \frac{[0.067/0.020 \text{ mol dm}^{-3}]\,[0.067/0.020 \text{ mol dm}^{-3}]}{[0.033/0.020 \text{ mol dm}^{-3}]\,[0.033/0.020 \text{ mol dm}^{-3}]} = 4.1$$

The units all cancel out. The volumes (0.020 dm³) also cancel out (in this case you didn't need to know the volume of the system). So $K_c = 4.1$ and K_c has no units.

> **Exam tip**
>
> The concentration of a solution is the number of moles of solute dissolved in 1 dm³ of solution. A square bracket around a formula is shorthand for 'concentration of that substance in mol dm⁻³'.

A reaction that has reached equilibrium at a given temperature will be a mixture of reactants and products. You can use the equilibrium expression to calculate the composition of this mixture.

Worked example

Calculating the composition of a reaction mixture

The reaction of ethanol and ethanoic acid is:

$$C_2H_5OH\,(l) \quad + \quad CH_3CO_2H\,(l) \quad \rightleftharpoons \quad CH_3CO_2C_2H_5\,(l) \quad + \quad H_2O\,(l)$$
$$\text{ethanol} \qquad\quad \text{ethanoic acid} \qquad \text{ethyl ethanoate} \qquad\quad \text{water}$$

You know that at equilibrium:

$$K_c = \frac{[CH_3CO_2C_2H_5\,(l)]\,[H_2O\,(l)]}{[CH_3CO_2H\,(l)]\,[C_2H_5OH\,(l)]}$$

Suppose that $K_c = 4.0$ at the temperature of our experiment. You want to know how much ethyl ethanoate you could produce by mixing one mol of ethanol and one mol of ethanoic acid. Set out the information as shown below:

Equation:	$C_2H_5OH\,(l)$ +	$CH_3CO_2H\,(l)$ $\rightleftharpoons$	$CH_3CO_2C_2H_5\,(l)$ +	$H_2O\,(l)$
	ethanol	ethanoic acid	ethyl ethanoate	water
At start:	1 mol	1 mol	1 mol	1 mol
At equilibrium:	$(1 - x)$ mol	$(1 - x)$ mol	x mol	x mol

You do not know how many moles of ethyl ethanoate will be produced, so you call this x. The equation tells us that x mol of water will also be produced. In doing so, x mol of both ethanol and ethanoic acid will be used up. So the amount of each of these remaining at equilibrium is $(1 - x)$ mol.

These figures are in moles, but you need concentrations in mol dm^{-3} to substitute in the equilibrium law expression. Suppose the volume of the system at equilibrium was V dm^{-3}. Then:

$$[C_2H_5OH(l)]_{eqm} = \frac{(1-x)}{V} \text{ mol dm}^{-3}$$

$$[CH_3CO_2H(l)]_{eqm} = \frac{(1-x)}{V} \text{ mol dm}^{-3}$$

$$[CH_3CO_2C_2H_5(l)]_{eqm} = \frac{x}{V} \text{ mol dm}^{-3}$$

$$[H_2O(l)]_{eqm} = \frac{x}{V} \text{ mol dm}^{-3}$$

These figures may now be put into the expression for K_c:

$$K_c = \frac{x/\cancel{V} \times x/\cancel{V}}{(1-x)/\cancel{V} \times (1-x)/\cancel{V}}$$

The V's cancel, so in this case you do not need to know the actual volume of the system.

> **Exam tip**
>
> If there are the same total number of entities in the products and reactants, the volume of the system will cancel out and you do not need to know it

$$4.0 = \frac{x \times x}{(1-x) \times (1-x)}$$

$$4.0 = \frac{x^2}{(1-x)^2}$$

Taking the square root of both sides, you get:

$$2 = \frac{x}{(1-x)}$$

$$2(1-x) = x$$

$$2 - 2x = x$$

$$2 = 3x$$

$$x = \frac{2}{3}$$

So $\frac{2}{3}$ mol of ethyl ethanoate and $\frac{2}{3}$ mol of water is produced if the reaction reaches equilibrium. The composition of the equilibrium mixture would be: ethanol $\frac{1}{3}$ mol, ethanoic acid $\frac{1}{3}$ mol, ethyl ethanoate $\frac{2}{3}$ mol, water $\frac{2}{3}$ mol.

Equilibrium constant K_p for gaseous equilibria

Many reversible reactions take place in the gas phase. These include many important industrial reactions such as the synthesis of ammonia and a key stage of the Contact process for making sulfuric acid. Gaseous equilibria also obey the equilibrium law. However we usually express their concentrations in a different way, using the idea of **partial pressure**.

Partial pressure

In a mixture of gases, each gas contributes to the total pressure. This contribution is called its partial pressure p and is the pressure that the gas would exert if it occupied the container on its own. The sum of the partial pressures of all the gases in a mixture is the total pressure. For example, air is a mixture of approximately 20% oxygen molecules and 80% nitrogen molecules and has a pressure (at sea level) of approximately 100 kPa (kilopascals).

> **Exam tip**
>
> The mole fraction of a gas, A, in a mixture is given by:
>
> $$\frac{\text{number of moles of gas A}}{\text{total number of moles of gas in the mixture}}$$

So the approximate partial pressure of oxygen in the air is 20 kPa and that of nitrogen is 80 kPa.

Mathematically, the partial pressure p of a gas in a mixture is given by its **mole fraction** multiplied by the total pressure.

partial pressure p of A = mole fraction of A × total pressure

Applying the equilibrium law to gaseous equilibria

An equilibrium constant can be found in the same way as for a reaction in solution. It is given the symbol K_p rather than K_c.

For a reaction $aA(g) + bB(g) \rightleftharpoons yY(g) + zZ(g)$

$$K_p = \frac{p^y Y(g)_{eqm}\, p^z Z(g)_{eqm}}{p^a A(g)_{eqm}\, p^b B(g)_{eqm}}$$

Note how this corresponds to the equilibrium law expressed in terms of concentration that you have seen earlier.

Worked example

Calculating partial pressure

This example shows how you can use the expression for K_p to calculate the composition of an equilibrium mixture

K_p is 0.020 for the reaction

$2HI(g) \rightleftharpoons H_2(g) + I_2(g)$

If the reaction started with pure HI, and the initial pressure of HI was 100 kPa, what would be the partial pressure of hydrogen when equilibrium is reached?

Set out the problem in the same way as when using K_c.

	$2HI(g)$	$\rightleftharpoons$	$H_2(g)$	$+$	$I_2(g)$
Start:	100 kPa		0 kPa		0 kPa
At eqm:	$(100 - 2x)$ kPa		x kPa		x kPa

The chemical equation tells us:

- that there will be the same number of moles of H_2 and I_2 at equilibrium, therefore $pH_{2eqm} = pI_{2eqm} = x$

- that for each mole of hydrogen (and of iodine) that is produced, two moles of hydrogen iodide are used up, so that if $pH_{2eqm} = x$, $pHI_{eqm} = (100 - 2x)$

$$K_p = \frac{pH_2(g)_{eqm}\, pI_2(g)_{eqm}}{(pHI(g)_{eqm})^2} \quad \text{(no units)}$$

Putting in the figures gives: $0.02 = \dfrac{x^2}{(100 - 2x)^2}$

Taking the square root of each side gives: $0.141 = \dfrac{x}{(100 - 2x)}$

$$0.141 \times (100 - 2x) = x$$

$$14.1 - 0.282x = x$$

$$14.1 = 1.282x$$

$$x = \frac{14.1}{1.282} = 10.99$$

$$pH_2(g) = 11 \text{ kPa (to 2 sf)}$$

The chemical equation tells us that $pI_2(g)$ must be the same as $pH_2(g)$ and that $pHI(g)$ must be $100 - (2 \times 11) = 78$ kPa.

Exam tip

Cancel the units of concentration (or pressure) in the equilibrium law expression to find the units of K_c (or K_p).

Figure 5 *Ammonia is made in this plant via a reversible reaction*

The effect of changing temperature and pressure on a gaseous equilibrium

Le Chatelier's principle applies to gaseous equilibria in the same way as to equilibria in solution. The only difference is that the partial pressure of reactants and products replace concentration.

So, for a reaction that is exothermic going left to right, increasing the temperature forces the equilibrium to the left, so that the reaction absorbs heat. In other words, increasing the temperature decreases K_p. So for the Haber process reaction:

$$3H_2(g) + N_2(g) \rightleftharpoons 2NH_3(g) \qquad \Delta H = -92 \text{ kJ mol}^{-1}$$

increasing the temperature decreases the yield of ammonia at equilibrium.

Increasing the pressure forces the equilibrium to move so as to reduce the total pressure, that is to the side with fewer molecules. So increasing the total pressure increases the yield of ammonia.

Changing the total pressure only affects the equilibrium position when there is a change in the total number of molecules on either side of the reaction. So for the equilibrium:

$$2HI(g) \rightleftharpoons H_2(g) + I_2(g)$$

pressure will have no effect on the equilibrium position.

Increasing the pressure on a gas phase reaction will increase the rate at which equilibrium is reached, as there will be more collisions between molecules. Increasing the temperature will also increase the rate at which equilibrium is reached, as will the use of a catalyst.

The effect of changing conditions on the equilibrium constant

Changing the *temperature* changes the value of the equilibrium constant, K_c. Whether K_c increases or decreases depends on whether the reaction is exothermic or endothermic. What happens is summarised in Table 1.

Table 1 *The effect of changing temperature on equilibria*

Type of reaction	Temperature change	Effect on K_c or K_p	Effect on products	Effect on reactants	Direction of change of equilibrium
endothermic	decrease	decrease	decrease	increase	moves left
endothermic	increase	increase	increase	decrease	moves right
exothermic	increase	decrease	decrease	increase	moves left
exothermic	decrease	increase	increase	decrease	moves right

Changing the concentration of the reactants or products has no effect on the value of the equilibrium constant.

Changing the pressure of the reactants or products has no effect on the value of the equilibrium constant.

Adding a catalyst has no effect on the value of the equilibrium constant.

K_c and the position of equilibrium

The size of the equilibrium constant K_c can tell us about the composition of the equilibrium mixture. The equilibrium expression is always of the general form:

$$\frac{[\text{products}]}{[\text{reactants}]}$$

So:

- If K_c is much greater than 1, products predominate over reactants and the equilibrium position is over to the right.
- If K_c is much less than 1, reactants predominate and the equilibrium position is over to the left.

Reactions where the equilibrium constant is greater than 10^{10} are usually regarded as going to completion. Reactions with an equilibrium constant of less than 10^{-10} are regarded as not taking place at all.

Equilibrium reactions in industry

A number of industrial processes involve reversible reactions. In these cases, the yield of the reaction is important and Le Chatelier's principle can be used to help find the best conditions for increasing it. However, yield is not the only consideration. Sometimes a low temperature would give the best yield but this would slow the reaction down. The costs of building and running a plant that operates at high temperatures and pressures must also be taken into account. In most cases, a compromise set of conditions is used. We will look at the industrial production of two important chemicals.

Sulfuric acid, H_2SO_4—the Contact process

Sulfuric acid is a vital industrial chemical; around 270 million tonnes is produced annually worldwide by the Contact process, and this figure is expected to increase. It is used in the manufacture of a large range of goods. It is made from sulfur, oxygen and water, and the key step is

$$2SO_2(g) + O_2(g) \rightleftharpoons 2SO_3(g) \qquad \Delta H = -196\,\text{kJ}\,\text{mol}^{-1}$$

catalysed by vanadium(V) oxide, V_2O_5.

Le Chatelier's principle predicts that the highest conversion to sulfur trioxide would be obtained at high pressure and low temperature. In fact, a temperature of 450°C (723 K) is used—at a lower temperature than this the reaction would be too slow. Only a slight increase over atmospheric pressure is used, because this gives sufficient conversion, and a high pressure plant is expensive to construct and run.

Ammonia, NH_3, and the Haber Process

Ammonia is another important chemical in industry. World production is over 140 million tonnes each year. Around 80% is used to make fertilisers like ammonium nitrate, ammonium sulfate, and urea. The rest is used to make synthetic fibres (including nylon), dyes, explosives, and plastics like polyurethane.

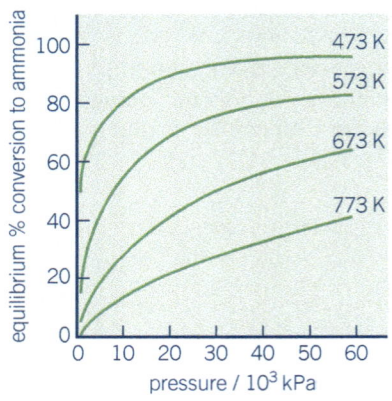

Figure 6 *Equilibrium % conversion of nitrogen and hydrogen to ammonia under different conditions*

Making ammonia

Nitrogen and hydrogen react together by a reversible reaction which, at equilibrium, forms a mixture of nitrogen, hydrogen, and ammonia:

$$N_2(g) + 3H_2(g) \rightleftharpoons 2NH_3(g) \qquad \Delta H = -92 \text{ kJ mol}^{-1}$$

The percentage of ammonia obtained at equilibrium depends on temperature and pressure, as shown in Figure 6. The graph shows that low temperature and high pressure would give close to 100% conversion. Low pressure and high temperature would give almost no ammonia.

The Haber process

The raw materials for the Haber process are air (which provides the nitrogen), water, and natural gas (methane, CH_4). These provide the hydrogen by the following reaction:

$$CH_4(g) + H_2O(g) \rightarrow CO(g) + 3H_2(g)$$

The nitrogen and hydrogen are fed into a converter in the ratio of 1 : 3 and passed over an iron catalyst.

Most plants run at a pressure of around 20 000 kPa (around 200 atmospheres) and a temperature of about 670 K. This is a lower pressure and a higher temperature than would give the maximum conversion.

Nitrogen and hydrogen flow continuously over the catalyst, so the gases do not spend long enough in contact with the catalyst to reach equilibrium. There is about 15% conversion to ammonia. The ammonia is cooled so that it becomes liquid and is piped off. Any nitrogen and hydrogen that is not converted into ammonia is fed back into the reactor.

The catalyst is iron in pea-sized lumps (to increase the surface area). It lasts about five years before it becomes poisoned by impurities in the gas stream and has to be replaced.

📖 The Haber process

Almost all ammonia is made by the Haber process, in which the reaction above is the key step. The process was developed by the German chemist Fritz Haber and the chemical engineer Carl Bosch in the early years of the 20th century. It allowed Germany to make explosives and fertilisers. This prolonged the First World War because, at that time, the source of nitrogen for these products was nitrates from South America. These could be blockaded by the navies of Britain and its allies.

Summary test 7.1

1 For each of the following statements about all equilibria, state whether it is true or false.
  a An equilibrium mixture always contains half reactants and half products.
  b At equilibrium the forward and the backward reactions come to a halt.
  c Equilibrium is only reached in a closed system.
  d Once equilibrium is reached the concentrations of the reactants and the products do not change.
2 State what can be said about the rates of the forward and the backward reactions when equilibrium is reached.
3 In which of the following reactions will the position of equilibrium be affected by changing the pressure? Explain your answers.
  a $2SO_2(g) + O_2(g) \rightleftharpoons 2SO_3(g)$
  b $CH_3CO_2H(aq) \rightleftharpoons CH_3CO_2^-(aq) + H^+(aq)$
  c $H_2(g) + CO_2(g) \rightleftharpoons H_2O(g) + CO(g)$
4 State the expression for the equilibrium constant for the following reactions in solution:
  a $A + B \rightleftharpoons C$
  b $2A + B \rightleftharpoons C$
  c $2A + 2B \rightleftharpoons 2C$
5 Deduce the units for K_c for question **4a** to **c**.

6 For the reaction between ethanol and ethanoic acid, at a different temperature to the worked example on page 100, the equilibrium mixture was found to contain 0.117 mol of ethanoic acid, 0.017 mol of ethanol, 0.083 mol ethyl ethanoate and 0.083 mol of water.
   a Calculate K_c.
   b Explain why you do not need to know the volume of the system to calculate K_c in this example.
   c State whether the equilibrium is further to the right or further to the left compared with the example on page 100.

7 Using Le Chatelier's principle, predict the effect of increasing: (i) the pressure and (ii) the temperature on the following reactions:
   a $2SO_2(g) + O_2(g) \rightleftharpoons 2SO_3(g)$ $\Delta H = -197$ kJ mol^{-1}
   b $N_2O_4(g) \rightleftharpoons 2NO_2(g)$ $\Delta H = +58$ kJ mol^{-1}
   c $H_2(g) + CO_2(g) \rightleftharpoons H_2O(g) + CO(g)$ $\Delta H = +40$ kJ mol^{-1}

8 $A(g) + B(g) \rightleftharpoons C(g) + D(g)$ represents an exothermic reaction and
   $$K_p = \frac{pC(g)\ pD(g)}{pA(g)\ pB(g)}.$$
   In the above expression, state what would happen to K_p:
   a if the temperature were decreased
   b if more A were added to the mixture
   c if a catalyst were added.

9 The reaction of ethanol with ethanoic acid produces ethyl ethanoate and water:
   $$C_2H_5OH(l) + CH_3COOH(l) \rightleftharpoons CH_3COOC_2H_5(l) + H_2O(l)$$
   A student suggested that the yield of ethyl ethanoate, $CH_3COOC_2H_5$, could be increased by removing the water as it was formed.
   Explain, using the idea of K_c, why this suggestion is sensible.

10 These questions are about reversible reactions. Give the correct term from **increases/decreases/does not change** to fill in the blank for each statement.
   a In an endothermic reaction, K_c _____ when the temperature is increased.
   b In an endothermic reaction, K_c _____ when the concentration of the reactants is decreased.
   c In an exothermic reaction, K_c _____ when the temperature is decreased.
   d In an exothermic reaction, K_c _____ when the concentration of the reactants is increased.
   e If a suitable catalyst is added to the reaction, K_c _____.

11 Use the graph in Figure 6 to find the equilibrium percentage conversion to ammonia at 20 000 kPa and 673 K.

12 Suggest why compromise conditions are used in manufacturing processes.

13 Explain how Le Chatelier's principle predicts that the highest conversion to ammonia in the equilibrium $N_2(g) + 3H_2(g) \rightleftharpoons 2NH_3(g)$ is obtained at
   a low temperature
   b high pressure.

7.2 Brønsted–Lowry theory of acids and bases

Learning outcomes

On these pages you will learn to:

- identify the common acids and alkalis
- use the Brønsted–Lowry theory of acids and bases
- explain the differences between strong and weak acids and bases
- explain pH and sketch titration curves
- describe neutralisation reactions
- explain the choice of indicator for an acid–base titration

Exam tip

Make sure you can write balanced symbol equations for the reactions of any of the acids in Table 2 with any of the bases.

You will need to know the names and formulae of some common acids and bases, shown in Table 2.

Table 2 *Some common acids and bases*

Acid	Formula	Base	Formula
hydrochloric	HCl	sodium hydroxide	NaOH
nitric	HNO_3	potassium hydroxide	KOH
sulfuric	H_2SO_4	ammonia	NH_3
ethanoic	CH_3COOH		

The reaction between an acid and a base to form a salt plus water is a neutralisation reaction.

For example:

hydrochloric acid + potassium hydroxide → potassium chloride + water

$$HCl(aq) + KOH(aq) \rightarrow KCl(aq) + H_2O(l)$$

We can write this equation ionically:

$$H^+(aq) + Cl^-(aq) + K^+(aq) + OH^-(aq) \rightarrow K^+(aq) + Cl^-(aq) + H_2O(l)$$

Notice that Cl^- and K^+ take no part in the reaction—they are spectator ions.

So cancelling them out, the reaction becomes:

$$H^+(aq) + OH^-(aq) \rightarrow H_2O(l)$$

This is the basis of all neutralisation reactions.

The Brønsted–Lowry theory

The Brønsted–Lowry description of acidity (developed in 1923 by Thomas Lowry and Johannes Brønsted independently) is the most useful current theory of acids and bases. It expands the idea of what is an acid or a base beyond the idea of neutralisation.

An acid is a substance that can donate a proton (H^+ ion) and a base is a substance that can accept a proton.

Proton transfer
Hydrogen chloride gas and ammonia gas react together to form ammonium chloride—a white ionic solid:

$$HCl(g) \quad + \quad NH_3(g) \quad \rightarrow \quad NH_4Cl(s)$$
hydrogen chloride   ammonia   ammonium chloride

Here, hydrogen chloride is acting as an acid by donating a proton to ammonia. Ammonia is acting as a base by accepting a proton. Acids and bases can only react in pairs—one acid and one base.

So, you could think of the reaction in these terms:

$$HCl(g) + NH_3(g) \rightarrow [NH_4^+Cl^-](s)$$
acid   base

Mixing bathroom cleaners

Bathroom cleaners come in essentially two types—bleach-based for removing stains and acid-based for removing limescale.

1 Limescale is made up of calcium carbonate, $CaCO_3$. Write the equation for the reaction of hydrochloric acid with calcium carbonate.

Most bathroom cleaners have a warning on the label not to mix them with other types of cleaner. The active ingredient in household bleach is chloric(I) acid (HClO), while acid-based cleaners contain hydrochloric acid (HCl). These react together to form chlorine gas.

$$HClO(aq) + HCl(aq) \rightleftharpoons Cl_2(g) + H_2O(aq)$$

Imagine you have put a large amount of bleach in the toilet bowl (so that chloric(I) acid is in excess) and then you add a squirt (say $50\,cm^3$) of acid-based cleaner of concentration $1\,mol\,dm^{-3}$. Assume the equilibrium is forced completely to the right.

2 How many moles of HCl have you added?

3 How many moles of chlorine would be produced?

4 What volume of chlorine is this?

5 Why would there be less chlorine gas in the bathroom than you have calculated in question 3?

This is a significant amount of chlorine and, considering that it was used as a poisonous gas in the First World War, something to be avoided.

Figure 7 *Bathroom cleaning products*

The pH scale

The acidity or alkalinity of a solution depends on the concentration of $H^+(aq)$ and is measured on the pH scale. Acids have pH values of less than 7 and alkalis values of greater than 7. A pH of 7 is neutral.

$$pH = -\log_{10}[H^+(aq)]$$

Pure water has a pH of 7, so $7 = -\log_{10}[H^+(aq)]$

Taking antilogs, $[H^+(aq)] = 10^{-7}\,mol\,dm^{-3}$

Since water dissociates:

$$H_2O(l) \rightleftharpoons H^+(aq) + OH^-(aq)$$

$$[H^+(aq)] = [OH^-(aq)] = 10^{-7}\,mol\,dm^{-3} \text{ in pure water}$$

How the pH scale was invented

Did you know that the pH scale was first introduced by a brewer? In 1909, the Danish biochemist Søren Sørenson was working for the Carlsberg company studying the brewing of beer. Brewing requires careful control of acidity to produce conditions in which yeast (which aids the fermentation process) will grow but unwanted bacteria will not. The concentrations of acid with which Sørenson was working were very small, such as one ten-thousandth of a mole per litre, and so he looked for a way to avoid using numbers such as 0.0001 (1×10^{-4}). Taking the $\log_{10}$ of this number gave -4, and for further convenience he took the negative off it, giving 4. So the pH scale was born.

Square brackets, [ ], mean the concentration in $mol\,dm^{-3}$.

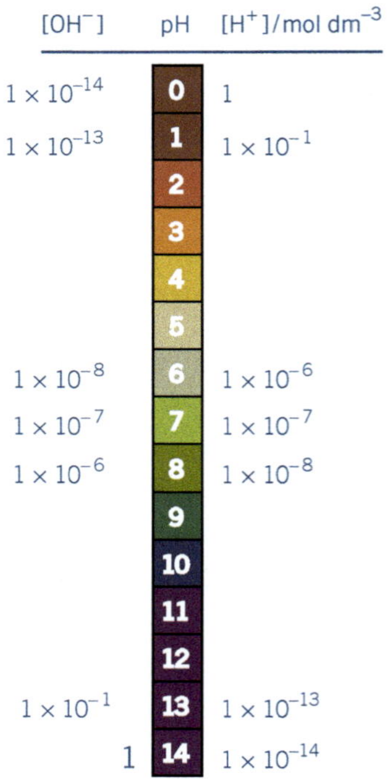

[OH⁻]	pH	[H⁺]/mol dm⁻³
1×10^{-14}	0	1
1×10^{-13}	1	1×10^{-1}
	2	
	3	
	4	
	5	
1×10^{-8}	6	1×10^{-6}
1×10^{-7}	7	1×10^{-7}
1×10^{-6}	8	1×10^{-8}
	9	
	10	
	11	
	12	
1×10^{-1}	13	1×10^{-13}
1	14	1×10^{-14}

Figure 8 *The pH scale*

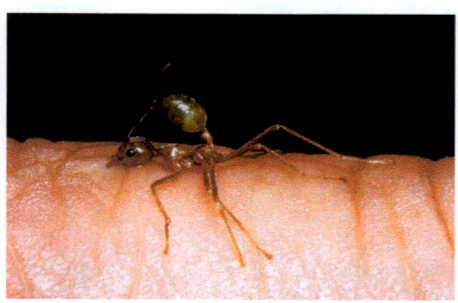

Figure 10 *Formic acid (methanoic acid) is quite concentrated when used as a weapon by the stinging ant and, although it is a weak acid, being sprayed with it can be a painful experience*

This expression is more complicated than simply stating the concentration of H⁺(aq). However, using the logarithm of the concentration does away with awkward numbers like 10⁻¹³, etc, which occur because the concentration of H⁺(aq) in most aqueous solutions is so small (see Figure 8). The minus sign makes almost all pH values positive (because the logs of numbers less than 1 are negative).

📖 Measuring pH

pH can be measured using an indicator paper or a solution, such as universal indicator. This is made from a mixture of dyes that change colour at different [H⁺(aq)]. This is fine for measurements to the nearest whole number, but for more precision a pH meter is used. A pH meter has an electrode which dips into a solution and produces a voltage related to [H⁺(aq)]. The pH readings can then be read directly on the meter or fed into a computer or data logger. This process can be used for continuous monitoring of chemical processes or medical procedures, for example.

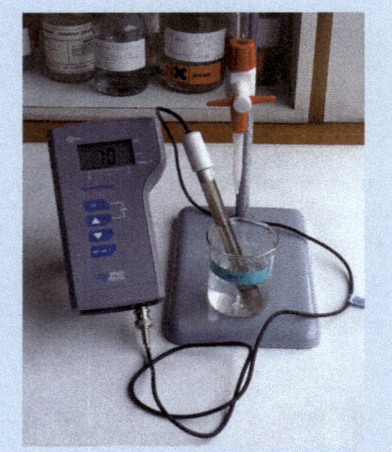

Figure 9 *Using a pH meter*

Strong and weak acids and bases

Acids that completely dissociate into ions in aqueous solutions are called **strong acids**. Hydrochloric acid is an example of a strong acid. In aqueous solution, it is completely dissociated.

$$HCl(g) + aq \rightarrow H^+(aq) + Cl^-(aq)$$

The word **strong** refers *only* to the extent of dissociation and not in any way to the concentration. So it is perfectly possible to have a very dilute solution of a strong acid.

The same arguments apply to bases. Strong bases are completely dissociated into ions in aqueous solutions. For example, sodium hydroxide is a strong base:

$$NaOH(aq) \rightarrow Na^+(aq) + OH^-(aq)$$

Weak acids

Many acids and bases are *only slightly* ionised (not fully dissociated) when dissolved in water. Ethanoic acid (the acid in vinegar, also known as acetic acid) is a typical example. In a 1 mol dm⁻³ solution of ethanoic acid, only about four in every thousand ethanoic acid molecules are dissociated into ions (so the degree of dissociation is $\frac{4}{1000}$. The rest remain dissolved as wholly covalently bonded molecules. In fact, an equilibrium is set up:

CH₃COOH(aq) ⇌	H⁺(aq) +	CH₃COO⁻(aq)
ethanoic acid	hydrogen ions	ethanoate ions
Before dissociation: 1000	0	0
At equilibrium: 996	4	4

Acids like this are called **weak acids**. Weak refers only to the degree of dissociation.

Compared with a strong acid of the same concentration, there will be far fewer ions in a solution of a weak acid. Its electrical conductivity will be significantly less.

For example, the pH of 1 mol dm⁻³ hydrochloric acid (a strong acid) is 0, while the pH of 1 mol dm⁻³ ethanoic acid (a weak acid) is 2.4.

Both acidic solutions will react with a reactive metal such as magnesium to produce a salt plus hydrogen. Ethanoic acid will react more slowly because of the reduced concentration of H^+ ions.

In a 5 mol dm⁻³ solution, ethanoic acid is still a weak acid; in a 10^{-4} mol dm⁻³ solution, hydrochloric acid is still a strong acid.

Weak bases

When ammonia dissolves in water, it forms an alkaline solution. The equilibrium lies well to the left and ammonia is weakly basic:

$$NH_3(aq) + H_2O(l) \rightleftharpoons NH_4^+(aq) + OH^-(aq)$$

Titrations

A titration is used to find the concentration of a solution by gradually adding to it a second solution with which it reacts. One of the solutions is of known concentration. To use a titration, you must know the equation for the reaction.

Titration procedure

The titration curves in Figure 13 on page 112 were determined using the apparatus shown in Figure 12a. Using a pipette, 25 cm³ acid was placed in the conical flask. The base was added 1 cm³ at a time from a burette and the pH recorded. At areas on the curve where the pH was changing rapidly, the experiment was repeated adding the base 0.1 cm³ at a time to find the shape of the curve more precisely.

pH changes during acid–base titrations

In an acid–base titration, an acid of known concentration is added from a burette to a measured amount of a solution of a base (an alkali) until an indicator shows that the base has been neutralised. Alternatively, the base is added to the acid until the acid is neutralised. You can then calculate the concentration of the alkali from the volume of acid used.

You can also follow a neutralisation reaction by measuring the pH with a pH meter (Figure 12b) in which case you do not need an indicator.

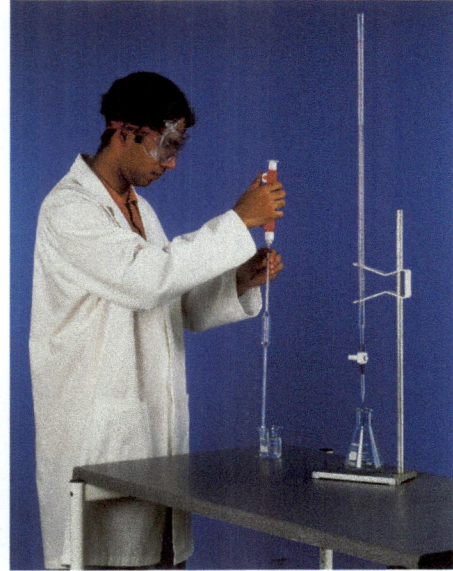

Figure 11 *An acid–base titration, to find the concentration of a base. A volumetric pipette is used to deliver an accurately measured volume of base of unknown concentration into the flask. The acid of known concentration is in the burette*

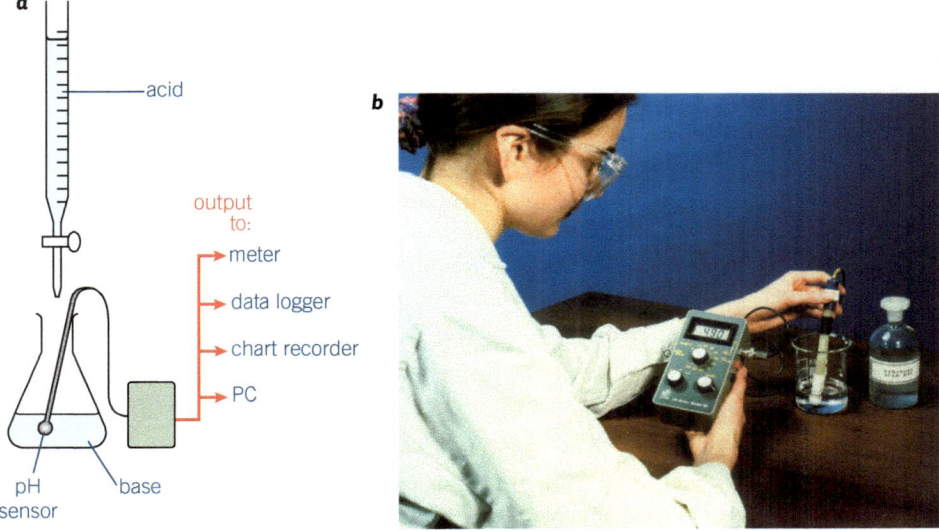

Figure 12 **a** *Apparatus to investigate pH changes during a titration* **b** *A pH meter*

Titration curves

Figure 13 shows the results obtained for four titrations using monoprotic acids (ie acids that donate a single H^+ ion). In each case, the base was added from the burette and the acid was accurately measured into a flask. The shape of each titration curve is typical for the type of acid–base titration.

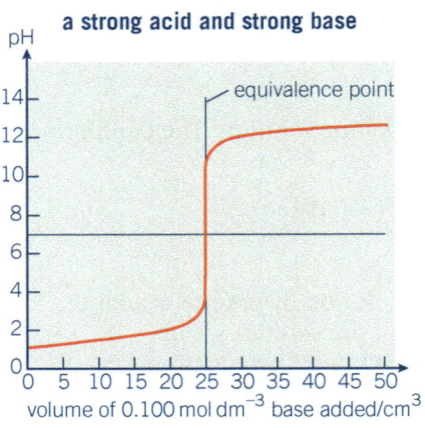

a strong acid and strong base

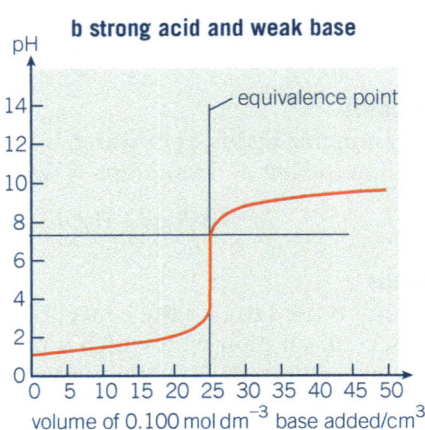

b strong acid and weak base

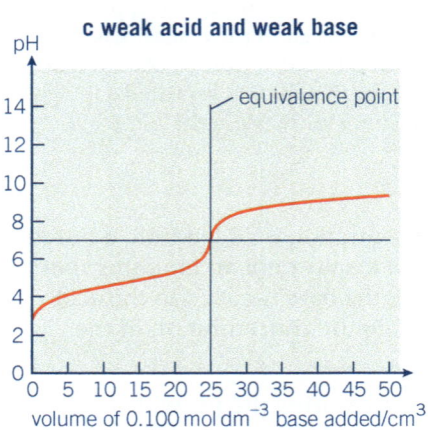

c weak acid and weak base

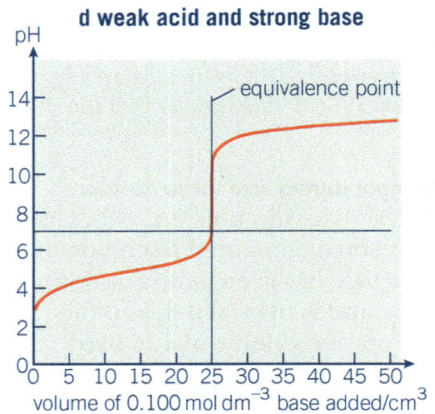

d weak acid and strong base

Figure 13 *Graphs of pH changes for titrations of different acids and bases*

The first thing to notice about these curves is that the pH does not change in a linear manner as the base is added. Each curve has almost horizontal sections where a lot of base can be added without changing the pH much. There is also a very steep portion of each curve, (except weak acid–weak base), where a single drop of base changes the pH by several units.

In a titration, the **equivalence point** is the point at which sufficient base has been added to just neutralise the acid (or vice-versa). In each of the titrations in Figure 13, the equivalence point is reached after $25.0\,cm^3$ of base has been added. However, the pH at the equivalence point is not always exactly 7.

In each case, except the weak acid–weak base titration, there is a large and rapid change of pH at the equivalence point (i.e., the curve is almost vertical) even though this is may not be centred on pH 7. This is relevant to the choice of indicator for a particular titration.

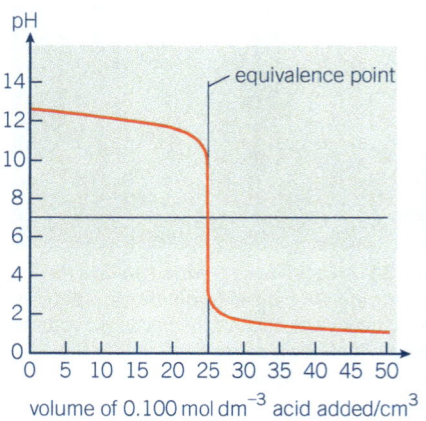

Figure 14 *Titration of a strong base–strong acid, adding $0.100\,mol\,dm^{-3}$ HCl(aq) to $25.0\,cm^3$ of $0.100\,mol\,dm^{-3}$ NaOH(aq)*

You can add the acid to the base for these pH curves and the shape will be flipped around the pH 7 line. For example, Figure 14 shows a strong acid–strong base curve.

In each case in Figure 13, the base was added to $25\,cm^3$ of $0.100\,mol\,dm^{-3}$ acid.

Worked example

A monoprotic acid

In a titration, the equivalence point is reached when 25.00 cm³ of 0.0150 mol dm⁻³ sodium hydroxide is neutralised by 15.00 cm³ hydrochloric acid. What is the concentration of the acid?

$$HCl(aq) + NaOH(aq) \rightarrow NaCl(aq) + H_2O(l)$$

The equivalence point shows that 15.0 cm³ hydrochloric acid of concentration A contains the same number of moles as 25.00 cm³ of 0.0150 mol dm⁻³ sodium hydroxide.

$$\text{number of moles in solution} = c \times \frac{V}{1000}$$

Where c is concentration in mol dm⁻³ and V is volume in cm³.

From the equation, number of moles HCl = number of moles NaOH

$$25.00 \times \frac{0.0150}{1000} = 15.00 \times \frac{A}{1000}$$
$$A = 0.025$$

So, the concentration of the acid is 0.0250 mol dm⁻³.

Worked example

A diprotic acid

In a titration, the equivalence point is reached when 20.00 cm³ of 0.0100 mol dm⁻³ sodium hydroxide is neutralised by 15.00 cm³ sulfuric acid. What is the concentration of the acid?

$$H_2SO_4(aq) + 2NaOH(aq) \rightarrow Na_2SO_4(aq) + 2H_2O(l)$$

The equivalence point shows that 15.00 cm³ sulfuric acid of concentration B contains the same number of moles of H⁺ ions as 20.00 cm³ of 0.0100 mol dm⁻³ sodium hydroxide contains OH⁻ ions.

$$\text{number of moles in solution} = c \times \frac{V}{1000}$$

Where c is concentration in mol dm⁻³ and V is volume in cm³.

$$\text{Number of moles of NaOH} = 20.00 \times \frac{0.0100}{1000} = \frac{0.2}{1000}$$

From the equation, number of moles H₂SO₄ = $\frac{1}{2}$ number of moles NaOH

So number of moles of H₂SO₄ = $\frac{0.1}{1000}$

$$\text{Number of moles of } H_2SO_4 = 15.00 \times \frac{B}{1000} = \frac{0.1}{1000}$$

So, the concentration, B, of the acid is 0.0067 mol dm⁻³.

Choice of indicators for titrations

Some common indicators are given in Table 3 with their colour changes.

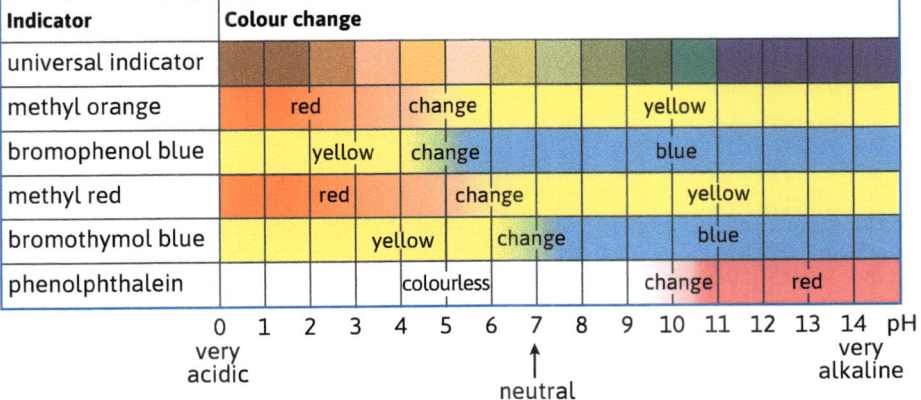

Table 3 *Some common indicators. Universal indicator is a mixture of indicators*

An acid–base titration uses an indicator to find the concentration of a solution of an acid or alkali. The equivalence point is the volume at which exactly the same number of moles of hydrogen ions (or hydroxide ions) has been added as there are moles of hydroxide ions (or hydrogen ions). The **end point** is the volume of alkali or acid added when the indicator just changes colour. Unless you choose the right indicator, the equivalence point and the end point may not always give the same answer.

A suitable indicator for a particular titration needs the following properties:

- The colour change must be sharp rather than gradual at the end point—no more than one drop of acid (or alkali) is needed to give a complete colour change. An indicator that changes colour gradually over several cubic centimetres would be unsuitable and would not give a sharp end point.
- The end point of the titration given by the indicator must be the same as the equivalence point, otherwise the titration will give the wrong answer.
- The indicator should give a distinct colour change, eg the colourless to pink change of phenolphthalein is easier to see than the red to yellow of methyl orange.
- Notice that the colour change of most indicators, shown in Table 3 on the previous page, takes place over a pH range of around two units. For this reason, not all indicators are suitable for all titrations. Universal indicator is not suitable for any titration because of its gradual colour changes.
- The following examples compare the suitability of two common indicators— phenolphthalein and methyl orange—for four different types of acid–base titration. In each case, the base is being added to the acid.

1 Strong acid–strong base, for example, hydrochloric acid and sodium hydroxide
Figure 15 is the graph of pH against volume of base added. The pH ranges over which two indicators change colour are shown. To fulfil the first two criteria above, the indicator must change within the vertical portion of the pH curve. Here either indicator would be suitable, but phenolphthalein is usually preferred because of its more easily seen colour change.

2 Weak acid–strong base titration, for example, ethanoic acid and sodium hydroxide
Methyl orange is not suitable (Figure 16). It does not change in the vertical portion of the curve and will change colour in the 'wrong' place and over the addition of many cubic centimetres of base. Phenolphthalein will change sharply at exactly 25 cm³, the equivalence point, and would therefore be a good choice.

Note

You can use Table 3 on the previous page to select other suitable indicators for a particular type of titration.

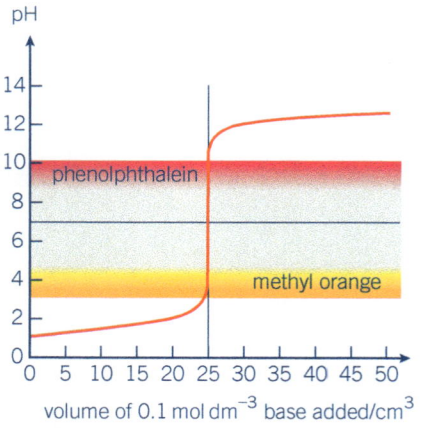

Figure 15 Titration of a strong acid–strong base, adding 0.1 mol dm⁻³ NaOH(aq) to 25 cm³ of 0.1 mol dm⁻³ HCl(aq)

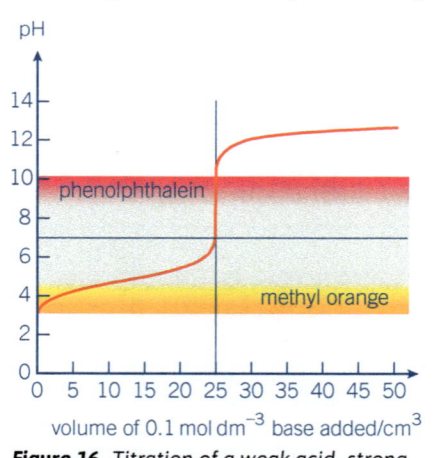

Figure 16 Titration of a weak acid–strong base, adding 0.1 mol dm⁻³ NaOH(aq) to 25 cm³ of 0.1 mol dm⁻³ CH₃COOH(aq)

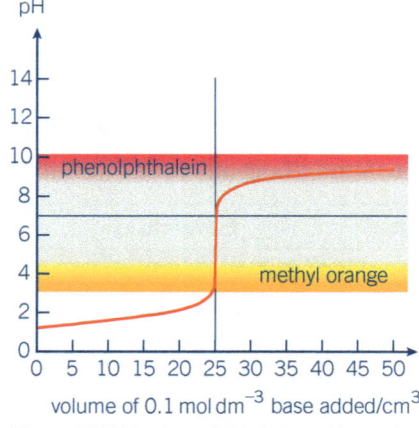

Figure 17 Titration of a strong acid–weak base, adding 0.1 mol dm⁻³ NH₃(aq) to 25 cm³ of 0.1 mol dm⁻³ HCl(aq)

3 Strong acid–weak base titration, for example, hydrochloric acid and ammonia
Here methyl orange will change sharply at the equivalence point but phenolphthalein would be of no use (Figure 17).

4 Weak acid–weak base, for example, ethanoic acid and ammonia

Here neither indicator is suitable (Figure 18). In fact, no indicator could be suitable as an indicator requires a vertical portion of the curve over two pH units at the equivalence point to give a sharp change.

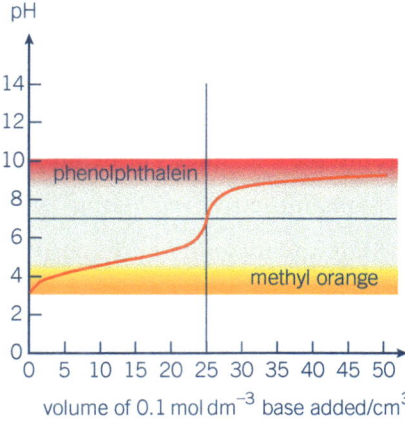

Figure 18 Titration of a weak acid–weak base, adding 0.1 mol dm^{-3} NH$_3$(aq) to 25 cm^3 of 0.1 mol dm^{-3} CH$_3$COOH(aq)

Summary test 7.2

1 Identify which reactant is an acid and which is a base in the following:
   a $HNO_3 + OH^- \rightleftharpoons NO_3^- + H_2O$
   b $CH_3COOH + H_2O \rightleftharpoons CH_3COO^- + H_3O^+$
2 In an acidic solution, [H$^+$] is 1×10^{-4} mol dm^{-3}. What is the pH?
3 What species are formed when the following bases accept a proton?
   a OH^-
   b NH_3
   c H_2O
   d Cl^-
4 25.0 cm^3 sodium hydroxide is neutralised by 15.0 cm^3 sulfuric acid, H$_2$SO$_4$, of concentration 0.100 mol dm^{-3}.
   a Write the equation for this reaction.
   b From the equation, determine how many moles of sulfuric acid will neutralise 1.00 mol of sodium hydroxide.
   c Determine how many moles of sulfuric acid are used in the neutralisation.
   d Determine the concentration of the sodium hydroxide.
5 The graph below shows two titration curves of two acids labelled A and B with a base.

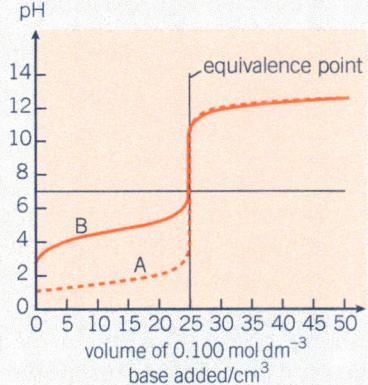

6 a State which curve represents:
   i a strong acid
   ii a weak acid.
   b State which one could represent:
   i ethanoic acid, CH$_3$COOH
   ii hydrochloric acid, HCl.
   c Was the base strong or weak? Explain your answer.
7 The indicator bromocresol purple changes colour between pH 5.2 and 6.8. Determine which of the following titration types it would be suitable for:
   a weak acid–weak base
   b strong acid–weak base
   c weak acid–strong base
   d strong acid–strong base
8 Determine which of the above titrations bromophenol blue would be suitable for (see Table 3 on page 113).

7 Exam-style questions

> Launch additional digital resources for the chapter

1 a State Le Chatelier's principle. *(2 marks)*

b Explain the term 'dynamic equilibrium'. *(1 mark)*

c State the meaning of the term 'closed system'. *(1 mark)*

2 Write K_c expressions for the following equations and deduce its units: *(6 marks)*

a $N_2(g) + 3H_2(g) \rightleftharpoons 2NH_3(g)$

b $2NH_3(g) \rightleftharpoons N_2(g) + 3H_2(g)$

c $2NO(g) + O_2(g) \rightleftharpoons 2NO_2(g)$

d $NO(g) + \frac{1}{2}O_2(g) \rightleftharpoons NO_2(g)$

3 Dinitrogen tetroxide is an important chemical in synthesis reactions. It forms an equilibrium mixture with nitrogen dioxide. The equilibrium constant (K_c) is $200\,mol^{-1}\,dm^3$ at 298 K.

$$2NO_2(g) \rightleftharpoons N_2O_4(g)$$

a Write an expression for the equilibrium constant (K_c) for this reaction. *(1 mark)*

b An equilibrium concentration of $0.02\,mol\,dm^{-3}$ $N_2O_4(g)$ was calculated at 298 K. Calculate the equilibrium concentration of $NO_2(g)$. *(2 marks)*

4 a Write K_p expressions for the following equations and deduce its units: *(4 marks)*

i $N_2(g) + 3H_2(g) \rightleftharpoons 2NH_3(g)$

ii $N_2O_4(g) \rightleftharpoons 2NO_2(g)$

b At 60°C and 1.0 atm N_2O_4 is 50% dissociated into NO_2. Calculate the value of K_p. *(2 marks)*

5 5 mol of ethanol, 6 mol of ethanoic acid, 6 mol of ethyl ethanoate and 4 mol of water were mixed in a sealed vessel and allowed to reach equilibrium at 15°C.

At equilibrium the vessel was found to contain 4 mol of ethanoic acid.

a Define the term Brønsted–Lowry acid. *(1 mark)*

b Ethanoic acid is a weak acid. Explain the difference between a strong and weak acid. *(2 marks)*

c Write an equation for the reaction between ethanol and ethanoic acid to form ethyl ethanoate and water. *(1 mark)*

d Write an expression for the equilibrium constant, K_c, for this reaction at 15°C. *(1 mark)*

e Calculate the number of moles of ethanol, ethyl ethanoate and water present in the equilibrium mixture. *(3 marks)*

f Use your answers to parts **d** and **e** to calculate the value of K_c for this reaction. *(1 mark)*

6 The Sabatier reaction involves the reaction of hydrogen with carbon dioxide in the presence of a nickel catalyst:

$$H_2(g) + CO_2(g) \rightleftharpoons H_2O(g) + CO(g)$$

a Write an expression for equilibrium constant, K_p, for this reaction. *(1 mark)*

b Explain why K_p has no units. *(1 mark)*

c The table below shows the value of K_p for the equilibrium at different temperatures:

Temperature	K_p
298	1.00×10^{-5}
900	6.03×10^{-1}
1300	2.82

i Describe the trend shown by the data in the table. *(1 mark)*

ii Use the above data to determine whether the forwards reaction is exothermic or endothermic. Explain your answer. *(3 marks)*

7 Carbonyl bromide can be formed according to the following equilibrium reaction:

$$CO(g) + Br_2(g) \rightleftharpoons COBr_2(g)$$

a Identify the species that undergoes oxidation in the above reaction. Explain your answer. *(2 marks)*

b Describe the effect of increasing pressure on the equilibrium yield of carbonyl bromide. *(3 marks)*

c Predict the effect of increasing the concentration of bromine gas on the position of the equilibrium. Give reasons for your answer. *(2 marks)*

d Describe the effect on the position of the equilibrium of adding a catalyst. Explain your answer. *(2 marks)*

e Write an expression for the equilibrium constant, K_c, for the reverse reaction. State the units of K_c. *(2 marks)*

8 a Write chemical formulae for the following acids:
  i hydrochloric ii nitric iii sulfuric.
  (3 marks)

 b Use your answer to part **a** to identify the species that is common to all acids and construct the simplest ionic equation to show the reaction of this species with hydroxide ions. State the type of reaction.
  (3 marks)

9 The table below describes the colours of some common indicators in both the acidic form (HIn) and their conjugate base (In⁻):

Indicator	Colour of HIn	Colour of In⁻
bromocresol green	yellow	blue
methyl red	red	Yellow
phenol red	yellow	red
phenolphthalein in ethanol	colourless	red

 a Construct a general equation for the equilibrium constant, K_a, for an indicator. *(1 mark)*

 b State the colour of methyl red in alkaline solution. Explain your answer. *(2 marks)*

10 Ammonia is manufactured by the Haber process, as detailed in the following reaction:

$N_2(g) + 3H_2(g) \rightleftharpoons 2NH_3(g)$ $\Delta H^\ominus = -92 \text{ kJ mol}^{-1}$

 a Use Le Chatelier's principle to explain why the chosen conditions of 250 atm and 500 K are considered to be a compromise. *(6 marks)*

 b Identify the catalyst used in the Haber process and state the effect of this catalyst on the amount of ammonia present at equilibrium. Explain your answer. *(3 marks)*

11 Saltpetre, commonly known as potassium nitrate, is a key component in the manufacture of gunpowder. It has the formula KNO_3.

 a Construct a balanced chemical equation to show how potassium nitrate can be formed from an acid–base reaction. *(2 marks)*

 b Identify the substance in part **a** that is a Brønsted–Lowry base. Explain your answer. *(2 marks)*

12 A student mixed 6.0 g of ethanoic acid (CH_3COOH) with 6.9 g of ethanol (C_2H_5OH). The system reached equilibrium. The equilibrium mixture contained 1.4 g of water as well as ethyl ethanoate ($CH_3COOC_2H_5$). The equation for the equilibrium is shown below:

$CH_3COOH(l) + C_2H_5OH(l) \rightleftharpoons CH_3COOC_2H_5(l) + H_2O(l)$

 a Calculate the amount in mol at the start of the reaction for:
  i ethanoic acid
  ii ethanol. *(4 marks)*

 b Calculate the amount in mol of water present in the equilibrium mixture. *(1 mark)*

 c Deduce the amount in mol of ethyl ethanoate present at equilibrium. *(1 mark)*

 d Calculate the amount in mol in the equilibrium mixture of:
  i Ethanoic acid
  ii Ethanol *(4 marks)*

 e Write an expression for K_c for the equilibrium. *(1 mark)*

 f Calculate K_c for the reaction, stating the units. *(3 marks)*

8.1 Rate of reaction

Kinetics is the study of the factors that affect rates of chemical reactions—how quickly they take place. There is a large variation in reaction rates. 'Popping' a test tube full of hydrogen is over in a fraction of a second. In contrast, the complete rusting away of an iron nail could take several years. Reactions can be speeded up or slowed down by changing the conditions.

Collision theory

For a reaction to take place between two particles, they must collide with enough energy to break bonds. The collision must also take place between the parts of the molecule that are going to react together—orientation is important. To get a lot of collisions you need a lot of particles in a small volume. For the particles to have enough energy to break bonds they need to be moving fast. So, for a fast reaction rate you need plenty of rapidly moving particles in a small volume.

Most collisions between molecules or other particles do not lead to reaction. They either do not have enough energy, or they are in the wrong orientation.

Factors that affect the rate of chemical reactions

The following factors will increase the rate of a reaction:

- **Increasing the temperature**. This increases the speed of the molecules, which in turn increases both their energy and also the frequency of collisions.
- **Increasing the concentration of a solution**. If there are more particles present in a given volume then collisions are more likely and the reaction rate would be faster—we say that the **frequency of collisions**. has increased. However, as a reaction proceeds, the reactants are used up and their concentration falls. So, in most reactions the rate of reaction drops as the reaction goes on.
- **Increasing the pressure of a gas reaction**. This has the same effect as increasing the concentration of a solution—there are more molecules or atoms in a given volume so collisions are more likely.
- **Increasing the surface area of solid reactants**. The greater the total surface area of a solid, the more of its particles are available for collisions with molecules in a gas or a liquid. This means that breaking a solid lump into smaller pieces increases the rate of its reaction because there are more sites for reaction.
- **Using a catalyst**. A catalyst is a substance that can change the rate of a chemical reaction without being chemically changed itself.

What is a reaction rate?

As a reaction, $A + 2B \rightarrow C$, takes place, the concentrations of the reactants A and B decrease with time and the concentration of product C increases with time. You could measure the concentration of A, B, or C with time and plot the results (Figure 1).

> **The rate of the reaction is defined as the change in concentration (of any of the reactants or products) with unit time.**

However, notice how different the curves are for A, B, and C. As [C] (the product) increases, [A] and [B] (the reactants) decrease. As the equation tells us, for every A that reacts there are two of B that react, so [B] decreases twice as fast as [A]. For this reason, when measuring the rate of a reaction it is important to state whether you are following the concentration of A, B, or C. Usually it is assumed that a rate is measured by following the concentration of a product(s), because the concentration of the product increases with time.

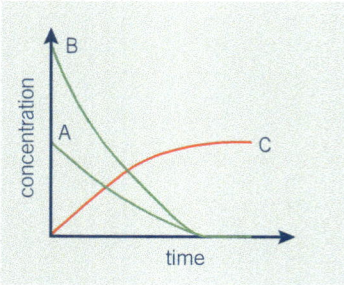

Figure 1 *Changes of concentration with time for A, B, and C*

The rate of reaction at any instant

You are often interested in the rate at a particular instant in time rather than over a period of time. To find the rate of change of [C] at a particular instant, draw a tangent to the curve at that time and then find its gradient (slope), as in Figure 2.

Reaction rates are measured in mol dm^{-3} s^{-1}.

$$\text{rate} = \frac{a}{b} = \frac{\text{change in concentration}}{\text{time}}$$

Experimental methods for measuring reaction rates

To measure a reaction rate, we need a method of measuring the concentration of a reactant or product as time goes on. The method chosen depends on the properties of the chosen substance—acid, base, coloured, oxidising agent, reducing agent etc—and the speed of the reaction. For example using a titration, which takes a few minutes, to measure a concentration would be suitable for a reaction that goes to completion in a few hours. However, it would not be suitable for a reaction which is over in a few seconds.

Sampling and titration

The reaction mixture is sampled by pipetting out a sample of the reaction mixture at intervals and carrying out a suitable titration (acid/base, or redox, for example). It may be possible to slow down or stop the reaction in the sample, for example by removing a catalyst, to allow more time for the titration. An example of a suitable reaction is that of iodine and propanone with an acid catalyst:

$$\overset{H^+}{I_2(aq) + CH_3COCH_3(aq) \rightarrow CH_2ICOCH_3(aq) + HI(aq)}$$

The reaction is complete in around 30 minutes. Samples can be taken every five minutes and added to sodium hydrogencarbonate solution to remove the catalyst and slow down the reaction. Then the samples can be titrated with sodium thiosulfate to measure the iodine concentration. The titration reaction is:

$$I_2(aq) + 2Na_2S_2O_3(aq) \rightarrow 2NaI(aq) + Na_2S_4O_6(aq)$$

using starch as an indicator.

Using a colorimeter

A colorimeter may be suitable if one of the reactants or products is coloured, ie it absorbs light. A colorimeter (see Figure 3) shines light of a wavelength absorbed by the coloured species through the reaction mixture onto a photocell. The readings can be recorded in real time by a data logger or PC and a concentration–time graph plotted immediately.

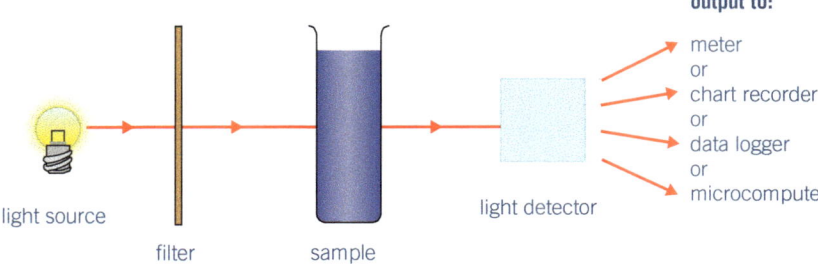

Figure 3 Using a colorimeter to measure reaction rates

The iodine/propanone reaction could also be investigated using a colorimeter as iodine is brown.

> **Exam tip**
>
> Square brackets around a chemical symbol mean its concentration in mol dm^{-3}.

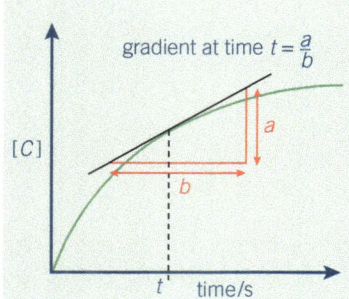

Figure 2 The rate of change of [C] at time t, is the gradient of the concentration–time graph at t

> **Exam tip**
>
> When planning a rate of reaction investigation, it is a good idea to select a property (the dependent variable) that is related to concentration and easy to measure.

> 📖 **The Beer–Lambert law**
>
> The absorbance of light by a coloured species is directly proportional to its concentration. This is called the Beer–Lambert law.

Measuring the volume of gas evolved

In a reaction such as:

$$Zn(s) + 2HCl(aq) \rightarrow ZnCl_2(aq) + H_2(g)$$

the gas evolved can be collected in a syringe—see Figure 4. The reaction is started by shaking the zinc into the acid.

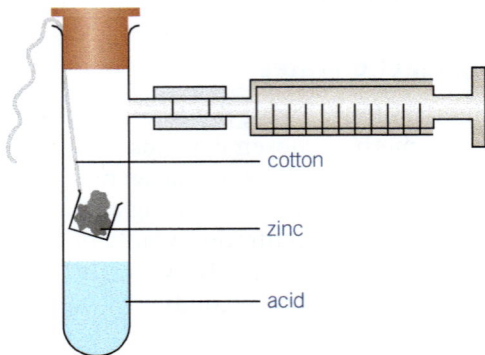

cotton

zinc

acid

Figure 4 *Measuring gas volume to monitor reaction rates*

Measuring conductivity of a solution

If the number of ions in a reaction changes over time, this will affect the conductivity of the solution. This can be measured as in Figure 5. Alternating current is used to avoid electrolysing the reaction mixture, which would affect the amount of ions present.

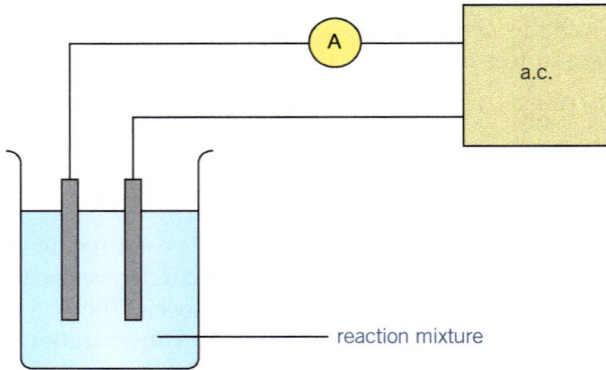

A

a.c.

reaction mixture

Figure 5 *Using a conductivity meter to monitor reaction rates*

The reaction $C_4H_9Br + H_2O \rightarrow C_4H_9OH + H^+ + Br^-$ could be followed by this method because more ions are produced as the reaction proceeds.

Worked example

Measuring a reaction rate

In the reaction between bromine and methanoic acid, the solution starts off brown (from the presence of bromine) and ends up colourless:

$$Br_2(aq) + HCO_2H(aq) \rightarrow 2Br^-(aq) + 2H^+(aq) + CO_2(g)$$

So, a colorimeter can be used to measure the decreasing concentration of bromine. The reaction is slow enough to enable the colorimeter to be read every half a minute and the measurements recorded. A computer or data logger could also be used to measure the readings, and this may be essential for faster reactions. The reaction can also be monitored by collecting the carbon dioxide gas.

Table 1 shows some typical results.

Table 1 *[Br₂] measured over time*

Time / s	$[Br_2]$ / mol dm^{-3}
0	0.0100
30	0.0090
60	0.0081
90	0.0073
120	0.0066
180	0.0053
240	0.0044
360	0.0028
480	0.0020
600	0.0013
720	0.0007

In order to find the reaction rate at different times, the results are plotted on a graph. You can then measure the gradients of the tangents at the times required. For example, at $t = 0$, 300 s, and 600 s (Figure 6).

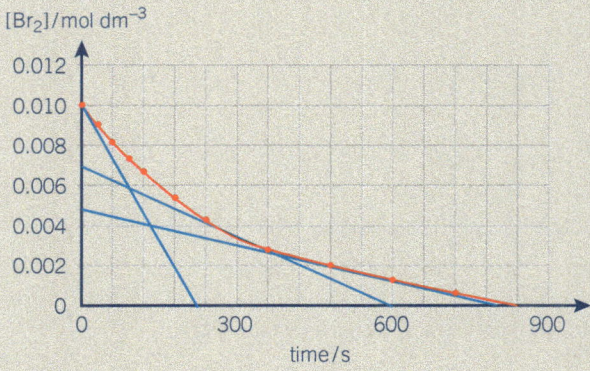

Figure 6 Finding the rate of reaction at t = 0, t = 300, and t = 600 s

At $t = 0\,s$, rate of reaction $= \dfrac{0.010}{240} = 0.000\,041\,6$ mol dm^{-3} s^{-1}

At $t = 300\,s$, rate of reaction $= \dfrac{0.0076}{540} = 0.000\,014$ mol dm^{-3} s^{-1}

At $t = 600\,s$, rate of reaction $= \dfrac{0.0046}{840} = 0.000\,005\,5$ mol dm^{-3} s^{-1}

Summary test 8.1

Answer the following questions about the reaction rate graph in Figure 7.

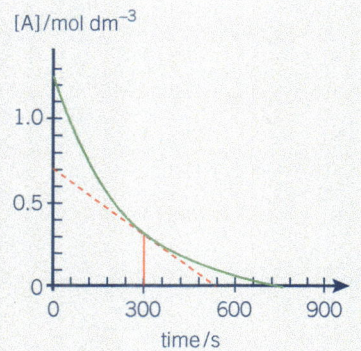

Figure 7

1. Is the concentration of a reactant or a product being plotted? Explain your answer.
2. The tangent to the curve at the time 300 seconds is drawn on the graph. Determine the gradient of the tangent. Remember to include units.
3. State what this gradient represents.
4. Without drawing tangents, state what can be said about the gradients of the tangents at time 0 seconds and time 600 seconds.
5. Explain your answer to question 4.
6. Suggest an experimental method of measuring the rate of:
   a. $CaCO_3(s) + 2HCl \rightarrow CaCl_2(aq) + H_2O(l) + CO_2(g)$
   b. $CuCl_2(aq) + Zn(s) \rightarrow ZnCl_2(aq) + Cu(s)$
   c. $C_2H_5Br + NaOH \rightarrow C_2H_5OH + NaBr$

Effect of temperature on reaction rates and the concept of activation energy

Learning outcomes

On these pages you will learn to:

- explain what activation energy means
- explain the effect of temperature change on the rate of a reaction

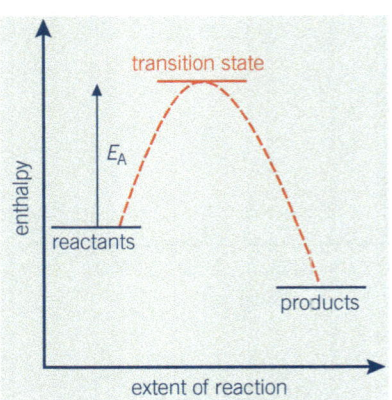

Figure 8 *An exothermic reaction with a large activation energy E_A*

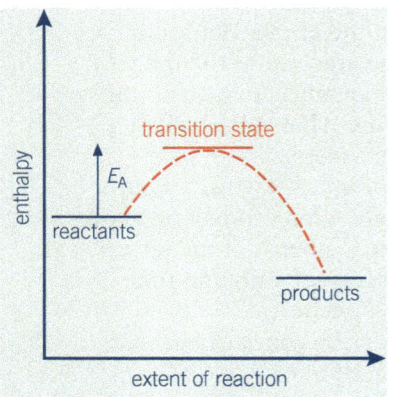

Figure 9 *An exothermic reaction with a small activation energy E_A*

Collision theory

Only a very small proportion of collisions actually result in a reaction.

Not all collisions between reactants will lead to a reaction. Those that do are called **effective collisions** and those that do not are called **non-effective collisions**. For a collision to result in a reaction, the molecules must have a certain minimum energy, enough to start breaking bonds. The minimum energy required for a collision to be effective and start the reaction is called the activation energy. It has the abbreviation E_A (see Section 5.1).

You can include the idea of activation energy on an enthalpy level diagram that shows the course of a reaction (see Figures 8 and 9).

The Maxwell–Boltzmann distribution

The particles in any gas (or solution) are all moving at different speeds. A few are moving slowly and a few are moving very fast, but most are somewhere in the middle. The energy of a particle depends on its speed, so the particles also have a range of energies. If you plot a graph of energy against the fraction of particles that have that energy, you end up with the curve shown in Figure 10. This particular shape is called the **Maxwell–Boltzmann distribution**—it tells us about the distribution of energy amongst the particles.

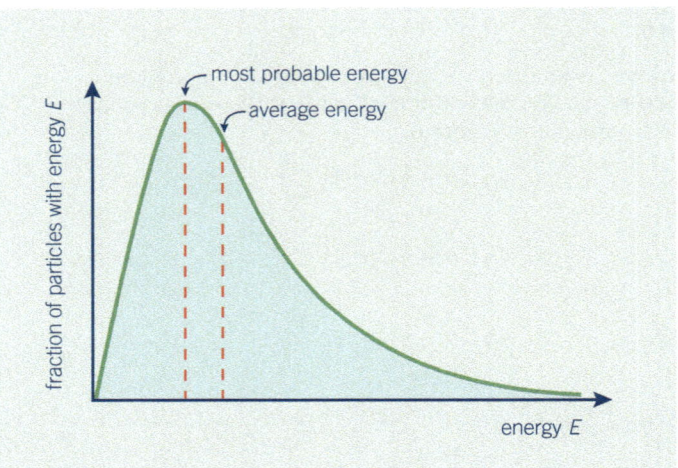

Figure 10 *The distribution of the energies of particles. The area under the graph represents the total number of particles*

- No particles have zero energy.
- Most particles have intermediate energies—around the peak of the curve.
- A few have very high energies (the right-hand side of the curve). In fact, there is no upper limit.
- Note also that the average energy is not the same as the most probable energy.

Activation energy E_A

If you mark the activation energy E_A on the Maxwell–Boltzmann distribution graph, Figure 11, then the area under the graph to the right of the activation energy line represents the number of particles with enough energy to react.

The need for the activation energy to be present before a reaction takes place explains why not all exothermic reactions occur spontaneously at room temperature.

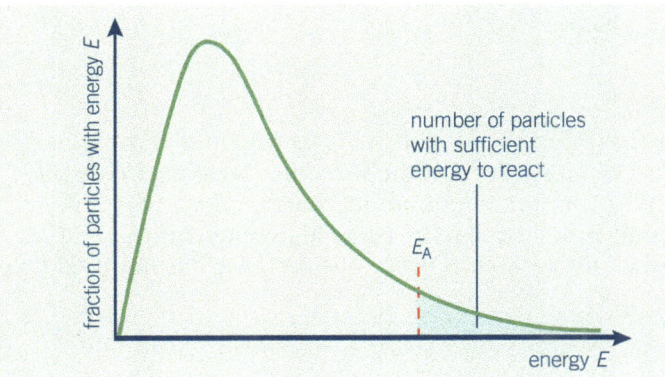

Figure 11 *Only particles with energy greater than E_A can react*

For example, fuels are mostly safe at room temperature, as in a petrol station. However, a small spark may provide enough energy to start the combustion reaction. The heat given out by the initial reaction is enough to provide the activation energy for further reaction. Similarly, the chemicals in a match head are quite stable until the activation energy is provided by friction.

Even the energy of a single spark can set off a reaction. This is why if you smell gas, you must not even turn on a light. An electric spark in the switch could produce enough energy to trigger an explosion.

The effect of temperature on reaction rate

The shape of the Maxwell–Boltzmann graph changes with temperature, as shown in Figure 12.

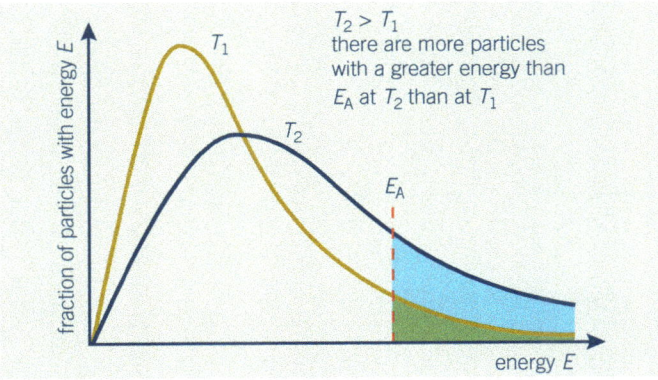

Figure 12 *The Maxwell–Boltzmann distribution of the energies of the same number of particles at two temperatures*

At higher temperatures the peak of the curve is lower and moves to the right. The number of particles with very high energy increases. The total area under the curve is the same for each temperature because it represents the total number of particles.

The shaded areas to the right of the E_A line represent the number of molecules that have greater energy than E_A at each temperature.

The graphs show that at higher temperatures more of the molecules have energy greater than E_A, so a there will be a higher percentage of effective collisions (ones which lead to reaction). This is why reaction rates increase with temperature. In fact, a small increase in temperature produces a large increase in the number of particles with energy greater than E_A.

Also, the total number of collisions in a given time increases a little as the particles move faster. However, this much less important to the rate of reaction than the increase in the number of effective collisions (those with energy greater than E_A).

1 Use Figure 13 to answer the following questions:

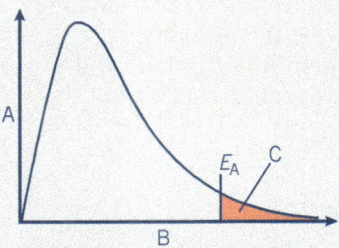

Figure 13 *The Maxwell–Boltzmann distribution of energies of particles at a particular temperature, with the activation energy, E_A, marked*

   a Identify the axis labelled A.
   b Identify the axis labelled B.
   c State what area C represents.
   d If the temperature is increased, describe what happens to the peak of the curve.
   e If the temperature is increased, describe what happens to E_A.

2 Give five factors that affect the speed of a chemical reaction.

Use the reaction profile below to answer questions 3 and 4:

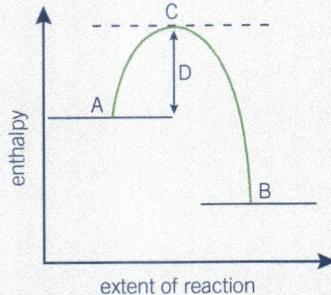

3 **a** Identify A.
   b Identify B.
   c Identify C.
   d Identify D.

4 **a** State whether the enthalpy level profile represents an endothermic or an exothermic reaction.
   b Explain your answer to part **a**.

8.3 Homogeneous and heterogeneous catalysts

Catalysts are substances that affect the rate of chemical reactions without being chemically changed themselves at the end of the reaction. Catalysts are usually used to speed up reactions so they are important in industry. It is cheaper to speed up a reaction by using a catalyst than by using high temperatures and pressures. This is true even if the catalyst is expensive, because it is not used up.

How catalysts work

Catalysts work because they provide a different pathway (or mechanism) for the reaction, one with a lower activation energy. Therefore, they reduce the activation energy of the reaction (the minimum amount of energy that is needed to start the reaction). You can see this on the enthalpy level diagrams in Figure 14.

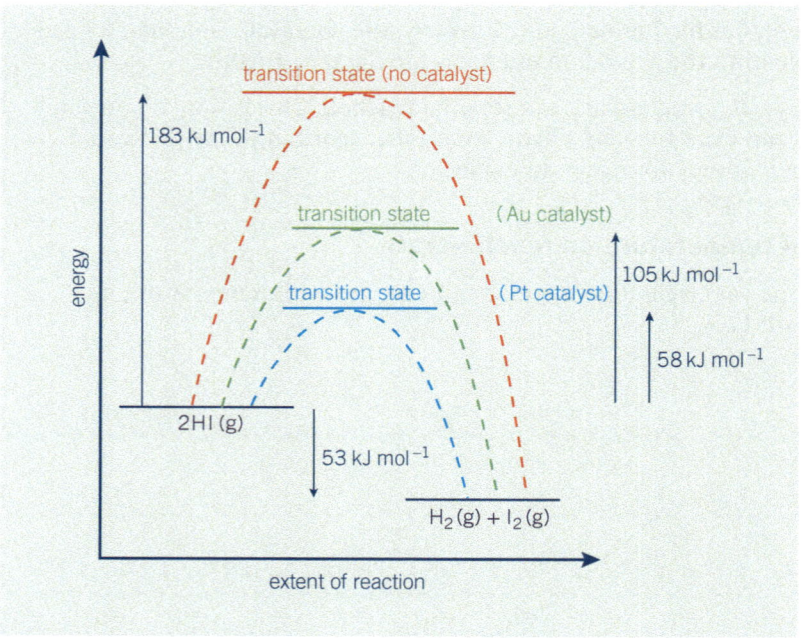

Figure 14 *The decomposition of hydrogen iodide with different catalysts*

For example, for the decomposition of hydrogen iodide:

$$2HI(g) \rightarrow H_2(g) + I_2(g)$$

$E_A = 183 \text{ kJ mol}^{-1}$ (without a catalyst)

$E_A = 105 \text{ kJ mol}^{-1}$ (with a gold catalyst)

$E_A = 58 \text{ kJ mol}^{-1}$ (with a platinum catalyst)

You can see what happens when you lower the activation energy if you look at the Maxwell–Boltzmann distribution curve in Figure 15.

The area that is shaded pink represents the number of effective collisions that can happen without a catalyst. The area shaded blue represents the additional molecules with energy equal to or above the activation energy. The area shaded blue, plus the area that is shaded pink, represents the number of effective collisions that can take place with a catalyst.

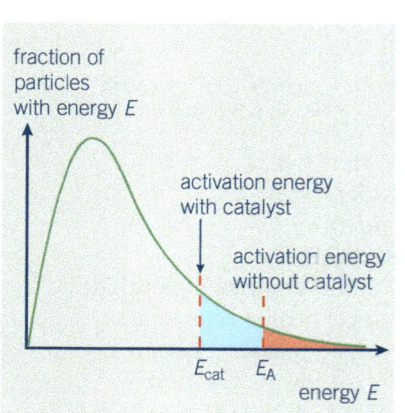

Figure 15 *With a catalyst the extra particles in the blue area react, in addition to the particles in the pink area*

Catalysts do not affect the enthalpy change of the reactions, nor do they affect the position of equilibrium in a reversible reaction (see Section 7.1).

Table 2 *Examples of catalysts*

Reaction	Catalyst	Use
$N_2(g) + 3H_2(g) \rightarrow 2NH_3(g)$ Haber process	iron	making fertilisers
$4NH_3 + 5O_2 \rightarrow 4NO + 6H_2O$ Ostwald process for making nitric acid	platinum and rhodium	making fertilisers and explosives
$H_2C=CH_2 + H_2 \rightarrow CH_3CH_3$ hardening of fats with hydrogen	nickel	making margarine
cracking hydrocarbon chains from crude oil	aluminium oxide and silicon dioxide zeolite	making petrol
catalytic converter reactions in car exhausts	platinum and rhodium	removing polluting gases
$H_2C=CH_2 + H_2O \rightarrow CH_3CH_2OH$ hydration of ethene to produce ethanol	H^+ absorbed on solid silica phosphoric acid, H_3PO_4	making ethanol – a fuel additive, solvent, and chemical feedstock
$CH_3CO_2H(l) + CH_3OH(l) \rightarrow CH_3CO_2CH_3(aq) + H_2O(l)$ esterification	H^+	making solvents

Many catalysts used in industry are transition metals or their compounds. Catalysts can be divided into two groups:

- heterogeneous
- homogeneous.

Heterogeneous catalysts

Heterogeneous catalysts are present in a reaction in a different phase (solid, liquid, or gas) than the reactants. They are usually present as solids, whilst the reactants may be gases or liquids. Their catalytic action occurs on the solid surface. The reactants pass over the catalyst surface, which remains in place so the catalyst is not lost and does not need to be separated from the products.

Homogeneous catalysts

This is when the catalyst is in the same phase as the reactant. For example, in the gas phase chlorine free radicals act as catalysts to destroy the ozone layer.

Summary test 8.3

1 The following questions refer to Figure 16.

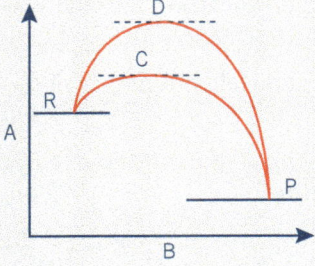

Figure 16 *A profile for a reaction with and without a catalyst*

    a Identify the labels A, B, C, R, and P.
    b State what the distances from D to R and from C to R represent.
    c State whether the reaction is exothermic or endothermic.
2 Identify from Table 2 above an example of:
    a heterogenous catalysis
    b homogenous catalysis.

8 Exam-style questions

 Launch additional digital resources for the chapter

1 The reaction between marble chips (calcium carbonate) and dilute hydrochloric acid was used by students to investigate the effect of surface area on reaction rate. The volume of carbon dioxide was collected for each experiment using a gas syringe. The students conducted two experiments:

Experiment 1	5 small marble chips (total mass = 2 g) and 50 cm³ of 0.1 mol dm⁻³ hydrochloric acid.
Experiment 2	1 large marble chip (total mass = 2 g) and 50 cm³ of 0.1 mol dm⁻³ hydrochloric acid.

a Write an equation for the reaction involved.
(2 marks)

b Draw and label a diagram of the apparatus used.
(3 marks)

c Explain why an excess of marble chips is used.
(1 mark)

d The results of the experiments are shown in the graph below:

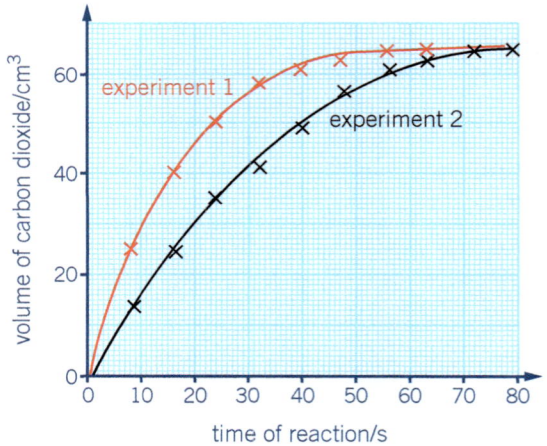

i Describe the how the rate of reaction changes in experiment 1. *(3 marks)*
ii Use your knowledge of collision theory to explain your answer to part **i**. *(2 marks)*

e Use the graph above to calculate the rate of reaction at 10 seconds for experiment 1, giving the units.
(3 marks)

f Explain why the initial rate of reaction is different for both experiments. *(3 marks)*

2 a Draw a graph with labelled axes showing the Maxwell–Boltzmann distribution for the energies of the molecules in a gas at 300K. Label this curve 'T1'.
(3 marks)

b Add to your answer to part **a** to include a curve for the same reaction at 320K, label your curve 'T2'.
(3 marks)

c Construct a labelled Maxwell–Boltzmann diagram to explain how a catalyst affects the rate of a chemical reaction. *(4 marks)*

3 a Define the term 'activation energy'. *(1 mark)*

b State the effect, if any, on each of the following changes to the activation energy, E_A, for a chemical reaction:
i increase in temperature
ii addition of a catalyst
iii reduction in concentration of reactants. *(3 marks)*

c Explain your answers to part **b**. *(3 marks)*

4 Some industrial processes require the use of a heterogeneous catalyst.

a Explain the term *heterogeneous catalyst*. *(1 mark)*

b Evaluate the different mechanisms of action of heterogeneous and homogeneous catalysts, giving an example of each. *(2 marks)*

5 A student conducted an experiment to investigate the rate of reaction between calcium carbonate and hydrochloric acid. 100 cm³ of 0.1 mol dm⁻³ hydrochloric acid was reacted with a small piece of calcium carbonate. A total of 120 cm³ of carbon dioxide was collected.

a Construct a balanced chemical equation for this reaction. *(2 marks)*

b Sketch a graph to show how the volume of carbon dioxide changed over time. *(2 marks)*

c Assuming an excess of calcium carbonate; on the same axes, sketch and label curves for the following:
i Curve A: 50 cm³ of 0.2 M hydrochloric acid with small pieces of calcium carbonate
ii Curve B: 50 cm³ of 0.1 M hydrochloric acid with small pieces of calcium carbonate
iii Curve C: 50 cm³ of 0.1 M hydrochloric acid with powdered calcium carbonate. *(3 marks)*

6 A chemist wanted to evaluate the purity of an antacid tablet by assaying the quantity of carbonate ions present. A crushed tablet was reacted with hydrochloric acid and the volume of carbon dioxide evolved was measured at 15 second time intervals. The results are given in the table below:

Time / s	0	15	30	45	60	75	90	105	120	135
Volume / cm³	0	68	135	190	235	265	280	285	285	285

   a Plot a graph using this data and draw a line of best fit. *(3 marks)*

   b Calculate the initial rate of this reaction *(3 marks)*

   c Use your graph to estimate the time taken for the reaction to be completed. *(1 mark)*

7 The diagram below shows how the volume of hydrogen changes with time in the reaction between magnesium and hydrochloric acid. Curve X is obtained when 1 g of magnesium ribbon reacts with 100 cm³ (excess) hydrochloric acid at 30°C.

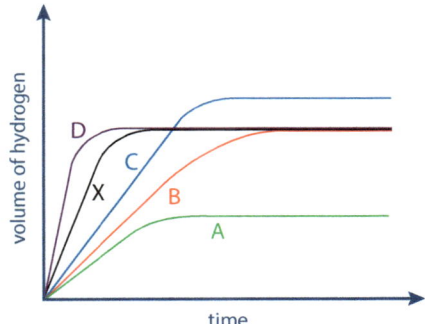

Which curve would you expect to obtain if:

   a 1 g of Mg ribbon reacts with 100 cm³ of the same acid at 50°C. *(1 mark)*

   b 1 g of Mg ribbon reacts with 100 cm³ of the same acid at 15°C. *(1 mark)*

   c 0.5 g of Mg ribbon reacts with 100 cm³ of the same acid at 30°C. *(1 mark)*

8 The diagram below shows the Maxwell-Boltzmann distribution curve for a sample of gas at a fixed temperature, T. The activation energy, E_A, for this reaction is also given.

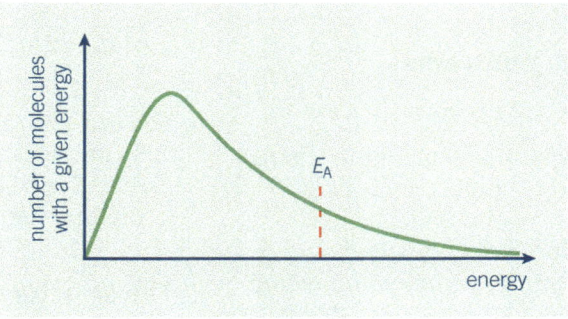

   a Copy the diagram above and sketch the distribution curve for a sample of gas that is at a temperature greater than T. Label this curve T_1. *(2 marks)*

   b State the effect of an increase in temperature on the rate of a chemical reaction. Explain your answer with references to part **a**. *(3 marks)*

   c State and explain the effect of the addition of a catalyst to this reaction. Label the diagram drawn in your answer to part **a** with $E_{A\,cat}$ to support your answer. *(4 marks)*

9 The water–gas shift reaction is used to produce hydrogen from the reaction of carbon monoxide and water according to the following reaction:

$$CO(g) + H_2O(g) \rightarrow H_2(g) + CO_2(g)$$

   a Explain why there is an increase in the rate of reaction between carbon monoxide and water with an increase in pressure, at constant temperature. *(2 marks)*

   b Define the term catalyst. *(2 marks)*

   c This reaction often uses a solid catalyst. Name this type of catalysis and give one reason why it is often used in the form of a powder. *(2 marks)*

Periodicity of physical properties of the elements in Period 3

The Periodic Table is a list of all the elements in order of increasing atomic number. You can predict the properties of an element from its position in the table. You can use the Periodic Table to explain the similarities of certain elements and the trends in their properties, in terms of their electronic arrangements.

The structure of the Periodic Table

The Periodic Table has been written in many forms including pyramids and spirals. The one we include on the inside back cover of this book has the usual layout. Some areas of the Periodic Table are given names. These are shown in Figure 1.

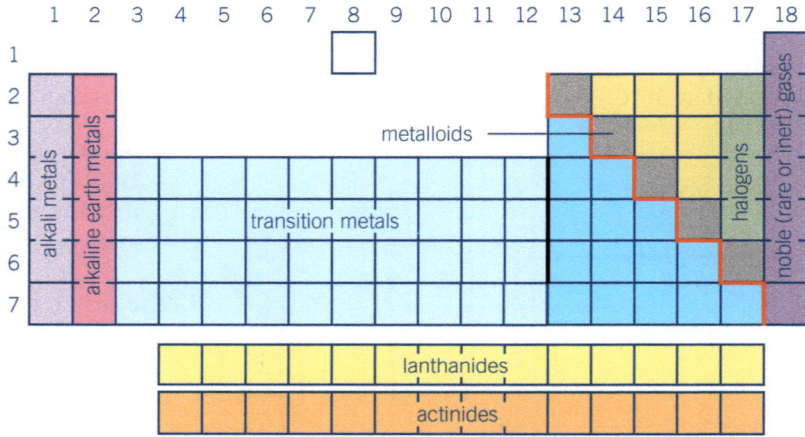

Figure 1 Named areas of the Periodic Table

Metals and non-metals

The red stepped line in Figure 1 (the 'staircase line') divides metals (on its left) from non-metals (on its right). Elements that touch this line, such as silicon, have a combination of metallic and non-metallic properties. They are called **metalloids** or **semi-metals**. Silicon, for example, is a non-metal but it looks quite shiny and conducts electricity, although not as well as a metal.

The s-, p-, d-, and f-blocks of the Periodic Table

Figure 2 shows the elements described in terms of their electronic arrangement.

Areas of the table are labelled s-block, p-block, d-block, and f-block.

- All the elements that have their highest energy electrons in s-orbitals are in the s-block, for example, sodium, Na ($1s^2 2s^2 2p^6 3s^1$).
- All the elements that have their highest energy electrons in p-orbitals are called p-block, for example, carbon, C ($1s^2 2s^2 2p^2$).
- All the elements that have their highest energy electrons in d-orbitals are called d-block, for example, iron, Fe ($1s^2 2s^2 2p^6 3s^2 3p^6 4s^2 3d^6$) and so on.

Strictly speaking, the transition metals and the d-block elements are not exactly the same. Scandium and zinc are not transition metals because they do not form any

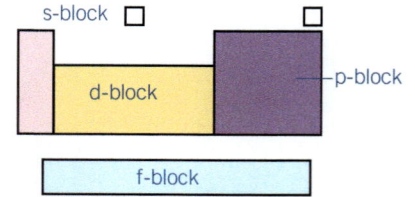

Figure 2 The s-block, p-block, d-block, and f-block areas of the Periodic Table

compounds in which they have partly filled d-orbitals, which is the characteristic of transition metals. You will learn more about the transition elements in Section 28.

Groups

A **group** is a vertical column of elements. The elements in the same group form a chemical 'family'—they have similar properties. Elements in the same group have the same number of electrons in the outer main level. The groups were traditionally numbered I–VII in Roman numerals plus zero for the noble gases, missing out the transition elements. It is now common to number them in ordinary numbers 1–18 including the transition metals.

Reactivity

In the s-block, elements (metals) get more reactive going down a group. To the right (non-metals), elements tend to get more reactive going up a group.

Transition elements are a block of relatively unreactive metals. This is where most of the useful metals are found.

Lanthanides are metals which are not often encountered. They all tend to form +3 ions in their compounds and have broadly similar reactivity.

Actinides are radioactive metals. Only thorium and uranium occur naturally in the Earth's crust in anything more than trace quantities.

Periods

Horizontal rows of elements in the Periodic Table are called **periods**. The periods are numbered starting from Period 1, which contains only hydrogen and helium. Period 2 contains the elements lithium to neon, and so on. There are trends in physical properties and chemical behaviour as you go across a period.

Periodicity of physical properties of elements in Period 3

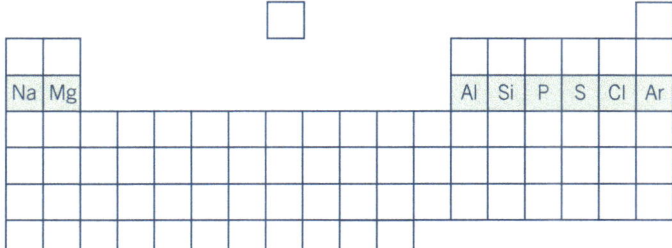

The Periodic Table reveals patterns in the properties of elements. For example, every time you go across a period you go from metals on the left to non-metals on the right. This is an example of **periodicity**. The word *periodic* means recurring regularly.

Periodicity is explained by the electron arrangements of the elements. In Period 3:

- Sodium, magnesium, and aluminium are metals. They have giant structures. They lose their outer electrons to form ionic compounds.
- Silicon has four electrons in its outer shell with which it forms four covalent bonds. The element has some metallic properties and is classed as a semi-metal.
- Phosphorus, sulfur, and chlorine are non-metals. They either accept electrons to form ionic compounds, or share their outer electrons to form covalent compounds.
- Argon is a noble gas—it has a full outer shell of electrons and is unreactive.

Atomic radii

Atomic radii tell us about the sizes of atoms. You cannot measure the radius of an isolated atom because there is no clear point at which the electron cloud density around it drops to zero. Instead half the distance between the centres of a pair of atoms is used, see Figure 4.

Even metals can form covalent molecules such as Na_2 in the gas phase. Since noble gases do not bond covalently with one another, they do not have covalent radii and so they are often left out of comparisons of atomic sizes.

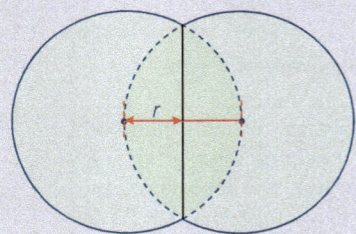

Figure 4 *Atomic radii are taken to be half the distance between the centres of a pair of atoms*

Exam tip

It is a common mistake to think that atoms increase in size as you cross a period. While the nuclei have more protons (and neutrons), the radius of the atom depends on the size of the electron shells.

Trends in atomic radii

The size of atoms is a periodic property. There are similar trends as you go across each period in the Periodic Table. In Period 3, all the outer electrons are in shell three and the nuclear charge increases across the period, drawing the electrons closer to the nucleus.

The size of the atoms decreases as we go across the period.

Atom	Na	Mg	Al	Si	P	S	Cl
Size of atom	2,8,1	2,8,2	2,8,3	2,8,4	2,8,5	2,8,6	2,8,7
Atomic (covalent) radius / nm	0.156	0.136	0.125	0.117	0.110	0.104	0.099
Nuclear charge	11+	12+	13+	14+	15+	16+	17+

Figure 3 *The sizes and electronic structures of the elements sodium to chlorine*

Trends in ionic radii

The ionic radii have a different pattern.

The radii of the metal ions are smaller than those of their parent atoms, because a whole outer shell has been lost. However, going across the period, the metal ions *decrease* in size.

When we reach the non-metal ions, there is an *increase* in ionic radius. The ionic radii of the non-metal ions are larger than their parent atoms, because they have gained electrons to form a full outer shell of electrons. Silicon does not form ions.

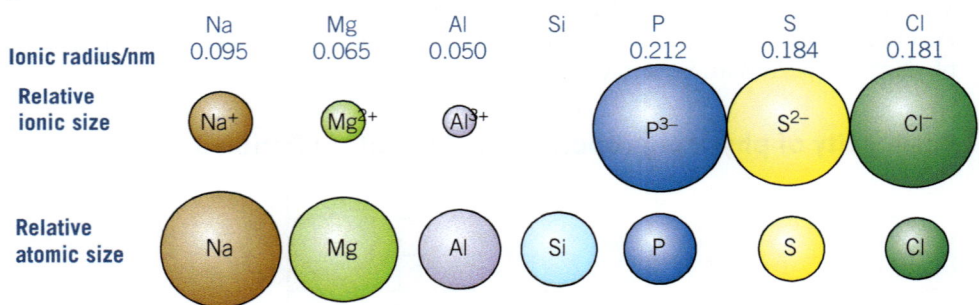

	Na	Mg	Al	Si	P	S	Cl
Ionic radius/nm	0.095	0.065	0.050		0.212	0.184	0.181

Figure 5 *Radii of the most stable ions of Na to Cl in Period 3*

Trends in melting and boiling points

Table 1 shows some trends across Period 3. (Similar trends are found in other periods.)

Table 1 *Some trends across Period 3*

Group	1	2	13	14	15	16	17	18
Element	sodium	magnesium	aluminium	silicon	phosphorus	sulfur	chlorine	argon
Electron arrangement	[Ne] 3s¹	[Ne] 3s²	[Ne] 3s² 3p¹	[Ne] 3s² 3p²	[Ne] 3s² 3p³	[Ne] 3s² 3p⁴	[Ne] 3s² 3p⁵	[Ne] 3s² 3p⁶
		s-block				p-block		
Classification		metals		semi-metal		non-metals		noble gas
Structure of element		giant metallic		giant molecular crystal (giant covalent)	molecular			atomic
					P_4	S_8	Cl_2	Ar
Melting point, T_m / K	371	922	933	1683	317 (white)	392 (monoclinic)	172	84
Boiling point, T_b / K	1156	1380	2740	2628	553 (white)	718	238	87

The trends in melting and boiling points are shown in Figure 6.

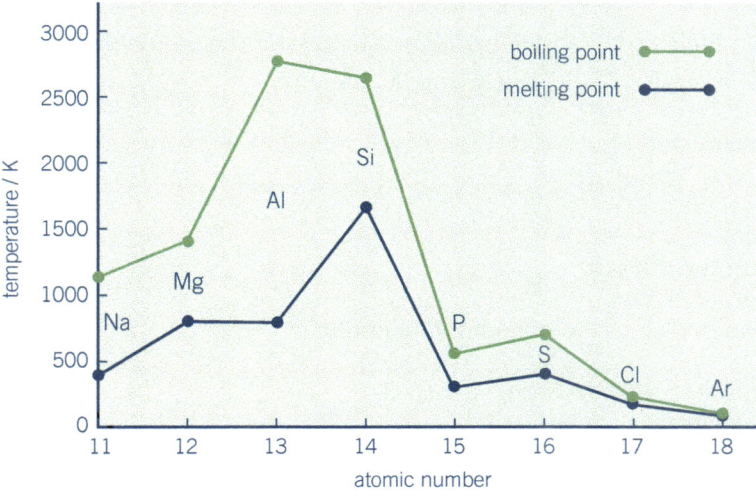

Figure 6 *Melting and boiling points of elements in Period 3*

There is a clear break in the middle of Figure 6 between elements with high melting points (on the left, with sodium, Na, in Group 1 as the exception) and those with low melting points (on the right). These trends are due to their structures.

- Giant structures (found on the left) tend to have high melting points and boiling points.
- Molecular or atomic structures (found on the right) tend to have low melting points and boiling points.

The melting points and boiling points of the metals increase from sodium to aluminium because of the strength of metallic bonding. As you go from left to right the charge on the ion increases so more electrons join the delocalised electron 'sea' that holds the giant metallic lattice together. So the melting point increases.

The melting points of the non-metals with molecular structures depend on the sizes of the van der Waals forces between the molecules. This in turn depends on the number of electrons in the molecule and how closely the molecules can pack together. As a result, the melting points of these non-metals are ordered: $S_8 > P_4 > Cl_2$. Silicon with its giant structure has a much higher melting point. Boiling points follow a similar pattern.

Trends in electrical conductivity

The metals (on the left) are all good conductors of electricity because of the sea of electrons which make up their structure. The non-metal elements on the right of the table do not conduct electricity because their bonding is covalent. Silicon in the middle is called a semi-conductor, like carbon above it in Period 2. It has no free electrons but some very limited conductivity when heated, because some of the electrons in the covalent bonds have enough energy to move.

> **Exam tip**
>
> The melting temperature of a substance is also the freezing temperature.

> **Exam tip**
>
> Remember that when a molecular substance melts, the covalent bonds remain intact but the van der Waals forces break.

> **Exam tip**
>
> You may need to look back at Chapter 4 to revise the properties of giant ionic, giant molecular and simple molecular structures.

Summary test 9.1

Use the Periodic Table on the inside back cover to answer the following:

1 From the elements, Br, Cl, Fe, K, Cs, and Sb, identify:
  a two elements
    i in the same period
    ii in the same group
    iii that are non-metals
  b one element
    i that is in the d-block
    ii that is in the s-block.

2 From the elements Tl, Ge, Xe, Sr, and W, identify:
  a a noble gas
  b the element described by Group 4, Period 4
  c an s-block element
  d a p-block element
  e a d-block element.

3 State what happens to the size of atoms as you go from left to right across a period Choose from 'increase', 'decrease', 'no change'.

4 a State what happens to the nuclear charge of the atoms as you go from left to right across a period.
  b Explain how this affects the size of the ions as you go across a period.

5 In Period 3 identify the element:
  a that most easily loses electron(s) when forming compounds
  b that most readily accepts electrons when forming compounds.

6 Identify the group where you find an element that exists as the following:
  a separate atoms
  b a giant molecular crystal (giant covalent).

7 A and B are both elements. Both conduct electricity—A well, B slightly. A melts at a low temperature, and B melts at a much higher temperature. Suggest the identity of A and B and explain how their bonding and structure account for their properties.

The reactions of the elements in Period 3 are all redox reactions. Every element starts with an oxidation number of zero. After it has reacted it has a positive or a negative oxidation number.

The reactivity of the metals *decreases* as we go across the period from left to right, because the number of electrons that need to be lost to form an ion increases.

The reactivity of the non-metals *increases* as we go across the period from left to right, because fewer electrons need to be gained to form a negative ion.

Reactions with water

Sodium and magnesium are the only metal elements in Period 3 that react with cold water. (Chlorine is the only non-metal that reacts with water.)

Sodium

The reaction of sodium with water is vigorous—the sodium floats on the surface of the water and fizzes rapidly, melting because of the heat energy released by the reaction. A strongly alkaline solution of sodium hydroxide is formed (pH 13–14). The oxidation number changes are shown here as small numbers above the following symbol equations:

$$\overset{0}{2Na(s)} + \overset{+1\ -2}{2H_2O(l)} \rightarrow \overset{+1\ -2+1}{2NaOH(aq)} + \overset{0}{H_2(g)}$$

Magnesium

The reaction of magnesium is very slow at room temperature—only a few bubbles of hydrogen are formed after some days. The resulting solution is less alkaline than in the case of sodium, because magnesium hydroxide is only sparingly soluble (pH around 10):

$$\overset{0}{Mg(s)} + \overset{+1\ -2}{2H_2O(l)} \rightarrow \overset{+2\ -2+1}{Mg(OH)_2(aq)} + \overset{0}{H_2(g)}$$

The reaction is much faster with heated magnesium and steam. It gives magnesium oxide and hydrogen:

$$\overset{0}{Mg(s)} + \overset{+1\ -2}{H_2O(g)} \rightarrow \overset{+2\ -2}{MgO(s)} + \overset{0}{H_2(g)}$$

All of these reactions are redox reactions, where the oxidation number of the metal *increases* and the oxidation number of some of the hydrogen atoms *decreases*.

The other elements in the period do not react easily with water or steam.

Reactions with oxygen

All the elements in Period 3 (except for argon) are relatively reactive. Their oxides can all be prepared by direct reaction of the element with oxygen. The reactions are exothermic.

Sodium

Sodium burns brightly in air (with a characteristic yellow flame) to form white sodium oxide:

$$\overset{0}{2Na(s)} + \overset{0}{\tfrac{1}{2}O_2(g)} \rightarrow \overset{+1\ -2}{Na_2O(s)}$$

Learning outcomes

On these pages you will learn to:

- describe the reactions of the Period 3 elements with oxygen, chlorine and water
- explain the variation of the oxidation number of the oxides and chlorides
- describe the acid/base behaviour of the oxides and their reactions with water
- describe the reactions of the chlorides with water
- interpret the trends in chemical reactions in terms of bonding and electronegativity
- suggest the type of chemical bonding in chlorides and oxides from their properties

Exam tip

The sodium oxide formed may have a yellowish appearance due to the production of some sodium peroxide, Na_2O_2.

Figure 7 *Magnesium burning in oxygen from the air*

Figure 8 *Aluminium burning in oxygen from the air. Powdered aluminium is being sprinkled into the flame*

Exam tip

The empirical formula of phosphorus pentoxide is P_2O_5. In the gas phase it forms molecules of P_4O_{10} and is sometimes referred to as phosphorus(V) oxide.

Figure 9 *Sulfur burning in oxygen*

Magnesium

A strip of magnesium ribbon burns in air with a bright white flame. The white powder that is produced is magnesium oxide. If burning magnesium is lowered into a gas jar of oxygen the flame is even more intense (Figure 7).

$$\text{magnesium} \quad + \quad \text{oxygen} \quad \rightarrow \quad \text{magnesium oxide}$$
$$0 \qquad\qquad\qquad 0 \qquad\qquad\qquad +2\ -2$$
$$2Mg(s) \quad + \quad O_2(g) \quad \rightarrow \quad 2MgO(s)$$

The oxidation numbers show how magnesium has been oxidised (its oxidation number has increased) and oxygen has been reduced (its oxidation number has decreased).

Aluminium

When aluminium powder is heated and then lowered into a gas jar of oxygen, it burns brightly to give aluminium oxide—a white powder. Aluminium powder also burns brightly in air (Figure 8).

$$\text{aluminium} \quad + \quad \text{oxygen} \quad \rightarrow \quad \text{aluminium oxide}$$
$$0 \qquad\qquad\qquad 0 \qquad\qquad\qquad +3\ -2$$
$$4Al(s) \quad + \quad 3O_2(g) \quad \rightarrow \quad 2Al_2O_3(s)$$

Aluminium is a reactive metal, but it is always coated with a strongly bonded surface layer of oxide – this protects it from further reaction. So, aluminium appears to be an unreactive metal and is used for many everyday purposes—saucepans, garage doors, window frames, and so on. Even if the surface is scratched, the exposed aluminium reacts rapidly with the air and seals off the surface.

Silicon

Silicon will also form the oxide if it is heated strongly in oxygen:

$$0 \qquad 0 \qquad +4\ -2$$
$$Si(s) + O_2(g) \rightarrow SiO_2(s)$$

Phosphorus

Red phosphorus must be heated before it will react with oxygen. White phosphorus spontaneously ignites in air and the white smoke of phosphorus pentoxide is given off. Red and white phosphorus are allotropes of phosphorus—the same element with the atoms arranged differently.

$$0 \qquad 0 \qquad +5\ -2$$
$$4P(s) + 5O_2(g) \rightarrow P_4O_{10}(s)$$

If the supply of oxygen is limited, phosphorus trioxide, P_2O_3, is also formed.

Sulfur

When sulfur powder is heated and lowered into a gas jar of oxygen, it burns with a blue flame to form the colourless gas sulfur dioxide (Figure 9). A little sulfur trioxide, SO_3, also forms, in which the oxidation number of the sulfur atom is +6.

$$\text{sulfur} + \text{oxygen} \rightarrow \text{sulfur dioxide}$$
$$0 \qquad 0 \qquad +4\ -2$$
$$S(s) + O_2(g) \rightarrow SO_2(g)$$

In all these redox reactions, the oxidation number of the Period 3 element increases and the oxidation number of the oxygen decreases (from 0 to −2 in each case). The oxidation number changes are shown as small numbers above the symbol equations above. The oxidation number of the Period 3 element in the oxide *increases* as you move from left to right across the period.

Reactions with chlorine

The metals sodium and magnesium both react vigorously when heated and plunged into chlorine gas, to give white fumes of the ionic chloride (see Figure 10). As before, oxidation numbers are given above each atom in the equations below.

$$\overset{0}{2Na(s)} + \overset{0}{Cl_2(g)} \rightarrow \overset{+1\ -1}{2NaCl(s)}$$

$$\overset{0}{Mg(s)} + \overset{0}{Cl_2(g)} \rightarrow \overset{+2\ -1}{MgCl_2(s)}$$

Aluminium reacts in a similar way. However, the chloride (which has the empirical formula $AlCl_3$) has the molecular formula Al_2Cl_6, which is covalently bonded (see Section 3.4) in the gas phase.

$$\overset{0}{4Al(s)} + \overset{0}{6Cl_2(g)} \rightarrow \overset{+3\ -1}{2Al_2Cl_6(s)}$$

Solid aluminium chloride forms a lattice structure, where layers of Al^{3+} ions alternate with layers of Cl^- ions.

The non-metals silicon, phosphorus and sulfur also combine vigorously with chlorine when heated, forming chlorides that are simple molecules:

$$\overset{0}{Si(s)} + \overset{0}{6Cl_2(g)} \rightarrow \overset{+4\ -1}{SiCl_4(l)}$$

$$\overset{0}{2P(s)} + \overset{0}{3Cl_2(g)} \rightarrow \overset{+3\ -1}{2PCl_3(l)} \quad \text{and} \quad \overset{0}{2P(s)} + \overset{0}{5Cl_2(g)} \rightarrow \overset{+5\ -1}{2PCl_5(s)}$$

$$\overset{0}{S(s)} + \overset{0}{Cl_2(g)} \rightarrow \overset{+2\ -1}{SCl_2(l)} \text{ (other chlorides of sulfur may be formed: } SCl_4 \text{ and } S_2Cl_2)$$

Figure 10 *Magnesium burning in chlorine*

The oxides and chlorides of Period 3

In all of their oxides and chlorides, Period 3 elements have a positive oxidation number. This is because oxygen and chlorine are the more electronegative elements in each compound. The oxidation number is the same as the number of outer electrons used in bonding in each compound.

Oxidation numbers of the oxides and chlorides

For both the chlorides and oxides of the Period 3 metal compounds (Na, Mg and Al), the oxidation number is the same as the charge on the ion. All the outer electrons are lost to form Na^+, Mg^{2+} and Al^{3+} ions, with the electron arrangement $1s^2\ 2s^2\ 2p^6$. From silicon onwards, the bonding in the compounds is covalent. Electrons are shared with the non-metal atoms to form a full outer 3-shell ($1s^2\ 2s^2\ 2p^6\ 3s^2\ 3p^6$). Notice that the 3-shell can take more than eight electrons and that this is the case in PCl_5, SO_2, and SO_3.

Reaction of the oxides with water

The overall trend is the formation of basic oxides on the left of Period 3 and acidic oxides on the right.

Aluminium oxide is considered to be **amphoteric**. This means that it can behave as a base or an acid ie it will react with both acids and bases:

As a base: $Al_2O_3(s) + 6HCl(aq) \rightarrow 2AlCl_3(aq) + 3H_2O(l)$

As a acid: $Al_2O_3(s) + 2NaOH(aq) + 3H_2O(l) \rightarrow 2NaAl(OH)_4$

$NaAl(OH)_4$ is called sodium aluminate.

> **Exam tip**
>
> The sum of the oxidation numbers in Al_2O_3 is zero, as it is in all compounds without a charge:
>
> $(2 \times 3) + (3 \times -2) = 0$

Basic oxides

Sodium and magnesium oxides are both bases.

Sodium oxide reacts with water to give sodium hydroxide solution – a strongly alkaline solution:

$$Na_2O(s) + H_2O(l) \rightarrow 2Na^+(aq) + 2OH^-(aq) \qquad \text{pH of solution} \approx 14$$

Magnesium oxide reacts with water to give magnesium hydroxide, which is sparingly soluble in water and produces a slightly alkaline solution:

$$MgO(s) + H_2O(l) \rightarrow Mg(OH)_2(s) \rightleftharpoons Mg^{2+}(aq) + 2OH^-(aq) \qquad \text{pH of solution} \approx 9$$

Aluminium hydroxide $(Al(OH)_3)$ is also amphoteric, as shown by its reaction with hydrochloric acid:

$$Al(OH)_3 + 3HCl \rightarrow AlCl_3 + 3H_2O$$

and also by its reaction with sodium hydroxide:

$$Al(OH)_3 + NaOH \rightarrow NaAl(OH)_4$$

Insoluble oxides

Aluminium oxide is insoluble in water.
Silicon dioxide is insoluble in water.

Acidic oxides

Non-metals on the right of the Periodic Table typically form acidic oxides. For example, phosphorus pentoxide reacts quite violently with water to produce an acidic solution of phosphoric(V) acid. This ionises, so the solution is acidic.

$$P_4O_{10}(s) + 6H_2O(l) \rightarrow 4H_3PO_4 (aq)$$

$H_3PO_4(aq)$ ionises in stages, the first being:

$$H_3PO_4(aq) \rightleftharpoons H^+ (aq) + H_2PO_4^-(aq)$$

Sulfur dioxide is fairly soluble in water and reacts with it to give an acidic solution of sulfuric(IV) acid (sulfurous acid). This partially dissociates producing H^+ ions, which cause the acidity of the solution:

$$SO_2(g) + H_2O(l) \rightarrow H_2SO_3(aq)$$

$$H_2SO_3(aq) \rightleftharpoons H^+(aq) + HSO_3^-(aq)$$

Sulfur trioxide reacts violently with water to produce sulfuric acid (sulfuric(VI) acid):

$$SO_3(g) + H_2O(l) \rightarrow H_2SO_4(aq) \rightarrow H^+(aq) + HSO_4^-(aq)$$

The overall pattern is:

- Metal oxides (on the left of the period) form alkaline solutions in water.
- Non-metal oxides (on the right of the period) form acidic solutions in water.
- Oxides in the middle of the period do not react.

Table 2 summarises these reactions.

Table 2 The oxides in water

Oxide	Bonding	Ions present after reaction with water	Acidity/alkalinity	Approx. pH (Actual values depend on concentration)
Na_2O	ionic	$Na^+(aq)$, $OH^-(aq)$	strongly alkaline	13–14
MgO	ionic	$Mg^{2+}(aq)$, $OH^-(aq)$	somewhat alkaline	10
Al_2O_3	covalent/ionic	insoluble, no reaction	–	7
SiO_2	covalent	insoluble, no reaction	–	7
P_4O_{10}	covalent	$H^+(aq) + H_2PO_4^-(aq)$	fairly strong acid	1–2
SO_2	covalent	$H^+(aq) + HSO_3^-(aq)$	weak acid	2–3
SO_3	covalent	$H^+(aq) + HSO_4^-(aq)$	strong acid	0–1

The behaviour of the oxides with water can be understood if you look at their bonding and structure (Table 3).

Table 3 *The trends in the bonding and structure of some of the oxides in Period 3*

	Na_2O	MgO	Al_2O_3	SiO_2	P_4O_{10}	SO_3	SO_2
T_m / K	1548	3125	2345	1883	573	290	200
Bonding	ionic	ionic	ionic/covalent	covalent	covalent	covalent	covalent
Structure	giant ionic	giant ionic	giant ionic	giant molecular crystal (giant covalent)	molecular	molecular	molecular

- Sodium and magnesium oxides, to the left of the Periodic Table, are composed of ions.
- Sodium oxide contains the oxide ion, O^{2-}, which is a very strong base (it strongly attracts protons) and so readily reacts with water to produce hydroxide ions. This forms a strongly alkaline solution.
- Magnesium oxide also contains oxide ions. However, its reaction with water produces a less alkaline solution than sodium oxide because it is less soluble than sodium oxide.
- Aluminium oxide is ionic but the bonding is too strong for the ions to be separated, partly because of the additional covalent bonding it has.
- Silicon dioxide is a giant molecular crystal (giant covalent) and water will not affect this type of structure.
- Phosphorus oxides and sulfur oxides are covalent molecules and react with water to form acid solutions.

General trend
Solutions of the oxides of the elements go from alkaline to acidic across the period.

This trend can be explained as follows.

Imagine each oxide reacts in water to form a compound E–O–H, where E represents any element. Whether this compound is an acid or an alkali depends on which bond E–O or O–H breaks.

If E–O breaks

$$E–O–H + aq \rightarrow E^+(aq) + OH^-(aq)$$

then the resulting solution is alkaline.

If O–H breaks

$$E–O–H + aq \rightarrow E–O^-(aq) + H^+(aq)$$

then the resulting solution is acidic.

Which bond breaks depends on the electronegativity of E. Electropositive elements (the metals) favour the first type of behaviour, where E^+ is formed. Electronegative elements (the non-metals) favour the second, where E is part of a negative ion.

Reaction of the chlorides with water
The ionic chlorides simply dissolve in water to form ionic solutions.

The non-metal chlorides all react with water (hydrolyse) to form acidic solutions containing H^+ ions and Cl^- ions:

$$AlCl_3(s) + 3H_2O(l) \rightarrow Al(OH)_3(s) + 3H^+(aq) + 3Cl^-(aq)$$

$$SiCl_4(l) + 2H_2O(l) \rightarrow SiO_2(s) + 4H^+(aq) + 4Cl^-(aq)$$

$$PCl_5(s) + 4H_2O(l) \rightarrow H_3PO_4(aq) + 5H^+(aq) + 5Cl^-(aq)$$

Table 4 summarises the reactions.

Table 4 *The chlorides in water*

Chloride	Bonding	Ions present after reaction with water	Acidity / alkalinity	Approx pH (actual values depend on concentration)
NaCl	ionic	$Na^+(aq)$, $Cl^-(aq)$	neutral	7
$MgCl_2$	ionic	$Mg^{2+}(aq)$, $Cl^-(aq)$	neutral	6.5–7
$AlCl_3$	covalent	$H^+(aq)$, $Cl^-(aq)$	acidic	2–3
$SiCl_4$	covalent	$H^+(aq)$, $Cl^-(aq)$	acidic	1
PCl_5	covalent	$H^+(aq)$, $Cl^-(aq)$	acidic	1
SCl_2	covalent	$H^+(aq)$, $Cl^-(aq)$	acidic	~1

Magnesium chloride hydrolyses to a small extent, so the pH of the resulting solution is slightly acidic.

Structure and bonding in oxides and chlorides

The ideas developed in Section 4.2 allow us to deduce the structure and bonding of the oxides and chlorides of the Period 3 elements. Those with low melting (and boiling) points have the structure of simple molecules and must therefore have covalent bonding. Those with high melting (and boiling) points must have giant structures. They could have either ionic or covalent bonding. The only way to be sure is to look at their conductivity when molten ionic compounds will conduct electricity and covalent compounds will not. Their conductivity in solution does not help, as many of the compounds react with water to form ions rather than just simply dissolving.

Summary test 9.2

1 Metals are shiny, conduct electricity, and react with acids to give hydrogen if they are reactive. Give three more properties not mentioned here.

2 Non-metals do not conduct electricity. Give two more properties typical of non-metals.

3 a State the oxidation number of sodium in all its compounds.
   b Deduce the oxidation number of oxygen in sodium peroxide, Na_2O_2. State what is unusual about this.
   c Demonstrate that the sum of the oxidation numbers in magnesium hydroxide is zero.

4 Deduce the oxidation number of sulfur in sulfur trioxide.

5 a Give an equation for the reaction of sodium oxide with water.
   b i State the oxidation number of sodium before and after the reaction.
     ii State whether the sodium been oxidised, reduced, or neither.

6 a State the ion that is responsible for the alkalinity of the solutions formed when sodium oxide and magnesium oxide react with water.
   b Give the range of pH values that represents an alkaline solution.

7 Phosphorus forms another oxide, P_4O_6.
   a Predict whether you expect it to react with water to form a neutral, acidic, or alkaline solution.
   b Explain your answer.
   c Give an equation for its reaction with water.

Chemical periodicity of other elements

Sections 9.1 and 9.2 have described and given examples of the periodicity of the properties of the elements in Period 3. Within a group of elements in the Periodic Table there are clear similarities. For example the Group 1 elements are all soft, low density metals that react readily with water, oxygen and chlorine to form ions with a single positive charge. Within this group there is a clear trend of increasing reactivity as we descend the group.

The Group 17 elements are all volatile non-metals which form compounds with metals where their ions have a single negative charge. Within this group there is a clear trend of decreasing reactivity as we descend the group.

Across periods, there is a change from metals on the left to non-metals on the right.

By combining your knowledge of both trends and similarities within the Periodic Table, it is possible to make predictions about the likely properties and reactions of other elements if you know their position in the Table. This was one of the successes of the work done by Dmitri Mendeleev, a Russian chemist.

It is possible to work out the likely properties of an element in the Periodic Table from its position.

 Mendeleev and the Periodic Table

The Periodic Table is one of the great achievements of science. It gives a logical structure to the properties of the 118 chemical elements now known. These are arranged in order of proton number (atomic number) in such a way as to show up similarities and trends in their properties, and to reflect their underlying electronic structure. Dmitri Mendeleev is credited as the 'father' of the Periodic Table, but when he first proposed it, he was aware of only around half this number. Furthermore, there was no knowledge of atomic structure. So he arranged the then known elements in order of relative atomic mass. Mendeleev's insight was to realise that there were elements still to be discovered and to leave gaps in his table to accommodate them.

Mendeleev used these trends and similarities to predict the properties of then-undiscovered elements. For many of the properties he used the data for the elements on either side of it in his Periodic Table. In this way, he predicted the properties of an element in Group 3 between aluminium and indium, which he called eka-aluminium. This element is now called gallium. His predictions for eka-aluminium are shown in Table 5.

Table 5 *Mendeleev's predictions of the properties of Gallium*

Property	Eka-aluminium	Gallium
atomic mass	68	69.723
formula of oxide	Ea_2O_3	Ga_2O_3
formula of chloride	Ea_2Cl_6 (volatile)	Ga_2Cl_6 (volatile)
density / g cm^{-3}	6.0	5.91
melting point / °C	low	29.76

Exam tip

The general pattern in the Periodic Table is for similarities within groups of elements, and changes in behaviour within periods of elements.

Summary test 9.3

1 a Predict the physical properties of the element caesium, Cs, from its position in the Periodic Table.
  b Give the likely formulae of its chloride, oxide and hydroxide.
2 A brown liquid element combines with strontium to form a salt. Predict where this element is likely to be found in the Periodic Table. Explain your answer.

9 Exam-style questions

(Launch additional digital resources for the chapter)

1 The table shows some data for the elements of Period 3:

	Na	Mg	Al	Si	P	S	Cl
Atomic radius / nm	0.186	0.160	0.143	0.117	0.110	0.104	0.099
Ionic radius / nm	0.095	0.065	0.05	–	–	0.184	0.181
First ionisation energy / kJ mol^{-1}	+494	+736	+577	+786	+1060	+1000	+1260

a Describe and explain the trend in atomic radii across the period from Na to Cl. *(4 marks)*

b Explain why the ionic radii of Cl$^-$ and S^{2-} are greater than the atomic radii. *(2 marks)*

c Describe the general trend in first ionisation energies across Period 3, giving reasons for your answer. *(4 marks)*

d i Give an equation for the first ionisation energy of aluminium. *(1 mark)*
   ii Explain why the first ionisation energy of aluminium deviates from the general trend described in part c. *(2 marks)*

2 The table below shows various properties of the chlorides of the elements in Period 2:

Formula of chloride	LiCl	BeCl$_2$	BCl$_3$	CCl$_4$	NCl$_3$	Cl$_2$O	CIF
State at 20°C	solid	solid	gas	liquid	liquid	gas	gas
Boiling point (°C)	1350	487	12	77	71	2	–101
Conductivity of aqueous solution	good	very poor	nil	nil	nil	nil	Nil
Structure	giant structures		simple molecular structures				

a Use your knowledge of structure and bonding to explain the differences in state and boiling point of the chlorides across period 2. *(4 marks)*

b Identify the type of bonding present in a molecule of lithium chloride. *(1 mark)*

c Explain why LiCl(s) does not conduct electricity, however LiCl(aq) is a good conductor of electricity. *(2 marks)*

d Describe and explain the trend in electronegativity across Period 2. *(3 marks)*

3 a Describe the structure and bonding of the oxides of the elements in Period 3 from Na to S. *(4 marks)*

b Describe the reaction of the oxides in part a with:
   i water *(3 marks)*
   ii dilute acid *(2 marks)*
   iii alkali *(2 marks)*

c Magnesium chloride is a crystalline white solid, whereas silicon tetrachloride is a volatile liquid. Use your knowledge of structure and bonding to explain these differences in melting point and boiling point. *(4 marks)*

4 The properties of elements Q and R are given below:

Element Q	Element R
i) Is soft and malleable ii) Floats on water iii) Has a melting point of less than 100°C iv) Forms an ionic hydride of formula QH	i) Has a density greater than 7 g cm^{-3} ii) Has a melting point greater than 2000K iii) Forms oxides of formulae RO, R$_2$O$_3$ and RO$_3$ iv) Forms compounds which are green, orange and violet

a State the deductions that you can make about Q and R from each piece of evidence. *(6 marks)*

b Predict the blocks of the Periodic Table in which elements Q and R are located. *(2 marks)*

c Suggest the identity of elements Q and R. *(2 marks)*

5 Consider the elements Na to Cl in Period 3.

a Which of these elements:
   i form cations? *(1 mark)*
   ii form a chloride of empirical formula, XCl$_3$? *(1 mark)*
   iii react together to form a compound of formula, XY? *(1 mark)*
   iv exist as diatomic molecules at room temperature? *(1 mark)*

b Describe the trend in boiling points across Period 3 for:
   i elements Na to Cl *(3 marks)*
   ii chlorides of Na to Cl. *(2 marks)*

6 The table below shows the melting point and electrical conductivity of five substances:

	Melting point / K	Electrical conductivity in solid state	Electrical conductivity in molten state
magnesium oxide, MgO	3173	poor	good
sodium chloride, NaCl	1081	poor	good
magnesium, Mg	923	good	good
carbon dioxide, CO_2	217	poor	poor
silicon(IV) oxide, SiO_2	1883	poor	poor

a Explain the difference in electrical conductivity between MgO(l) and MgO(s). *(2 marks)*

b Explain why the melting point of MgO is considerably higher than that of NaCl. *(2 marks)*

c Use your knowledge of structure and bonding to explain why the electrical conductivity of magnesium is good in both solid and liquid states. *(1 mark)*

d Carbon and silicon are both Group 4 elements. Explain why their oxides have very different melting points. *(4 marks)*

7 a State and explain the general trend in first ionisation energies of the elements in Period 3 from sodium to argon. *(4 marks)*

b Give the identity of an element that deviates from the general trend in first ionisation energy across Period 3. Explain why this occurs. *(3 marks)*

c Describe the general trend in first ionisation energies across Period 3, giving reasons for your answer. *(4 marks)*

d The successive ionisation energies of an element X are given in the table below:

Ionisation number	1	2	3	4	5	6	7	8
Ionisation energy / kJ mol^{-1}	786	1580	3230	4360	16090	19800	23780	29290

Deduce, with reasons, the identity of element X using the data from the table above to support your response. *(3 marks)*

e Give the identity of the Period 3 element that has the highest melting point. Give reasons for your answer using your knowledge of structure and bonding. *(4 marks)*

8 The oxides of Period 3 exhibit a range of different properties.

a Construct a balanced chemical equation for the reaction that occurs when sodium metal is heated in oxygen. *(1 mark)*

b Suggest two observations that are made when carrying out the reaction described in part **a**. *(2 marks)*

c Construct a balanced chemical equation, and suggest one observation made during the formation of phosphorus(V) oxide from heating phosphorus in the presence of oxygen. *(2 marks)*

d The melting points of the oxides of sodium, magnesium, silicon(IV) and phosphorus(V) are given in the table below:

Compound	Melting point / K
sodium oxide	1405
magnesium oxide	3125
silicon(IV) oxide	1986
phosphorus(v) oxide	613

i Explain the differences in melting point between the oxides of sodium and magnesium. *(2 marks)*

ii Explain the differences in melting point between the oxides of silicon and phosphorus. *(3 marks)*

141

10 Group 2

Similarities and trends in the properties of the Group 2 metals, magnesium to barium, and their compounds

Learning outcomes

On these pages you will learn to:

- describe the reactions of the Group 2 elements with oxygen, water and dilute acids
- describe the reactions of the oxides, hydroxides and carbonates with water and dilute acids
- describe the thermal decomposition of the nitrates and carbonates
- interpret and make predictions from patterns in physical and chemical properties
- describe patterns in solubility of the hydroxides and sulfates

The elements in Group 2 are sometimes called the **alkaline earth metals**. This is because their oxides and hydroxides are alkaline. Like Group 1, they are s-block elements. They are similar in many ways to Group 1 but they are less reactive. Beryllium is not typical of the group and is not considered here.

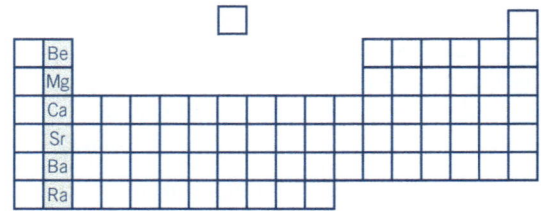

Exam tip

The untypical behaviour of beryllium is due to the very small size of the Be^{2+} ion, which has only a single shell of electrons.

The physical properties of the Group 2 elements, magnesium to barium

A summary of some of the physical properties of the elements from magnesium to barium is given in Table 1 below. Trends in properties are shown by the arrows, which show the direction of increase. In all their reactions, the metals get more reactive going down the group.

The chemical reactions of the Group 2 elements, magnesium to barium

Exam tip

The only oxidation states adopted by Group 2 elements are zero and +2.

Oxidation is *loss of electrons* so in all their reactions the Group 2 metals are oxidised. The metals go from oxidation state 0 to oxidation state +2. These are redox reactions. The metals are all acting as reducing agents.

Reaction with water
With water you see a trend in reactivity—the metals get more reactive going down the group. These are also redox reactions.

Table 1 *The physical properties of Group 2, magnesium to barium*

	Atomic number Z	Electron arrangement	Metallic radius / nm	First + second IEs / kJ mol^{-1}	Melting point T_m / K	Boiling point T_b / K	Density ρ / g cm^{-3}
magnesium, Mg	12	[Ne] 3s^2	0.160	736 + 1450 = 2186	922	1380	1.74
calcium, Ca	20	[Ar] 4s^2	0.197	590 + 1150 = 1740	1112	1757	1.54
strontium, Sr	38	[Kr] 5s^2	0.215	550 + 1064 = 1614	1042	1657	2.60
barium, Ba	56	[Xe] 6s^2	0.224	503 + 965 = 1468	998	1913	3.51

The basic reaction is as follows, where M is any Group 2 metal:

$$\overset{0}{M(s)} + \overset{+1-2}{2H_2O(l)} \rightarrow \overset{+2-2+1}{M(OH)_2(aq)} + \overset{0}{H_2(g)}$$

Magnesium reacts very slowly with cold water but rapidly with steam to form an alkaline oxide and hydrogen.

$$Mg(s) + H_2O(g) \rightarrow MgO(s) + H_2(g)$$

Calcium reacts in the same way but more vigorously, even with cold water. Strontium and barium react more vigorously still.

Reaction with oxygen

All the metals react vigorously when heated in air or in oxygen

$$2M(s) + O_2(g) \rightarrow 2MO(s)$$

to give oxides containing M^{2+} ions.

Strontium and barium also form peroxides containing the O_2^{2-} ion, eg

$$Sr(s) + O_2 \rightarrow SrO_2(s)$$

Reaction with dilute acids

All the metals react with dilute hydrochloric acid to give a chloride salt and hydrogen. The reaction becomes more vigorous as we descend the group because the outer electrons are further from the nucleus and therefore more easily lost. For example:

$$Mg(s) + 2HCl(aq) \rightarrow MgCl_2(aq) + H_2(g)$$

With sulfuric acid, sulfates are formed. For example:

$$Mg(s) + H_2SO_4(aq) \rightarrow MgSO_4(s) + H_2(g)$$

However, as we descend the group from calcium, the reaction soon stops, because the sulfates become increasingly insoluble and coat the metal, protecting it from the acid.

Reactions of the Group 2 oxides and hydroxides with water and dilute acids

Reaction with water

In general the oxides react with water to form a hydroxide:

$$MO(s) + H_2O(l) \rightarrow M(OH)_2(s)$$

This reaction is slow for magnesium oxide but vigorous for the rest of the Group 2 metal oxides.

The hydroxides then dissolve in water to form alkaline solutions:

$$M(OH)_2(s) + aq \rightarrow M^{2+}(aq) + 2OH^-(aq)$$

The solubility of the hydroxides *increases* on descending the group. Magnesium hydroxide is only slightly soluble and forms a weakly alkaline solution.

Reaction with dilute acids

The oxides and hydroxides are neutralised by dilute acids, forming salts, eg

$$MgO(s) + 2HCl(aq) \rightarrow MgCl_2(aq) + H_2O(l)$$

$$MgO(s) + H_2SO_4(aq) \rightarrow MgSO_4(aq) + H_2O(l)$$

Extension

Uses of Group 2 hydroxides

Magnesium hydroxide can be used in indigestion remedies to neutralise excess stomach acid, which causes heartburn, indigestion, and gas.

Calcium hydroxide is sometimes called slaked lime and is used to treat acidic soil. Most plants have an optimum level of acidity or alkalinity in which they thrive. For example, grass prefers a pH of around 6. If the soil has a pH much below this, grass will not grow as well as it could. Crops such as wheat, corn, oats, and barley prefer soil that is nearly neutral.

As before, with the oxides of calcium, strontium and barium, the sulfates that form initially are insoluble, so the reaction will slow down and stop.

$$Mg(OH)_2(s) + 2HCl(aq) \rightarrow MgCl_2(aq) + 2H_2O(l)$$

$$Mg(OH)_2(s) + H_2SO_4(aq) \rightarrow MgSO_4(aq) + 2H_2O(l)$$

Reactions of the Group 2 carbonates with water and with dilute acids

Reaction with water
The carbonates of all the Group 2 metals are insoluble in water.

Reaction with dilute acids
The carbonates of all the Group 2 metals react with dilute hydrochloric acid to form carbon dioxide, water and the chloride salt, eg

$$CaCO_3(s) + 2HCl(aq) \rightarrow CaCl_2(aq) + H_2O(l) + CO_2(g)$$

The reaction of magnesium carbonate with sulfuric acid produces a soluble sulfate:

$$MgCO_3(s) + H_2SO_4(aq) \rightarrow MgSO_4(aq) + H_2O(l) + CO_2(g)$$

However, the reaction of calcium carbonate with sulfuric acid produces insoluble calcium sulfate. So the reaction slows down and stops.

The sulfates of barium and strontium are also insoluble, so the same thing happens.

The thermal decomposition of the Group 2 carbonates and nitrates

The carbonates decompose to the oxide and carbon dioxide, eg

$$MgCO_3(s) \rightarrow MgO(s) + CO_2(g)$$

As we descend the group, the carbonates are more stable. So higher temperatures are required to bring about decomposition.

The nitrates all decompose to form brown nitrogen dioxide gas and oxygen, eg

$$2Ca(NO_3)_2(s) \rightarrow 2CaO(s) + 4NO_2(g) + O_2(g)$$

Again higher temperatures are required to bring about decomposition as we descend the group.

These trends are related to the size of the metal ion and are explained in Section 27.1.

The solubilities of the Group 2 metal hydroxides and sulfates

There are clear trends in the solubilities of the hydroxides and the sulfates.

Hydroxides
As we go down the group the hydroxides become more soluble. They are all white solids.

- Magnesium hydroxide, $Mg(OH)_2$, is almost insoluble.
- Calcium hydroxide, $Ca(OH)_2$, is sparingly soluble. A solution of $Ca(OH)_2$ is used as limewater.
- Strontium hydroxide, $Sr(OH)_2$, is more soluble.
- Barium hydroxide, $Ba(OH)_2$, dissolves to produce a strongly alkaline solution:

$$Ba(OH)_2(s) + aq \rightarrow Ba^{2+}(aq) + 2OH^-(aq)$$

Sulfates

The solubility trend in the sulfates is exactly the opposite to the trend in the hydroxides. The Group 2 sulfates become *less soluble* going down the group. Barium sulfate is virtually insoluble. These trends in solubility are also explained in Section 27.1.

The insolubility of barium sulfate is used in a simple test for sulfate ions in solution. The solution is first acidified with nitric or hydrochloric acid. Then barium chloride solution is added. If sulfate ions are present, a white precipitate of barium sulfate is formed:

$$Ba^{2+}(aq) + SO_4^{2-}(aq) \rightarrow BaSO_4(s)$$

(Acid is added to remove any carbonate ions (as carbon dioxide). This is because barium carbonate is also a white insoluble solid, and would be indistinguishable from barium sulfate.)

Summary test 10.1

1 a State the oxidation number of all Group 2 elements in their compounds.
   b Explain your answer.
2 Explain why it becomes easier to form +2 ions going down Group 2.
3 Explain why this is a redox reaction:

$$Ca + Cl_2 \rightarrow CaCl_2$$

4 Give the equation for the reaction of calcium with water. Include the oxidation state of each element.
5 Predict how the reaction of strontium with water would compare with those of the following. Explain your answers.
   a calcium
   b barium
6 Radium is below strontium in Group 2. Predict how the solubilities of the following compounds would compare with the other members of the group. Explain your answers.
   a radium hydroxide
   b radium sulfate

There are exam-style questions to test your knowledge of the material in this chapter at the end of Chapter 12.

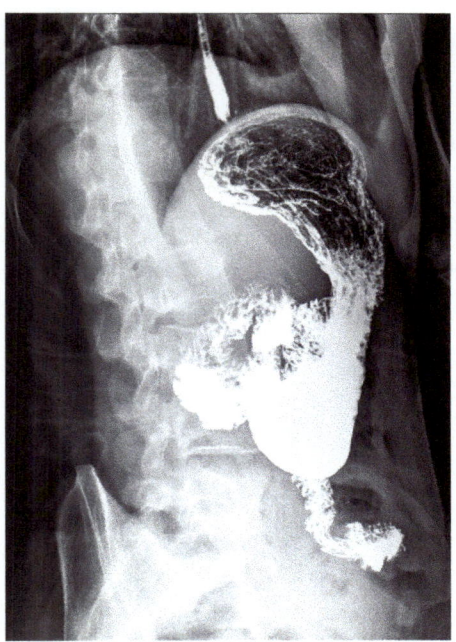

Figure 1 *A barium meal highlights a patient's esophageal cancer in this X-ray*

Physical properties of the Group 17 elements

Learning outcomes

On these pages you will learn to:

- describe and explain the physical properties of the halogens
- describe and explain the trend in bond strength

Group 17, the halogens, on the right-hand side of the Periodic Table, is made up of non-metals. As elements they exist as diatomic molecules, F_2, Cl_2, Br_2, and I_2, called the halogens. (Astatine is rare and radioactive.)

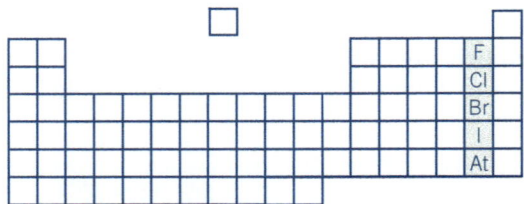

The halogens vary in appearance, as shown in Figure 1. At room temperature, fluorine is a pale yellow gas, chlorine a greenish gas, bromine a red-brown liquid, and iodine a black solid – they get darker and denser going down the group.

All the halogens have a characteristic 'swimming-pool' smell.

Figure 1 *Fluorine, chlorine, bromine and iodine in their gaseous states*

The bond strengths of the halogens

The general pattern for the bond energies of the halogens is for them to get smaller as we descend the group. This is because the shared electrons in the halogen–halogen bond get further from the nuclei of the atoms (ie the atomic radii increase) while 'feeling' the same nuclear charge (after allowing for shielding of the inner shells).

Table 1 *Bond energies for fluorine, chlorine, bromine, and iodine*

Bond	Bond energy / kJ mol^{-1}
F–F	158
Cl–Cl	242
Br–Br	193
I–I	151

A number of the properties of fluorine are untypical. Many of these untypical properties are because the F–F bond is unexpectedly weak, compared with the trend for the rest of the halogens (see Table 1). The small size of the fluorine atom leads to repulsion between non-bonding electrons as they are so close together, and the bond is readily broken.

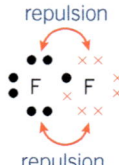

The physical properties of fluorine, chlorine, bromine, and iodine are shown in Table 2.

There are some clear trends shown by the red arrows.

Table 2 *The physical properties of Group 17, fluorine to iodine*

Halogen	Proton number, Z	Electron arrangement	Electronegativity	Atomic (covalent) radius / nm	Melting point T_m / K	Boiling point T_b / K
fluorine	9	[He] $2s^2\ 2p^5$	4.0	0.071	53	85
chlorine	17	[Ne] $3s^2\ 3p^5$	3.0	0.099	172	238
bromine	35	[Ar] $3d^{10}\ 4s^2\ 4p^5$	2.6	0.114	266	332
iodine	53	[Kr] $4d^{10}\ 5s^2\ 5p^5$	2.5	0.133	387	457

Volatility

Volatility is the ease with which the elements turn into gases. Fluorine and chlorine are gases at room temperature. Bromine is a volatile liquid and iodine sublimes (turns directly from solid to gas). So the elements get less volatile as we descend the group. This is another way of stating that the melting and boiling points increase as we go down the group.

We can explain this because the halogen molecules attract each other solely by London dispersion forces (instantaneous dipole–induced dipole forces)—see Section 3.6. These forces increase with proton number because there are more electrons in the atom to create an instantaneous dipole.

> **Exam tip**
>
> Electronegativity is the power of an atom to attract electrons to itself.

> **Summary test 11.1**
>
> 1 Predict the properties of astatine compared with the other halogens in terms of:
>   a physical state at room temperature, including colour
>   b size of atom
>   c electronegativity.
> 2 Explain your answers to question **1**.
> 3 a Use the data in Table 2 to give a rough estimate of the boiling point of astatine.
>   b Explain why you would expect the boiling point of astatine to be the largest.

11.2 The chemical properties of the halogen elements and the hydrogen halides

Learning outcomes

On these pages you will learn to:

- describe and explain the oxidising behaviour of the halogens
- describe and explain the reactions of the halogens and their hydrogen halides
- describe and explain the thermal stabilities of the hydrogen halides

Exam tip

Remember **OIL RIG**:

Oxidation **I**s **L**oss of electrons.

Reduction **I**s **G**ain of electrons.

Halogens usually react by gaining electrons to become negative ions, with a charge of –1. These reactions are redox reactions—halogens are oxidising agents and are themselves reduced. For example:

$$Cl_2 + 2e^- \xrightarrow{\text{gain of electrons}} 2Cl^-$$

The oxidising ability of the halogens increases going up the group.

Fluorine is one of the most powerful oxidising agents known.

fluorine chlorine bromine iodine

← increasing oxidising power ——

The reactions of the halogen elements

Halogens will react with metal halides in aqueous solution in such a way that the halide in the compound will be displaced by a more reactive halogen but not by a less reactive one. This is called a **displacement reaction**. For example:

$$\overset{0}{Cl_2}(aq) + 2Na\overset{-1}{Br}(aq) \rightarrow \overset{0}{Br_2}(aq) + 2Na\overset{-1}{Cl}(aq)$$

This is also a redox reaction in which chlorine acts as an oxidising agent. The ionic equation for this reaction is:

$$Cl_2(aq) + \cancel{2Na^+(aq)} + 2Br^-(aq) \rightarrow Br_2(aq) + \cancel{2Na^+(aq)} + 2Cl^-(aq)$$

The sodium ions are spectator ions—they take no part in the reaction.

$$Cl_2(aq) + 2Br^-(aq) \rightarrow Br_2(aq) + 2Cl^-(aq)$$

The two colourless starting materials react to produce the red-brown colour of bromine.

The chlorine acts as an oxidising agent by removing electrons from Br^- and oxidising $2Br^-$ to Br_2 (the oxidation number of the bromine increases from –1 to 0).

Exam tip

A halogen will always oxidise a halide ion below it in the Periodic Table.

Table 3 *The oxidation of a halide by a halogen*

	F^-	Cl^-	Br^-	I^-
F_2	–	yes	yes	yes
Cl_2	no	–	yes	yes
Br_2	no	no	–	yes
I_2	no	no	no	–

In general, a halogen will always oxidise a halide ion below it in the Periodic Table (see Table 3).

You cannot investigate fluorine in an aqueous solution because it reacts with water.

The extraction of bromine from sea water

The oxidation of a halide by a halogen is the basis of a method for extracting bromine from sea water. Sea water contains small amounts of bromide ions which can be oxidised by chlorine to produce bromine:

$$Cl_2(aq) + 2Br^-(aq) \rightarrow Br_2(aq) + 2Cl^-(aq)$$

Extraction of iodine from kelp

Iodine was discovered in 1811. It was extracted from kelp, which is obtained by burning seaweed. Some iodine is still produced in this way. Salts such as sodium chloride, potassium chloride, and potassium sulfate are removed from the kelp by washing with water. The residue is then heated with manganese dioxide and concentrated sulfuric acid and iodine is liberated:

$$2I^- + MnO_2 + 4H^+ \rightarrow Mn^{2+} + 2H_2O + I_2$$

1 Is the reaction an oxidation or a reduction of the iodide ion? Explain your answer.
2 Find out why table salt often has potassium iodide added to it.

Reactions of the halogens with hydrogen

All the halogens will react directly with hydrogen to form hydrogen halides:

$$X_2 + H_2(g) \rightarrow 2HX(g)$$

where X represents any halogen.

In each case, the halogen oxidises the hydrogen from oxidation number zero to +1.

There is a clear trend of decreasing reactivity as we descend the group.

- Fluorine and hydrogen react explosively even at low temperature and in the dark.
- Chlorine and hydrogen react explosively at room temperature when exposed to ultraviolet light.
- Bromine and hydrogen react slowly and require a catalyst.
- Iodine vapour and hydrogen react to form an equilibrium mixture.

$$I_2(g) + H_2(g) \rightleftharpoons 2HI(g)$$

So fluorine is the best oxidising agent and iodine is the worst.

Bond energies

The trend in reactivity can be explained by using the bond energies in Table 4.

Table 4 Bond energies of the hydrogen halides and the halogens

Halogen halide	Bond energy / kJ mol^{-1}	Halogen	Bond energy / kJ mol^{-1}
HF	562	F–F	158
HCl	431	Cl–Cl	242
HBr	366	Br–Br	193
HI	299	I–I	151

The H–H bond energy is 436 kJ mol^{-1}.

For example:

$$F_2(g) + H_2(g) \rightarrow 2HF(g)$$

Showing the bonds:

$$F-F + H-H \rightarrow 2H-F$$

To break the fluorine and hydrogen bonds we have to *put in* the bond energy of F–F ($158\,kJ\,mol^{-1}$) + the bond energy of H–H ($436\,kJ\,mol^{-1}$) = $594\,kJ\,mol^{-1}$

The energy *given out* when hydrogen fluoride is made is 2 × bond energy of H–F = 2 × $562\,kJ\,mol^{-1}$ = $1124\,kJ\,mol^{-1}$

The difference is $1124 - 594 = 530\,kJ\,mol^{-1}$

More energy is given out than is put in, so $\Delta H = -530\,kJ\,mol^{-1}$

A similar calculation for the formation of H–I gives $\Delta H = -11\,kJ\,mol^{-1}$

So, in the reactions of the halogens with hydrogen, fluorine is the most reactive element. Hydrogen fluoride, with the strongest bond, is the most stable hydrogen halide.

Thermal decomposition

The reverse of these reactions can be investigated by placing a red-hot wire or glass rod into the halogen halide gas. This provides the energy to start the reaction:

$$2HX \rightarrow H_2 + X_2$$

No reaction is observed with HF or with HCl.

HBr decomposes to produce a small amount of red-brown bromine gas.
HI decomposes rapidly to form large amounts of violet iodine vapour.

These trends can be explained in terms of the strength of the H–X covalent bond. This decreases as we descend the group, as shown in Table 4 (on the previous page). So the activation energy needed to start the reaction by breaking the H–X bond decreases.

Summary test 11.2

1 a State which of the following mixtures would react:
  i $Br_2(aq) + NaCl(aq)$
  ii $Cl_2(aq) + NaI(aq)$
  b Explain your answers.
  c Give the complete equation for the mixture that reacts.
2 Use bond energies to calculate the value of ΔH for the formation of H–Cl and H–Br from their elements. Using the values for H–F and H–I calculated above, explain the trend.

Halide ions can act as reducing agents. In these reactions the halide ions lose (give away) electrons and become halogen molecules. There is a definite trend in their reducing ability. This is linked to the size of the ions. *The larger the ion, the more easily it loses an electron.* This is because the electron is lost from the outer shell, which is further from the nucleus. So as the ion gets larger, the attraction to the outer electron is less.

$$\text{------ increasing reducing power } \longrightarrow$$

	F^-	Cl^-	Br^-	I^-
Ionic radius / nm	0.133	0.180	0.195	0.215

This trend can be seen in the reactions of solid sodium halides with concentrated sulfuric acid.

The reactions of sodium halides with concentrated sulfuric acid

Solid sodium halides react with concentrated sulfuric acid. The products are different and reflect the reducing powers of the halide ions shown above.

Sodium chloride (solid)

In this reaction, drops of concentrated sulfuric acid are added to solid sodium chloride. Steamy fumes of hydrogen chloride are seen. The solid product is sodium hydrogensulfate.

The reaction is:

$$NaCl(s) + H_2SO_4(l) \rightarrow NaHSO_4(s) + HCl(g)$$

This is not a redox reaction because no oxidation number has changed. The chloride ion is too weak a reducing agent to reduce the sulfur (oxidation number = +6) in sulfuric acid. It is an acid–base reaction.

$$\underset{NaCl(s)}{\overset{+1\ -1}{}} + \underset{H_2SO_4(l)}{\overset{+1+6-2}{}} \rightarrow \underset{NaHSO_4(s)}{\overset{+1+1+6-2}{}} + \underset{HCl(g)}{\overset{+1-1}{}}$$

This reaction can be used to prepare hydrogen chloride gas which, because of this reaction, was once called *salt gas*.

Sodium bromide (solid)

In this case, you will see steamy white fumes of hydrogen bromide and brown fumes of bromine. Colourless sulfur dioxide is also formed.

Two reactions occur.

First sodium hydrogensulfate and hydrogen bromide are produced (in a similar acid–base reaction to sodium chloride):

$$NaBr(s) + H_2SO_4(l) \rightarrow NaHSO_4(s) + HBr(g)$$

However, bromide ions are strong enough reducing agents to reduce the sulfuric acid to sulfur dioxide. The oxidation number of the sulfur is reduced from +6 to +4 and that of the bromine increases from −1 to 0.

$$\underset{2HBr(g)}{\overset{-1}{}} + \underset{H_2SO_4(l)}{\overset{+6}{}} \rightarrow \underset{SO_2(g)}{\overset{+4}{}} + 2H_2O(l) + \underset{Br_2(l)}{\overset{0}{}}$$

This is a redox reaction. The reactions are exothermic and some of the bromine vaporises.

Learning outcomes

On these pages you will learn to:

- describe and explain the behaviour of halide ions as reducing agents
- describe and explain the reactions of the halide ions

Exam tip

Remember that the reactions take place between *solid* halide salts and *concentrated* sulfuric acid.

Exam tip

A similar reaction to that of sodium chloride occurs with sodium fluoride to produce hydrogen fluoride, an extremely dangerous gas that will etch glass. The fluoride ion is an even weaker reducing agent than the chloride ion.

Sodium iodide (solid)

In this case you see steamy fumes of hydrogen iodide, the black solid of iodine, and also the bad egg smell of hydrogen sulfide gas is present. Colourless sulfur dioxide is also evolved. Yellow solid sulfur may also be seen.

Several reactions occur. Hydrogen iodide is produced in an acid–base reaction as before:

$$NaI(s) + H_2SO_4(l) \rightarrow NaHSO_4(s) + HI(g)$$

Iodide ions are better reducing agents than bromide ions, so they reduce the sulfur in sulfuric acid even further (from +6 to zero and –2). Sulfur dioxide, sulfur, and hydrogen sulfide gas are produced. For example:

$$\underset{-1}{8H^+} + \underset{}{8I^-} + \underset{+6}{H_2SO_4(l)} \rightarrow \underset{-2}{H_2S(g)} + 4H_2O(l) + \underset{0}{4I_2(s)}$$

During the reduction from +6 to –2, the sulfur passes through oxidation number 0 and some yellow, solid sulfur may be seen.

The reduction occurs in a series of steps:

$$\mathbf{1} \quad \underset{-1}{2HI(g)} + \underset{+6}{H_2SO_4(l)} \rightarrow \underset{+4}{SO_2(g)} + \underset{0}{I_2(s)} + 2H_2O(l)$$

$$\mathbf{2} \quad \underset{-1}{4HI(g)} + \underset{+4}{SO_2(g)} \rightarrow \underset{0}{S(s)} + \underset{0}{2I_2(s)} + 2H_2O(l)$$

$$\mathbf{3} \quad \underset{-1}{2HI(g)} + \underset{0}{S(s)} \rightarrow \underset{-2}{H_2S(s)} + \underset{0}{I_2(s)}$$

Adding these equations and cancelling species that occur on both sides of the arrows give the overall equation shown above.

The reaction of metal halides with silver ions

All metal halides (except fluorides) react with the silver ions in aqueous silver nitrate, to form a precipitate of the insoluble silver halide. For example:

$$Cl^-(aq) + Ag^+(aq) \rightarrow AgCl(s)$$

(Silver fluoride does not form a precipitate because it is soluble in water.)

This reaction can be used to identify the halide present in a solution of the halide, if there are no other ions present that could react with the silver ions.

Step 1

Dilute nitric acid HNO_3 or ($H^+(aq) + NO_3^-(aq)$) is first added to the halide solution to remove any soluble carbonate, $CO_3^{2-}(aq)$, or hydroxide, $OH^-(aq)$ impurities:

Removing the carbonate ions:

$$CO_3^{2-}(aq) + 2H^+(aq) + 2NO_3^-(aq) \rightarrow CO_2(g) + H_2O(l) + 2NO_3^-(aq)$$

Removing the hydroxide ions:

$$OH^-(aq) + H^+(aq) + NO_3^-(aq) \rightarrow H_2O(l) + NO_3^-(aq)$$

Exam tip

Sulfuric acid or hydrochloric acid cannot be used as an alternative to nitric acid. Sulfuric acid would give a precipitate of silver sulfate and hydrochloric acid a precipitate of silver chloride. Either of these would invalidate the test.

Exam tip

Silver hydroxide in fact is converted into silver oxide—a brown precipitate:

$$2AgOH \rightarrow Ag_2O + H_2O$$

These would otherwise interfere with the test by forming insoluble precipitates:

$$2Ag^+(aq) + CO_3^{2-}(aq) \rightarrow Ag_2CO_3(s)$$

silver carbonate

$$Ag^+(aq) + OH^-(aq) \rightarrow AgOH(s)$$

silver hydroxide

Step 2

Then a few drops of silver nitrate solution are added and the halide precipitate forms.

The reaction can be used as a test for halides because you can tell from the colour of the precipitate which halide has formed, see Table 5. The colours of silver bromide and silver iodide are similar. However, if you add a few drops of concentrated ammonia solution the silver bromide dissolves but silver iodide does not.

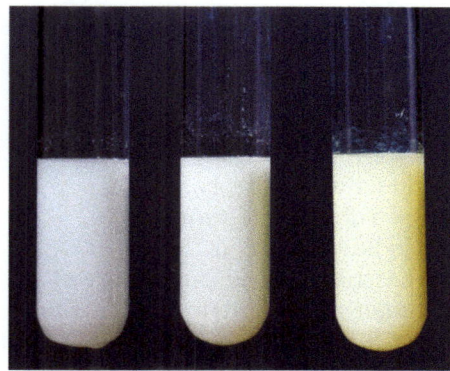

Figure 2 *The colours of the silver halides: (from left to right) AgCl, AgBr, AgI*

Table 5 *Tests for halides*

Halide	silver fluoride	silver chloride	silver bromide	silver iodide
Colour	no precipitate	white ppt	cream ppt	pale yellow ppt
Further tests		dissolves in dilute ammonia	dissolves in concentrated ammonia	insoluble in concentrated ammonia

Summary test 11.3

1 The reaction between concentrated sulfuric acid and solid sodium fluoride is not usually carried out in the laboratory.
   a State how the reducing power of the fluoride ion compares with the other halide ions.
   b Explain why you would predict this.
   c Give a balanced symbol equation for the reaction between concentrated sulfuric acid and sodium fluoride.
   d Is this a redox reaction? Explain your answer.
2 A few drops of silver nitrate were added to an acidified solution, to show the presence of sodium bromide.
   a State what you would see.
   b Give the equation for the reaction.
   c State what would happen if you now added a few drops of concentrated ammonia solution.
   d Explain why an acid is added to sodium bromide solution initially.
   e Neither hydrochloric nor sulfuric acid may be used to acidify the solution. Explain why this is so.
   f Explain why this test cannot be used to find out if fluoride ions are present.

Learning outcomes

On these pages you will learn to:

- describe and explain the reactions of chlorine with sodium hydroxide
- describe and explain the use of chlorine in purifying water

Chlorine is a poisonous gas and was notoriously used as such in the First World War. However, it is soluble in water and in this form has become an essential part of our lives in the treatment of water, both for drinking and in swimming pools.

Reaction with water

Chlorine reacts with water in a reversible reaction to form chloric(I) acid, HClO, and hydrochloric acid, HCl:

$$\overset{0}{Cl_2(g)} + H_2O(l) \rightleftharpoons \overset{+1}{HClO(aq)} + \overset{-1}{HCl(aq)}$$

In this reaction, the oxidation number of one of the chlorine atoms increases from 0 to +1, and the oxidation number of the other chlorine atom decreases from 0 to −1.

This type of redox reaction, where the oxidation numbers of some atoms of the same element increase and others decrease, is called **disproportionation**.

Figure 3 *This testing kit monitors the pH and chlorine levels of the swimming pool*

Water purification

The reaction above takes place when chlorine is used to purify water for drinking and in swimming pools, to prevent life-threatening diseases. Chloric(I) acid is an oxidising agent and kills bacteria by oxidation. It is also a bleach.

In sunlight, a different reaction occurs:

$$2Cl_2(g) + 2H_2O(l) \rightarrow 4HCl(aq) + O_2(g)$$
$$\text{pale green} \qquad \text{colourless}$$

Chlorine is rapidly lost from swimming pool water in sunlight, so shallow pools need frequent addition of chlorine.

An alternative to the direct chlorination of swimming pools is to add solid sodium (or calcium) chlorate(I). This dissolves in water to form chloric(I) acid, HClO(aq), in a reversible reaction:

$$NaClO(s) + H_2O \rightleftharpoons Na^+(aq) + OH^-(aq) + HClO(aq)$$

In alkaline solution, this equilibrium moves to the left and the HClO is removed as ClO⁻ ions. To prevent this happening, swimming pools need to be kept slightly acidic.

However, this is carefully monitored and the water never gets acidic enough to corrode metal components and affect swimmers.

Reaction with sodium hydroxide

Chlorine reacts with cold, dilute sodium hydroxide to form sodium chlorate(I), NaClO. This is an oxidising agent and the active ingredient in household bleach. This is also a disproportionation reaction—see the oxidation numbers above the relevant species:

$$\overset{0}{Cl_2}(g) + 2NaOH(aq) \rightarrow Na\overset{+1}{Cl}O(aq) + Na\overset{-1}{Cl}(aq) + H_2O(l)$$

The other halogens behave similarly.

With hot, concentrated sodium hydroxide, a different reaction occurs which is also a disproportionation. The product is called sodium chlorate(V) or just sodium chlorate.

$$3\overset{0}{Cl_2}(g) + 6NaOH(aq) \rightarrow Na\overset{+5}{Cl}O_3(aq) + 5Na\overset{-1}{Cl}(aq) + 3H_2O(l)$$

Similar reactions occur for bromine and iodine.

Exam tip

Sodium chlorate(I) is also known as sodium hypochlorite.

Summary test 11.4

1 a Add the oxidation numbers for the elements Na, O and H in the following two reactions:

$$\overset{0}{Cl_2}(g) + 2NaOH(aq) \rightarrow Na\overset{+1}{Cl}O(aq) + Na\overset{-1}{Cl}(aq) + H_2O(l)$$

and

$$3\overset{0}{Cl_2}(g) + 6NaOH(aq) \rightarrow Na\overset{+5}{Cl}O_3(aq) + 5Na\overset{-1}{Cl}(aq) + 3H_2O(l)$$

b Explain what you notice.

2 Deduce the oxidation numbers of the chlorine atoms in the reaction:

$$Cl_2(g) + H_2O(l) \rightleftharpoons HClO(aq) + HCl(aq)$$

State the type of reaction this is.

3 In sunlight, the concentration of chlorine in a swimming pool can drop. Explain what will happen to the concentration of HClO. (Hint: Use Le Chatelier's principle.)

4 Chlorine reacts with water when exposed to sunlight:

$$2Cl_2(g) + 2H_2O(l) \rightarrow 4HCl(aq) + O_2(g)$$

a Deduce the oxidation numbers of each element before and after this reaction.

b Identify the oxidising agent and the reducing agent in this reaction.

There are exam-style questions to test your knowledge of the material in this chapter at the end of Chapter 12.

12.1 Nitrogen and sulfur

Both of the elements nitrogen and sulfur are important in everyday life, in both positive and negative ways.

The unreactivity of nitrogen

Nitrogen is an unreactive gas at the top of Group 15 of the Periodic Table. It has five electrons in its outer shell ($1s^2$, $2s^2$, $2p^3$).

Nitrogen exists as N_2 molecules in which the two nitrogen atoms are held together by a triple bond ($N\equiv N$). The dot-and-cross diagram is shown in Figure 1.

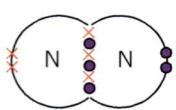

Figure 1 *The dot-and-cross diagram of the bonding in the N_2 molecule*

Looking at the orbital picture, Figure 2, the triple bond is formed from:

- the overlap of a 2p-orbital on each nitrogen atom along the axis of the molecule, to form a σ molecular orbital, and
- the overlap of two 2p-orbitals on each nitrogen atom to form π-orbitals above and below the axis of the molecule.

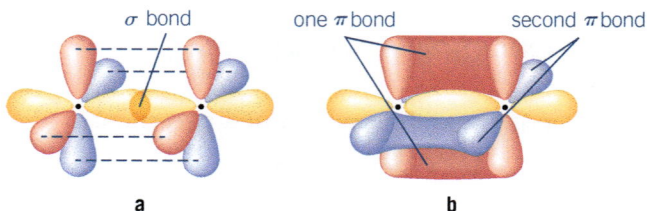

Figure 2 *The bonding orbitals in nitrogen, N_2*

The resulting bond is one of the strongest chemical bonds known, with a bond energy of 944 kJ mol^{-1}. As both of the atoms in the bond are the same, the bond is completely non-polar and is therefore not attacked by positively- or negatively-charged reagents. These two factors explain why nitrogen is chemically very inert.

Ammonia, NH_3

The dot-and-cross diagram for the ammonia molecule is shown in Figure 3.

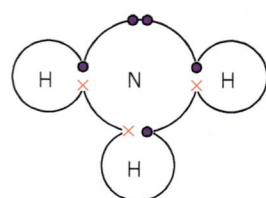

Figure 3 *The dot-and-cross diagram for the bonding in the NH_3 molecule*

The orbital picture shows three σ bonds with hydrogen atoms. These are formed by the overlap of sp³ hybrid orbitals (see Section 3.4) on the nitrogen atom with s-orbitals on each hydrogen atom (see Figure 4).

Notice that both diagrams show a lone pair of electrons on the nitrogen atom. This lone (non-bonding) pair of electrons is very important in the chemistry of ammonia.

If the lone pair is donated to a hydrogen ion (a proton) we get the ammonium ion (NH_4^+). The ammonia molecule is then acting as a Brønsted–Lowry base (see Section 7.2). The shape of the ammonium ion is a perfect tetrahedron as all four bonds are equivalent.

$$NH_3(aq) + H^+(aq) \rightarrow NH_4^+(aq)$$

The ammonium ion can form the positive ion in a salt (in the same way as a metal ion). For example in ammonium chloride:

$$NH_3(aq) + HCl(aq) \rightleftharpoons NH_4^+(aq) + Cl^-(aq)$$

This equilibrium can be displaced to the left by adding a strong base such as sodium hydroxide to remove the hydrochloric acid. So ammonia can be displaced from its salts by reaction with a base. For example:

$$NH_4NO_3(aq) + NaOH(aq) \rightarrow NaNO_3(aq) + NH_3(aq) + H_2O$$

Nitrogen oxides

At high temperatures and with a spark, nitrogen will react with oxygen to form nitrogen monoxide, NO:

$$N_2(g) + O_2(g) \rightarrow 2NO(g)$$

This occurs naturally during thunderstorms. It also happens in combustion processes when fuels burn in air, for example in internal combustion engines.

Further oxidation of NO by oxygen in the air leads to the formation of nitrogen dioxide:

$$2NO(g) + O_2(g) \rightarrow 2NO_2$$

Pollution from nitrogen oxides

A mixture of nitrogen monoxide and nitrogen dioxide is often referred to as NO_x. NO_x is a pollutant:

- It reacts in moist air (oxygen and water) to form nitric acid—this is a component of acid rain.
- It can cause breathing problems such as asthma.
- It contributes to photochemical smogs when it reacts with unburnt hydrocarbons from fuels.

NO_x can be removed from car exhaust gases by using a catalytic converter (see Figure 5). Platinum and rhodium in the catalytic converter catalyse the following reactions:

$$\text{carbon monoxide} + \text{nitrogen oxides} \rightarrow \text{carbon dioxide} + \text{nitrogen}$$

$$\text{hydrocarbons} + \text{nitrogen oxides} \rightarrow \text{carbon dioxide} + \text{nitrogen} + \text{water}$$

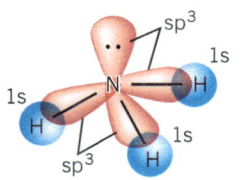

Figure 4 *The bonding orbitals in ammonia, NH₃*

Exam tip

Because of the lone pair of electrons, the ammonia molecule is a triangular pyramid, not trigonal.

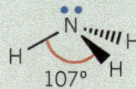

Exam tip

The formation of the ammonium ion is an example coordinate (dative) bonding. The lone pair of electrons on the nitrogen atom is used to form a bond with a H⁺ ion (a proton). See Section 3.4 for more on coordinate bonding.

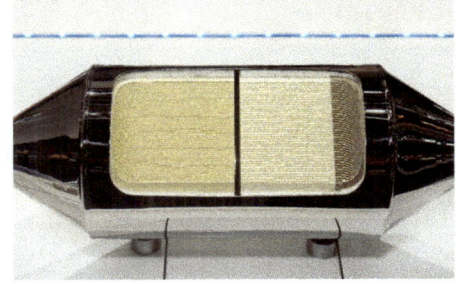

Figure 5 *A catalytic converter containing an alloy of platinum and rhodium*

Typical balanced equations for reactions that take place in a catalytic converter include:

$$CO(g) + NO(g) \rightarrow CO_2(g) + \tfrac{1}{2}N_2(g)$$

$$2CO(g) + NO_2(g) \rightarrow 2CO_2(g) + \tfrac{1}{2}N_2(g)$$

$$C_8H_{18}(g) + 25NO(g) \rightarrow 8CO_2(g) + 9H_2O(g) + 12\tfrac{1}{2}N_2(g)$$

$$C_8H_{18}(g) + 25NO_2(g) \rightarrow 16CO_2(g) + 18H_2O(g) + 12\tfrac{1}{2}N_2(g)$$

Carbon monoxide, resulting from the incomplete combustion of fuels, is toxic. The products are less polluting than the reactants, although carbon dioxide is a greenhouse gas.

Photochemical smogs

Photochemical smogs are a form of pollution formed at ground level, particularly in cities where motor vehicles produce NO_x and also unburned hydrocarbons from their fuels. These primary pollutants can react together to form a mixture of secondary pollutants, including ozone (O_3) and compounds called peroxyacetyl nitrates (PANs). This occurs particularly in large, low-lying cities where there are many vehicles and the wind cannot blow away the polluting gases.

PANs have the general formula:

where R represents a variety of hydrocarbons (compounds of hydrogen and carbon) derived from unburned vehicle fuel. They are *lachrymators*, which means they cause eye irritation, a bit like tear gas. Photochemical smogs can also cause breathing difficulties and heart problems.

Sulfur and acid rain

Many fossil fuels—in particular coal, which is still used in many power stations—contain sulfur. They produce sulfur dioxide (SO_2) when the fuel is burned. This, too, reacts with oxygen and water in the atmosphere. It forms sulfuric acid, another component of acid rain. The reaction is catalysed by NO_x.

When the fuel is burned:

$$S(s) + O_2(g) \rightarrow SO_2(g)$$

In the atmosphere, sulfur dioxide is oxidised to sulfur trioxide:

$$2SO_2(g) + O_2(g) \rightarrow 2SO_3(g)$$

This is then converted to sulfuric acid:

$$SO_3(g) + H_2O(l) \rightarrow H_2SO_4(l)$$

The oxidation of sulfur dioxide is catalysed by NO:

$$2NO(g) + O_2(g) \rightarrow 2NO_2(g) \qquad \text{reaction 1}$$

$$SO_2(g) + NO_2(g) \rightarrow SO_3(g) + NO(g) \qquad \text{reaction 2}$$

Reaction 2 converts sulfur dioxide to sulfur trioxide. Reaction 1 then converts nitrogen monoxide back to nitrogen dioxide. So the nitrogen oxides effectively act as a catalyst.

Figure 6 *A cloud of smog above Mexico City is pictured here. The brown colouration is due to the presence of NO_2 formed from photochemical reactions*

Adding the two reactions shows that no nitrogen oxide is used up in the process:

$$2NO(g) + O_2(g) \rightarrow 2NO_2(g) \qquad \text{reaction 1}$$

$$2SO_2(g) + 2NO_2(g) \rightarrow 2SO_3(g) + 2NO(g) \qquad \text{reaction 2}$$

$$\overline{2NO(g) + O_2(g) + 2SO_2(g) + 2NO_2(g) \rightarrow 2NO_2(g) + 2SO_3(g) + 2NO(g)}$$

So

$$O_2(g) + 2SO_2(g) \rightarrow 2SO_3(g)$$

Summary test 12.1

1 Figure 7 shows a way of preparing ammonia.

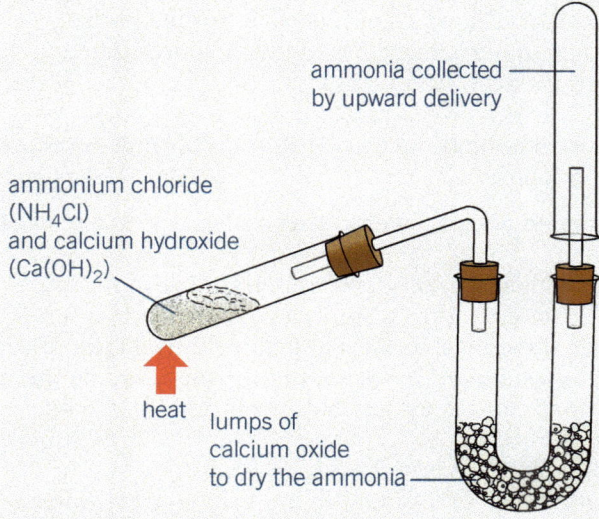

Figure 7 *Making ammonia on a small scale*

   a Give a balanced equation for the reaction that occurs.
   b Explain why the ammonia has to be dried.
   c Give a balanced equation for the drying reaction.
   d Concentrated sulfuric acid is a more efficient drying agent than calcium oxide. Explain why it is not suitable here.
   e Ammonia is collected by upward delivery because it is less dense than air. Explain why ammonia is less dense than air.

2 a State which of these reactions is/are redox reactions. Explain your answer using oxidation numbers.

$$S + O_2(g) \rightarrow SO_2(g)$$

$$2SO_2(g) + O_2(g) \rightarrow 2SO_3(g)$$

$$SO_3(g) + H_2O(g) \rightarrow H_2SO_4(l)$$

   b If 64 tonnes of sulfur dioxide are released into the atmosphere by a power station, calculate how many tonnes of sulfuric acid will be produced eventually. Use the equation:

$$SO_2(g) + \frac{1}{2} O_2(g) + H_2O(l) \rightarrow H_2SO_4(l)$$

Launch additional digital resources for the chapter

1 State the general trend in the following properties descending Group 2, giving reasons for your answers:

a atomic radius (1 mark)

b first ionisation energy (1 mark)

c relative strength as a reducing agent (1 mark)

2 This question is about the reactions of Group 2 elements and their compounds:

a Barium reacts vigorously with water.
  i Construct a balanced chemical equation for this reaction. (2 marks)
  ii Ba^{2+} ions are toxic. Explain why barium sulfate is safe for patients to use in medical imaging. (2 marks)

b Magnesium reacts with dilute hydrochloric acid to form magnesium chloride solution and hydrogen gas.
  i Construct a balanced chemical equation for this reaction. (2 marks)
  ii Predict the observations on adding barium to dilute hydrochloric acid and name the products formed. (2 marks)

c Magnesium carbonate reacts with dilute hydrochloric acid at room temperature to form magnesium oxide, water and a gas.
  i Identify the gas formed in this reaction. (1 mark)
  ii Construct a balanced equation for this reaction. (2 marks)

3 Describe the general trend in the thermal stability of Group 2 carbonates as you descend the group. Explain your answer. (3 marks)

4 A student investigated the reaction of concentrated sulfuric acid with separate samples of solid sodium chloride, solid sodium bromide and solid sodium iodide. Observations for the resulting reactions are shown in the table below:

Halide	Observations
Sodium chloride	Acidic gas produced
Sodium bromide	Acidic gas produced as well as a little red-brown gas
Sodium iodide	Acidic gas produced as well as a purple gas

a Identify the gas formed in the reaction with sodium chloride and construct an equation for this reaction. (2 marks)

b i Identify the gas formed in the reaction with sodium bromide. (1 mark)
  ii Construct an equation to show how the red-brown gas is made from the acidic gas that forms beforehand. (2 marks)
  iii Use you answer to part ii to deduce whether bromine is oxidised or reduced. Use oxidation numbers to explain your answer. (2 marks)

c Explain why no coloured gas is formed in the reaction of sodium chloride with concentrated sulfuric acid. (1 mark)

5 a Describe the trend in volatility of chlorine, bromine and iodine. (1 mark)

b Explain your answer to part a. (2 marks)

6 Three halide ion solutions were found with missing labels. Describe how a student could perform a series of simple test-tube reactions to deduce the identity of each solution, stating the observations made. Assume these solutions contain either chloride, bromide or iodide ions. (6 marks)

7 Chlorine and its compounds are used to treat water to make it safe to drink.

a Construct an equation to show the equilibrium that occurs when chlorine is added to water forming chloric(I) acid (HOCl) and hydrochloric acid. (2 marks)

b Chloric acid behaves as a weak acid. Write an equation to show this. (2 marks)

c In acidic conditions chloric(I) acid behaves as an oxidising agent. Using the equation below, explain how this occurs.

$2HOCl(aq) + 2H^+(aq) + 2e^- \rightleftharpoons Cl_2(g) + 2H_2O(l)$

d Explain how HOCl molecules make water safe to drink. (1 mark)

8 Ammonia and chlorine react in the gas phase to form nitrogen and ammonium chloride.

a Construct a balanced chemical equation for this reaction. (1 mark)

b Explain why ammonia reacts as both a base and a reducing agent in this reaction. (1 mark)

9 This question is about the oxides of nitrogen and sulfur.

a Construct a balanced chemical equation to show how nitrogen monoxide is made from its elements at high temperatures in vehicle engines. *(2 marks)*

b Write an equation to show the equilibrium formed in the atmosphere when nitrogen dioxide and nitrogen monoxide are interconverted. *(1 mark)*

c In the atmosphere, nitrogen dioxide catalyses the conversion of sulfur dioxide to sulfur trioxide. Explain the catalytic role of nitrogen dioxide in this process using a balanced equation. *(2 marks)*

d Construct balanced equations to show how sulfur trioxide and nitrogen dioxide form acid rain. *(2 marks)*

e Describe two problems caused by acid rain. *(2 marks)*

f Nitrogen monoxide is removed from the exhaust gases of vehicles using catalytic converters.
  i Construct a balanced equation for the reaction of nitrogen monoxide and carbon monoxide in a catalytic converter. *(1 mark)*
  ii Using your answer to part **i**, state and explain which substances are oxidised and reduced in this reaction. *(2 marks)*
  iii Identify a metal used in a catalytic converter and suggest why these are usually recycled. *(2 marks)*

10 Magnesium is an element found in the s-block of the Periodic Table.

a State the full electron configuration for a magnesium ion, Mg^{2+}, and explain why magnesium is an s-block element. *(2 marks)*

b State and explain the difference in the values for the second ionisation energies of magnesium and sodium. *(3 marks)*

c Give the formula of the Group 2 hydroxide, from Mg to Ba, that is the least soluble in water. *(2 marks)*

11 A salt, X, containing a Group 2 metal was observed to be a white crystalline solid that was soluble in water. The following tests were conducted and observations noted:

Test 1: Formed a white precipitate upon addition of sulfuric acid.

Test 2: Formed a white precipitate upon addition of magnesium nitrate solution.

Evaluate the physical properties of X and experimental observations above to suggest a chemical formula for X. Give ionic equations, including state symbols, for each test to support your answer. *(6 marks)*

12 This question is about the elements of Group 17 from fluorine to iodine:

a Define the term electronegativity. *(1 mark)*

b State and explain the trend in electronegativity as you descend Group 17. *(3 marks)*

c Identify the element that is the weakest reducing agent. Explain your answer. *(3 marks)*

d Describe an experiment that would allow a student to distinguish between aqueous solutions of sodium chloride and sodium bromide. State what would be observed for each substance. *(4 marks)*

e Explain why it is necessary to acidify silver nitrate solution when testing for halide ions. *(1 mark)*

13 a Explain why traces of sulfur dioxide are emitted from oil-burning furnaces. *(2 marks)*

b Give the name of the process used to removed sulfur dioxide gas from fossil fuel power station emissions, and construct a balanced equation to show this could be achieved using one of the following substances.

$$CaCl_2 \qquad Ca(OH)_2 \qquad CaSO_4 \quad \textit{(3 marks)}$$

c Draw a 'dot-and-cross' diagram to explain why molecules of sulfur dioxide are described as non-linear (or V-shaped). *(2 marks)*

14 The first step in the manufacture of nitric acid from ammonia involves the exothermic oxidation of ammonia to nitrogen monoxide (NO) and steam. This is a reversible reaction.

a Construct a balanced equation for the reaction of ammonia with oxygen to form nitrogen monoxide and steam. *(2 marks)*

b Predict, qualitatively, the conditions of temperature and pressure to obtain the maximum yield of nitrogen monoxide in the equilibrium mixture. Give reasons for your answer. *(4 marks)*

An introduction to organic chemistry

Formulae, functional groups and the naming of organic compounds

Learning outcomes

On these pages you will learn to:

- define the term hydrocarbon
- identify functional groups
- interpret general, structural, displayed and skeletal formulae, and use these to deduce molecular formulae
- name organic compounds

Organic chemistry is the chemistry of carbon compounds. Life on our planet is based on carbon, and *organic* means to do with living things. Nowadays, many carbon-based materials, like plastics and drugs, are made synthetically and there are large industries based on synthetic materials. There are far more compounds of carbon known than those of all the other elements put together, well over 10 million.

What is special about carbon?

Carbon can form rings and very long chains, which may be branched. This is because:

- A carbon atom has four electrons in its outer shell, so it forms four covalent bonds.
- Carbon–carbon bonds are relatively strong (350 kJ mol^{-1}) and non-polar.

The carbon–hydrogen bond is also strong (410 kJ mol^{-1}) and relatively non-polar. **Hydrocarbon** chains form the skeleton of most organic compounds—see Figures 1, 2, and 3.

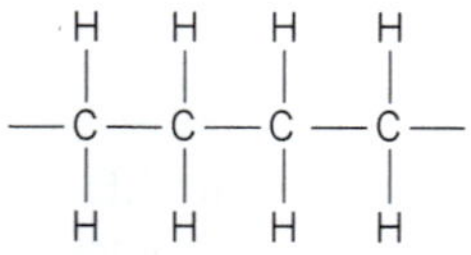

Figure 1 *Part of a straight hydrocarbon chain*

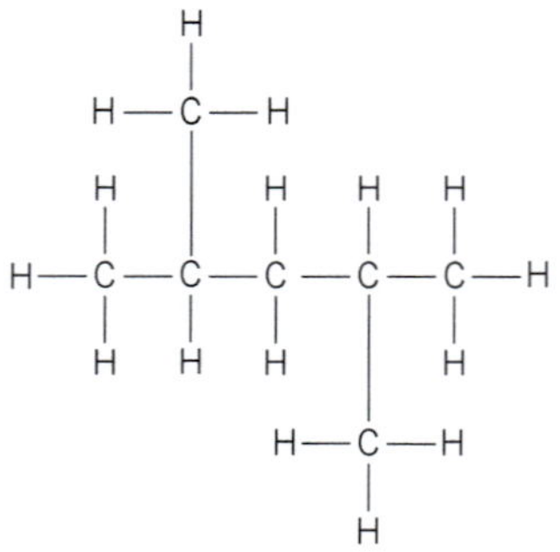

Figure 2 *A branched hydrocarbon chain*

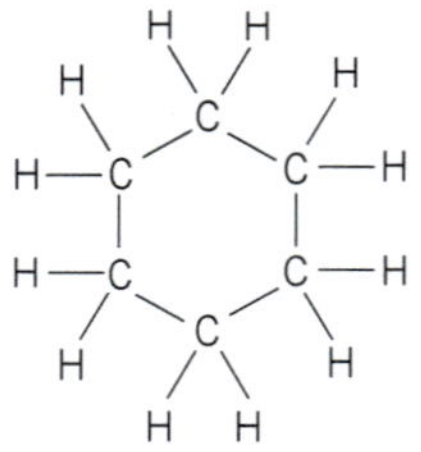

Figure 3 *A hydrocarbon ring (described as 'cyclic')*

The simplest organic compounds are hydrocarbons, which contain carbon and hydrogen only. The simplest hydrocarbons are **alkanes**, which have no reactive groups (called **functional groups**). More complex organic compounds are based on hydrocarbons with one (or more) functional groups attached.

Functional groups

The hydrocarbon skeleton of organic compounds is relatively unreactive, so functional groups dictate the chemical and physical properties of organic compounds. Functional groups may be attached to hydrocarbon chains of different lengths to form chemical 'families' called **homologous series**.

Table 1 shows the main types of functional groups along with examples of their names and different ways of expressing their formulae.

Table 1 *The main functional groups. R indicates an unspecified organic group, usually an alkyl group with the general formula C_nH_{2n+1}*

Class of compound	Name of functional group	Structural formula of functional group	Displayed formula	Skeletal formula	Name
alkene	alkene	R₂C=CR₂ (R on all four positions)	H—C—C=C (with H's)	(skeletal)	propene
halogenoalkane (primary, secondary and tertiary)	halogen	R—X	H—C—C—C—X (with H's)	(skeletal with X)	1-chloropropane (when X is chlorine)
alcohol (primary, seconday and tertiary)	hydroxy	R—OH	H—C—C—C—O—H (with H's)	(skeletal with OH)	propan-1-ol
aldehyde	carbonyl	R–C=O with H	H—C—C—C=O (with H)	(skeletal with OH)	propanal
ketone	carbonyl	R, R'–C=O	H—C—C—C—H (with O)	(skeletal with O)	propanone
carboxylic acid	carboxyl	R—C with =O and OH	H—C—C—C with =O and O—H	(skeletal with O, OH)	propanoic acid
ester	ester	R—O—C—R with =O	H—C—O—C—C—C—H (with O, H's)	(skeletal with O)	methyl propanoate
primary amine	amine	R—NH₂	H—C—C—C—N (with H's)	(skeletal with NH₂)	propylamine
nitrile	nitrile	R—C≡N	H—C—C—C≡N (with H's)	(skeletal with N)	propanenitrile

Bonding in carbon compounds

In *all* stable carbon compounds, carbon forms four covalent bonds and has eight electrons in its outer shell. It can do this by forming bonds in different ways.

- By forming four single bonds as in methane:

163

- By forming two single bonds and one double bond as in ethene:

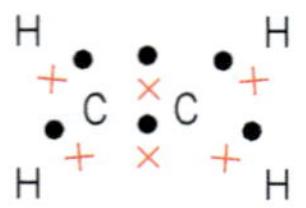

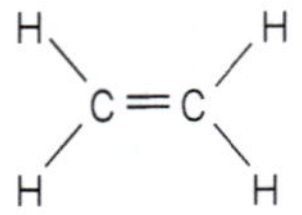

- By forming two single bonds and one double bond as in ethyne:

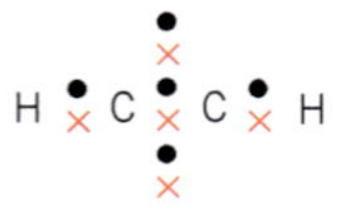

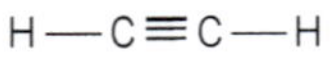

Hydrocarbons without double or triple bonds are referred to as **saturated**, meaning that no extra hydrogen can be added. Hydrocarbons with double or triple bonds are called **unsaturated**.

Types of formulae

Because of the complexity of organic compounds, several types of formulae are used. Each type is useful in different situations. You may want to know about the way the atoms are arranged within the molecule, or just the number of each atom present. There are different ways of doing this.

The molecular formula (general formula)

The molecular formula is the formula that shows the actual number of atoms of each element in the molecule. It is found from:

- the empirical formula
- the relative molecular mass of the empirical formula
- the relative molecular mass of the molecule.

The displayed formula

This shows every atom and every bond in the molecule:

— is a single bond

= is a double bond

≡ is a triple bond

For ethene, C_2H_4, the displayed formula is:

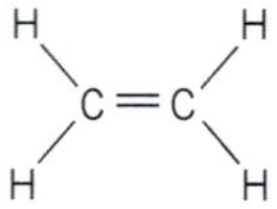

For ethanol, C_2H_6O, the displayed formula is:

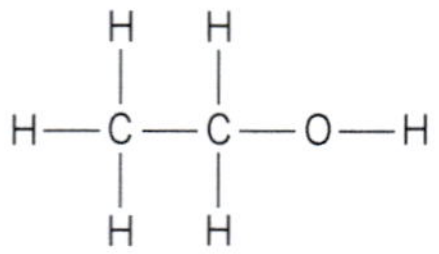

The structural formula

This shows the arrangement of atoms in a molecule in a simplified form, without showing all the bonds.

Each carbon is written separately, with the atoms or groups that are attached to it.

is written CH_3CH_3

Worked example

Empirical formula

The empirical formula is a formula that shows the simplest ratio of the atoms of each element present in a compound. For example, ethane:

3.00 g of ethane contains 2.40 g of carbon, $A_r = 12.0$, and 0.60 g of hydrogen, $A_r = 1.0$. What is its empirical formula?

$$\text{number of moles of carbon} = \frac{2.40}{12.0} = 0.20 \text{ mol of carbon}$$

$$\text{number of moles of hydrogen} = \frac{0.60}{1.0} = 0.60 \text{ mol of hydrogen}$$

Dividing through by the smaller number (0.20) gives C : H as 1 : 3

So the empirical formula of ethane is CH_3.

The mass of carbon plus the mass of hydrogen adds up to the mass of ethane, so no other element is present.

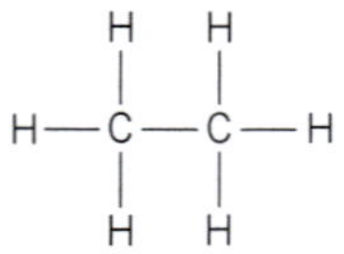

is written $CH_3CH_2CH_2OH$

Branches in the carbon chains are shown in brackets:

is written $CH_3CH(CH_3)CH_3$

Skeletal formulae

With more complex molecules, displayed structural formulae become time-consuming to draw. In skeletal notation the carbon atoms are not drawn at all. Straight lines represent carbon–carbon bonds and carbon atoms are assumed to be where the bonds meet. Neither hydrogen atoms nor C–H bonds are drawn. Each carbon is assumed to form enough C–H bonds to make a total of four bonds (counting double bonds as two).

would be written

would be written

would be written

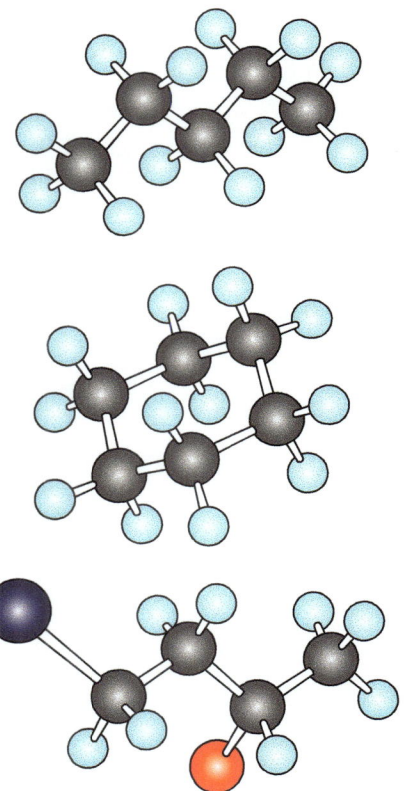

Figure 4 The 3D structures of the molecules on the left

The choice of type of formula to use depends on the circumstances and the type of information you need to give. Notice that skeletal formulae give a rough idea of the bond angles. In an unbranched alkane chain these are 109.5°.

Three-dimensional structural formulae

These attempt to show the three-dimensional structure of the molecule. Bonds coming out of the paper are shown by wedges ✔ and bonds going into the paper by dotted lines ⠒⠒.

So methane would be represented as

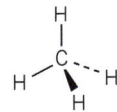

Extension

From inorganic to organic

Organic compounds were originally thought to be produced by living things only. This was disproved by Friedrich Wöhler in 1828. He made urea (an organic compound found in urine) from ammonium cyanate (an inorganic compound).

$$NH_4^+(NCO)^- \rightarrow (NH_2)_2CO$$
ammonium cyanate   urea

He reported to a fellow chemist

"I cannot, so to say, hold my chemical water and must tell you that I can make urea without thereby needing to have kidneys, or anyhow, an animal, be it human or dog".

Worked example

Molecular formula

The empirical formula of ethane is CH_3 and this group of atoms has a relative molecular mass of 15.0.

The relative molecular mass of ethane is 30.0, which is 2×15.0. So there must be two units of the empirical formula in every molecule of ethane.

The molecular formula is therefore $(CH_3)_2$ or C_2H_6.

Using different types of formula

The different types of formula present information in different ways. It is important to be able to translate the information from one type to another.

1,4-dichlorobutane has the skeletal formula

To translate this into a displayed formula you must remember that:

- Each change in angle of the skeleton represents a carbon atom.
- Each carbon atom is bonded to enough hydrogen atoms so that it is making four bonds

This gives the displayed formula below:

$$Cl-\underset{\underset{H}{|}}{\overset{\overset{H}{|}}{C}}-\underset{\underset{H}{|}}{\overset{\overset{H}{|}}{C}}-\underset{\underset{H}{|}}{\overset{\overset{H}{|}}{C}}-\underset{\underset{H}{|}}{\overset{\overset{H}{|}}{C}}-Cl$$

To translate this to a structural formula, each carbon atom is written separately along with the atoms it is bonded to. This gives the structural formula below.

$$CH_2ClCH_2CH_2CH_2Cl$$

The molecular formula groups the atoms of each element together:

$$C_4H_8Cl_2$$

The empirical formula is the simplest whole number ratio of the elements.

$$C_2H_4Cl$$

Naming organic compounds

Functional groups

In general, organic compounds are made of a hydrocarbon chain (or ring) with one or more reactive groups attached. These reactive groups are called **functional groups** and react in the same way whatever the length of the hydrocarbon chain. So if you learn the reactions of one alcohol, for example, you can apply this knowledge to any alcohol.

Homologous series

A **homologous series** is a family of organic compounds with the same functional group but different carbon chain lengths.

- Members of a homologous series have a general formula. For example, the alkanes are C_nH_{2n+2} and the alkenes, with one double bond, are C_nH_{2n}.
- Each member of the series differs from the next by CH_2.
- The length of the carbon chain has little effect on the chemical reactivity of the functional group and there are trends in physical properties such as boiling point.

- The length of the carbon chain affects physical properties, like melting point, boiling point, and solubility. Melting points and boiling points increase by a small amount as the number of carbon atoms in the chain increases. This is because the intermolecular forces increase. In general, small molecules are gases, and larger ones are liquids or solids.
- Chain branching generally reduces melting points because the molecules pack together less well.

The IUPAC system

Many organic compounds have everyday names, but chemists use a system devised by the International Union of Pure and Applied Chemistry (IUPAC). IUPAC is an international organisation of chemists that develops standards to be used by chemists throughout the world. The IUPAC naming system is rather like a universal language of chemistry. The systematic names tell us about the structures of the compounds rather than just their formula. The basic principles are covered below.

Roots

A systematic name has a root that tells us the longest unbranched hydrocarbon chain or ring, see Table 2.

The syllable after the root tells us whether there are any double bonds.

-*ane* means no double bonds. For example, ethane:

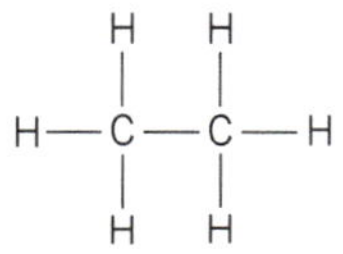

has two carbon atoms and no double bond.

Table 2 *The first six roots used in naming organic compounds*

Number of carbons	Root
1	meth
2	eth
3	prop
4	but
5	pent
6	hex

-*ene* means there is a double bond. For example, ethene:

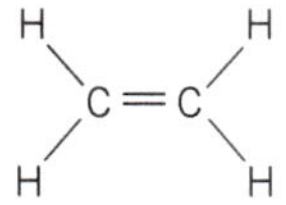

has two carbon atoms and one double bond.

Prefixes and suffixes

Prefixes and suffixes describe the changes that have been made to the root molecule.

- Prefixes are added to the beginning of the root.

For example, side chains are shown by a prefix, whose name tells us the number of carbons:

methyl CH_3- ethyl C_2H_5-
propyl C_3H_7- butyl C_4H_9-

For example:

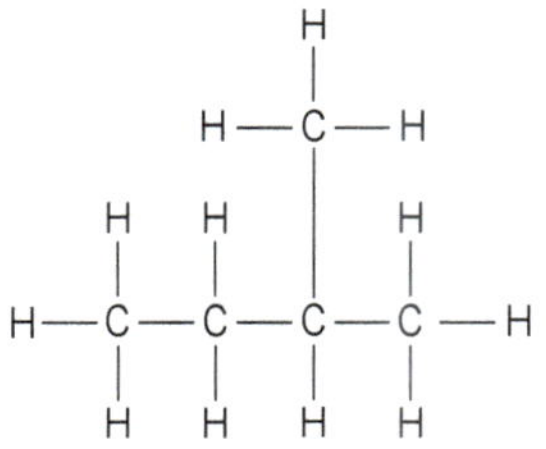

is called *methyl*butane. The longest unbranched chain is four carbons long, which gives us butane (as there are no double bonds) and there is a side chain of one carbon, a methyl group.

> **Exam tip**
>
> Ring shaped molecules are described as **cyclic**.

Hydrocarbon ring molecules have the additional prefix *cyclo*. So the compound below would be named cyclohexane:

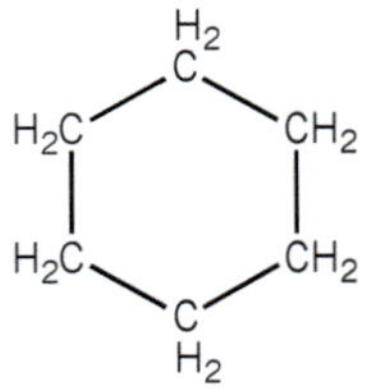

- Suffixes are added to the end of the root.

For example, alcohols, –OH, have the suffix *-ol*, as in methanol, CH_3OH.

Table 3 shows the suffixes of some common functional groups.

> **Exam tip**
>
> R is used to symbolise an alkyl group. If there is more than one different R group, they are written R′, R″ and so on.

***Table 3** The suffixes of some functional groups*

Family	General functional group	Suffix	Example
Alkanes	C_nH_{2n+2}	-ane	Ethane, CH_3CH_3
Alkenes	R–CH=CH–R	-ene	Propene, $CH_3CH{=}CH_2$
Halogenoalkanes	R–X (X = F, Cl, Br, I)	none	Chloromethane, CH_3Cl
Alcohols	R–OH	-ol	Ethanol, CH_3CH_2OH
Aldehydes	RCHO	-al	Ethanal, CH_3CHO
Ketones	RCOR′	-one	Propanone, CH_3COCH_3
Carboxylic acids	RCOOH	-oic acid	Ethanoic acid, CH_3COOH
Acid chlorides	RCOCl	-oyl chloride	Ethanoyl chloride, CH_3COCl
Amides	$RCONH_2$	-amide	Ethanamide, CH_3CONH_2
Amines	$R–NH_2$	-amine	Methylamine, CH_3NH_2
Amino acids	$H_2NCHRCOOH$	none	Aminoethanoic acid, H_2NCH_2COOH
Ester	RCOOR′	none	Ethyl ethanoate, $CH_3COOC_2H_5$
Nitrile	R–C≡N	-nitrile	Ethanenitrile, $CH_3C{\equiv}N$

Note that the halogenoalkanes are named using a prefix (fluoro-, chloro-, bromo-, iodo-) rather than a suffix. R is often used to represent a hydrocarbon chain (of any length). Think of it as representing the rest of the molecule.

Examples

In bromoethane, *eth* indicates that the molecule has a chain of two carbon atoms, *ane* that it is has no double or triple bonds, and *bromo* that one of the hydrogen atoms of ethane is replaced by a bromine atom.

bromoethane

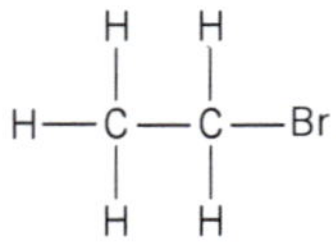

In propene, *prop* indicates a chain of three carbon atoms, and *ene* that there is one C=C (double bond).

propene

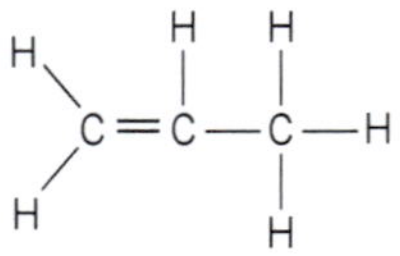

In methanol, *meth* indicates a single carbon, *an* that there are no double bonds, and *ol* that there is an OH group (an alcohol).

methanol 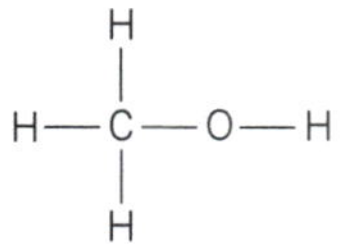

Molecules with more than one functional group or side chain

A molecule may have more than one functional group. For example:

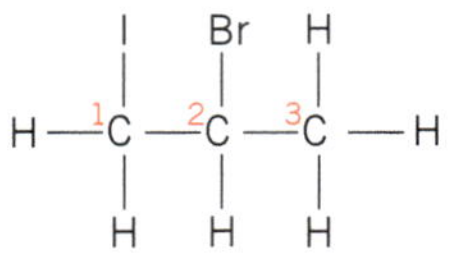 2-bromo-1-iodopropane

Even though iodine is on carbon 1 and bromine is on carbon 2, *bromo* is written before *iodo*. This is because the substituting groups are put in alphabetical order (rather than in the numerical order of the functional groups).

You can show that you have more than one of the same substituting group by adding prefixes as well as functional groups. *di-*, *tri-*, and *tetra-* mean two, three, and four, respectively.

So, 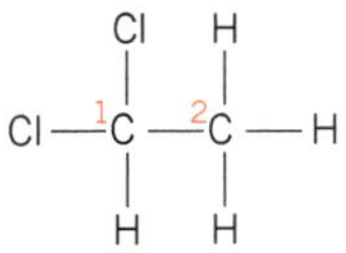 is called 1,1-dichloroethane

and 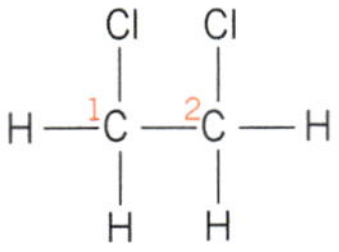 is called 1,2-dichloroethane.

Exam tip

In chemical names, commas are used to separate strings of numbers. A dash separates numbers from words.

Summary test 13.1

1 A compound comprising only carbon and hydrogen, in which 4.8 g of carbon combine with 1.0 g of hydrogen, has a relative molecular mass of 58.

a Calculate how many moles of carbon there are in 4.8 g.
b Calculate how many moles of hydrogen there are in 1.0 g.
c Give the empirical formula of this compound.
d Give the molecular formula of this compound.
e Sketch the structural formula and the skeletal formula of the compound that has a straight chain.
f Sketch the displayed formula and the skeletal formula of the compound that has a branched chain.

2 Give the name of each of the following.

a $CH_3CH_2CH_2Cl$
b $CH_3CH_2CH_2CH_2CH_3$
c $CH_3CH_2CH{=}CHCH_3$
d $CH_3CH_2CH_2CH(CH_3)CH_3$

3 Sketch the displayed formulae for:

a methylbutanone
b but-2-ene
c 2-chlorohexane
d but-1-ene.

Learning outcomes

On these pages you will learn to:

- describe the different types of organic reactions
- use the terminology associated with organic reaction mechanisms

Exam tip

Bonds between atoms of the same element are non-polar and break to give free radicals.

Exam tip

Polar bonds tend to break to give a positive ion and a negative ion.

Exam tip

Positive ions move towards the cathode in electrolysis and so are called **cations**. Negative ions move towards the anode and so are called **anions**.

Exam tip

Molecules with a double or triple bond are described as **unsaturated**, because they can have extra atoms added to them. Molecules with single bonds only are called **saturated**.

For a reaction to take place, chemical bonds in the reactants must break. The amount of energy needed to break a chemical bond is called the **bond energy** (see Section 5.1).

Bond breaking

Looking at the energies of the bonds in an organic compound can identify the weakest bond. This is often the bond most likely to break.

A single covalent bond consists of a pair of electrons shared between two atoms. When a bond between two atoms of the same element is broken, one of the electrons is likely to go to each of the atoms in the bond. This leaves each of the atoms with an unpaired electron. These species are usually highly reactive and are called **free radicals**, or just radicals. The process is called **homolytic fission** of the bond and is often brought about by ultraviolet light.

The free radicals are written with a dot to indicate the unpaired electron. For example:

$$Br-Br \rightarrow Br\cdot + Br\cdot$$

The three steps of a typical reaction involving free radicals are discussed in detail in Section 14.1. In brief they are:

- initiation—formation of free radicals by homolytic bond breaking
- propagation—where a free radical reacts to give a product and another free radical
- termination—where two free radicals react to pair their unpaired electrons.

When a polar bond breaks, the shared electrons will both go to the more electronegative atom. It will form a negative ion (an anion). The less electronegative atom loses control of the bonding electrons and becomes a positive ion (a cation). For example:

$$CH_3Cl \rightarrow CH_3^+ + Cl^-$$

This is called **heterolytic fission**.

Positive ions where the positive charge is located on a carbon atom are called **carbocations** (or carbonium ions).

Types of reagent

An ion with a positive charge or a molecule with an area of positive charge will attack negatively charged areas (such as a double bond) on another organic molecule. Such a species is called an **electrophile** (literally electron-lover), for example H^+.

An ion with a negative charge or a molecule with an area of negative charge will attack positively charged areas on another organic molecule. Such a species is called a **nucleophile** (literally nucleus-lover), for example $^-$:OH. A nucleophile has lone pair of electrons with which it can form a covalent bond with an electron-deficient species.

Types of reaction

In principle there are three types of reaction—**addition**, **substitution** and **elimination**. For example:

Addition

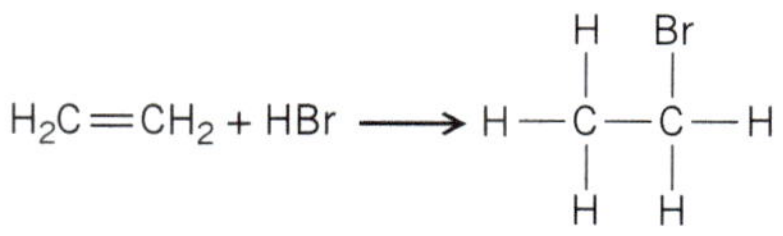

$$H_2C{=}CH_2 + HBr \longrightarrow$$

Substitution

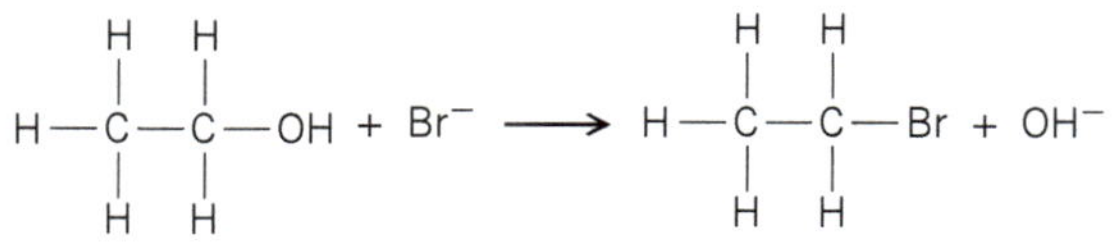

Elimination

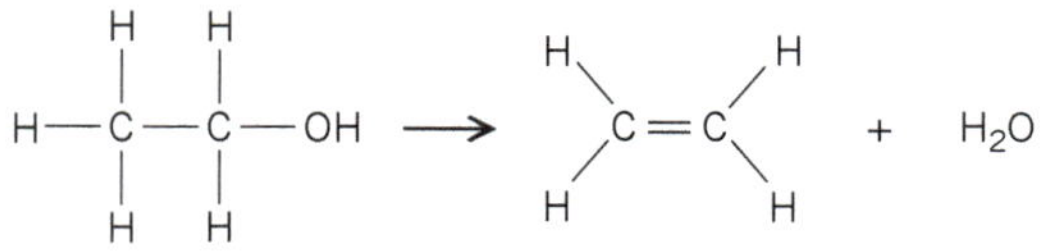

Two other, more specific, terms are sometimes used.

- If an elimination reaction eliminates water (as in the third example above) it may be called a **condensation** reaction.
- A **hydrolysis** reaction is one in which water is a reactant.

The type of reagent and the type of reaction are often used together. So we may refer to a **free radical substitution** reaction or an **electrophilic addition** reaction or **nucleophilic substitution** (or addition).

Redox processes

Oxidation numbers are not so useful in organic chemistry. It is clearer here to define oxidation as the addition of oxygen and reduction as the addition of hydrogen. We use the symbol [O] to represent a single atom of oxygen from the oxidising agent, and [H] to represent a single atom of hydrogen from the reducing agent. For example:

$$CH_3CH_2OH + 2[O] \rightarrow CH_3COOH + H_2O$$

and

$$CH_3CHO + 2[H] \rightarrow CH_3CH_2OH$$

Reaction mechanisms and 'curly arrows'

We can often explain what happens in organic reactions by considering the movement of electrons. As electrons are negatively charged, they tend to move from areas of high electron density to more positively-charged areas.

For example, a lone pair of electrons will be attracted to the positive end of a polar bond, written as $C^{\delta+}$. The movement of a pair of electrons is shown by a curly arrow. The arrow starts from a lone pair of electrons or from a covalent bond, and it moves towards a positively charged area of a molecule to form a new bond. Or an electron pair can move from a bond towards a more electronegative species to become a lone pair on a negatively charged ion. For example:

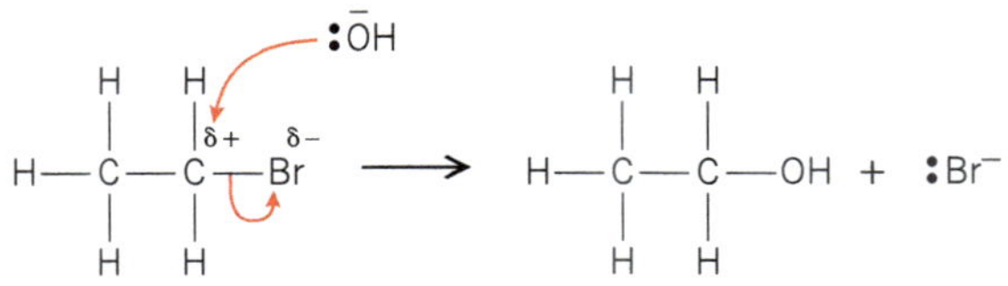

Learning outcomes

On these pages you will learn to:

- describe and predict molecule shapes and bond angles
- describe hybridisation in organic molecules
- explain the formation of σ and π bonds

The hydrocarbon chain of an organic molecule may be straight (unbranched), branched or cyclic.

The principles governing the shapes of organic molecules are the same as those described in Section 3.5, ie the **electron pair repulsion theory**. This says that each pair of electrons around an atom will repel all the other electron pairs, including lone (non-bonding) pairs, so that they become as far apart as possible from each other.

Lone pairs are pulled closer to the nucleus than bonding pairs and will therefore repel more strongly.

Double and triple bonds have four and six electrons respectively, so they also repel more strongly than bonding pairs.

To work out the shape of an organic molecule you need to first draw the dot-and-cross diagram. For example:

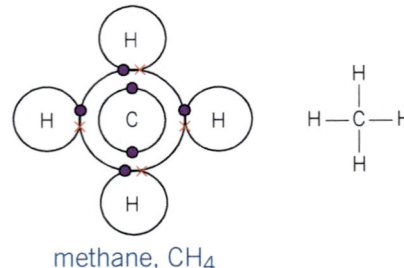

methane, CH_4

So methane has four bonding pairs of electrons and adopts the shape of a regular tetrahedron.

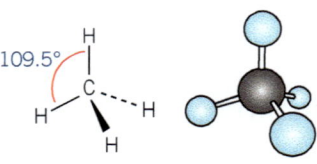

Hybridisation

A more sophisticated picture of the bonding in organic molecules uses the idea of hybrid molecular orbitals, as shown in Section 3.4.

Remember that four orbitals in the outer shell of a carbon atom (one s-orbital and three p-orbitals) can mix together to form **hybrid orbitals**, a process called **hybridisation**. These hybrid orbitals can form bonds by overlapping with orbitals on other atoms.

In carbon, the orbitals involved in the outer shell can mix in three ways, as shown in Figure 5.

Notice that the four original orbitals always give rise to a total of four hybrid orbitals. The notation sp, sp^2 and sp^3 indicates the proportions of the original in the hybrid. So sp^3 is 25% s and 75% p, for example.

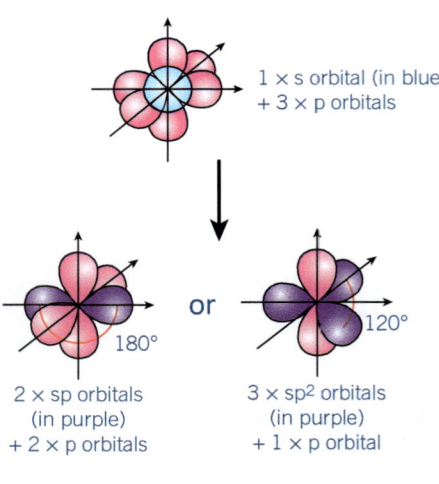

1 × s orbital (in blue)
+ 3 × p orbitals

2 × sp orbitals
(in purple)
+ 2 × p orbitals

180°

3 × sp² orbitals
(in purple)
+ 1 × p orbital

120°

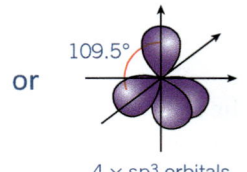

109.5°

4 × sp³ orbitals

Figure 5 Formation of hybrid orbitals—three possibilities

The shapes of the hybrids are as follows:

- sp orbitals are at 180° to each other, ie in a straight line.
- sp^2 orbitals are at 120° to each other, ie in a flat trigonal shape.
- sp^3 orbitals are at 109.5° to each other, ie pointing to the corners of a tetrahedron.

The bonding in ethane, C_2H_6

The ethane molecule is shown in Figure 6. Both carbon atoms are hybridised sp^3.

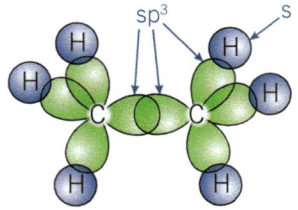

Figure 6 *The bonding in ethane*

The C–C bond is formed by overlap of two sp^3 orbitals to form a σ molecular orbital. Each C–H bond is formed by overlap of an sp^3 orbital on a carbon atom with an s orbital on hydrogen to form a σ molecular orbital. The H–C–H and C–C–H angles are all 109.5°.

The bonding in ethene, C_2H_4

In ethene, each carbon is sp^2 hybridised. This leaves an unhybridised p orbital—see Figure 7.

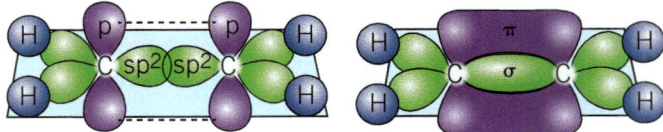

Figure 7 *The bonding in ethene, $H_2C=CH_2$*

The four C–H bonds are formed by the overlap of sp^2 orbitals on carbon with an s orbital on hydrogen, to form σ molecular orbitals. The C–C bond is in two parts:

- a σ orbital formed by the overlap of two sp^2 orbitals
- a π orbital formed by the overlap of the p-orbitals on each carbon atom.

The H–C–H and H–C–C angles are approximately 120°. However, because the four electrons in the double bond repel more than the two electrons in the single bonds, the H–C–H angles are about 118° and the H–C–C angles are about 121°. The molecule as a whole is flat—often described a **planar**—because all six atoms are in the same plane.

The bonding in ethyne H–C≡C–H

In ethyne the carbon atoms are hybridised sp with two p-orbitals remaining. The C–H bonds are formed by the overlap of sp-orbitals on the carbon atoms with s-orbitals on the hydrogens. The carbon–carbon triple bond consists of a σ bond formed by overlap of sp-orbitals, and two π bonds formed by overlap of the p-orbitals.

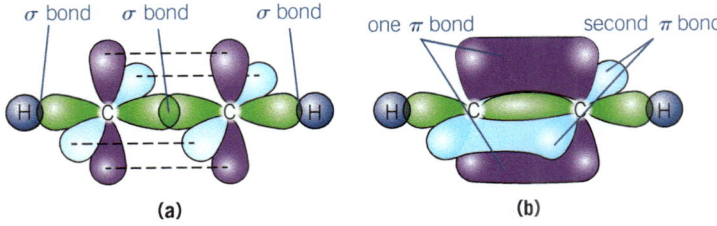

(a)   (b)

Figure 8 *The bonding in ethyne, HC≡CH*

Summary test 13.3

1 The C–C bond energy in ethane is 350 kJ mol⁻¹ and the C=C bond energy in ethene is 610 kJ mol⁻¹. Explain what this suggests about the strength of a carbon–carbon π bond compared with a carbon–carbon σ bond.
2 Suggest why the H–C–H bonds are slightly less than 120°.
3 Explain why the ethyne molecule (H–C≡C–H) is exactly linear.

173

13.4 Isomerism: structural and stereoisomerism

Learning outcomes

On these pages you will learn to:

- describe and explain chain, positional and functional group isomerism
- describe and explain geometrical and optical isomerism
- identify a chiral centre in a molecule

Exam tip

To understand isomerism it is very helpful to use molecular models—even modelling clay and matchsticks will do.

Isomers are molecules with the same molecular formula but a different arrangement of atoms in space.

Structural isomerism—chain, positional and functional group isomers

Chain isomers

If a long chain hydrocarbon has a side chain, you need to say where the side chain is located on the main chain. For example, methylpentane could refer to:

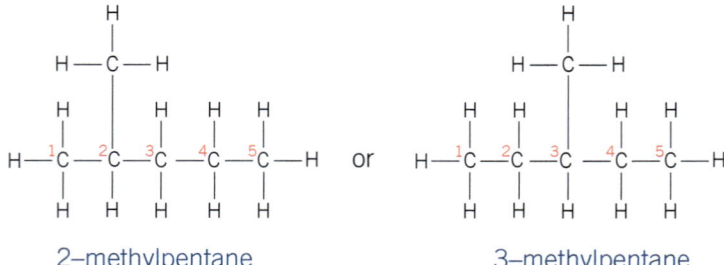

2–methylpentane   3–methylpentane

This is called **chain isomerism**. A number (sometimes called a **locant**) is used to tell us the position of any branching in a chain (and the position of any functional group).

Positional isomers

Positional isomers vary in the position of a functional group on the hydrocarbon chain. The molecules below are a pair of positional isomers:

CH₃—CH₂—CH₂—Br   CH₃—CH—CH₃
                                          |
                                          Br

1-bromopropane   2-bromopropane

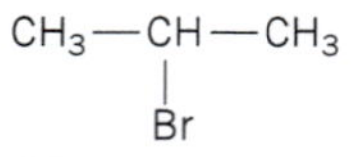

Take care, though. Both molecules below are 1-bromopropane. The right-hand one is not 3-bromopropane because the smallest possible number is always used.

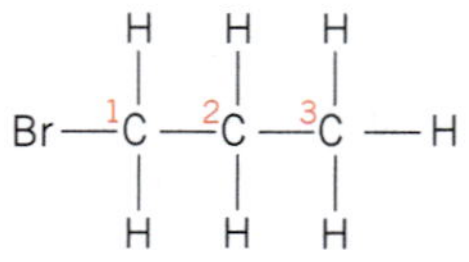

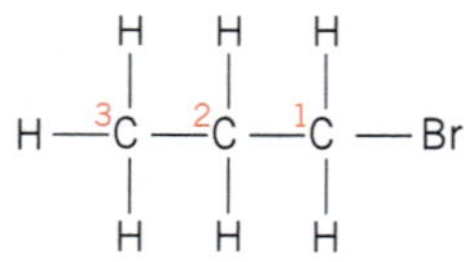

1-bromopropane may also be represented by either of the structural formulae below, because all the hydrogens on carbon 1 are equivalent.

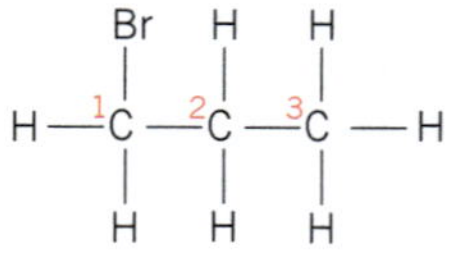

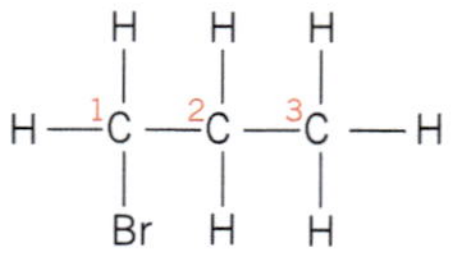

Functional group isomerism

The same molecular formula may contain totally different functional groups. This is **functional group isomerism.**

For example, C_2H_6O could represent:

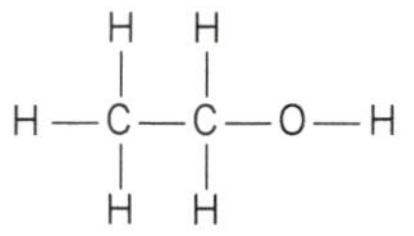

ethanol (an alcohol)   or   methoxymethane (an ether)

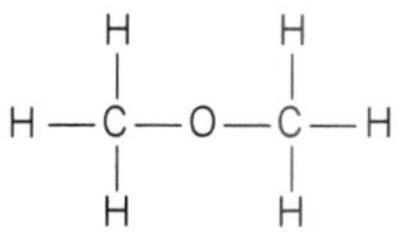

C_3H_6O could represent:

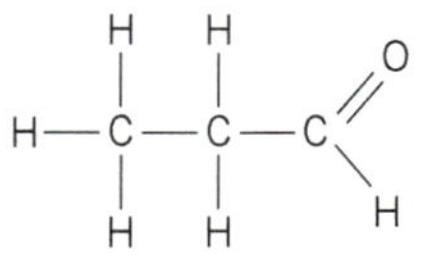

propanal
(an aldehyde)   or   propanone
(a ketone)

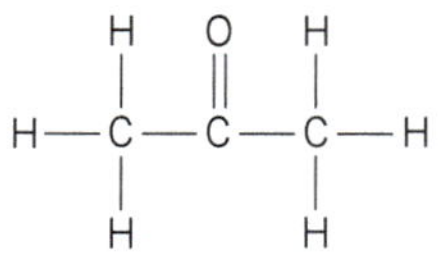

$C_4H_8O_2$ could represent:

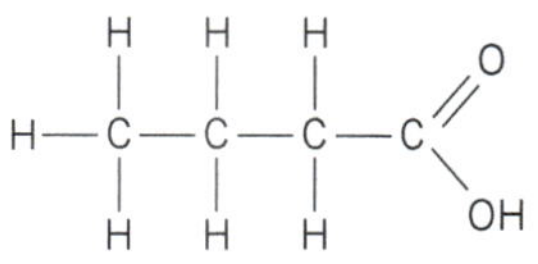

butanoic acid
(a carboxylic acid)   methyl propanoate
(an ester)

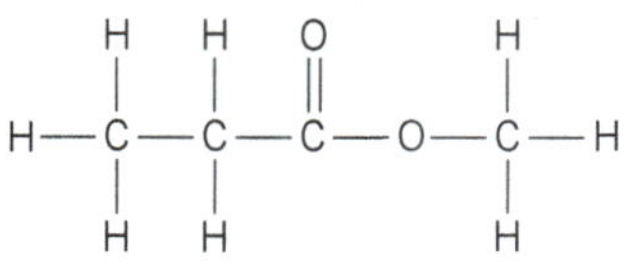

Stereoisomerism

Stereoisomerism is where two (or more) compounds have the same structural formula. They differ in the arrangement of the bonds in space. There are two types:

- **geometrical isomerism**, also called *cis–trans* isomerism (or *E–Z* isomerism)
- **optical isomerism**.

Geometrical isomerism

Geometrical isomerism (also known as *cis–trans* isomerism) tells us about the positions of substituents at either side of a carbon–carbon double bond. (It is also known as *E–Z* isomerism.)

Two substituents may either be

- on the same side of the bond, *cis* (*Z*)
- on opposite sides of the bond, *trans* (*E*)—see Figure 9.

cis-1,2-dichloroethene   trans-1,2-dichloroethene

Figure 9 *Groups cannot rotate about a double bond so these are a pair of isomers*

Cis- and *trans*-isomers are separate compounds and are not easily converted from one to the other. There is no rotation possible around the double bond, as this would involve breaking the π part of the bond (which consists of a σ plus a π orbital)—see Figure 7 in Section 13.3.

Where there is a double bond, no rotation is possible and two isomers can be distinguished. Substituted groups joined by a single bond can rotate around the single bond, so there are no isomers (Figure 10).

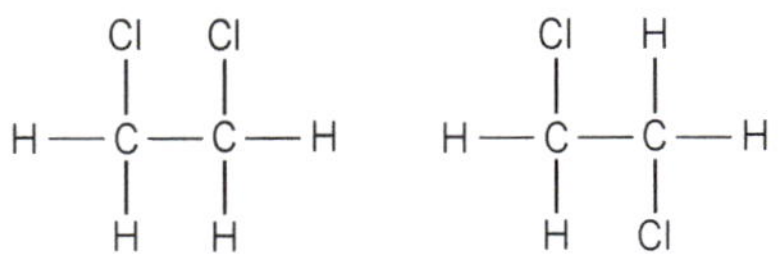

Figure 10 *Groups can rotate around a single bond. These are representations of the same molecule and are not isomers*

Extension

E–Z is from the German entgegen (opposite—*trans*) and zusammen (together—*cis*).

Exam tip

Each carbon atom in the C=C bond must be bonded to two different atoms for *cis–trans* isomerism to occur.

Extension

Geometric isomerism in dienes
In the case of dienes (molecules with two carbon–carbon double bonds), each C=C may be *cis* or *trans*. This makes a possible four isomers—*cis-cis, cis-trans, trans-cis and cis-cis*. The molecule below illustrates this. The isomer shown is *trans-cis*.

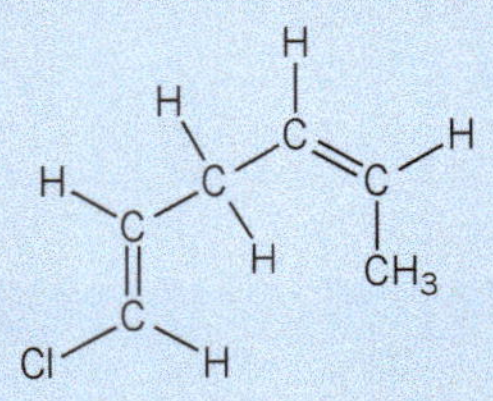

The left hand C=C is *trans* and the right hand one is *cis*. Try drawing the other three isomers of this molecule.

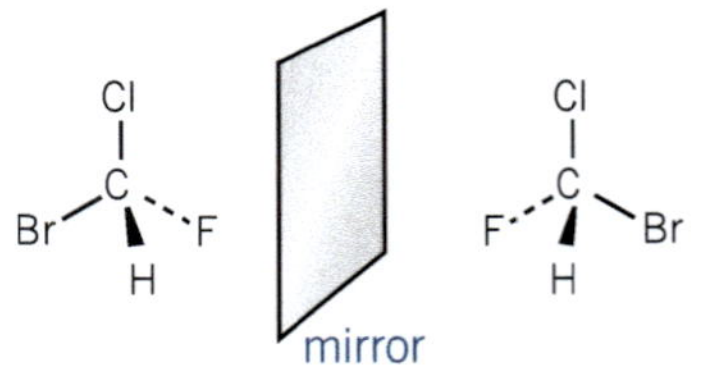

Optical isomerism

Optical isomers occur when there are four different substituents attached to one carbon atom. This results in two isomers that are non-superimposable mirror images of one another, but are not identical. For example, bromochlorofluoromethane exists as two mirror image forms:

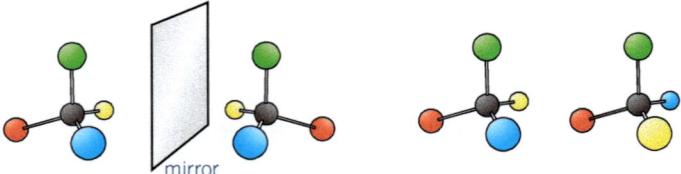

***Figure 11** Bromochlorofluoromethane has a pair of mirror isomers which are not identical*

The ball and stick models of bromochlorofluoromethane in Figure 11 may help you to see that these are not identical.

Imagine rotating one of the molecules about the C–Cl bond (pointing upwards) until the two bromine atoms (in red) are in the same position.

The positions of the hydrogen (blue) and fluorine atoms (yellow) will not match—you cannot superimpose one molecule onto the other.

This is just like a pair of shoes. A left shoe and a right shoe are mirror images, but they are not identical—they cannot be superimposed.

Pairs of molecules like this are called optical isomers, because they differ in the way they rotate the plane of polarisation of **polarised light**—either clockwise ((+)-isomer) or anticlockwise ((–)-isomer).

> **Exam tip**
>
> Pairs of optical isomers can only be distinguished on paper using 3D representations like those above. ▲ represents a bond coming out of the paper. ⋯ represents a bond going into the paper.

> **Exam tip**
>
> Drawn structures of optical isomers should show a tetrahedral arrangement of groups around a central carbon atom as in the structures above.

Chirality

Optical isomers are said to be **chiral**—this means 'handed', as in left- and right-handed. The two isomers are called a pair of **enantiomers**. The carbon bonded to the four different groups is called the **chiral centre** or the **asymmetric carbon atom**. It is often indicated on formulae by *. You can easily pick out a chiral molecule because it contains at least one carbon atom that has four different groups attached to it.

- All α-amino acids—except aminoethanoic acid (glycine), the simplest one (Figure 12)—have a chiral centre. For example, the chiral centre of α-aminopropanoic acid (2-aminopropanoic acid) is:

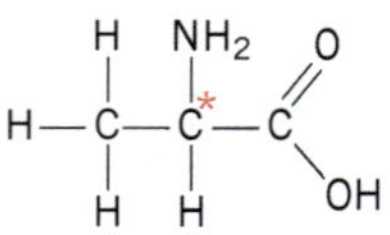

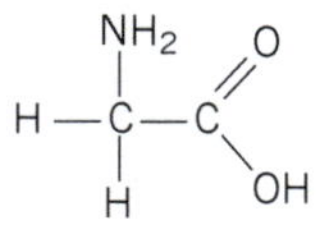

glycine

***Figure 12** Aminoethanoic acid (glycine)*

- 2-hydroxypropanoic acid (non-systematic name lactic acid) is also chiral. Although the chiral carbon is bonded to two other carbon atoms, these carbons are part of different groups and you must count the whole group.

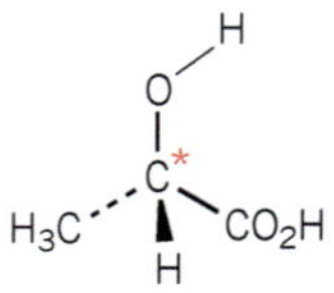

Optical isomerism happens because of the three-dimensional structures of the isomers. So it can only be shown by three-dimensional representations or models.

> **Exam tip**
>
> More complex molecules may have two or more chiral centres.

Extension

Light consists of vibrating electric and magnetic fields. You can think of it as waves with vibrations occurring in all directions at right angles to the direction of motion of the light wave. If the light passes through a special filter, called a **polaroid** (as in polaroid sunglasses) all the vibrations are cut out except those in one plane, for example, the vertical plane (Figure 13).

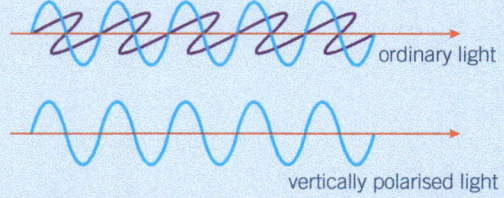

ordinary light

vertically polarised light

Figure 13 *Polarised light*

The light is now vertically polarised and it will be affected differently by different optical isomers of the same substance.

Optical rotation can be measured using a **polarimeter** (Figure 15).

1 Polarised light is passed through two solutions of the same concentration, each containing a different optical isomer of the same substance.
2 One solution will rotate the plane of polarisation through a particular angle, clockwise. This is the (+)-isomer.
3 The other will rotate the plane of polarisation by the same angle, anticlockwise. We call this the (−)-isomer.

(There are several other systems in use for distinguishing pairs of isomers: as well as (+) and (−) you may see R and S, D and L, or d and l.)

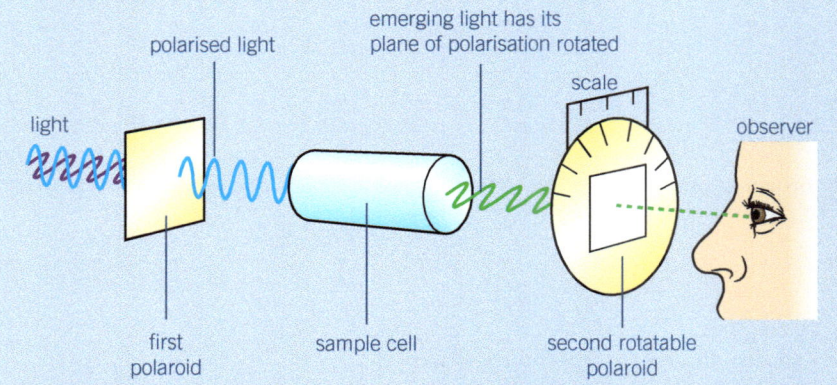

emerging light has its plane of polarisation rotated

polarised light

scale

light

observer

first polaroid

sample cell

second rotatable polaroid

Figure 15 A polarimeter measures rotation of the plane of polarisation of polarised light

Figure 14 *The lenses of polarised sunglasses filter sunlight in the same way a polaroid filters light in a polarimeter*

Exam tip

When determining if a molecule has a chiral centre, it may help if you draw in all the hydrogen atoms that are not shown explicitly in the skeletal formula.

Worked example

Identifying isomers

Optical isomers

In principle this is straightforward—you look for a carbon with four different groups attached and this is your chiral centre. However in some cases it can be tricky. Here are some examples:

1 Does the following structure have a chiral centre?

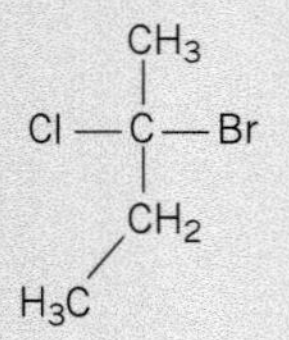

At first sight you might say 'no', because the central carbon atom has a Cl atom, a Br atom and two C atoms bonded to it. However, you must look at the whole of any groups attached. These are Cl, Br, CH_3 and C_2H_5—four different groups. So the molecule is chiral.

2 Does the molecule below have chiral centre?

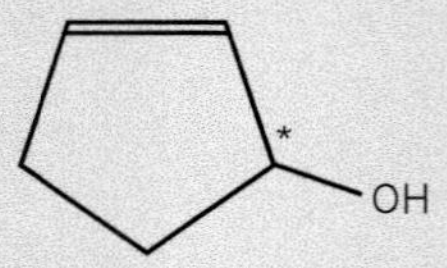

First remember that in a skeletal formula each carbon atom has enough H atoms to make four bonds (counting a double bond as two). So the starred C atom has a H atom bonded to it that is not drawn here. It is still bonded to an O atom, a H atom and two C atoms. However, you need to work backwards from the starred atom. Going clockwise you come first to a CH_2 group. But going anticlockwise, you come first to a carbon atom which is forming a double bond—these are different groups, so the starred carbon is chiral.

3 Does the compound below have a chiral centre?

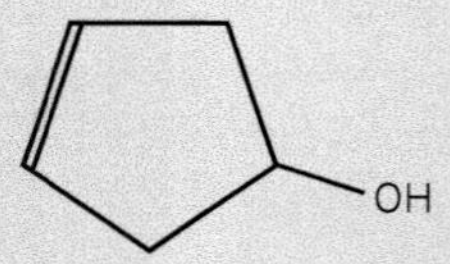

The carbon atom bonded to the OH group also has a hydrogen atom. However, the other groups are the same whichever way you go around the ring, so this carbon atom is not chiral.

Cis–trans isomers

You need two conditions:

1 a carbon–carbon double bond so that there is restricted rotation
2 two different groups on one side of the double bond and two different groups on the other.

Which of the pairs of molecules below can have *cis–trans* isomerism?

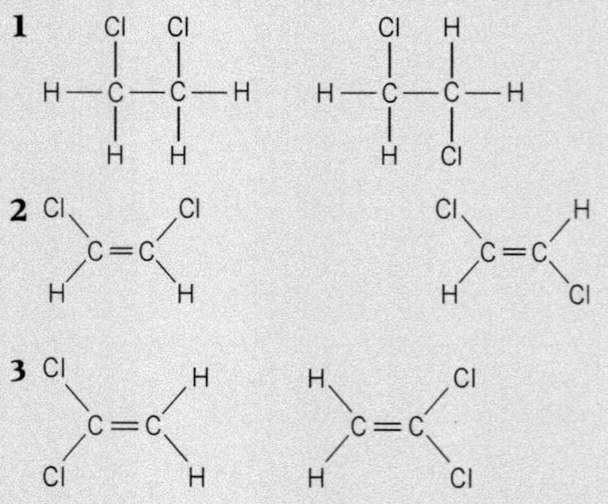

Pair 1 are different representations of the same molecule. Rotation is possible about the C—C single bond.

Pair 2 are *cis–trans* isomers. There is no rotation about the C=C double bond and there are two different groups on one side of the double bond and two different groups on the other.

Pair 3 are different representations of the same molecule. Rotating the left-hand structure 180° clockwise will give the right-hand structure.

Numbers of isomers
As the size of molecules increases, the number of possible isomers increases rapidly. Methane, ethane and propane have no isomers, butane has two (see below) and pentane has three. Decane has 75!

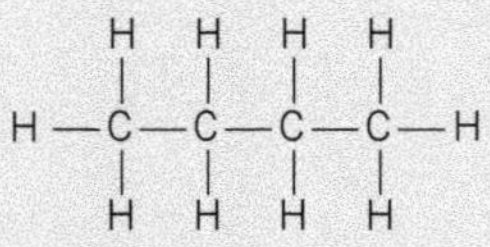

butane 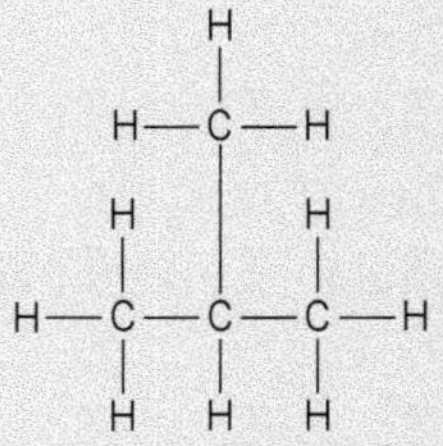   methylpropane

Cholesterol has eight chiral centres and 256 different optical isomers.

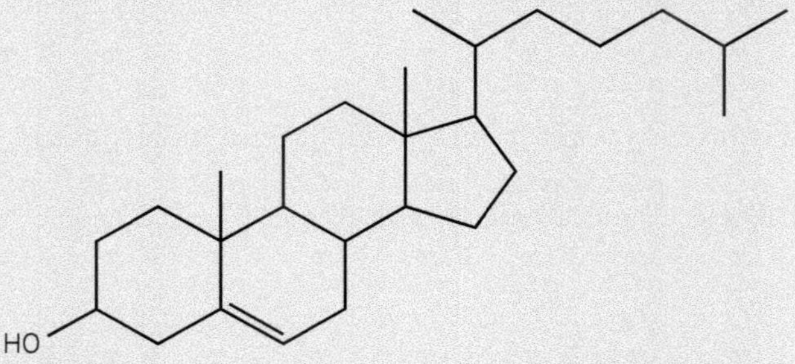

cholesterol

Summary test 13.4

1 a Give the name of this:

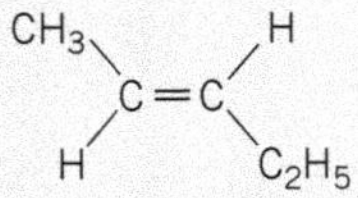

b State the name of its geometrical isomer.

2 State which of the following compounds show optical isomerism:

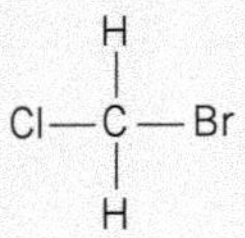

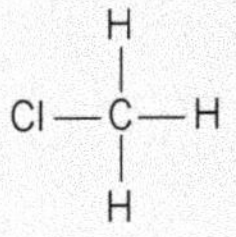

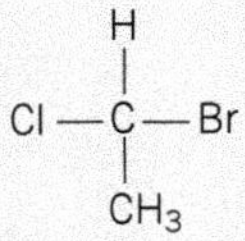

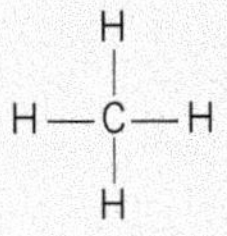

3 Identify the chiral centre on this molecule using a *:

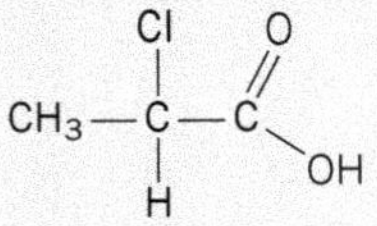

Draw the two optical isomers using 3D representations.

When drawing isomers you must take care not to draw the same molecule twice from different angles.

4 The molecule below has two chiral centres. Identify them both.

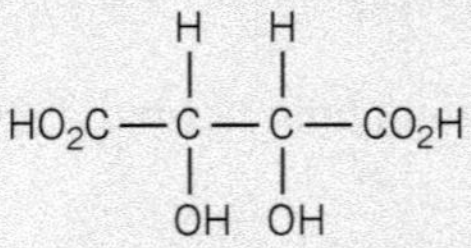

$$HO_2C - \underset{\underset{OH}{|}}{\overset{\overset{H}{|}}{C}} - \underset{\underset{OH}{|}}{\overset{\overset{H}{|}}{C}} - CO_2H$$

5 State the type of structural isomerism that is shown by the following pairs of molecules. Choose from: A = functional groups at different points, B = different functional groups, C = chain branching.

a $CH_3CH_2OCH_3$ and $CH_3CH_2CH_2OH$

b $CH_3CH_2CH_2OH$ and $CH_3CH(OH)CH_3$

c $CH_3CH_2CH_2CH_2CH_3$ and $CH_3CH(CH_3)CH_2CH_3$

6 a Give the displayed and structural formulae for all the five isomers of hexane, C_6H_{14}.

b Give the names of these isomers.

7 State which of these molecules can show *cis–trans* isomerism.

A $CH_2{=}CH_2$

B $CH_3{-}CH_3$

C $RCH{=}CH_2$

D $RCH{=}CHR$

8 Identify as many chiral centres as you can on the cholesterol molecule (on the previous page).

9 Sketch (and identify) the three isomers of pentane.

> ⬚ **Launch additional digital resources for the chapter**

1 Alkanes are simple hydrocarbons within a homologous series with the general formula C_nH_{2n+2}.

a Define 'hydrocarbon'. *(1 mark)*

b Explain what is a 'homologous series'. *(2 marks)*

c Define the molecular formula of an alkane containing 18 carbon atoms. *(1 mark)*

d Explain why compounds in a homologous series have similar chemical properties. *(1 mark)*

2 State the homologous series and draw the displayed formulae for the following compounds:

a $CH_3CH_2CH_2OH$

b $CH_3CH_2COCH_3$

c CH_3CH_2Cl

d CH_3CH_2COOH

e $CH_3CH=CHCH_3$

f H_2NCH_2COOH

g $CH_3CH_2OCH_2CH_2CH_3$ *(7 marks)*

3 a Write the full structural formulae of all the isomers of the following, stating which type of isomerism is involved:
  i C_3H_7Cl
  ii C_6H_{14}
  iii $C_2H_3Cl_2Br$ *(3 marks)*

b Write the structural and molecular formulae for the following alkanes:
  i ethylcyclohexane
  ii 1,2-dimethylcyclopentane
  iii 2,2,3-trimethylbutane
  iv 3,4-diethyl-2,2-dimethylheptane *(8 marks)*

4 In methane, CH_4, the carbon atom has four hybrid sp^3 orbitals.

a State the number of s orbitals and p orbitals that merge to form one sp^3 hybrid orbital. *(1 mark)*

b In methane, each carbon sp^3 orbital overlaps with one orbital in a hydrogen atom. Give the name of this hydrogen atom orbital.
  A 1s   **B** 2s   **C** 2p   **D** 3s   *(1 mark)*

c In methane, what type of bond forms between a carbon atom and a hydrogen atom?
  A α   **B** β   **C** σ   **D** π   *(1 mark)*

d Explain why a methane molecule is tetrahedral. *(1 mark)*

5 Consider the following compounds:

A hex-2-ene, $CH_3CH_2CH_2CH=CHCH_3$

B hex-1-ene, $CH_3CH_2CH_2CH_2CH=CH_2$

C hexane, $CH_3CH_2CH_2CH_2CH_2CH_3$

D cyclohexane,

$$H_2C \overset{\displaystyle CH_2}{\underset{\displaystyle CH_2}{<}} \overset{\displaystyle CH_2}{\underset{\displaystyle CH_2}{>}} CH_2$$

State which compound(s):

a Would decolourise bromine in the absence of sunlight.

b Would react with chlorine, but only when heated or exposed to light.

c Would absorb 1 mole of hydrogen per mole of the compound in the presence of a nickel catalyst.

d Has *cis* and *trans*-isomers.

e Are unsaturated. *(5 marks)*

6 Butene (C_4H_4) is able to exist as two positional isomers in which only one is a stereoisomer.

a Define the term 'stereoisomer'. *(2 marks)*

b Give the displayed formula of the isomer of butene that shows stereoisomerism. Name this type of isomerism and explain how it occurs. *(4 marks)*

c For each of the following compounds, state whether you would expect them to show *cis–trans* isomerism. For the compounds that show *cis–trans* isomerism, draw and name the structures of all possible isomers.

(4 marks)

14.1 Alkanes

Alkanes are saturated hydrocarbons. They are used as fuels and lubricants and as the starting materials for making other compounds. They are not normally synthesised as they are usually derived from crude oil by fractional distillation.

The general formula

Alkanes may be unbranched chains, branched chains, or rings. The general formula for all chain alkanes is C_nH_{2n+2}.

Unbranched chains

Unbranched chains are often called straight chains although the C–C–C angle is 109.5°. This means that the chains are not literally straight. In an unbranched alkane, each carbon atom has two hydrogen atoms, except the end carbons which each have one extra.

For example, pentane, C_5H_{12}:

$CH_3CH_2CH_2CH_2CH_3$

displayed   structural

Branched chains

For example, methylbutane, C_5H_{12}, which is an isomer of pentane:

$CH_3CH_2CH(CH_3)CH_3$

displayed   structural

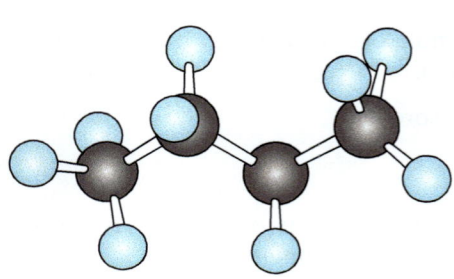

Figure 1 A 3D representation of an unbranched chain alkane, butane. Butane has four carbons in its chain

Ring alkanes

Ring alkanes have the general molecular formula C_nH_{2n} because the end hydrogens are not required.

Naming alkanes

Straight chains

Alkanes are named from the root, which tells us the number of carbon atoms. They have the suffix -*ane*, denoting an alkane, see Table 1.

Branched chains

When you are naming a hydrocarbon with a branched chain, you must first find the longest unbranched chain. This can sometimes be a bit tricky (see the example below). This gives the root name. Then name the branches or side chains as prefixes—methyl-, ethyl-, propyl-, and so on. Finally, add locants—these are numbers that say which carbon atoms the side chains are attached to.

Table 1 Names of the first six straight chain alkanes

methane	CH_4
ethane	C_2H_6
propane	C_3H_8
butane	C_4H_{10}
pentane	C_5H_{12}
hexane	C_6H_{14}

Worked example

Both the hydrocarbons below are the same, though they may seem different at first sight.

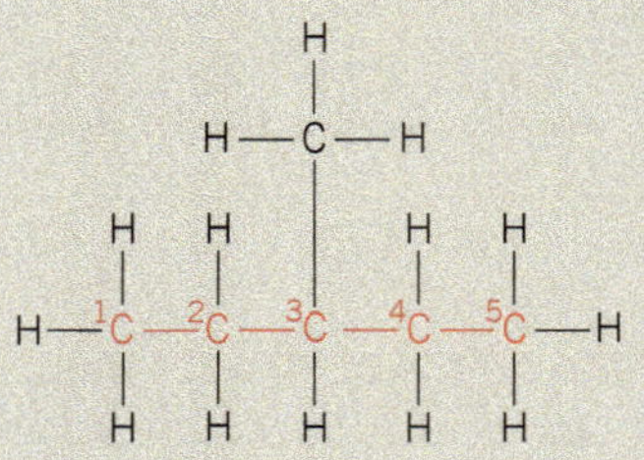

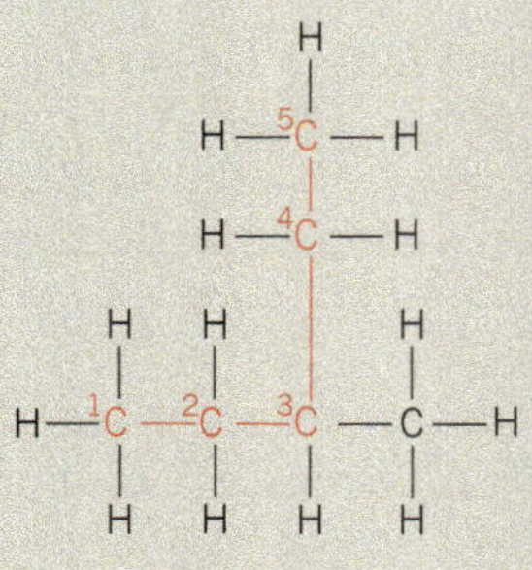

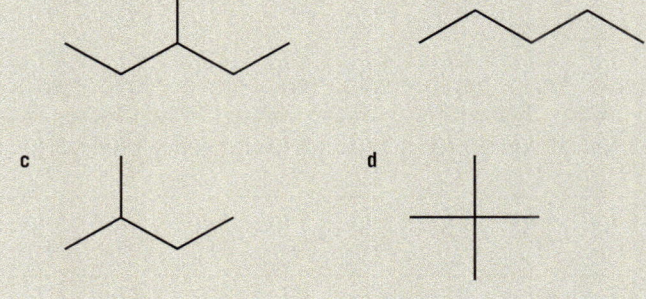

In both, the longest unbranched chain (in red) is five carbons, so the root is pentane. The only side chain has one carbon so it is methyl-. It is attached at carbon 3 so the full name is 3-methylpentane.

Which of the following is the skeletal formulae for 3-methylpentane?

a

b

c

d

Answer: a

Isomerism

Methane, ethane, and propane have no isomers. The number of possible chain isomers then increases with the number of carbons in the alkane. Butane, with four carbons, has two isomers; and pentane has three isomers:

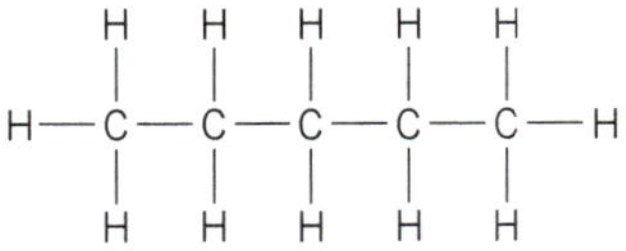

pentane 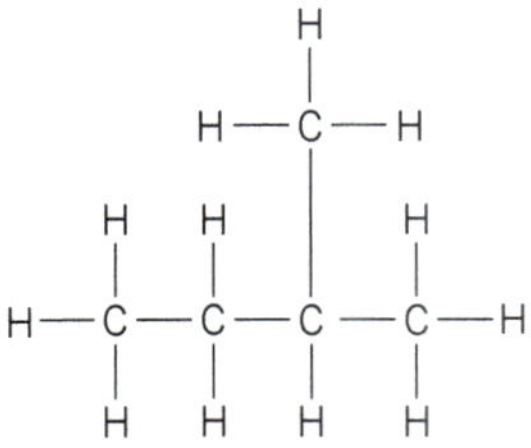   methylbutane 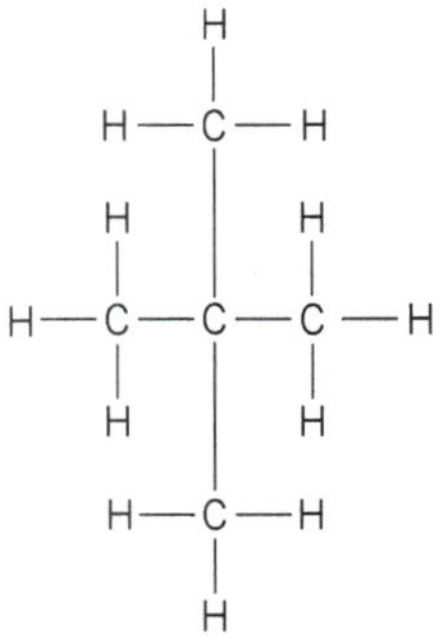   2,2-dimethylpropane

The number of isomers rises rapidly with chain length. Decane, $C_{10}H_{22}$, has 75 and $C_{30}H_{62}$ has over 4 billion.

Figure 2 *Campingaz is a mixture of propane and butane. Polar expeditions use special gas mixtures with a higher proportion of propane, because butane is liquid below 272 K (–1°C)*

increasing chain length

Figure 3 *The effect of increasing chain length on the physical properties of alkanes*

Physical properties

Physical properties are those that do not involve chemical reactions.

Polarity

Alkanes are almost non-polar because the electronegativities of carbon (2.5) and hydrogen (2.1) are so similar. As a result, the only intermolecular forces between their molecules are weak London dispersion forces, and the larger the molecule, the stronger the London dispersion forces.

Boiling points

This increasing intermolecular force is why the boiling points of alkanes increase as the chain length increases. The shorter chains are gases at room temperature. Pentane, with five carbons, is a liquid with a low boiling point of 309 K (36°C). At a chain length of about 18 carbons, the alkanes become solids at room temperature. The solids have a waxy feel.

Alkanes with branched chains have lower melting points than straight chain alkanes with the same number of carbon atoms. This is because they cannot pack together as closely as unbranched chains and so the London dispersion forces are not so effective.

Solubility

Alkanes are insoluble in water. This is because water molecules are held together by hydrogen bonds which are much stronger than the van der Waal's forces that act between alkane molecules. However, alkanes do mix with other relatively non-polar liquids.

Synthesising alkanes

In the laboratory, alkanes may be made by hydrogenating **alkenes** using hydrogen gas and a catalyst such as platinum or nickel at a temperature of around 150°C. For example with ethene:

$$CH_2{=}CH_2 \ + \ H_2 \ \xrightarrow{\text{Ni}} \ CH_3{-}CH_3$$

A similar reaction is used to hydrogenate oils—which contain long carbon chains which have carbon–carbon double bonds – to make margarine. This is known as hardening oils as the products have higher melting points (are more solid) than the starting materials.

Long chain hydrocarbons can be **cracked** – heated with a catalyst – to produce a mixture of shorter chain alkanes and also alkenes. In the laboratory this is done by heating the vapour of the long chain alkane such as liquid paraffin and passing it over a catalyst of aluminium oxide or even porcelain chips (see Figure 4). This reaction will produce a mixture of shorter chain alkanes and alkenes.

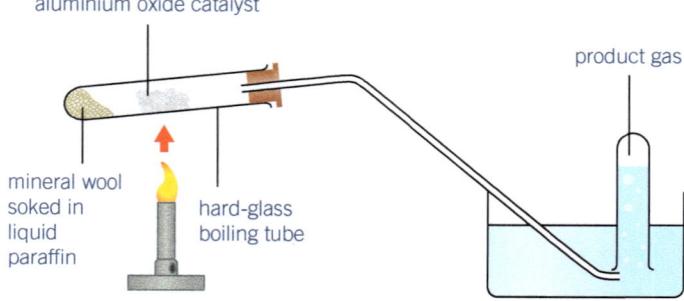

Figure 4 *Cracking an alkane*

For example, cracking decane might give octane and ethene:

Reactions of alkanes

Alkanes are generally unreactive. They have strong C–C and C–H bonds only. The bond energy of C–C is $350\,kJ\,mol^{-1}$ and that of C–H is $410\,kJ\,mol^{-1}$. This means that the bonds are hard to break. They are also relatively non-polar, so they are not attacked by charged reagents. This means that alkanes have only three important reactions—combustion, cracking and free radical substitution with halogens.

Combustion

The shorter chain alkanes burn completely in a plentiful supply of oxygen to give carbon dioxide and water. For example:

$$CH_4(g) + 2O_2(g) \rightarrow CO_2(g) + 2H_2O(l) \qquad \Delta H = -890 \text{ kJ mol}^{-1}$$

$$C_2H_6(g) + 3\tfrac{1}{2}O_2(g) \rightarrow 2CO_2(g) + 3H_2O(l) \qquad \Delta H = -1559.7 \text{ kJ mol}^{-1}$$

Combustion reactions give out heat. They have large negative enthalpies of combustion. The more carbons present, the greater the heat output. For this reason they are important as fuels. Alkanes store a large amount of energy for a small amount of weight. For example, octane produces approximately 48 kJ of energy per gram when burnt, which is about twice the energy output per gram of coal. Examples of alkane fuels include:

- methane (the main component of natural gas)
- propane (camping gas)
- butane (Calor gas)
- petrol (a mixture of hydrocarbons of approximate chain length C_8)
- paraffin (a mixture of hydrocarbons of chain lengths C_{10} to C_{18}).

Incomplete combustion

In a limited supply of oxygen, the poisonous gas carbon monoxide, CO, is formed. For example, with propane:

$$C_3H_8(g) + 3\tfrac{1}{2}O_2(g) \rightarrow 3CO(g) + 4H_2O(l)$$

This is called **incomplete combustion**.

With even less oxygen, carbon (soot) is produced. For example, when a Bunsen burner is used with a closed air hole, the flame is yellow and a black sooty deposit appears on the apparatus. Incomplete combustion often happens with longer chain hydrocarbons, which need more oxygen to burn compared with shorter chains.

Environmental problems

Burning alkanes as fuel has several environmental consequences. Carbon dioxide is a greenhouse gas which causes global warming. The production of carbon monoxide by incomplete combustion has been mentioned above. There is also the possibility of unburned hydrocarbons entering the environment. Finally, at the temperatures found in internal combustion engines, nitrogen and oxygen from the air will combine to form oxides of nitrogen, NO and NO_2 (together called NO_x). These can combine with unburned alkanes to form a photochemical smog (as discussed in Section 12.1). Modern vehicles have catalytic converters in their exhaust systems which contain a honeycomb coated with platinum and rhodium metals. This catalyses the reaction of these pollutants to less-polluting products.

> **Exam tip**
>
> The general equation for the complete combustion of an alkane with the general formula C_nH_{2n+2} is
>
> $$C_nH_{2n+2} + \frac{3n+1}{2}O_2$$
> $$\downarrow$$
> $$nCO_2 + (n+1)H_2O$$
>
> For alkenes of formula C_nH_{2n} it is
>
> $$C_nH_{2n} + \frac{3n}{2}O_2$$
> $$\downarrow$$
> $$nCO_2 + nH_2O$$

$$2CO(g) \quad + \quad 2NO(g) \quad \rightarrow \quad N_2(g) \quad + \quad 2CO_2(g)$$

carbon monoxide   nitrogen oxide   nitrogen   carbon dioxide

hydrocarbons + nitrogen oxides → nitrogen + carbon dioxide + water

For example, $C_8H_{18} + 25NO \rightarrow 12\frac{1}{2}N_2 + 8CO_2 + 9H_2O$

The reactions take place on the surface of the catalyst, on the layer of platinum and rhodium metals, converting the hydrocarbons into carbon dioxide and water.

Extension

Fractional distillation of crude oil

Crude oil is at present the world's main source of organic chemicals. It is called a fossil fuel because it was formed millions of years ago by the breakdown of plant and animal remains at the high pressures and temperatures deep below the Earth's surface. Because it forms very slowly, it is effectively non-renewable.

Crude oil is a mixture mostly of alkanes, both unbranched and branched. Crude oils from different sources have different compositions.

The composition of a typical crude oil is given in Table 2.

Table 2 *The composition of a typical crude oil*

	Gases	Petrol	Naphtha	Kerosene	Gas oil	Fuel oil and wax
Approximate boiling point / K	310	310–450	400–490	430–523	590–620	above 620
Chain length	1–5	5–10	8–12	11–16	16–24	25+
Percentage present	2	8	10	14	21	45

Crude oil contains small amounts of other compounds dissolved in it. These come from other elements in the original plants and animals the oil was formed from, for example, some contain sulfur. These produce sulfur dioxide, SO_2, when they are burnt. This is one of the causes of acid rain – sulfur dioxide reacts with oxygen high in the atmosphere to form sulfur trioxide. This reacts with water in the atmosphere to form sulfuric acid.

To convert crude oil into useful products you have to separate the mixture. This is done by heating it and collecting the fractions that boil over different ranges of temperatures. Each fraction is a mixture of hydrocarbons of similar chain length and therefore similar properties, see Figure 5.

The process is called **fractional distillation** and it is done in a fractionating tower.

- The crude oil is first heated in a furnace.
- A mixture of liquid and vapour passes into a tower that is cooler at the top than at the bottom.
- The vapours pass up the tower via a series of trays containing bubble caps until they arrive at a tray that is sufficiently cool (at a lower temperature than their boiling point). Then they condense to liquid.
- The mixture of liquids that condenses on each tray is piped off.
- The shorter chain hydrocarbons condense in the trays nearer to the top of the tower, where it is cooler, because they have lower boiling points.
- The thick residue that collects at the base of the tower is called tar or bitumen. It can be used for road surfacing but, as supply often exceeds demand, this fraction is often further processed to give more valuable products

1 Draw the displayed formula and structural formula of hexane.
2 In which of the crude oil fractions named in Table 2 is hexane most likely to be found?
3 What is fractional distillation and how is it different from distillation?
4 Give the names of two gases produced in fractional distillation.

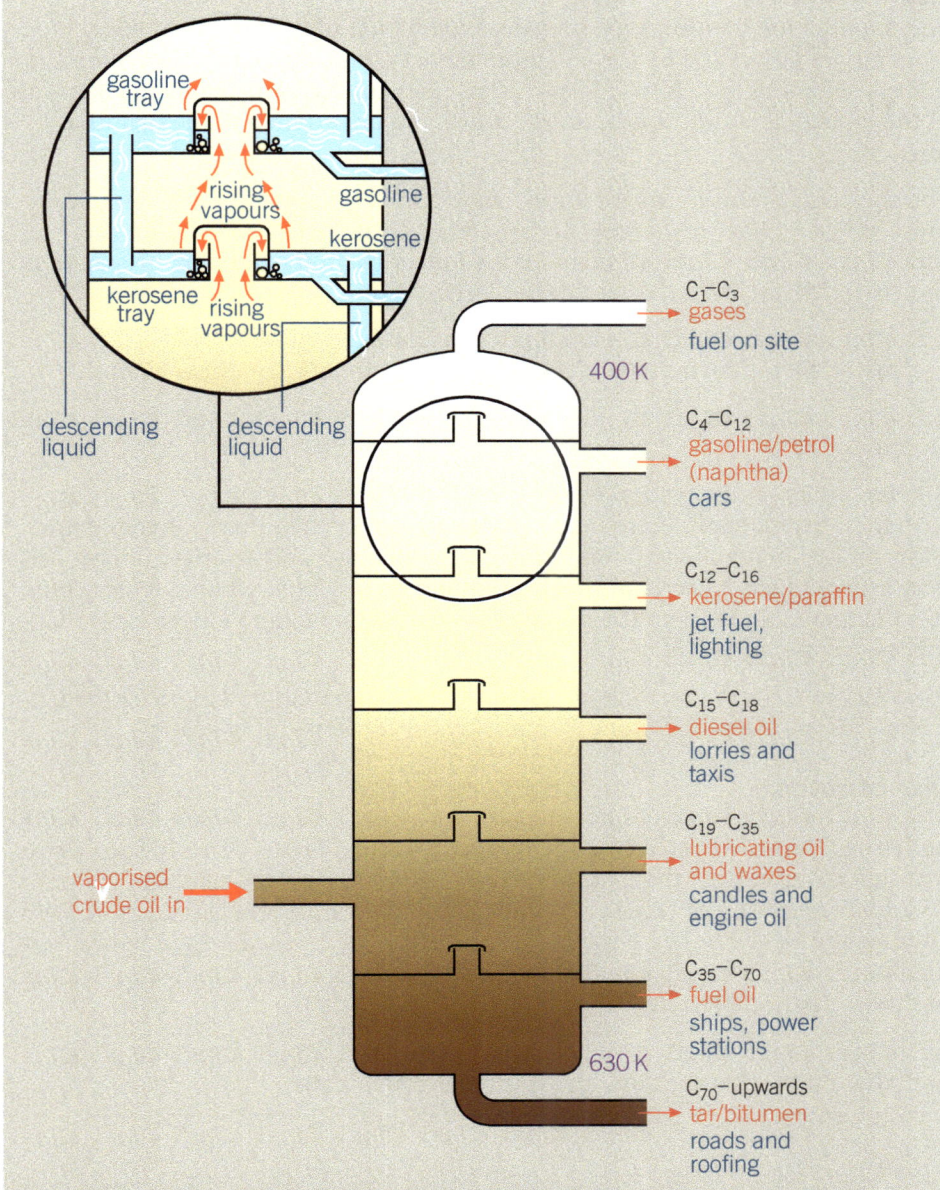

gasoline
tray

rising
vapours

gasoline

kerosene

kerosene
tray

rising
vapours

descending
liquid

descending
liquid

C_1–C_3
gases
fuel on site

400 K

C_4–C_{12}
gasoline/petrol
(naphtha)
cars

C_{12}–C_{16}
kerosene/paraffin
jet fuel,
lighting

C_{15}–C_{18}
diesel oil
lorries and
taxis

vaporised
crude oil in

C_{19}–C_{35}
lubricating oil
and waxes
candles and
engine oil

C_{35}–C_{70}
fuel oil
ships, power
stations

630 K

C_{70}–upwards
tar/bitumen
roads and
roofing

Figure 5 *The fractional distillation of crude oil. The chain length ranges are approximate*

Figure 6 *Crude oil is separated into fractions by distillation in cylindrical towers typically 8 m in diameter and 40 m high. Oil refineries vary but a typical one might process 3.5 million tonnes of crude oil per year*

Figure 7 *A range of products obtained from crude oil*

Industrial cracking

The naphtha fraction from the fractional distillation of crude oil is in huge demand, for petrol and by the chemical industry. The longer chain fractions are not as useful and therefore of lower value economically. Most crude oil has more of the longer chain fractions than is wanted and not enough of the naphtha fraction.

The shorter chain products (of lower relative molecular mass) are economically more valuable than the longer chain material. To meet the demand for the shorter chain hydrocarbons, many of the longer chain fractions are broken into shorter lengths (cracked). This has two useful results:

- shorter, more useful chains are produced, especially petrol
- some of the products are alkenes, which are more reactive than alkanes.

Petrol is a mixture of mainly alkanes containing between four and twelve carbon atoms.

Alkenes are used as chemical feedstock, which means they supply industries with the starting materials to make different products. They are converted into a huge range of other compounds, including polymers and a variety of products from paints to drugs. Perhaps the most important alkene is ethene, which is the starting material for poly(ethene) (also called polythene) and a wide range of other everyday materials.

Alkanes are very unreactive and harsh conditions are required to break them down. There are a number of different ways of carrying out cracking.

Thermal cracking

Thermal cracking involves heating alkanes to a high temperature, 700–1200 K, under high pressure, up to 7000 kPa. Carbon–carbon bonds break in such a way that one electron from the pair in the covalent bond goes to each carbon atom. So initially two shorter chains are produced, each ending in a carbon atom with an unpaired electron. These fragments are called **free radicals**. Free radicals are highly reactive intermediates. They react in a number of ways to form a variety of shorter chain molecules.

As there are not enough hydrogen atoms to produce two alkanes, one of the new chains must have a C=C, and is therefore an alkene:

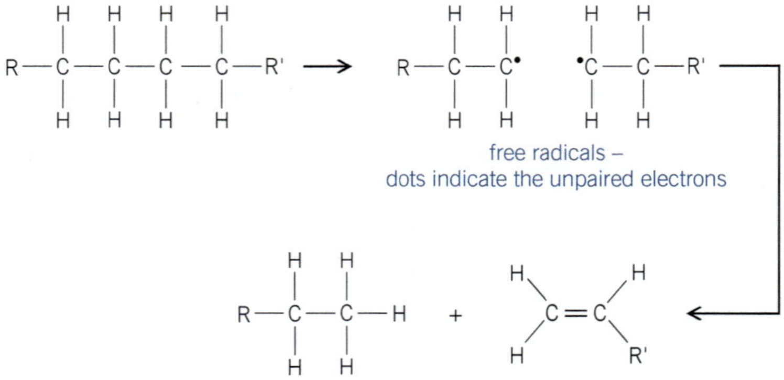

free radicals –
dots indicate the unpaired electrons

Figure 8 *Thermal cracking*

Any number of carbon–carbon bonds may break and the chain does not necessarily break in the middle. Hydrogen may also be produced. Thermal cracking tends to produce a high proportion of alkenes. To avoid too much decomposition (ultimately to carbon and hydrogen) the alkanes are kept in these conditions for a very short time, typically one second. The equation in Figure 8 shows cracking of a long chain alkane to give a shorter chain alkane and an alkene. The chain could break at any point.

Catalytic cracking

Catalytic cracking takes place at a lower temperature (approximately 720 K) and lower pressure (but more than atmospheric), using a zeolite catalyst, consisting of silicon dioxide and aluminium oxide (aluminosilicates). Zeolites have a honeycomb structure with an enormous surface area. They are also acidic. This form of cracking is used mainly to produce motor fuels. The products are mostly branched alkanes, cycloalkanes (rings), and aromatic compounds (compounds based on the six-carbon benzene ring with delocalised bonding, see Section 30.1).

Free radical substitution of alkanes

When you put a mixture of an alkane and a halogen into bright sunlight, or shine a photographer's lamp onto the mixture, the alkane and the halogen will react to form a halogenoalkane. The ultraviolet component of the light starts the reaction. Alkanes do not react with halogens in the dark at room temperature.

For example, if you put a mixture of hexane and a little liquid bromine into a test tube and leave it in the dark, it stays red-brown (the colour of bromine). However, if you shine ultraviolet light onto it, the mixture becomes colourless and misty fumes of hydrogen bromide appear.

A substitution reaction has taken place. One or more of the hydrogen atoms in the alkane has been replaced by a bromine atom and hydrogen bromide is given off as a gas. The main reaction is:

$$C_6H_{14}(g) + Br_2(l) \rightarrow C_6H_{13}Br(l) + HBr(g)$$
$$\text{hexane} \quad \text{bromine} \quad \text{bromohexane} \quad \text{hydrogen bromide}$$

Bromohexane is a halogenoalkane.

Chain reactions

The reaction above is called a free radical substitution. It is chain reaction which takes place in three stages—initiation, propagation, and termination.

The reaction between any alkane and a halogen has this mechanism.

For example, ethane and chlorine:

$$C_2H_6(g) + Cl_2(g) \rightarrow C_2H_5Cl(g) + HCl(g)$$

Initiation

- The first, or initiation, step of the reaction is breaking the Cl—Cl bond to form two chlorine atoms.
- The chlorine molecule absorbs the energy of a single quantum of ultraviolet (UV) light. The energy of one quantum of UV light is greater than the Cl—Cl bond energy, so the bond will break.
- Since both atoms are the same, the Cl–Cl bond breaks homolytically—one electron goes to each chlorine atom.
- This results in two separate chlorine atoms, written Cl•. They are called free radicals. The dot is used to show the unpaired electron.

1 Complete the word equation for one possibility for the thermal cracking of decane.

decane → hexane + ...

2 In the laboratory cracking of alkanes, state how you can tell that the products have shorter chains than the starting materials.

3 Explain why we would not crack octane industrially.

4 Explain how the temperature required for cracking can be reduced.

5 Give two economic reasons for cracking long chain alkanes.

6 Give the name of the alkane $CH_3CH_2CH(CH_3)CH_3$ and sketch its displayed formula.

7 Sketch the displayed formula and structural formula of 2-methylhexane.

8 Give the name of an isomer of 2-methylhexane that has a straight chain.

9 State which of the two isomers in question 8 will have the higher melting point? Explain your answer.

10 Give an equation for:
 a the complete combustion of propane
 b the incomplete combustion of ethane to carbon monoxide.

11 Explain why you should not use a camping gas stove inside a small tent.

12 State which stage of a free-radical reaction of bromine with methane is represented by the following.
 a $Br\bullet + Br\bullet \rightarrow Br_2$
 b $CH_4 + Br\bullet \rightarrow CH_3\bullet + HBr$
 c $\bullet CH_3 + Br_2 \rightarrow CH_3Br + Br\bullet$
 d $Br_2 \rightarrow 2Br\bullet$

$$Cl–Cl \xrightarrow{\text{UV light}} 2Cl\bullet$$

• Free radicals are highly reactive.
• The C–H bond in the alkane needs more energy to break than is available in a quantum of ultraviolet radiation. So this bond does not break.

Propagation

This takes place in two stages:

1 The chlorine free radical takes a hydrogen atom from ethane to form hydrogen chloride, a stable compound. This leaves an ethyl free radical, $\bullet C_2H_5$.

$$Cl\bullet + C_2H_6 \rightarrow HCl + \bullet C_2H_5$$

2 The ethyl free radical is also very reactive and reacts with a chlorine molecule. This produces another chlorine free radical and a molecule of chloroethane—a stable compound.

$$\bullet C_2H_5 + Cl_2 \rightarrow C_2H_5Cl + Cl\bullet$$

The effect of these two steps is to produce hydrogen chloride, chloroethane, and a new Cl• free radical. This free radical is ready to react with more ethane and repeat the two steps. This is the 'chain' part of the chain reaction. These steps may take place thousands of times before the free radicals are destroyed in the termination step.

Termination

Termination is the step in which the free radicals are removed. This can happen in any of the following three ways:

$Cl\bullet + Cl\bullet \rightarrow Cl_2$	Two chlorine free radicals react together to give chlorine.
$\bullet C_2H_5 + \bullet C_2H_5 \rightarrow C_4H_{10}$	Two ethyl free radicals react together to give butane.
$Cl\bullet + \bullet C_2H_5 \rightarrow C_2H_5Cl$	A chlorine free radical and an ethyl free radical react together to give chloroethane.

Notice that in every case, two free radicals react to form a stable compound with no unpaired electrons.

Other products of the chain reaction

Other products are formed as well as the main ones, chloroethane and hydrogen chloride.

• Some butane is produced at the termination stage, as shown above.
• Dichloroethane may be made at the propagation stage, if a chlorine radical reacts with some chloroethane that has already formed.

$$C_2H_5Cl + Cl\bullet \rightarrow \bullet C_2H_4Cl + HCl$$

followed by $\bullet C_2H_4Cl + Cl_2 \rightarrow C_2H_4Cl_2 + Cl\bullet$

• With longer-chain alkanes there will be many isomers formed because the Cl• can replace any of the hydrogen atoms.
• Chain reactions are not very useful because they produce such a mixture of products. They will also occur without light at high temperatures.

Alkenes

Alkenes are unsaturated hydrocarbons. They are made of carbon and hydrogen only and have one or more carbon–carbon double bonds. This means that alkenes have fewer than the maximum possible number of hydrogen atoms. The double bond makes them more reactive than alkanes because of the high concentration of electrons (high electron density) between the two carbon atoms. Ethene, the simplest alkene, is the starting material for a large range of products, including polymers such as polythene, PVC, polystyrene, and PET fabric, as well as products like antifreeze and paints. Alkenes are produced in large quantities when crude oil is thermally cracked.

The general formula

The homologous series of alkenes with one double bond has the general formula C_nH_{2n}.

How to name alkenes

The suffix -ene indicates the presence of a carbon-carbon double bond. There cannot be a C=C bond if there is only one carbon. So, the simplest alkene is ethene, $CH_2=CH_2$ followed by propene, $CH_3CH=CH_2$. With longer chains, a locant is needed to indicate where the double bond is situated.

The shape of alkenes

Ethene is a planar (flat) molecule. This makes the angles between each bond roughly 120°.

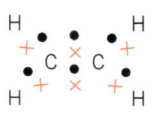 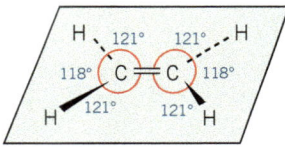

Unlike the C–C bonds in alkanes, there is no rotation about the double bond. This is because of the way a double bond is made. Any molecules where a hydrogen atom in ethene has been replaced by another atom or group will have the same flat shape around the carbon–carbon double bond.

Why a double bond cannot rotate

In a C=C double bond, as well as a normal C–C single bond there is a p-orbital (which contains a single electron) on each carbon. These two p-orbitals overlap to form an orbital with a cloud of electron density above and below the single bond, see Figure 9 and Figure 10. This is called a π-orbital and its presence means the bond cannot rotate. This is sometimes called **restricted rotation**.

p-orbitals

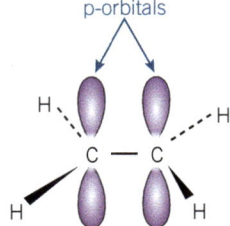

two p-orbitals produce

π-orbital

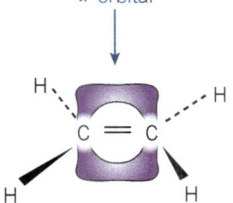

the π-orbital in ethene

Figure 9 *The double bond in ethene*

Learning outcomes

On these pages you will learn to:

- describe how alkenes are produced
- describe the reactions of alkenes
- describe the test for a C=C double bond
- describe the mechanism of electrophilic addition
- explain the inductive effect of alkyl groups
- describe the characteristics of addition polymerisation

Exam tip

Remember, when writing a systematic name, the groups are listed alphabetically.

Exam tip

The H–C–H angle is slightly less than 120° because the group of four electrons in the C=C double bond repels more strongly than the groups of two in the C–H single bonds.

Exam tip

The C–C single bond is formed by the overlap of two p-orbitals along the line joining the two carbon atoms. The orbital formed is called a σ-orbital.

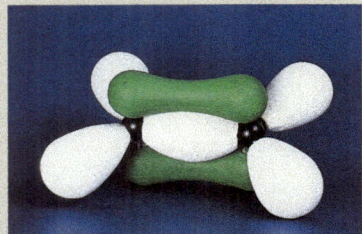

Figure 10 *Model of ethene showing orbitals*

Isomers

Alkenes with more than three carbons can form different types of isomers and they are named according to the IUPAC system, using the suffix -ene to indicate a double bond.

As well as chain isomers like those found in alkanes, alkenes can form two types of isomer that involve the double bond:

- positional isomers
- geometrical isomers.

Positional isomers

These are isomers with the double bond in different positions – between a pair of neighbouring carbon atoms in different positions in the carbon chain.

but-2-ene   but-1-ene

The longer the carbon chain, the more possibilities there will be and therefore the greater the number of isomers.

Geometrical isomers

Geometrical isomerism is a form of stereoisomerism. The two stereoisomers have the same structural formula but the bonds are arranged differently in space. It occurs only around C=C double bonds. For example, but-2-ene, above, can exist as shown below.

$\longrightarrow$

The isomer in which both –CH_3 groups are on the same side of the double bond is called *cis*-but-2-ene. The isomer where they are on opposite sides is called *trans*-but-2-ene. This type of isomerism is often called *cis-trans* isomerism.

Exam tip

Compounds such as *cis*-but-2-ene, which have two substituents on the same side of the double bond, are also called *Z*-isomers. Those with two substituents on opposite sides, like *trans*-but-2-ene, are also called *E*-isomers.

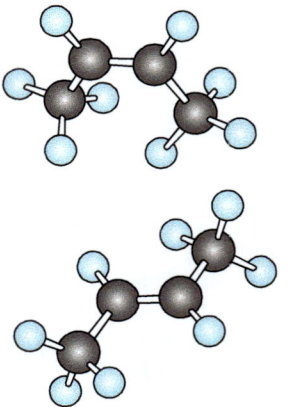

Figure 11 *3D representations of the cis- and trans-isomers of but-2-ene*

Bond energies

Remember that a C=C bond consists of a σ-bond and a π-bond.

1 Use the bond energies C–C = 350 kJ mol^{-1} and C=C = 610 kJ mol^{-1} to calculate the strength of the π part of the bond alone.

2 Explain why the π part of the bond is weaker than the σ part.

Physical properties of alkenes

The double bond does not greatly affect properties such as boiling and melting points. London dispersion forces are the only intermolecular forces that act between the alkene molecules. This means that the physical properties of alkenes are very similar to those of the alkanes. The melting and boiling points increase with the number of carbon atoms present. Alkenes are not soluble in water.

Synthesising alkenes

Alkenes can be produced:

- by cracking long chain alkanes as described in Section 14.1
- by dehydrating an **alcohol** (see Section 16.1)
- by eliminating a hydrogen halide from a halogenoalkane (see Section 15.1).

a Cracking alkanes—the alkane vapour is passed over a heated catalyst:

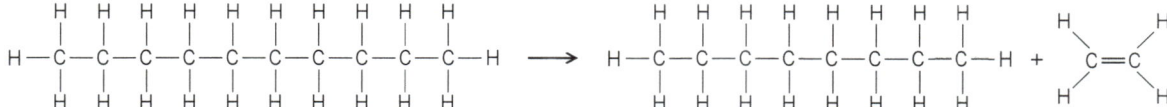

b Dehydrating an alcohol—the alcohol vapour is either passed over a heated catalyst:

$$CH_3-CH_2-OH \xrightarrow{Al_2O_3} CH_2{=}CH_2 + H_2O$$

or heated with concentrated sulfuric acid:

$$CH_3-CH_2-OH \xrightarrow{conc.\ H_2SO_4} CH_2{=}CH_2 + H_2O$$

c Eliminating a hydrogen halide from a halogenoalkane—the halogenoalkane is heated with a solution of sodium hydroxide in ethanol:

$$\underset{\overset{|}{Br}}{CH_3CHCH_3} + NaOH \longrightarrow CH_2{=}CHCH_3 + NaBr + H_2O$$

Reactions of alkenes

The double bond makes a big difference to the reactivity of alkenes compared with alkanes. The bond enthalpy for C–C is $350\,kJ\,mol^{-1}$ and for C=C it is $610\,kJ\,mol^{-1}$. So you might predict that alkenes would be less reactive than alkanes. In fact alkenes are *more* reactive than alkanes.

The C=C forms an electron-rich area in the molecule, which can easily be attacked by positively charged reagents. These reagents are called **electrophiles** (electron-liking). They are electron pair acceptors. An example of a good electrophile is the H^+ ion. As alkenes are unsaturated they can undergo addition reactions.

So most of the reactions of alkenes are electrophilic additions.

Electrophilic addition reactions

The reactions of alkenes are typically electrophilic additions. The four electrons in the carbon–carbon double bond make it a centre of high electron density. Electrophiles are attracted to it and can form a bond by using two of the four electrons in the carbon–carbon double bond (of the four electrons, the two that are in a π-bond, see above).

The mechanism is always essentially the same:

1 The electrophile is attracted to the double bond.
2 Electrophiles are positively charged and accept a pair of electrons from the double bond. The electrophile may be a positively charged ion or have a positively charged area.
3 A positive ion (a carbocation) is formed.
4 A negatively charged ion forms a bond with the carbocation.

See how the examples below fit this general mechanism.

> **Exam tip**
>
> Learn the definition: an electrophile is an electron pair acceptor.

> **Exam tip**
>
> Remember that cations are positively charged.

Reaction with hydrogen halides

Hydrogen halides, HCl, HBr, and HI, add across the double bond to form a halogenoalkane. This reaction occurs readily at room temperature. For example:

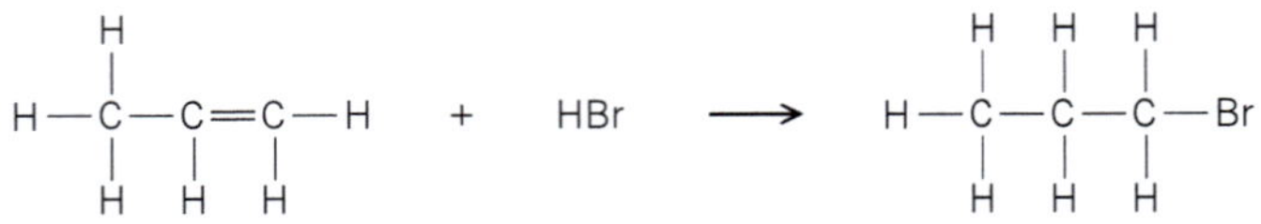

propene   hydrogen bromide   1-bromopropane

- Bromine is more electronegative than hydrogen, so the hydrogen bromide molecule is polar, $H^{\delta+}-Br^{\delta-}$.
- The electrophile is the $H^{\delta+}$ of the $H^{\delta+}-Br^{\delta-}$.
- The $H^{\delta+}$ of HBr is attracted to the C=C bond because of the double bond's high electron density.
- One of the pairs of electrons from the C=C forms a bond with the $H^{\delta+}$ to form a positive ion (called a **carbocation**). At the same time the electrons in the $H^{\delta+}-Br^{\delta-}$ bond are drawn towards the $Br^{\delta-}$.
- The bond in hydrogen bromide breaks heterolytically. Both electrons from the shared pair in the bond go to the bromine atom because it is more electronegative than hydrogen, leaving a Br^- ion.
- The Br^- ion attaches to the positively charged carbon of the carbocation forming a bond with one of its electron pairs.

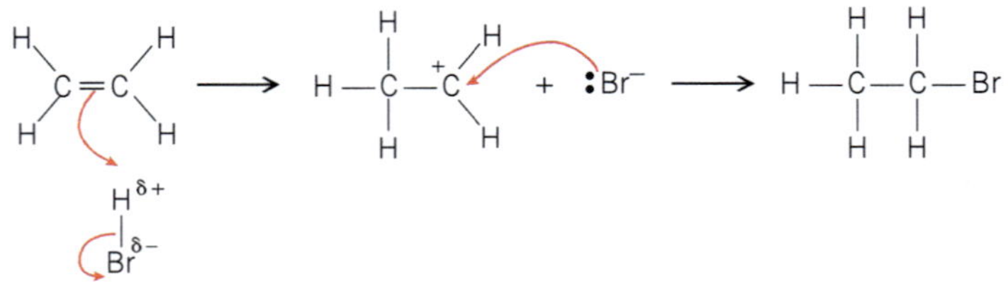

Exam tip

Heterolytic bond breaking means that when a covalent bond breaks, both electrons go to one of the atoms involved in the bond and none go to the other. This results in the formation of a negative ion and a positive ion.

Asymmetrical alkenes

When hydrogen bromide adds to ethene, bromoethane is the only possible product.

However, when the double bond is not exactly in the middle of the chain, there are two possible products—the bromine of the hydrogen bromide could bond to either of the carbon atoms of the double bond.

For example, propene could produce:

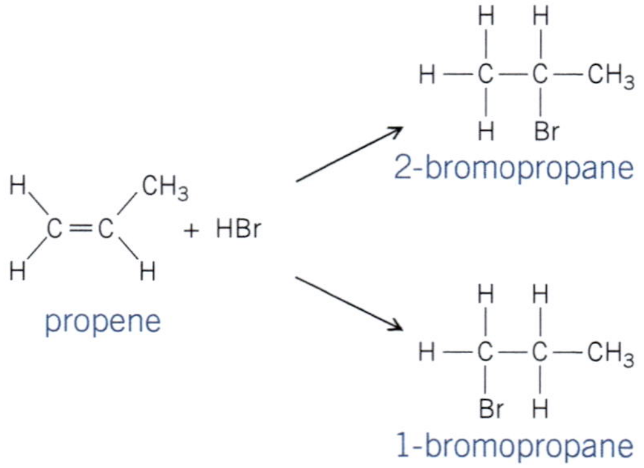

Exam tip

There is simple way to work out the product. When hydrogen halides add on to alkenes, the hydrogen adds on to the carbon atom which already has the most hydrogens. This is called Markovnikov's rule.

2-bromopropane

propene + HBr

1-bromopropane

In fact, the product is almost entirely 2-bromopropane.

To explain this, you need to know that alkyl groups, for example $-CH_3$ or $-C_2H_5$, have a tendency to release electrons. This is known as a **positive inductive effect** and is sometimes represented by an arrow along their bonds to show the direction of the release.

This electron-releasing effect tends to stabilise the positive charge of the intermediate carbocation. The more alkyl groups there are attached to the positively charged carbon atom, the more stable the carbocation is. So, a positively charged carbon atom which has three alkyl groups (called a **tertiary carbocation**) is more stable than one with two alkyl groups (a **secondary carbocation**,) which is more stable than one with just one alkyl group (a **primary carbocation**), see Figure 12.

The product will tend to come from the more stable carbocation.

So, the two possible carbocations when propene reacts with HBr are:

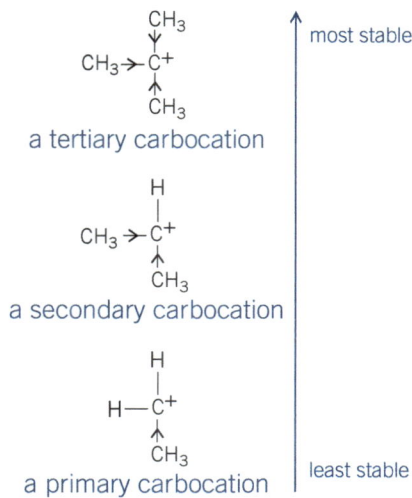

Figure 12 *Stability of primary, secondary, and tertiary carbocations*

a secondary carbocation   a primary carbocation
(more stable, product   (less stable)
formed from this)

|Br⁻

2-bromopropane

The secondary carbocation is more stable because it has two methyl groups releasing electrons towards the positive carbon. The majority of the product is formed from this.

Reaction of alkenes with halogens
Alkenes react rapidly with chlorine gas, or with solutions of bromine and iodine in an organic solvent, to give dihalogenoalkanes.

The halogen atoms add across the double bond.

In this case the halogen molecules act as electrophiles:

- At any instant, a bromine (or any other halogen) molecule is likely to have an instantaneous dipole, $Br^{\delta+}$–$Br^{\delta-}$. (An instant later, the dipole could be reversed $Br^{\delta-}$–$Br^{\delta+}$.) The $\delta+$ end of this dipole is attracted to the electron-rich double bond in the alkene—the bromine molecule has become an electrophile.
- The electrons in the double bond are attracted to the $Br^{\delta+}$. They repel the electrons in the Br–Br bond and this strengthens the dipole of the bromine molecule.

Exam tip

The instantaneous dipole $Br^{\delta+}$–$Br^{\delta-}$ is also induced when a bromine molecule collides with the electron-rich double bond.

- Two of the electrons from the double bond form a bond with the $Br^{\delta+}$ and the other bromine atom becomes a Br⁻ ion. This leaves a carbocation, in which the carbon atom that is not bonded to the bromine has the positive charge.
- The Br⁻ ion now forms a bond with the carbocation.

195

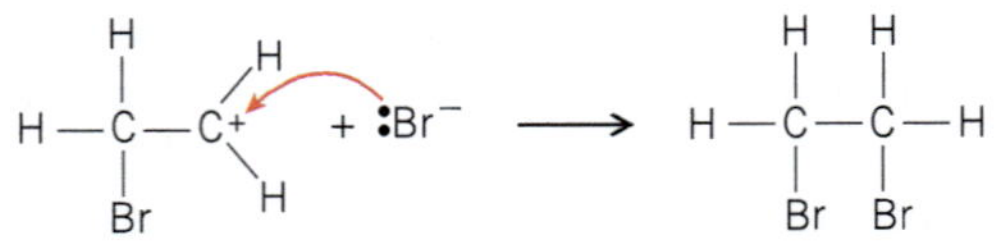

So the addition takes place in two steps:

1 formation of the carbocation by electrophilic addition
2 rapid reaction with a negative ion.

The test for a double bond

This addition reaction is used to test for a carbon–carbon double bond. When a few drops of bromine solution, sometimes called bromine water (which is reddish-brown), are added to an alkene, the solution is decolourised because the products are colourless.

Reaction with water

Water also adds on across the double bond in alkenes. The reaction is used industrially to make alcohols and is carried out with steam, at a suitable temperature and pressure, using an acid catalyst such as phosphoric acid, H_3PO_4.

$$CH_2=CH_2(g) + H_2O(g) \rightarrow CH_3CH_2OH(g)$$

Reaction with hydrogen

This reaction takes place at around 150°C with a nickel catalyst:

$$CH_2=CH_2 \ + \ H_2 \ \xrightarrow{\text{Ni}} \ CH_3-CH_3$$

Oxidation reactions

Combustion is, of course, an oxidation reaction

$$CH_2=CH_2(g) + 3O_2(g) \rightarrow 2CO_2(g) + 2H_2O(l)$$

This reaction is of little practical importance, as the alkenes are too useful to be burned as fuels.

Oxidising an alkene using an oxidising agent such as potassium manganate(VII) (potassium permanganate) can take place in two ways. With cold, dilute acidified potassium manganate(VII), a **diol** is formed.

$$CH_2=CH_2 \ + \ H_2O \ + \ [O] \ \rightarrow \ CH_2OHCH_2OH$$
$$\text{ethene} \qquad\qquad\qquad\qquad \text{ethane-1,2-diol}$$

Notice that the double bond has disappeared but the two carbon atoms are still bonded with a single bond. Notice also the convention of using [O] to represent the oxidising agent.

With hot, concentrated acidified potassium manganate(VII) the C=C breaks completely and the two carbon-containing fragments are oxidised to carbonyl compounds (**aldehydes** or **ketones**) and then further to **carboxylic acids**. This can form the basis of a method for finding the position of the double bond in an alkene molecule.

For example, oxidising hex-2-ene initially produces ethanal (2 carbon atoms) and butanal (4 carbon atoms)

$$CH_3CH=CHCH_2CH_2CH_3 + 2[O] \rightarrow CH_3CHO + CH_3CH_2CH_2CHO$$

So we can deduce where the double bond was located.

Exam tip

The carbocation will react with any nucleophile that is present. In aqueous solution, such as bromine water, water reacts with the carbocation, forming some CH_2BrCH_2OH, 2-bromoethanol.

Further oxidation of ethanal gives ethanoic acid, and further oxidation of butanal gives butanoic acid.

$$CH_3CHO + [O] \rightarrow CH_3COOH$$

$$CH_3CH_2CH_2CHO + [O] \rightarrow CH_3CH_2CH_2COOH$$

Addition polymerisation

Addition polymers are made from a monomer or monomers with a carbon–carbon double bond (alkenes). The monomer has the general formula:

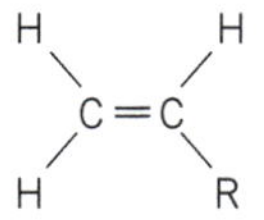

When the monomers **polymerise**, the double bond opens and the monomers bond together to form a backbone of carbon atoms as shown:

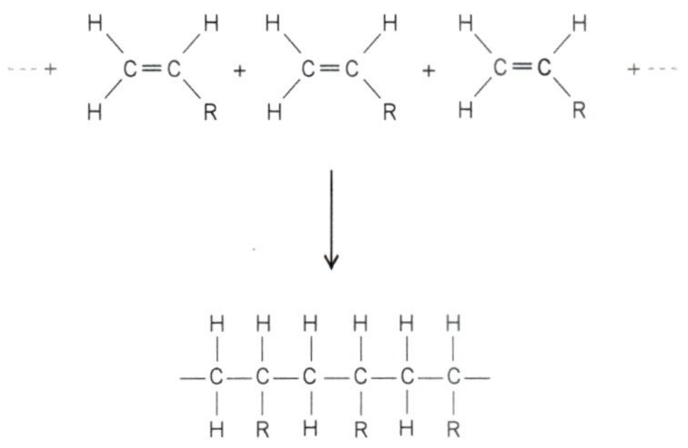

This may also be represented by equations such as:

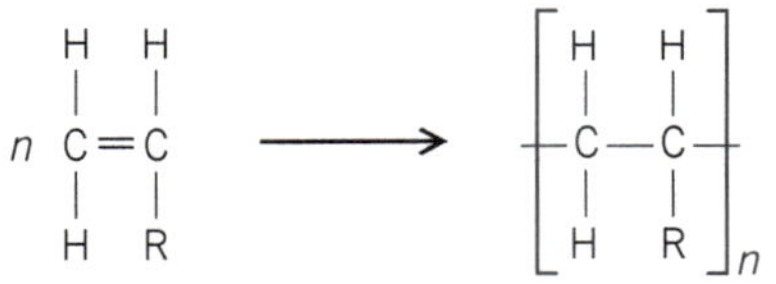

R may be an alkyl group such as CH_3 in which case the polymer is poly(propene):

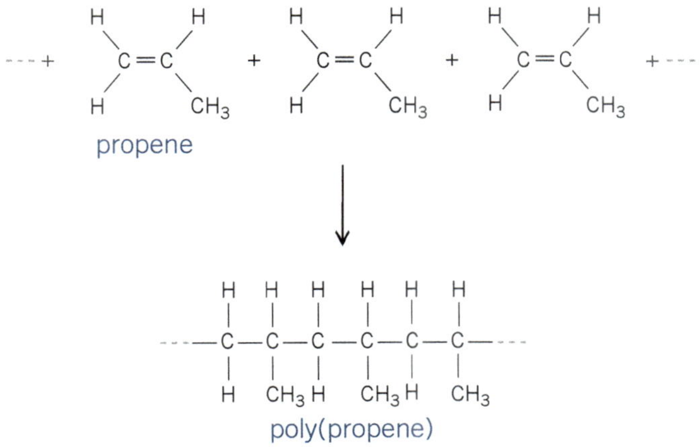

propene

poly(propene)

In another example, ethene polymerises to form poly(ethene):

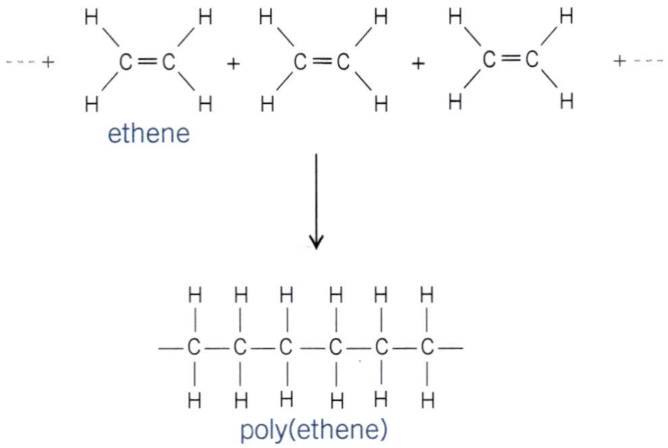

ethene

poly(ethene)

More details about addition polymerisation are in Chapter 20.

Summary test 14.2

1 Give the name of $CH_3CH=CHCH_2CH_2CH_3$

2 Sketch the structural formula for hex-1-ene.

3 There are six isomeric pentenes. Sketch their displayed formulae.

4 State which of these attacks the double bond in an alkene. Choose from a, b, or c.

  A electrophiles

  B nucleophiles

  C alkanes

5 State which of these best describes the double bond in an alkene. Choose from A, B, C or D.

  A electron-rich

  B positively charged

  C electron-deficient

  D acidic

6 Give the equation for the complete combustion of propene.

7 State which of the following are typical reactions of alkenes.

  A electrophilic additions

  B electrophilic substitutions

  C nucleophilic substitutions

8 a Give the two possible products of the reaction between propene and hydrogen bromide.

  b State which the main product is.

  c Explain why this product is more likely.

9 Give the product of the reaction between ethene and hydrogen chloride.

10 State which of the following is the test for a carbon–carbon double bond.

  A Forms a white precipitate with silver nitrate.

  B Turns limewater milky.

  C Decolourises bromine solution.

11 Suggest the product (or products) of the oxidation of cyclohexene, 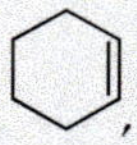 , with hot concentrated acidified potassium manganate(VII).

Launch additional digital resources for the chapter

1 Chloromethane (CH_3Cl) is synthesised from the reaction of methane with chlorine gas. This is an example of a free radial substitution reaction where chlorine gas forms chlorine radicals.

a Define the term 'free radical' and state the type of bond fission taking place when a radical species is formed. *(2 marks)*

b Explain with the aid of equations, how chloromethane is synthesised. Give the names of each stage and reaction conditions where appropriate. *(5 marks)*

c Following the synthesis of chloromethane, a by-product with the formula CH_2Cl_2 was also identified. Explain with the aid of equations how this occurs. *(2 marks)*

2 An alkene, X, reacts with hot, concentrated acidified potassium manganate (VII) ions. The products of the reaction are propanoic acid and pentanoic acid.

a Give the structural formula of alkene X. *(1 mark)*

b Draw the displayed formula of the product formed when alkene X reacts with cold, dilute acidified manganate(VII) ions. *(1 mark)*

3 From the reaction scheme below:

a State the reagents and conditions represented by B, D and F. *(3 marks)*

b Give the formulae for the compounds represented by A, C and E. *(3 marks)*

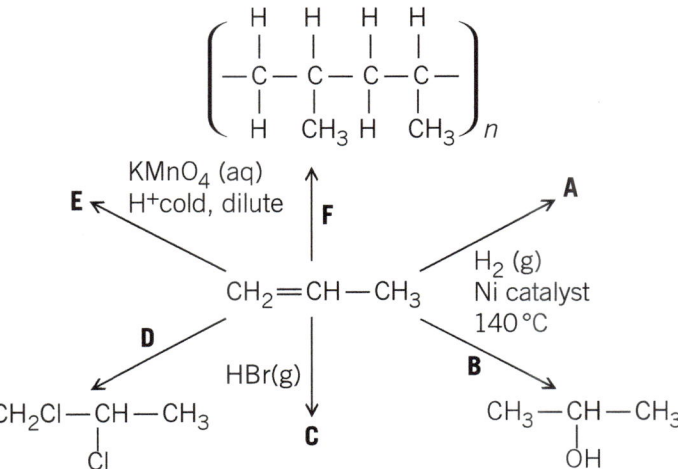

4 This question is about the reactions of three alkenes with hydrogen bromide.

a Ethene reacts with hydrogen bromide to make bromoethane.

i State the conditions required for this reaction. *(1 mark)*

ii Give a mechanism for this reaction. Name this type of mechanism. *(4 marks)*

b Propene reacts with hydrogen bromide to make two possible products.

i Give the displayed formulae and names of the two products. *(2 marks)*

ii Identify from your answer to part **i**, which of the two products is the major product. Give reasons for your answer. *(3 marks)*

c Hydrogen bromide reacts with the alkene shown below to make two products.

$$CH_3$$
$$\overset{|}{C}$$
$$H_2C \diagdown \quad \diagup CH_3$$

i State the IUPAC name for the substance shown by the above formula. *(1 mark)*

ii Give the displayed formula of the major product of the reaction between hydrogen bromide and the alkene given above. Name this product. *(2 marks)*

5 Dodecane, $C_{12}H_{26}$, is a major component of crude oil.

a Construct a balanced equation to show the complete combustion of dodecane. *(1 mark)*

b Give the empirical formula of dodecane. *(1 mark)*

c Dodecane can be cracked to produce octane which is used as a fuel for vehicles and so has a greater demand. Cracking dodecane also produces butene.

i State one condition required for cracking of dodecane to occur. *(1 mark)*

ii Construct an equation to show the cracking of one molecule of dodecane. *(2 marks)*

iii Suggest one use for the butene produced as a by-product. *(1 mark)*

d When the octane produced is combusted in vehicle engines, the exhaust gases are also found to contain NO and NO_2.

i Octane does not contain any nitrogen. Suggest where the nitrogen component of the NO exhaust gases comes from. *(1 mark)*

ii Catalytic converters are used to remove harmful substances from vehicle exhaust emissions. Give an equation to show how NO is removed from exhaust gases. *(2 marks)*

iii Give an example of an element found in a catalytic converter. *(1 mark)*

15.1 Halogenoalkanes

Not many halogenoalkanes occur naturally but they are the basis of many synthetic compounds. Some examples of these are PVC (used to make drainpipes), Teflon (the non stick coating on pans), and a number of anaesthetics and solvents. Halogenoalkanes have an alkane skeleton with one or more halogen (fluorine, chlorine, bromine, or iodine) atoms in place of hydrogen atoms.

The general formula

The general formula of a halogenoalkane with a single halogen atom is $C_nH_{2n+1}X$, where X is the halogen. This is often shortened to R–X.

How to name halogenoalkanes
- The prefixes fluoro-, chloro-, bromo-, and iodo- tell us which halogen is present.
- Locants are used, if needed, to show on which carbon the halogen is bonded:

1-chloropropane   1-iodopropane   2-bromo-2-methylpropane

The prefixes di-, tri-, tetra-, and so on, are used to show how many atoms of each halogen are present.

- When a compound contains different halogens they are listed in alphabetical order, not in order of the number of the carbon atom to which they are bonded. For example:

is 3-chloro-2-iodopentane not 2-iodo-3-chloropentane. (C is before I in the alphabet.)

Bond polarity

Halogenoalkanes have a C–X bond. This bond is polar, $C^{\delta+}–X^{\delta-}$, because halogens are more electronegative than carbon. The electronegativities of carbon and the halogens are shown in Table 1. Notice that as you go down the group, the bonds get less polar.

Physical properties of halogenoalkanes

Solubility
- The polar $C^{\delta+}–X^{\delta-}$ bonds are not polar enough to make the halogenoalkanes soluble in water.
- The main intermolecular forces of attraction are dipole–dipole attractions and London dispersion forces.
- Halogenoalkanes mix with hydrocarbons so they can be used as dry-cleaning fluids and to remove oily stains. (Oil is a mixture of hydrocarbons.)

Table 1 *Electronegativities of carbon and the halogens*

Element	Electronegativity
carbon	2.5
fluorine	4.0
chlorine	3.0
bromine	2.6
iodine	2.5

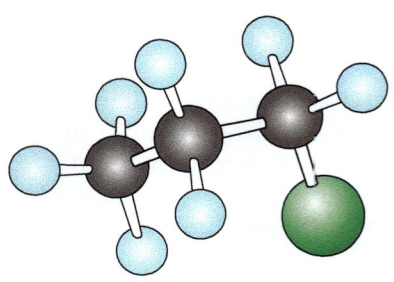

Figure 1 3D representation of 1-chloropropane, a halogenoalkane

Melting and boiling points

The boiling point depends on the number of carbon atoms and halogen atoms.

- Boiling point increases with increased chain length.
- Boiling point increases going down the halogen group.

Both these effects are caused by increased van der Waals forces, because the larger the molecules the greater the number of electrons (and therefore the larger the van der Waals forces).

As in other homologous series, increased branching of the carbon chain will tend to lower the melting point.

Halogenoalkanes have higher boiling points than alkanes with similar chain lengths because they have higher relative molecular masses and they are more polar.

Exam tip

Branched chain molecules do not pack together well in the solid state and are therefore easier to melt.

How the halogenoalkanes react—the reactivity of the C–X bond

When halogenoalkanes react it is almost always the C–X bond that breaks. There are two factors that determine how readily the C–X bond reacts. These are:

- the $C^{\delta+}$–$X^{\delta-}$ bond polarity
- the C–X bond energy.

Bond polarity

The halogens are more electronegative than carbon so the bond polarity will be $C^{\delta+}$–$X^{\delta-}$. This means that the carbon bonded to the halogen has a partial positive charge—it is electron deficient. So it can be attacked by reagents that are electron-rich or have electron-rich areas. These are called nucleophiles. A **nucleophile** is an electron pair donor.

The polarity of the C–X bond would predict that the C–F bond would be the most reactive. It is the most polar, so the $C^{\delta+}$ has the most positive charge and is therefore most easily attacked by a nucleophile. This argument would make the C–I bond least reactive because it is the least polar.

Exam tip

Nucleophiles are reagents that have a lone pair of electrons capable of forming a bond with $C^{\delta+}$. They may be negative ions or neutral molecules with an area of $\delta-$ charge.

Bond energies

C–X bond energies are listed in Table 2. The bonds get weaker going down the group. Fluorine is the smallest atom of the halogens and the shared electrons in the C–F bond are strongly attracted to the fluorine nucleus. This makes a strong bond. Going down the group, the shared electrons in the C–X bond get further and further away from the halogen nucleus, so the bond becomes weaker.

The bond energies would predict that iodo-compounds, with the weakest bonds, are the most reactive, and fluoro-compounds, with the strongest bonds, are the least reactive. Experiments confirm that reactivity increases going down the group. This means that bond energy is a more important factor than bond polarity.

Table 2 Carbon–halogen bond energies

Bond	Bond energy / kJ mol^{-1}
C–F	467
[C–H	410]
C–Cl	340
C–Br	280
C–I	240

Classification of halogenoalkanes

The halogen atom may occur in one of three positions on a hydrocarbon chain. If it occurs at the end of the chain, it is called a **primary** (1°) halogenoalkane. If it occurs in the body of the chain, it is called a **secondary** (2°) halogenoalkane. If it occurs at a branch in the chain, it is called a **tertiary** (3°) halogenoalkane.

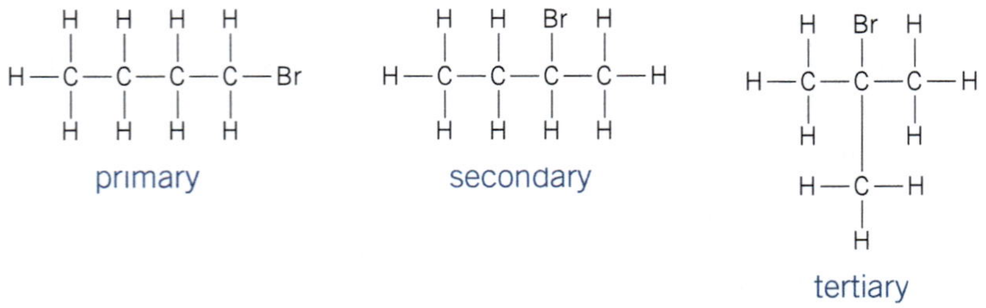

primary   secondary   tertiary

In a tertiary halogenoalkane, the α-carbon has no hydrogen atoms attached, a secondary halogenoalkane has one hydrogen atom attached and a primary halogenoalkane has two (or three in the case of CH_3X).

Synthesising halogenoalkanes

Halogenoalkanes can be produced by:

a free radical substitution of alkanes with chlorine or bromine in ultraviolet light, for example: $C_6H_{14} + Br_2 \rightarrow C_6H_{13}Br + HBr$ (see Section 14.1 for more details)

b addition of a halogen or a hydrogen halide across the double bond in an alkene at room temperature, for example:

$CH_2{=}CH_2 + HBr \rightarrow CH_3CH_2Br$ (see Section 14.2 for more details)

c substitution of the OH group of an alcohol with a halogen.

Substitution of the OH group of an alcohol by a halogen
Alcohols react with hydrogen halides such as hydrogen bromide, HBr, to substitute the halogen for the OH group, forming a halogenoalkane. The rate of the reaction varies with the halogen in the order HI > HBr > HCl. A strong acid catalyst such as sulfuric acid is used. The hydrogen halide is generated in the reaction flask by the reaction of an alkali metal halide, for example:

$$2KBr(s) + H_2SO_4 \rightarrow K_2SO_4(s) + 2HBr(g)$$

$$CH_3CH_2OH + HBr \rightarrow CH_3CH_2Br + H_2O$$

Other reagents which bring about halogen substitution are:

- phosphorus trichloride, PCl_3, when heated
- phosphorus pentachloride, PCl_5
- a mixture of red phosphorus and bromine (or iodine)
- thionyl chloride, $SOCl_2$.

Example equations are given below:

$$3CH_3CH_2CH_2OH + PCl_3 \rightarrow 3CH_3CH_2CH_2Cl + H_3PO_3$$

$$CH_3CH_2CH_2OH + PCl_5 \rightarrow CH_3CH_2CH_2Cl + POCl_3 + HCl$$

$$2P + 3I_2 \rightarrow 2PI_3$$

followed by:

$$3CH_3CH_2OH + PI_3 \rightarrow 3CH_3CH_2I + H_3PO_3$$

$$CH_3CH_2CH_2OH + SOCl_2 \rightarrow CH_3CH_2CH_2Cl + HCl + SO_2$$

Reactions of halogenoalkanes

Most reactions of organic compounds take place via a series of steps. You can often predict these steps by thinking about how electrons are likely to move. This can help you understand why reactions take place as they do and can save a great deal of memorising. The reactions of halogenoalkanes are usually nucleophilic substitution.

Exam tip

Remember that the free radical substitution of alkanes is a chain reaction that proceeds via three steps: initiation, propagation, and termination.

Nucleophiles

Nucleophiles are reagents that attack and form bonds with positively- or partially positively-charged carbon atoms.

- A nucleophile is either a negatively-charged ion or has an atom with a $\delta-$ charge.
- A nucleophile has a lone (unshared) pair of electrons which it can use to form a covalent bond.
- The lone pair is situated on an electronegative atom.

So, in organic chemistry a nucleophile is a species that has a lone pair of electrons with which it can form a bond by donating its electrons to an electron deficient carbon atom. Some common nucleophiles are:

- the hydroxide ion, $^-$:OH
- ammonia, :NH_3, where the electronegative nitrogen atom has a $\delta-$ charge
- the cyanide ion, $^-$:CN
- the water molecule, :OH_2, where the electronegative oxygen atom has a $\delta-$ charge.

They will each replace the halogen in a halogenoalkane. These reactions are called **nucleophilic substitutions** and they all follow essentially the same reaction mechanism.

A reaction mechanism describes a route from reactants to products via a series of theoretical steps. These may involve short-lived intermediates.

Figure 2 *Applications of halogenoalkanes*

Nucleophilic substitution

The general equation for nucleophilic substitution, using :Nu^- to represent any negatively charged nucleophile and X to represent a halogen atom, is:

$$
\begin{array}{c}
\text{H} \\
| \\
\text{R—C—X} \\
| \\
\text{H}
\end{array}
+ \text{ :Nu}^- \longrightarrow
\begin{array}{c}
\text{H} \\
| \\
\text{R—C—Nu} \\
| \\
\text{H}
\end{array}
+ \text{ :X}^-
$$

Reaction mechanisms and curly arrows

Curly arrows are used to show how electron pairs move in organic reactions. These are shown here in red for clarity. You can write the above reaction as:

$$
\begin{array}{c}
\text{:Nu}^- \\
\text{H} \\
| \\
\text{R—C}^{\delta+}\text{—X}^{\delta-} \\
| \\
\text{H}
\end{array}
\longrightarrow
\begin{array}{c}
\text{H} \\
| \\
\text{R—C—Nu} \\
| \\
\text{H}
\end{array}
+ \text{ :X}^-
$$

The lone pair of electrons of a nucleophile is attracted towards a partially positively charged carbon atom. A curly arrow starts at a lone pair of electrons and moves towards $C^{\delta+}$.

The lower curly arrow shows the electron pair in the C–X bond moving to the halogen atom, X, and making it a halide ion, X^-, which has a lone pair of elecctrons. The halide ion is called the **leaving group**.

The rate of substitution depends on the halogen. Fluorine compounds are unreactive due to the strength of the C–F bond. Then, going down the group, the rate of reaction increases as the C–X bond strength decreases.

Examples of nucleophilic substitution reactions

All these reactions are similar. Remember the basic pattern, shown above. Then work out the product with a particular nucleophile. This is easier than trying to remember the separate reactions.

Halogenoalkanes with aqueous sodium (or potassium) hydroxide

The nucleophile is the hydroxide ion, $^-$:OH.

Exam tip
The ethanol is acting as a co-solvent.

This reaction occurs very slowly at room temperature. To speed up the reaction it is necessary to warm the mixture. Halogenoalkanes do not mix with water, so ethanol is used as a solvent in which the halogenoalkane and the aqueous sodium (or potassium) hydroxide both mix. This is called a **hydrolysis reaction**.

The overall reaction is:

$$R–X + OH^- \rightarrow ROH + X^-$$

so an alcohol, ROH, is formed.

For example:

$$C_2H_5Br \quad + \quad OH^- \quad \longrightarrow \quad C_2H_5OH \quad + \quad Br^-$$

bromoethane   ethanol

This is the mechanism:

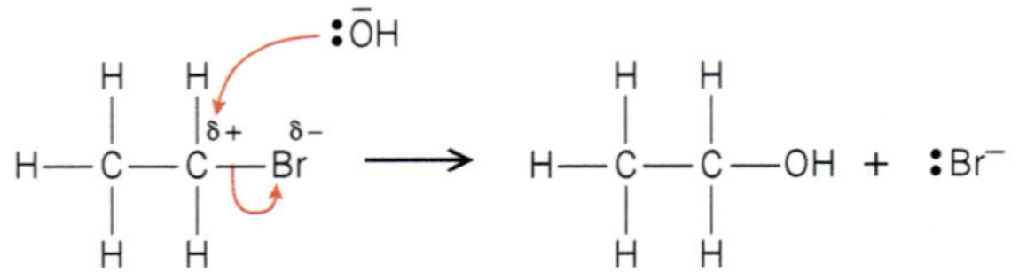

The rate of the reaction depends on the strength of the carbon–halogen bond: C–F > C–Cl > C–Br > C–I (see Table 2 above). So fluoroalkanes do not react at all, while iodoalkanes react rapidly.

Halogenoalkanes with cyanide ions

When halogenoalkanes are warmed with an aqueous ethanolic solution of potassium cyanide, nitriles are formed. The nucleophile is the cyanide ion, $^-$:CN.

The reaction is:

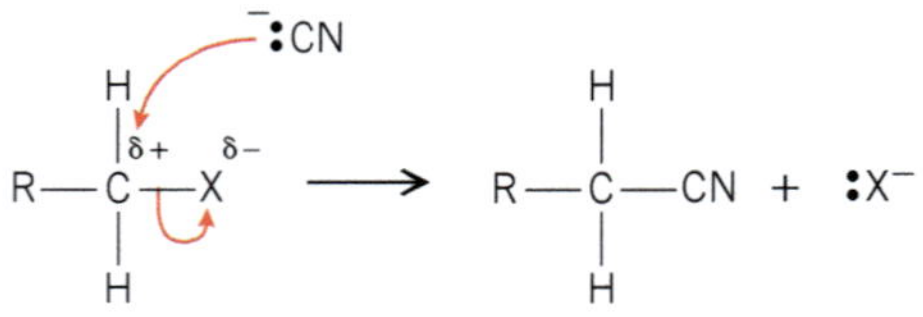

Exam tip
Nitriles have the functional group –C≡N. They are named from the number of carbon atoms and the suffix nitrile. The carbon of the –CN group is counted as part of the root, so CH₃CH₂CN is propanenitrile, not ethanenitrile.

The product is called a **nitrile**. It has one extra carbon in the chain than the starting halogenoalkane. This reaction is often useful if you want to make a product that has one carbon more than the starting material.

Halogenoalkanes with ammonia

The nucleophile is ammonia, :NH$_3$.

The reaction of halogenoalkanes with an excess concentrated solution of ammonia in ethanol is carried out under pressure. The reaction produces an **amine**, RNH$_2$.

$$R—X + 2NH_3 \rightarrow RNH_2 + NH_4X$$

Exam tip
Primary amines have the functional group –NH₂. They are named with the suffix amine attached to the appropriate side chain stem, rather than the usual root name. So C₂H₅NH₂ is ethylamine, not ethanamine.

Ammonia is a nucleophile because it has a lone pair of electrons that it can donate (although it has no negative charge) and the nitrogen atom has a δ– charge.

Because ammonia is a neutral nucleophile, a proton, H^+, must be lost to form the neutral product, called an amine. The H^+ ion reacts with a second ammonia molecule to form an NH_4^+ ion.

This is the mechanism:

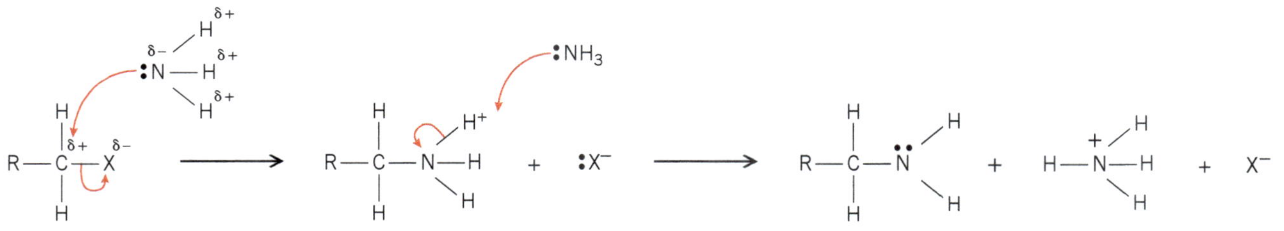

The mechanisms of nucleophilic substitution

Nucleophilic substitution reactions can take place in two ways. These are called S_N2 and S_N1.

S_N2

The nucleophile forms a bond with the $C^{\delta+}$ of the halogenoalkane using its lone pair of electrons. *At the same time*, the halogen–carbon bond breaks. Both electrons go to the halogen atom, which leaves as a negative ion. This ion is the leaving group. An intermediate species called an **activated complex** exists briefly.

The movement of the electron pairs is shown by the 'curly arrows' in the scheme below.

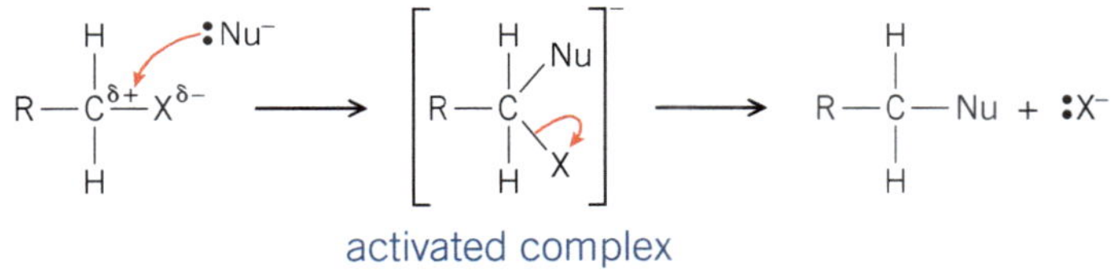

activated complex

The rate of an S_N2 reaction depends on the concentration of two species—the halogenoalkane *and* the nucleophile. So S_N2 is short for *substitution, nucleophilic* in which *two* species are involved in the key step on which the reaction rate depends.

S_N1

Here the carbon–halogen bond breaks first. Both electrons go to the halogen atom, which leaves as a halide ion. This is the slowest step of the reaction and it leaves a positive ion called a **carbocation**. This ion is then attacked by the nucleophile—this is a rapid step because it takes place between two oppositely-charged ions. The overall rate of the reaction depends only on the concentration of the halogenoalkane. Once the carbocation is formed, it is immediately attacked by the nucleophile.

a carbocation

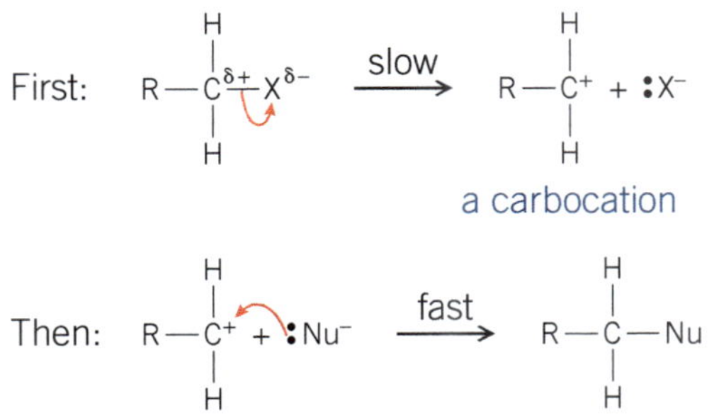

So S_N1 is short for *substitution, nucleophilic* in which *one* species is involved in the key step on which the reaction rate depends.

> **Exam tip**
>
> With excess halogenoalkane, secondary and tertiary amines may also be formed, along with quaternary ammonium salts (see Section 34.1)

> **Exam tip**
>
> The slowest step in a reaction mechanism is called the rate-determining step.

S_N1 or S_N2?

Which mechanism actually occurs depends on the halogenoalkane. The key factor is the stability of the carbocation that is formed as an intermediate in the S_N1 mechanism. The more stable this ion is, the more likely it is that the S_N1 mechanism will occur.

Exam tip

The inductive effect is shown by an arrow on the bond in the direction of electron-release.

Remember that alkyl groups such as CH_3 and C_2H_5 have an electron-releasing effect called the **inductive effect** (see Section 14.2). This effect tends to stabilise the positive charge on the intermediate carbocation. In a primary halogenoalkane, there is only one alkyl group attached to the positively-charged carbon atom. In secondary halogenoalkanes, there are two alkyl groups, and in tertiary halogenoalkanes there are three (see Figure 3). So tertiary halogenoalkanes are more likely to react by the S_N1 mechanism and primary halogenoalkanes via S_N2. Secondary halogenoalkanes react by a mixture of both mechanisms.

Extension

In tertiary halogenoalkanes, the three R groups attached to the α-carbon block the attack of a nucleophile. This means the S_N2 reaction is unlikely.

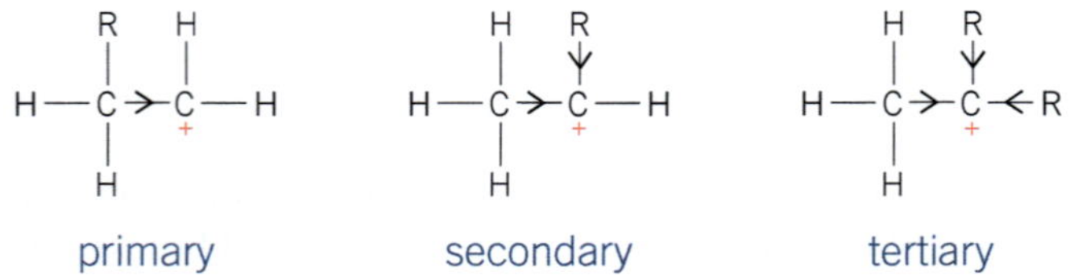

***Figure 3** Inductive effects (shown by the arrows) in primary, secondary and tertiary carbocations*

The uses of nucleophilic substitution

Nucleophilic substitution reactions are useful because they are a way of introducing new functional groups into organic compounds. Halogenoalkanes can be converted into alcohols, amines, and nitriles. These in turn can be converted to other functional groups (see Figure 4).

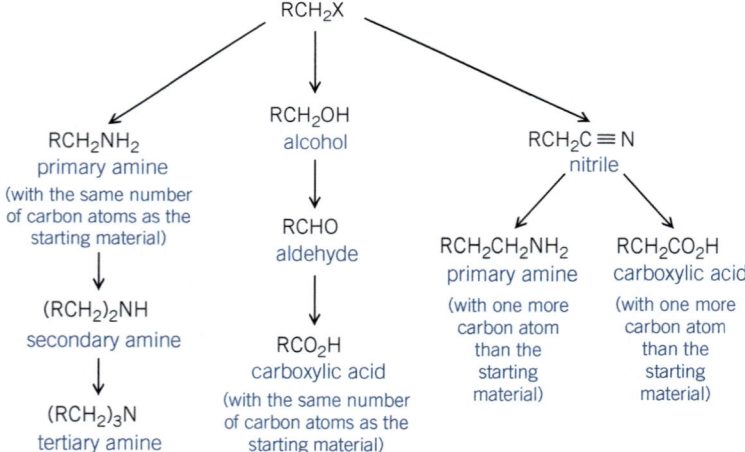

***Figure 4** Uses of nucleophilic substitution*

Elimination reactions of halogenoalkanes

Halogenoalkanes typically react by nucleophilic substitution. But, under different conditions they react by elimination. A hydrogen halide is eliminated from the molecule, leaving a double bond in its place so that an alkene is formed.

OH⁻ ion acting as a base

You saw in earlier that the OH⁻ ion, from aqueous sodium or potassium hydroxide, is a nucleophile and its lone pair will attack a halogenoalkane at $C^{\delta+}$ to form an alcohol.

Under different conditions, the OH⁻ ion can act as a base, removing an H⁺ ion from the halogenoalkane. In this case, it is an elimination reaction rather than a substitution. In the example below, bromoethane reacts with potassium hydroxide to form ethene. A molecule of hydrogen bromide, HBr, is eliminated then the hydrogen bromide reacts with the potassium hydroxide. The reaction produces ethene, potassium bromide, and water.

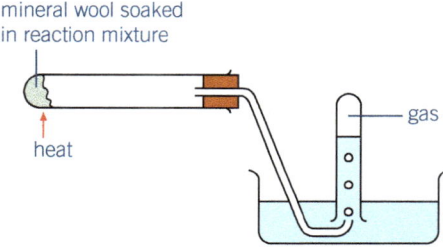

The conditions of reaction

The sodium (or potassium) hydroxide is dissolved in ethanol and mixed with the bromoethane. *There is no water present.* The mixture is heated. The experiment can be carried out using the apparatus shown in Figure 5.

mineral wool soaked
in reaction mixture

gas

heat

Figure 5 *Apparatus for elimination of hydrogen bromide from bromoethane*

The product is ethene. Ethene burns and also decolourises bromine solution, showing that it has a C=C bond.

The mechanism of elimination

Hydrogen bromide is eliminated as follows. The curly arrows show the movement of electron pairs:

- The OH⁻ ion uses its lone pair to form a bond with one of the hydrogen atoms on the carbon next to the C–Br bond. These hydrogen atoms are very slightly δ+.
- The electron pair from the C–H bond now becomes part of a carbon–carbon double bond.
- The bromine takes the pair of electrons in the C–Br bond and leaves as a bromide ion (the leaving group).

This reaction is a useful way of making molecules with carbon–carbon double bonds.

Extension

CFCs

CFCs (chlorofluorocarbons) were introduced in the 1930s by an American engineer, Thomas Midgley, for use in refrigerators. He famously demonstrated their non-toxicity and non-flammability to a scientific conference by breathing in a lungful and exhaling it to extinguish a lighted candle. It was not until long after Midgley's death that it was realised that CFCs were involved in the depletion of the ozone layer because they release chlorine atoms in the stratosphere.

Chemists have developed less harmful replacements for CFCs. Initially these were HCFCs (hydrochlorofluorocarbons), which contain hydrogen, carbon, fluorine, and chlorine. One example is $CHFCl_2$. These decompose more easily than CFCs due to their C–H bonds, and the chlorine atoms are released lower in the atmosphere where they do not contribute to the destruction of the ozone layer.

The so-called second generation replacements are called HFCs (hydrofluorocarbons) such as CHF_2CF_3. These contain no chlorine and therefore do not damage the ozone layer. They are not wholly free of environmental problems though, and chemists are working on third generation compounds. Some are considering reverting to refrigerants such as ammonia, which were used before the advent of CFCs.

1 Draw a 3D representation of the formula of $CHFCl_2$. What shape is this molecule? Does it have any isomers? Explain your answer.
2 What sort of formula is CHF_2CF_3? Draw its displayed formula. What is its molecular formula?

Isomeric products

In some cases, a mixture of isomeric elimination products is possible.

Substitution or elimination?

As the hydroxide ion will react with halogenoalkanes as a nucleophile or as a base, there is competition between substitution and elimination. In general, a mixture of an alcohol and an alkene is produced. For example:

The reaction that predominates depends on three factors:

- the reaction conditions (aqueous or ethanolic solution)
- the temperature
- the type of halogenoalkane (primary, secondary, or tertiary).

The conditions of the reaction

- Hydroxide ions, warmed, dissolved in water (aqueous), favour substitution.
- Hydroxide ions at high temperature, dissolved in ethanol, favour elimination.

The type of halogenoalkane

- Primary halogenoalkanes tend to react by substitution.
- Tertiary halogenoalkanes tend to react by elimination.
- Secondary halogenoalkanes react by both substitution and elimination.

Relative strength of the C–X bond

The reactivities of the halogenoalkanes to substitution of the halogen depend on the strength of the C–X bond. Carbon–halogen bond energies were listed in Table 2 earlier in this chapter. The bonds get weaker going down the group. The stronger the bond, the slower the reaction.

You can demonstrate this by reacting different halogenoalkanes with aqueous silver nitrate using ethanol as a solvent in which all components dissolve. The water acts as a weak nucleophile:

$$R–X + H_2O \rightarrow R–OH + H^+ + X^-$$

As soon as a significant concentration of X^- is produced, it reacts with the silver nitrate to form a precipitate:

$$X^-(aq) + AgNO_3(aq) \rightarrow AgX(s) + NO_3^-(aq)$$

The precipitates form in the order I^- faster than Br^-, faster than Cl^-, confirming that the key factor is the C–X bond energy.

Identifying the halogen

This reaction can also be used to determine which halogen is present by examining the precipitate. Cl^- forms a white precipitate, which is soluble in dilute ammonia. Br^- forms a cream precipitate and I^- a yellowish one (see Figure 6). The latter two can be distinguished by their solubility in concentrated ammonia. AgBr dissolves in concentrated ammonia, while AgI does not—see Section 11.3.

Figure 6 *Precipitates formed from the silver nitrate test; left to right, they are silver chloride, silver bromide, and silver iodide*

Summary test 15.1

1 These questions are about the following halogenoalkanes:
   i $CH_3CH_2CH_2CH_2I$
   ii $CH_3CHBrCH_3$
   iii $CH_2ClCH_2CH_2CH_3$
   iv $CH_3CH_2CHBrCH_3$
   a Sketch the displayed formula for each halogenoalkane and mark the polarity of the C–X bond.
   b Identify each halogenoalkane.
   c Predict which of them would have the highest boiling point and explain your answer.
2 Explain why the halogenoalkanes get less reactive going up the halogen group.
3 This equation represents the hydrolysis of a halogenoalkane by sodium hydroxide solution:

$$R–X + OH^- \rightarrow ROH + X^-$$

   a Explain why the reaction is carried out in ethanol.
   b Identify the nucleophile.
   c Explain why this is a substitution.
   d Identify the leaving group.
   e State which would have the fastest reaction: R–F, R–Cl, R–Br, or R–I.
4 **a** Starting with bromoethane, state which nucleophile will produce a product with three carbon atoms.
   b Give the equation for this, using curly arrows to show the mechanism of the reaction.
   c Identify the product.
5 In elimination reactions of halogenoalkanes, state which of the following the OH– group is acting as.
   A a base
   B an acid
   C a nucleophile
   D an electrophile
6 **a** Identify the two possible products when 2-bromopropane reacts with hydroxide ions.
   b Explain how could you show that one of the products is an alkene.
   c Give the mechanism (using curly arrows) of the reaction that is an elimination.
7 Use the electron arrangements of fluorine and of chlorine to suggest why the C–F bond is stronger than the C–Cl bond.

⎡⎡ **Launch additional digital resources for the chapter**

1 Dichloromethane, CH_2Cl_2 was used as an early anaesthetic. It was discovered that further substitution by chlorine produced more potent anaesthetics and hence Chloroform, $CHCl_3$ became more commonly used.

  a Name the type of reaction that is used to produce chloroform from dichloromethane.   *(1 mark)*

  b Explain with the aid of balanced equations, how further substitution of dichloromethane can produce chloroform. Name this step in the reaction.   *(3 marks)*

  c Give the IUPAC name for chloroform.   *(1 mark)*

2 Methanol can be synthesised from bromomethane. Suggest a mechanism for the formation of methanol from bromomethane. Name the type of reaction and state the conditions required.   *(4 marks)*

3 Organic chemists often use nitrile synthesis as a means of increasing carbon chain length in synthesis reactions.

  a Suggest a reagent and the reaction conditions used to synthesise a nitrile from a halogenoalkane.   *(2 marks)*

  b Bromoethane is used in the synthesis of a nitrile with the structural formula CH_3CH_2CN. Give the displayed formula and IUPAC name for this nitrile.   *(2 marks)*

4 2-methylpropan-2-ol is used as a fuel additive to increase the octane rating of performance fuels. A student investigated the synthesis of 2-methylpropan-2-ol from a brominated halogenoalkane X and potassium hydroxide solution. The rate of reaction was found to be independent of hydroxide ion concentration.

  a Give the IUPAC name for halogenoalkane X.   *(1 mark)*

  b Using the information above, suggest a mechanism for the formation of 2-methyl propan-2-ol from hydrocarbon X. Name this type of reaction.   *(3 marks)*

  c Using your answer to part **b**, explain why the rate of reaction is independent of hydroxide ion concentration. Comment of the relative stability of any species involved.   *(2 marks)*

  d Predict the effect on the rate of reaction for the formation of 2-methylpropan-2-ol if halogenoalkane X were to contain chlorine instead of bromine. Explain your answer.   *(3 marks)*

5 Identify the compounds A to D and the reagents P and Q in the reaction scheme below: *(6 marks)*

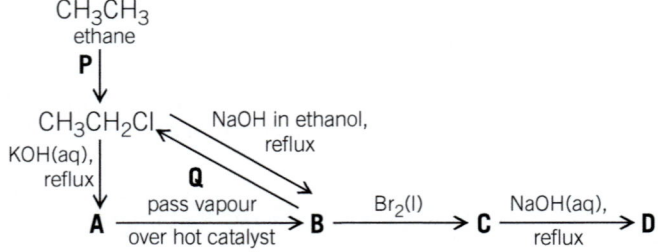

6 Halogenoalkanes can undergo either nucleophilic substitution reactions or elimination reactions when they react with hydroxide ions, such as in NaOH(aq).

  a OH^- acts as a nucleophile in nucleophilic substitution reactions. Define the term 'nucleophile'.   *(1 mark)*

  b Explain how hydroxide ions behave when heated under reflux with 1-chloropropane, $CH_3CH_2CH_2Cl$ in ethanolic conditions. Name the type of reaction and identify the homologous series of compounds formed under these conditions.   *(3 marks)*

  c Using your answer to part **b**, draw a mechanism for this reaction.   *(3 marks)*

7 Ethylamine, $CH_3CH_2NH_2$, can be synthesised by nucleophilic substitution from the reaction of chloroethane with ammonia.

  a State the conditions required for this reaction.   *(1 mark)*

  b Suggest a mechanism for this reaction.   *(4 marks)*

  c A small quantity of diethylamine is produced as a by-product of this reaction. Give the displayed formula for diethylamine.   *(1 mark)*

8 Describe how a student might carry out a series of simple test tube reactions to differentiate between a selection of halogenoalkanes that may contain either chlorine, bromine and iodine atoms. Include at least one ionic equation in your answer. *(6 marks)*

9 Suggest the reaction conditions needed to synthesise a halogenoalkane from an alcohol. *(1 mark)*

10 The reaction of propene with hydrogen bromide produces two organic products.

  a Give the IUPAC name and displayed formula for the minor product. *(2 marks)*

  b Suggest a mechanism for the formation of the major product and explain why this is the preferred product. *(4 marks)*

11 Consider the following compounds:
  I CCl_3F **II** CF_3CF_3
  III CH_3CH_2Cl **IV** $CH_3CHBrCH_3$

  a Name each compound. *(4 marks)*

  b Identify which of the above compounds would react most readily with aqueous sodium hydroxide. *(1 mark)*

  c Identify which of the above compounds would react least readily with aqueous sodium hydroxide. *(1 mark)*

  d Which compound(s) would undergo an elimination reaction when heated under reflux with NaOH in ethanol? *(1 mark)*

  e State which compound would be most suitable for use as a refrigerant and describe one problem caused by this compound if it escapes into the atmosphere. Explain why this occurs for this type of compound. *(4 marks)*

12 This question concerns the hydrolysis of three different halogenoalkanes:

Four drops of each halogenoalkane were added separately to three separate tubes standing in a water bath at 60°C. Each tube contains 1 cm^3 of 0.1 mol dm^{-3} silver nitrate solution. The results are as follows:

Compound	Observation
1-chlorobutane	Slight cloudiness after three minutes. Still only slightly cloudy after 15 minutes.
1-bromobutane	Slightly cloudy after one minute, opaque after three minutes, coagulation and precipitation after six minutes.
1-iodobutane	Immediately opaque, yellow precipitate within first minute.

  a Give the formula of the precipitate formed in each reaction. *(3 marks)*

  b Explain why a precipitate is formed in each reaction. *(1 mark)*

  c Each reaction involves a substitution reaction occurring between a halogenoalkane and a nucleophile. Give the formula of the nucleophile involved. *(1 mark)*

  d Identify which halogenoalkane undergoes substitution most readily, explain your answer with reference to the relative strength of the carbon-halogen bond. *(3 marks)*

13 Nucleophilic substitution reactions occur via an S_N1 or S_N2 mechanism.

  a Describe the difference between an S_N1 and an S_N2 mechanism. *(2 marks)*

  b Identify the fastest step of an S_N1 mechanism. Explain your answer. *(2 marks)*

  c Reactions that take place in aqueous conditions usually proceed via an S_N1 mechanism. Give a reason why this occurs. *(2 marks)*

  d In an S_N2 reaction, which of the following species would be the most reactive? Give reasons for your answer.

  i CH_3Cl **ii** CH_3Br **iii** CH_3I **iv** CH_3F *(2 marks)*

16.1 Alcohols

Learning outcomes

On these pages you will learn to:

- describe the reactions used to form alcohols
- describe the reactions of alcohols, including combustion, substitution to give halogenoalkanes, the reaction with sodium, oxidation to carbonyl compounds and carboxylic acids, dehydration to alkenes, and reactions with carboxylic acids to form esters
- classify alcohols as primary, secondary or tertiary and learn how to distinguish them
- describe the test for a $CH_3CH(OH)-$ group in an alcohol
- explain the relative acidities of water and ethanol

Alcohols have the functional group –OH attached to a hydrocarbon chain. They are relatively reactive. The general formula of an alcohol is $C_nH_{2n+1}OH$. This is often shortened to ROH.

How to name alcohols

The name of the functional group (the –OH group) is normally given by the suffix -ol. (The prefix hydroxy- is used if some other functional groups are present.)

$$H-\underset{\underset{H}{|}}{\overset{\overset{H}{|}}{C}}-\underset{\underset{H}{|}}{\overset{\overset{H}{|}}{C}}-O-H$$

ethanol

With chains longer than ethanol, you need a number to show where the –OH group is.

propan-1-ol   propan-2-ol

If there is more than one –OH group, di-, tri-, tetra-, and so on are used to say how many –OH groups there are and numbers to say where they are located.

butane-1,4-diol   propane-1,2,3-triol

Propane-1,2,3-triol is also known as glycerol. It may be obtained from the fats and oils found in living organisms.

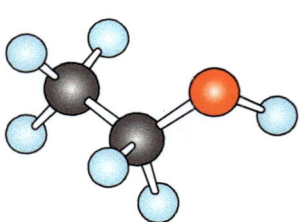

Figure 1 A 3D representation of ethanol, an alcohol

Shape

In alcohols, the oxygen atom has two bonding pairs of electrons and two lone pairs, as it does in the water molecule. The C–O–H angle is about 105° because the 109.5° angle of a perfect tetrahedron is 'squeezed down' by the presence of the lone pairs. These two lone pairs will repel each other more than the pairs of electrons in a covalent bond.

104.5°

Classification of alcohols

Alcohols are classified as primary (1°), secondary (2°), or tertiary (3°) according to how many other groups (R) are bonded to the carbon that has the –OH group.

Primary alcohols

In a primary alcohol, the carbon with the –OH group has one R group (and therefore two hydrogen atoms).

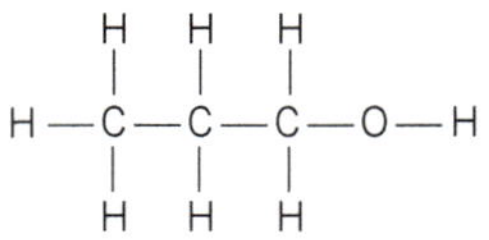

propan-1-ol is a primary alcohol

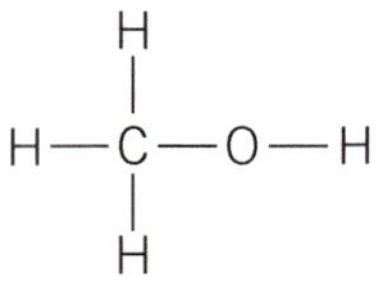

methanol, where the carbon has no R groups is counted as a primary alcohol

A primary alcohol has the –OH group at the end of a chain.

Secondary alcohols

In a secondary alcohol, the –OH group is attached to a carbon with two R groups (and therefore one hydrogen atom).

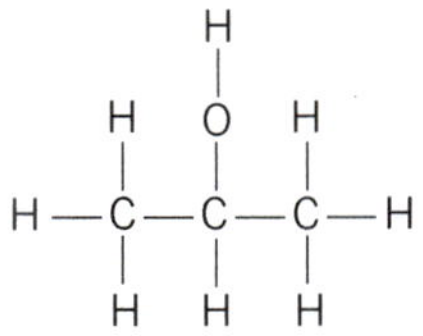

propan-2-ol is a secondary alcohol

A secondary alcohol has the –OH group in the body of the chain.

Tertiary alcohols

Tertiary alcohols have three R groups attached to the carbon that is bonded to the –OH (so this carbon has no hydrogen atoms).

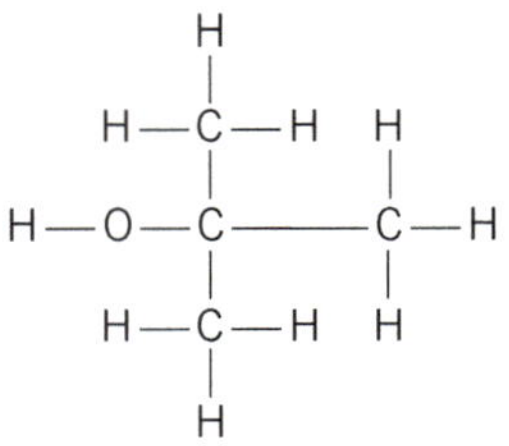

2-methylpropan-2-ol is a tertiary alcohol

A tertiary alcohol has the –OH group at a branch in the chain.

Polyols

Alcohols with more than one –OH group are called polyols. They are named -diol if they have two –OH groups (for example ethane-1,2-diol, $CH_2(OH)CH_2OH$) and -triol if they have three –OH groups.

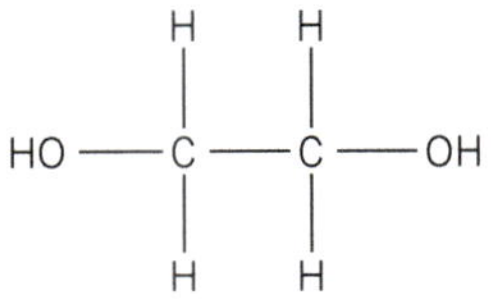

ethane-1,2-diol is a polyol; specifically, a diol

Physical properties

The –OH group in alcohols means that hydrogen bonding occurs between the molecules. This is the reason that alcohols have higher melting and boiling points than alkanes of similar relative molecular mass.

The –OH group of alcohols can hydrogen bond to water molecules, but the non-polar hydrocarbon chain cannot. This means that the alcohols with short hydrocarbon chains are soluble in water because the hydrogen bonding predominates. In longer-chain alcohols the non-polar hydrocarbon chain dominates and the alcohols become insoluble in water.

Exam tip

Halogenoalkanes may be classified as primary, secondary, or tertiary in the same way.

Extension

Antifreeze

Ethane-1,2-diol is the main ingredient in most antifreezes. Antifreezes are added to the water in the cooling systems of motor vehicles, so that the resulting coolant mixture does not freeze at winter temperatures.

1. Draw the structural formula of ethane-1,2-diol and indicate where hydrogen bonding with water may take place.
2. Suggest why solutions of ethane-1,2-diol have lower freezing temperatures than pure water.

Extension

The reactivity of alcohols

How an organic molecule reacts depends, among other things, on the strength and polarity of the bonds within it.

1. Draw a displayed formula for ethanol and mark on it the bond energy of each bond (the amount of energy required to break it) and the electronegativity of each atom. Use the table of bond energies at the back of the book. Table 1 in Section 3.1 shows the electronegativities of various elements.
2. Use your diagram and the electronegativity values to explain why:
   - a the hydrocarbon skeleton remains intact in most reactions of ethanol
   - b the typical reactions of ethanol are nucleophilic substitutions in which the –OH group is lost.

Synthesising alcohols

Alcohols can be made by a number of reactions:

1   Addition of water (steam) across the double bond of an alkene at high temperature and pressure using a catalyst of phosphoric acid, H_3PO_4. This is a hydration reaction, and how ethanol is made industrially.

$$CH_2{=}CH_2 + H_2O \xrightarrow[60\ atm]{330°C} CH_3CH_2OH$$

2   Oxidation of an alkene by cold, dilute acidified potassium manganate(VII) ($KMnO_4$), to form a diol. For example:

This is ethene-1,2-diol (ethylene glycol), used in anti-freeze.

[O] comes from the oxidising agent, $KMnO_4$.

3   Substitution of the halogen of a halogenoalkane by –OH using hot sodium (or potassium) hydroxide in aqueous solution. For example:

$$CH_3CH_2Br + OH^- \rightarrow CH_3CH_2OH + Br^-$$

4   Reduction of an **aldehyde** (or **ketone**) with $LiAlH_4$ or $NaBH_4$. Aldehydes give primary alcohols and ketones give secondary alcohols. For example:

$$CH_3CHO + 2[H] \rightarrow CH_3CH_2OH \text{ (a primary alcohol)}$$

$$CH_3COCH_3 + 2[H] \rightarrow CH_3CH(OH)CH_3 \text{ (a secondary alcohol)}$$

[H] comes from the reducing agent, $LiAlH_4$ or $NaBH_4$.

5   Reduction of a **carboxylic acid** by $LiAlH_4$ or $NaBH_4$, or by heating with $H_2(g)$ and a nickel catalyst. For example:

$$CH_3COOH + 4[H] \rightarrow CH_3CH_2OH + H_2O$$

6   Hydrolysis of an **ester** using hot dilute acid or alkali, For example:

$$CH_3COOCH_3 + H_2O \rightarrow CH_3COOH + CH_3OH$$

Each of these reactions is discussed in more detail in the relevant section.

The reactions of alcohols

Combustion

Alcohols burn completely to produce carbon dioxide and water if there is excess oxygen available. (Otherwise there is incomplete combustion and carbon monoxide or even carbon (soot) is produced.) This is the equation for the complete combustion of ethanol:

$$C_2H_5OH(l) + 3O_2(g) \rightarrow 2CO_2(g) + 3H_2O(l)$$

Ethanol is often used in fuel mixtures, for example in picnic stoves.

Figure 2 *Alcohol-burning stove*

Dehydration

Alcohols can be dehydrated with excess hot concentrated sulfuric acid or by passing their vapours over a heated aluminium oxide catalyst. An alkene is formed. For example, propan-1-ol is dehydrated to propene:

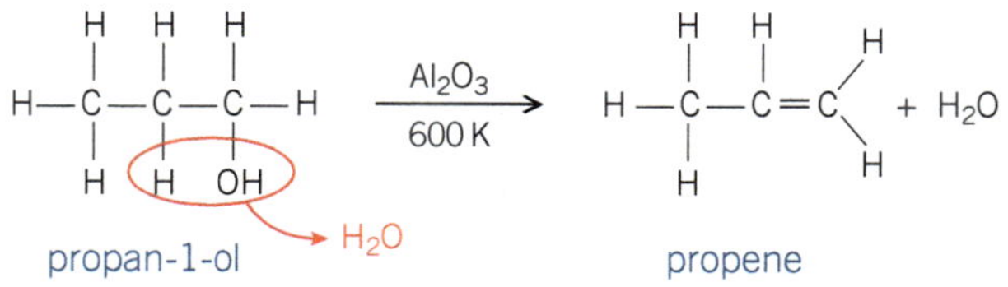

***Figure 3** Dehydration of propan-1-ol*

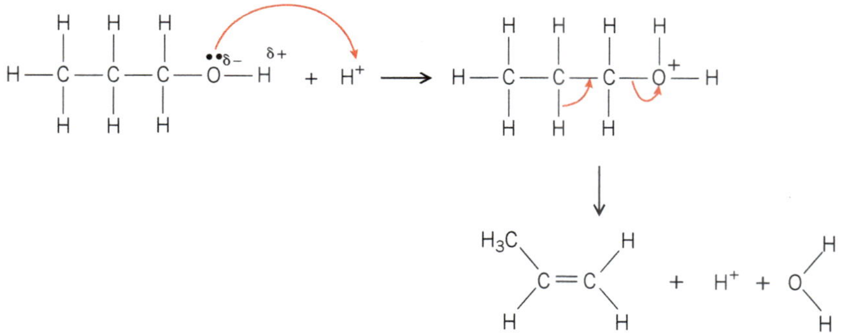

***Figure 4** The mechanism*

Phosphoric(V) acid is an alternative dehydrating agent.

Extension

Isomeric alkenes

Dehydration of longer chain or branched alcohols may produce a mixture of alkenes, including ones with *cis–trans* isomers, see Section 13.4.

For example, with butan-2-ol there are three possible products: but-1-ene, *cis*-but-2-ene, and *trans*-but-2-ene.

1 Name the two isomeric alkanes that are formed using the *E/Z* notation.

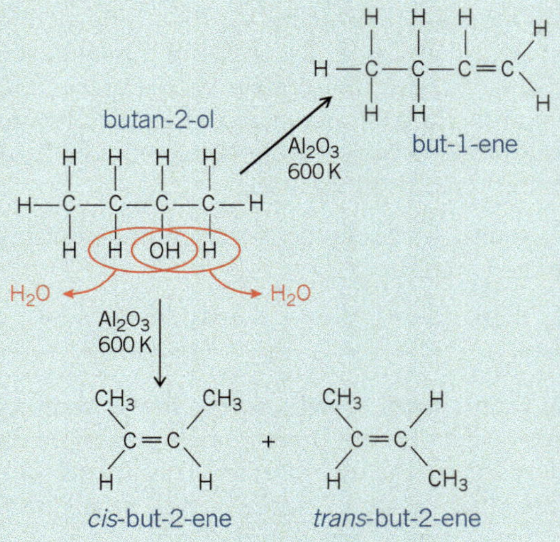

Exam tip

In order to be dehydrated an alcohol must have a hydrogen atom on a carbon next to the –OH group.

Oxidation

Combustion is usually complete oxidation. Alcohols can also be oxidised gently and in stages. Primary alcohols are oxidised to aldehydes, RCHO. Aldehydes can be further oxidised to carboxylic acids, RCOOH. For example:

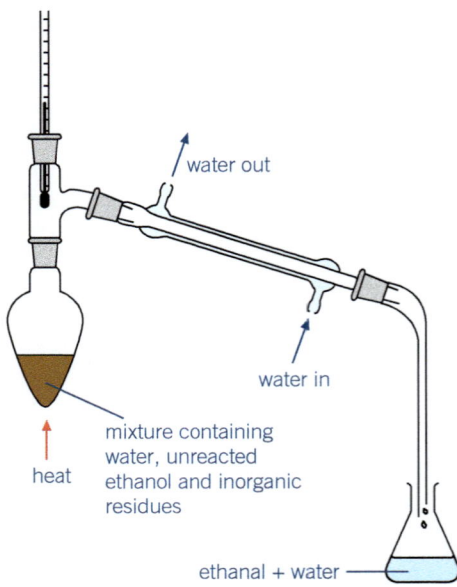

Figure 5 *Apparatus for distilling ethanal from the reaction mixture*

Secondary alcohols are oxidised to ketones, R_2CO. Ketones are not oxidised further.

propan-2-ol   propanone (a ketone)

Tertiary alcohols are not easily oxidised. This is because oxidation would need a C–C bond to break, rather than a C–H bond (which is what happens when an aldehyde is oxidised). Ketones are not oxidised further for the same reason.

Many aldehydes and ketones have pleasant smells.

The experimental details

A solution of potassium dichromate, acidified with dilute sulfuric acid, is often used to oxidise alcohols to aldehydes and ketones. It is the oxidising agent. In the reaction, the orange dichromate(VI) ions are reduced to green chromium(III) ions.

Tertiary alcohols are not oxidised by dichromate(VI) ions. This can be used as a method to distinguish them from primary and secondary alcohols.

To oxidise ethanol (1° alcohol) to ethanal—an aldehyde

Dilute acid and less potassium dichromate(VI) than is needed for complete oxidation to carboxylic acid are used. The mixture is heated gently in apparatus like that shown in Figure 5, with the receiver cooled in ice to reduce evaporation of the product. Ethanal (boiling temperature 294 K, 21°C) vaporises as soon as it is formed and distils off. This stops it from being oxidised further to ethanoic acid. Unreacted ethanol remains in the flask.

The notation [O] is used to represent oxygen from the oxidising agent. The reaction is given by the equation:

$$CH_3CH_2OH(l) \quad + \quad [O] \quad \rightarrow \quad CH_3CHO(g) \quad + \quad H_2O(l)$$

ethanol   ethanal

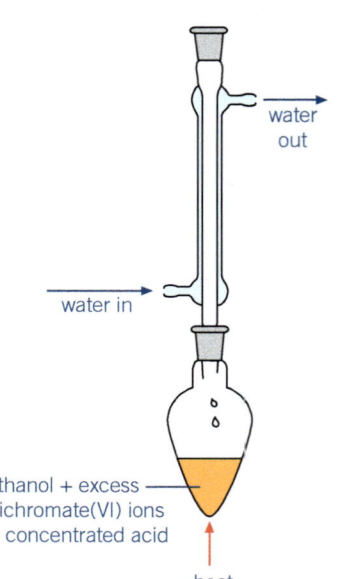

Figure 6 *Reflux apparatus for oxidation of ethanol to ethanoic acid*

To oxidise ethanol (1° alcohol) to ethanoic acid—a carboxylic acid

Concentrated sulfuric acid and more than enough potassium dichromate(VI) is used for complete reaction (the dichromate(VI) is in excess). The mixture is refluxed in the apparatus shown in Figure 6. **Reflux** means that vapour condenses and drips back into the reaction flask.

While the reaction mixture is refluxing, any ethanol or ethanal vapour will condense and drip back into the flask until, eventually, it is all oxidised to the acid. After refluxing for around 20 minutes, you can distil off the ethanoic acid (boiling temperature 391 K, 118°C), along with any water, by rearranging the apparatus to that shown in Figure 5.

Using [O] to represent oxygen from the oxidising agent, the equation is:

$$CH_3CH_2OH(l) \quad + \quad 2[O] \quad \rightarrow \quad CH_3COOH(g) \quad + \quad H_2O(l)$$

ethanol   ethanoic acid

Notice that twice as much oxidising agent is used in this reaction compared with the oxidation to ethanal.

Oxidising a secondary alcohol to a ketone

Secondary alcohols are oxidised to ketones by acidified dichromate. You do not have to worry about further oxidation of the ketone.

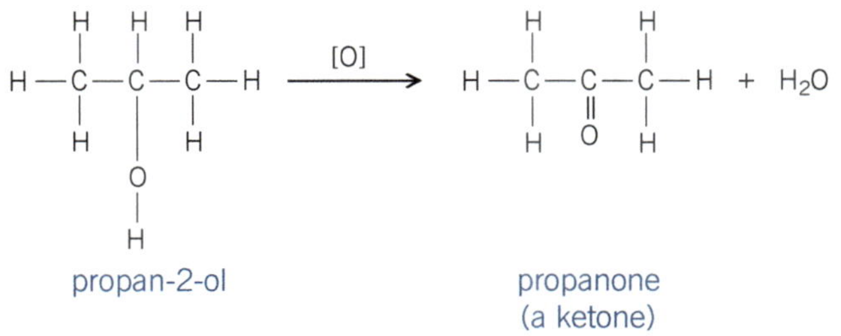

propan-2-ol   propanone
                               (a ketone)

Reaction with sodium

Dropping a small piece of metallic sodium into ethanol produces bubbles of hydrogen and forms the ionic compound sodium ethoxide:

$$2CH_3CH_2OH + 2Na \rightarrow H_2 + 2(CH_3CH_2O^-Na^+)$$

Other alcohols undergo similar reactions, where the alcohol is behaving as an acid whose hydrogen is displaced by a metal.

Notice that it is only the hydrogen bonded to the oxygen that is involved. This is because this bond is polarised $O^{\delta-}-H^{\delta+}$. The reaction is slower than the reaction of sodium with water, showing that ethanol is a weaker acid than water.

📖 Why are alcohols weaker acids than water?

To answer this we need to look at the following equations where a H^+ ion is being lost:

$$CH_3CH_2OH \rightleftharpoons CH_3CH_2O^- + H^+ \quad (1)$$

$$H_2O \rightleftharpoons {}^-OH + H^+ \quad (2)$$

Alkyl groups, such as the ethyl group CH_3CH_2-, are slightly electron-releasing (this is called the inductive effect). This places extra negative charge on the oxygen atom in ethanol and also strengthens the O–H bond. This makes it more difficult for the H of the OH group to be released as a H^+ ion than is the case in water. So water releases a H^+ ion more readily than ethanol does and is therefore a stronger acid.

Substitution of the –OH group by a halogen

Alcohols react with hydrogen halides such as hydrogen bromide, HBr, to form a halogenoalkane. The halogen is substituted for the –OH group.

The rate of the reaction varies with the halogen, in the order HI > HBr > HCl. A strong acid catalyst such as sulfuric acid is used.

The hydrogen halide is generated in the reaction flask by the reaction of an alkali metal halide, for example:

$$2KBr(s) + H_2SO_4 \rightarrow K_2SO_4(s) + 2HBr(g)$$

The acid catalyst works by first donating a proton (a H^+ ion) to the alcohol—we say the alcohol has been protonated:

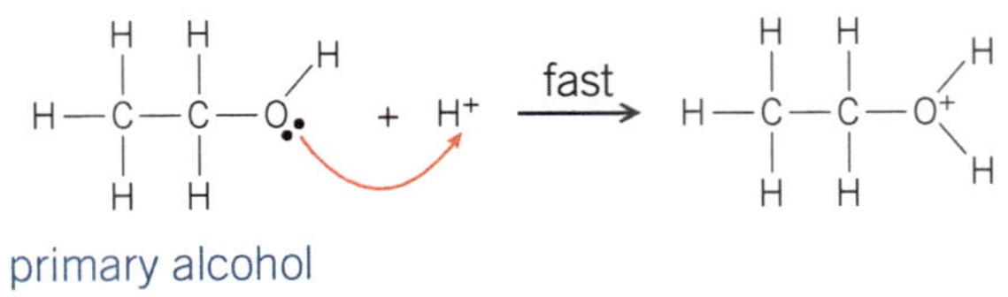

primary alcohol

Some of the positive charge resides on the carbon next to the –OH group. In the case of a primary alcohol, this is then attacked by the nucleophile Br⁻, followed by the loss of a molecule of water as the leaving group.

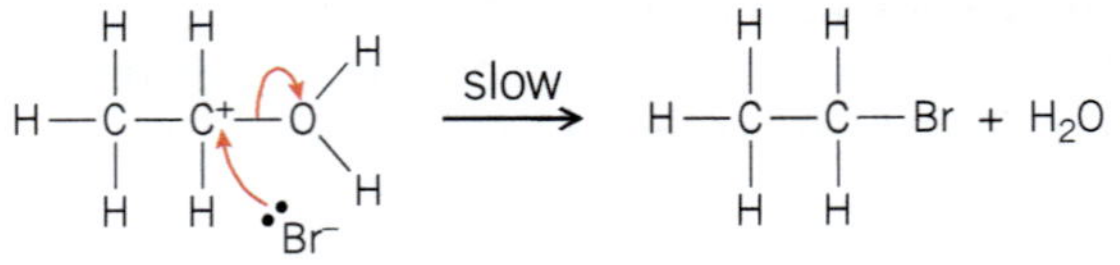

This is a nucleophilic substitution (S_N2) reaction. It is essentially the reverse of the substitution reaction described in Section 15.1 for halogenoalkanes with OH⁻.

Tertiary alcohols undergo a similar reaction via an S_N1 mechanism. A molecule of water is first lost from the protonated alcohol to form a carbocation, which is then rapidly attacked by Br⁻.

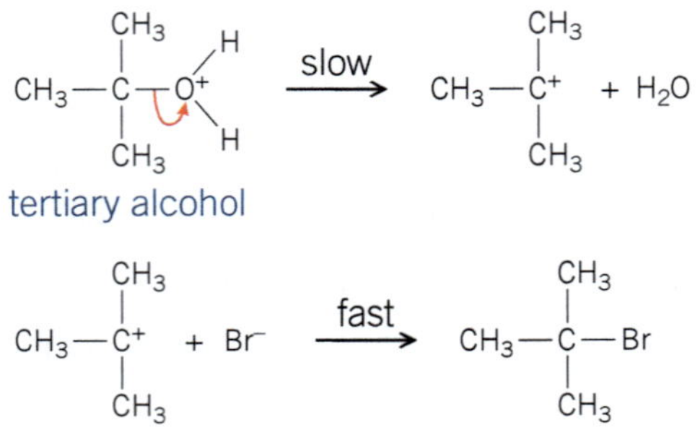

tertiary alcohol

Tertiary alcohols react by the S_N1 mechanism (due to the stabilisation of the carbocation by the inductive effect of three alkyl groups) and primary alcohols react by the S_N2 mechanism.

The overall reaction is:

$$ROH + HBr \rightarrow RBr + H_2O$$

Other reagents which bring about halogen substitution are:

- phosphorus trichloride, PCl_3, and heat
- phosphorus pentachloride, PCl_5
- a mixture of phosphorus and bromine (or iodine)
- thionyl chloride, $SOCl_2$.

These are all nucleophilic substitution reactions. The reagent first generates a halide ion which acts as the nucleophile.

Typical equations are:

$$3CH_3CH_2OH + PCl_3 \rightarrow 3CH_3CH_2Cl + H_3PO_3$$

$$CH_3CH_2OH + PCl_5 \rightarrow CH_3CH_2Cl + POCl_3 + HCl$$

$$2P + 3Br_2 \rightarrow 2PBr_3$$

followed by:

$$3CH_3CH_2OH + PBr_3 \rightarrow 3CH_3CH_2Br + H_3PO_3$$

And similarly for iodine:

$$CH_3CH_2OH + SOCl_2 \rightarrow CH_3CH_2Cl + SO_2 + HCl$$

Formation of esters

Alcohols will react with carboxylic acids, such as ethanoic acid, CH_3COOH, to form esters (see Section 18.2 for more details about esters). A concentrated sulfuric acid catalyst is used but the reaction does not go to completion:

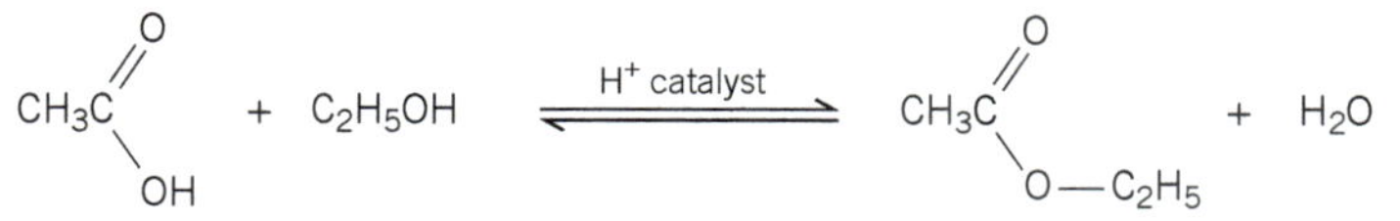

The sulfuric acid has two functions. It protonates the carboxylic acid, making it more susceptible to nucleophilic attack by the oxygen atom of the alcohol. It is also a good dehydrating agent so it removes water from the reaction mixture and forces the equilibrium to the right by Le Chatelier's principle (see Section 7.1).

Esters can also be formed by the reaction of alcohols with acyl chlorides (see Section 33.3). These are more reactive than carboxylic acids and no strong acid catalyst is needed. For example:

$$CH_3COCl + CH_3CH_2OH \rightarrow CH_3COOCH_2CH_3 + HCl$$

The tri-iodomethane reaction

This reaction occurs only with alcohols containing the group:

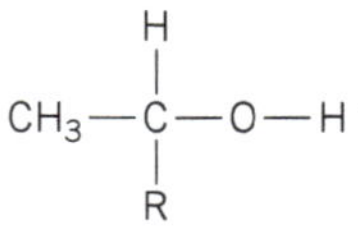

where R is a H atom or an alkyl group.

So it will work with ethanol and propan-2-ol, but not with propan-1-ol.

The alcohol is warmed with a mixture of iodine and sodium hydroxide. The formation of a yellow precipitate of solid tri-iodomethane indicates that the group above is present.

The overall reaction is as follows:

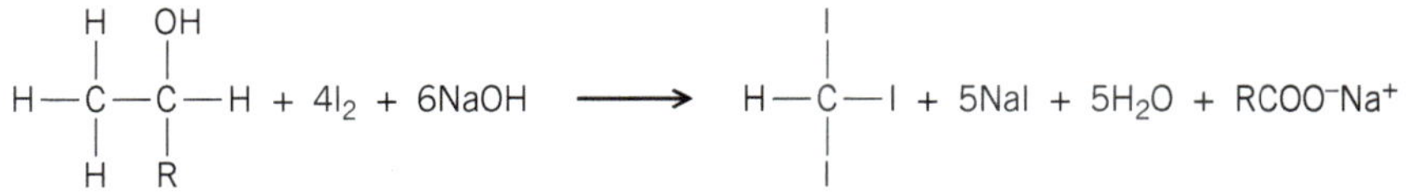

Summary test 16.1

1 Sketch the displayed formula and name the alcohol $C_2H_5CH(OH)CH_3$.

2 Identify these alcohols as primary, secondary, or tertiary:
  a butan-2-ol
  b 2-methylpentan-2-ol
  c methanol

3 Explain why the C–O–H angle in alcohols is less than 109.5°.

4 State what happens in each case when the following alcohols are oxidised as much as possible by acidified potassium dichromate:
  a a primary alcohol
  b a secondary alcohol

5 Explain why a tertiary alcohol is not oxidised by the method outlined in question 4.

6 Describe the difference between distilling and refluxing.

7 Give the equation for the elimination of water from ethanol and identify the product.

8 State the possible products of dehydrating pentan-2-ol.

9 Sketch the displayed formulae of propan-2-ol to show that it will undergo the tri-iodomethane reaction.

There are exam-style questions to test your knowledge of the material in this chapter at the end of Chapter 18.

17.1 Aldehydes and ketones

Aldehydes and **ketones** are compounds which contain the carbonyl group.

The carbonyl group consists of a carbon–oxygen double bond: $\diagdown C{=}O$

The group is also present in carboxylic acids and derivatives (see Chapter 18).

In aldehydes, the carbon bonded to the oxygen (the carbonyl carbon) has at least one hydrogen atom bonded to it, so the general formula of an aldehyde is:

$$\begin{array}{c} R \diagdown \\ C{=}O \\ H \diagup \end{array}$$

This is sometimes written as RCHO.

In ketones, the carbonyl carbon has two organic groups, which can be represented by R and R′, so the formula of a ketone is:

$$\begin{array}{c} R' \diagdown \\ C{=}O \\ R \diagup \end{array}$$

The R groups in both aldehydes and ketones may be alkyl or aryl (ie ones based on a benzene ring, see Section 29.3).

How to name aldehydes and ketones

Aldehydes are named using the suffix *-al*. The carbon of the aldehyde functional group is counted as part of the carbon chain of the root. So:

$H{-}C\diagup^{O}_{\diagdown H}$ or HCHO is methanal and $H{-}C{-}C\diagup^{O}_{\diagdown H}$ or CH_3CHO is ethanal.

The aldehyde group can only occur at the end of a chain, so a numbering system is not needed to show its location.

Ketones are named using the suffix *-one*. In the same way as aldehydes, the carbon atom of the ketone functional group is counted as part of the root. So the simplest ketone:

$$H{-}\overset{H}{\underset{H}{C}}{-}\overset{O}{\underset{}{C}}{-}\overset{H}{\underset{H}{C}}{-}H \text{ or } CH_3COCH_3, \text{ is called propanone.}$$

No ketone with fewer than three carbon atoms is possible.

You do not need to number the carbon in propanone or in butanone:

$$H{-}\overset{H}{\underset{H}{C}}{-}\overset{H}{\underset{H}{C}}{-}\overset{O}{\underset{}{C}}{-}\overset{H}{\underset{H}{C}}{-}H,$$

$CH_3CH_2COCH_3$, because the carbonyl group can only be in one position. With larger numbers of carbon atoms, numbers are needed to locate the carbonyl group on the chain. For example, pentanone could be pentan-3-one, $CH_3CH_2COCH_2CH_3$, or pentan-2-one, $CH_3COCH_2CH_2CH_3$.

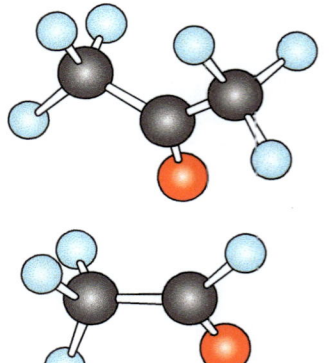

Figure 1 *3D representations of propanone (a ketone) and ethanal (an aldehyde)*

Physical properties of carbonyl compounds

The carbonyl group is strongly polar, $C^{\delta+}=O^{\delta-}$, so there are permanent dipole–dipole forces between the molecules. These forces mean that boiling points are higher than those of alkanes of comparable relative molecular mass, but not as high as those of alcohols where **hydrogen bonding** can occur between the molecules (Table 1).

Table 1 Boiling point data

Name	Formula	M_r	T_b / K
butane	$CH_3CH_2CH_2CH_3$	60	273
propanone	CH_3COCH_3	58	359
propan-1-ol	$CH_3CH_2CH_2OH$	60	370

Solubility in water

Shorter chain aldehydes and ketones mix completely with water because hydrogen bonds form between the oxygen of the carbonyl compound and water (Figure 2). As the length of the carbon chain increases, carbonyl compounds become less soluble in water.

Methanal, HCHO, is a gas at room temperature. Other short chain aldehydes and ketones are liquids, with characteristic smells (propanone, sometimes known as acetone, is often found in nail varnish remover).

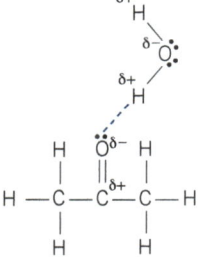

Figure 2 Hydrogen bonding between propanone and water

The reactivity of carbonyl compounds

The C=O bond in carbonyl compounds is strong (Table 2) and you might think that the C=O bond would be the least reactive bond. However, almost all reactions of carbonyl compounds involve the C=O bond.

This is because the big difference in electronegativity between carbon and oxygen makes the $C^{\delta+}=O^{\delta-}$ strongly polar. So, nucleophilic reagents can attack the $C^{\delta+}$. Also, since they contain a double bond, carbonyl compounds are **unsaturated** and addition reactions are possible.

In fact, the most typical reactions of the carbonyl group are **nucleophilic additions**.

Table 2 Comparison of bond strengths

Bond	Mean bond enthalpy / kJ mol^{-1}
C=O	740
C=C	610
C–O	360
C–C	350

Synthesising carbonyl compounds

Carbonyl compounds can be produced by the oxidation of alcohols using acidified potassium dichromate(VI) or acidified potassium manganate(VII) as the oxidising agent.

Primary alcohols are oxidised to aldehydes, for example:

$$CH_3CH_2OH + [O] \rightarrow CH_3CHO + H_2O$$

Secondary alcohols are oxidised to ketones, for example:

$$CH_3CH(OH)CH_3 + [O] \rightarrow CH_3COCH_3 + H_2O$$

These reactions are discussed in more detail in Section 16.1.

Reactions of aldehydes and ketones

Many of the reactions of carbonyl compounds are nucleophilic addition reactions.

They also undergo redox reactions.

Nucleophilic addition reactions

Representing the nucleophile as $:Nu^-$, the general reaction is:

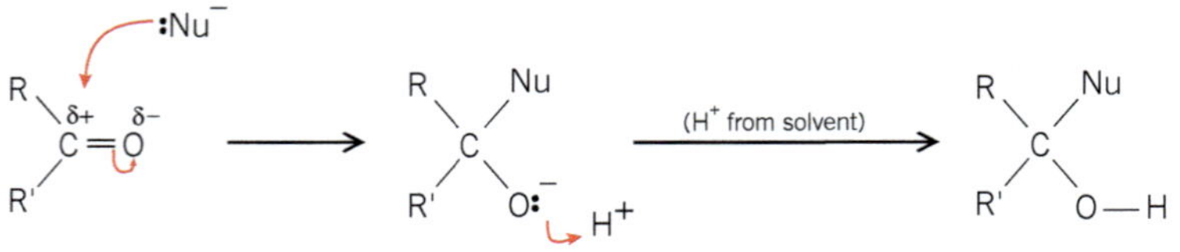

The addition of hydrogen cyanide is a good example of a nucleophilic addition.

Addition of hydrogen cyanide

Sodium cyanide or potassium cyanide is used as a source of cyanide ions followed by the addition of dilute hydrochloric acid. You will not carry out this reaction in the laboratory because of the toxic nature of the CN^- ion.

The products are called **hydroxynitriles**. (This is an example where the –OH group is named using the prefix hydroxy- rather than the suffix -ol.) Hydroxynitriles are useful in synthesis because both the –OH and –CN groups are reactive and can be converted into other functional groups. Here the nucleophile is $:CN^-$.

With a ketone:

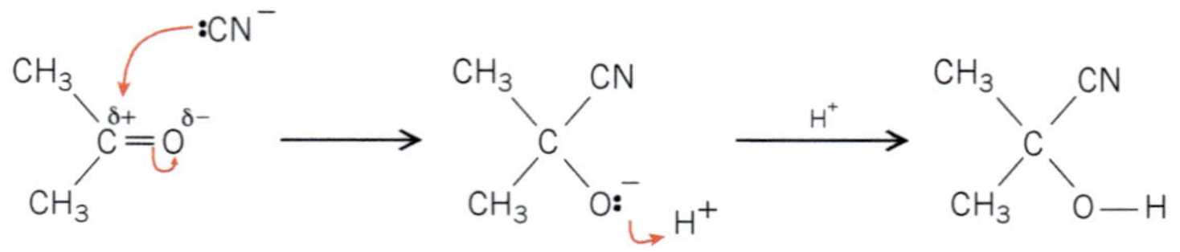

Or with an aldehyde:

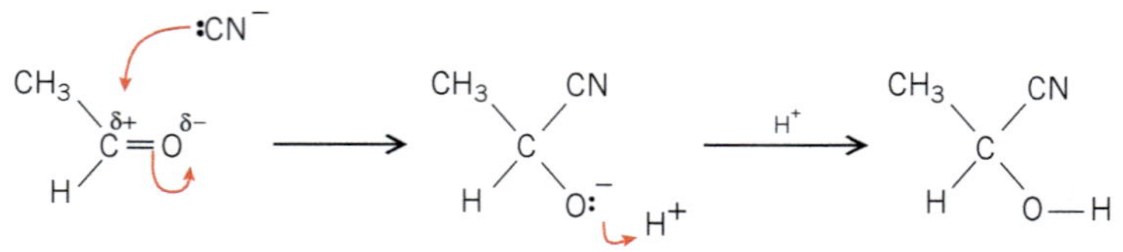

The overall balanced equation for the reaction with an aldehyde is:

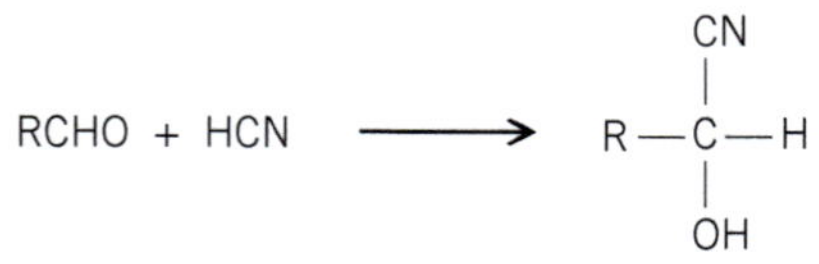

This reaction is important in **organic synthesis** because it increases the length of the carbon chain by one carbon.

This reaction will produce a racemic mixture of two optical isomers (enantiomers) when carried out with an aldehyde or an unsymmetrical ketone, because the $:CN^-$ ion may attack from above or below the flat C=O group (Figure 3).

Exam tip

A racemic mixture contains equal amounts of two enantiomers and has no effect on polarised light. Optical activity is discussed in Sections 13.4 and 29.4.

Exam tip

Nucleophiles have a lone pair with which to attack the $C^{\delta+}$. Some nucleophiles are negatively charged, and others use the negative end of a dipole to attack $C^{\delta+}$.

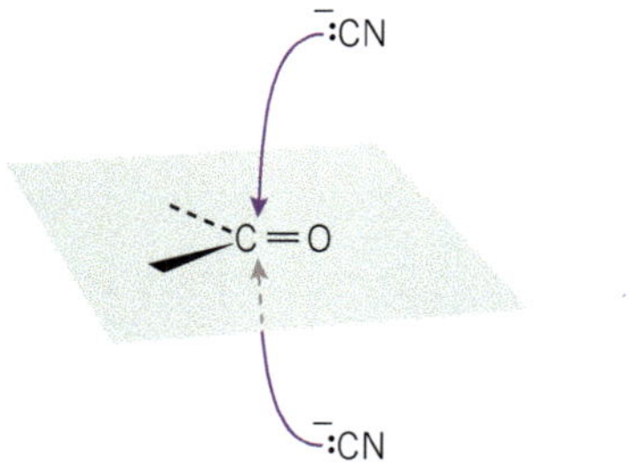

Figure 3 *The $:CN^-$ ion may attack from above or below the C=O group*

Exam tip

In theory, hydrogen cyanide could be used to generate the nucleophile. However, this is toxic and, being a gas, it is hard to stop it escaping into the laboratory.

Redox reactions

Oxidation

Aldehydes can be oxidised to carboxylic acids. Remember that [O] is used to represent the oxidising agent, for example:

$$CH_3CHO + [O] \rightarrow CH_3COOH$$

One oxidising agent commonly used is acidified (with dilute sulfuric acid) potassium dichromate(VI), $K_2Cr_2O_7/H^+$.

Ketones cannot be oxidised easily to carboxylic acids because, unlike aldehydes, a C–C bond must be broken. Stronger oxidising agents break the hydrocarbon chain of the ketone molecule resulting in a shorter chain molecule, carbon dioxide, and water.

Distinguishing aldehydes from ketones

Weak oxidising agents can oxidise aldehydes but not ketones. This is the basis of two tests to distinguish between them.

Fehling's test

Fehling's solution is made from a mixture of two solutions—Fehling's A which contains the Cu^{2+} ion and is therefore coloured blue, and Fehling's B which contains an alkali and a complexing agent.

- When an aldehyde is warmed with Fehling's solution, a brick red precipitate of copper(I) oxide is produced—the copper(I) oxidises the aldehyde to a carboxylic acid, and is itself reduced to copper(II), Figure 4.
- Ketones give no reaction to this test.

Figure 4 When an aldehyde is warmed with Fehling's solution, the blue colour will turn green then a brick-red precipitate forms

The silver mirror test (Tollens' reagent)

Tollens' reagent contains the complex ion $[Ag(NH_3)_2]^+$ which is formed when aqueous ammonia is added to an aqueous solution of silver nitrate.

- When an aldehyde is warmed with Tollens' reagent, metallic silver is formed. Aldehydes are oxidised to carboxylic acids by Tollens' reagent. The Ag^+ is reduced to metallic silver. A silver mirror will be formed on the inside of the test tube (which has to be spotlessly clean).

$$RCHO + [O] \rightarrow RCOOH \qquad \text{The aldehyde is oxidised.}$$

$$[Ag(NH_3)_2]^+ + e^- \rightarrow Ag + 2NH_3 \qquad \text{The silver is reduced.}$$

- Ketones give no reaction to this test.

Reduction

Many reducing agents will reduce both aldehydes and ketones to alcohols. One such reducing agent is lithium tetrahydridoaluminate(III) (lithium aluminium hydride, $LiAlH_4$). This generates the nucleophile $:H^-$, the hydride ion. Sodium tetrahydridoborate(III) (sodium borohydride, $NaBH_4$) will bring about the same reduction.

The H^- ion reduces $C^{\delta+}=O^{\delta-}$ but not C=C, as it is repelled by the high electron density in the C=C bond but is attracted to the $C^{\delta+}$ of the C=O bond.

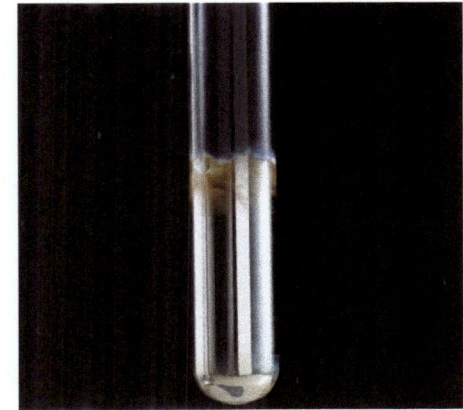

Figure 5 The reaction of aldehydes with Ag^+ ions was once used as a method of silvering mirrors

Reducing an aldehyde

Aldehydes are reduced to primary alcohols by the following mechanism in which H⁻ acts as a nucleophile:

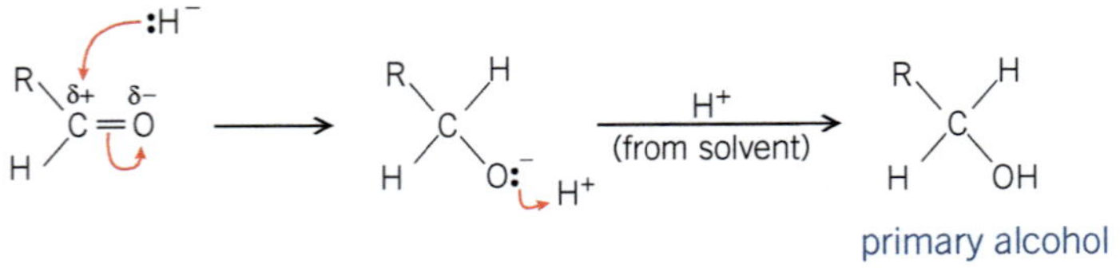

[H] is used to represent reduction in equations.

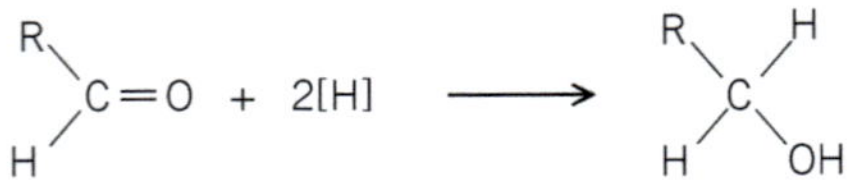

Reducing a ketone

Ketones are reduced to secondary alcohols in a similar way.

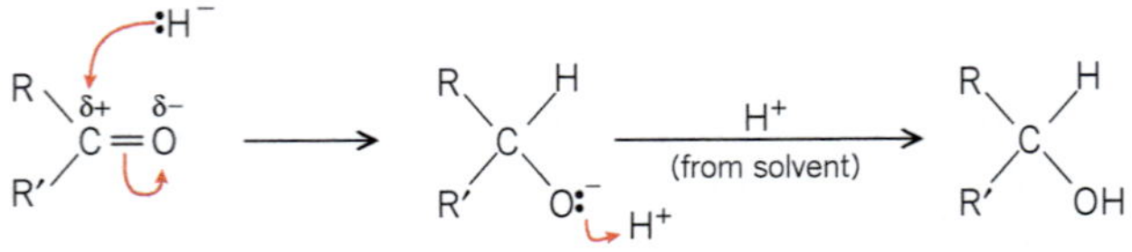

Using [H]:

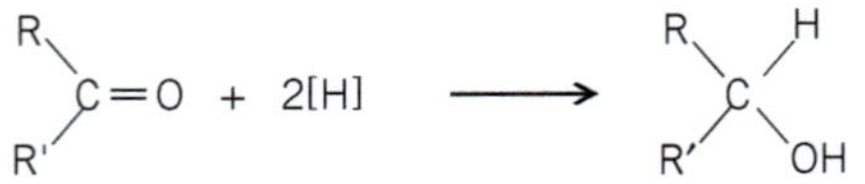

These reactions are **nucleophilic addition** reactions (because the H⁻ ion is a nucleophile).

Detecting the presence of carbonyl compounds

The presence of carbonyl compounds can be deduced from the formation of an orange precipitate with the reagent 2,4-dinitrophenylhydrazine (sometimes called 2,4-DNPH or Brady's reagent). The reaction is:

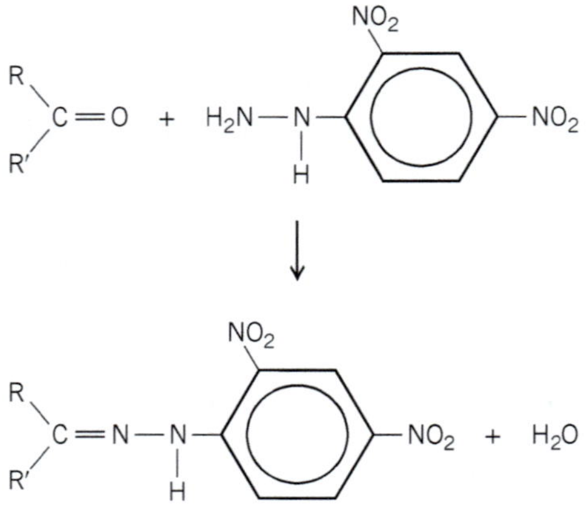

The product is called a 2,4-dinitrophenylhydrazone. The test can also be used to identify the actual carbonyl compound. The precipitate is washed and dried and its melting point measured. By comparing the melting point with a table in a database, the identity of the actual carbonyl compound can be found.

The tri-iodomethane reaction

We have seen this reaction in Section 16.1 where it was used with alcohols containing the group:

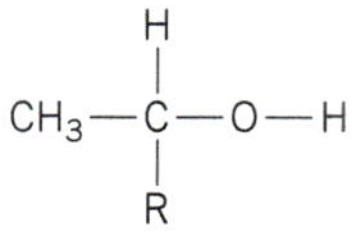

$$CH_3 - \underset{\underset{R}{|}}{\overset{\overset{H}{|}}{C}} - O - H$$

It will also give a positive result with carbonyl compounds containing the group:

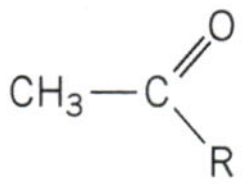

$$CH_3 - C \underset{R}{\overset{O}{\diagup}}$$

ie ethanal (where R is H), and methyl ketones such as propanone.

The carbonyl compound is warmed with a mixture of iodine and sodium hydroxide. Formation of a yellow precipitate of solid tri-iodomethane indicates that CH_3CO-, is present.

$$CH_3 - C \underset{R}{\overset{O}{\diagup}} + 3I_2 + 4OH^- \longrightarrow CHI_3 + RCOO^- + 3I^- + 3H_2O$$

> There are exam-style questions to test your knowledge of the material in this chapter at the end of Chapter 18.

Summary test 17.1

1 Give the name of the following compounds:

a

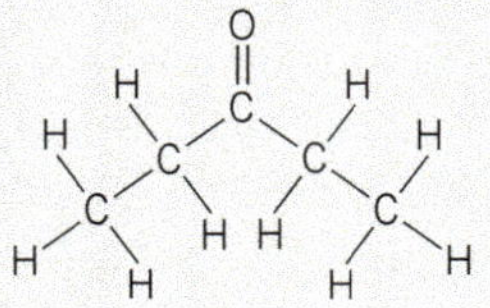

b

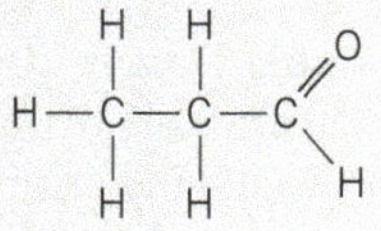

2 Explain why:
   a No ketone with fewer than three carbons is possible.
   b No numbering system is needed in the ketone butanone.
   c No numbering is ever needed to locate the position of the C=O group when naming aldehydes.
3 Explain why there are no hydrogen bonds between propanone molecules.
4 Explain why hydrogen bonds can form between propanone and water molecules.
5 State which of the following is a nucleophile:
   H^+, Cl^-, $Cl\cdot$, H^-
6 Sodium tetrahydridoborate(III) generates the nucleophile :H^- and converts aldehydes and ketones to alcohols.
   a State if you would you expect this reagent to reduce:

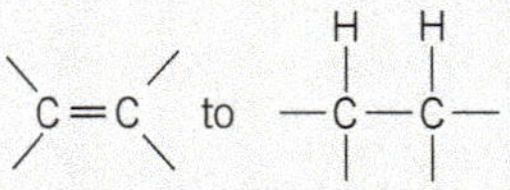

   $$\overset{\diagdown}{\underset{\diagup}{C}} = \overset{\diagdown}{\underset{\diagup}{C}} \quad to \quad -\underset{|}{\overset{\overset{H}{|}}{C}} - \underset{|}{\overset{\overset{H}{|}}{C}} -$$

   b Explain your answer.
   c Predict the product when sodium tetrahydridoborate(III) reacts with:

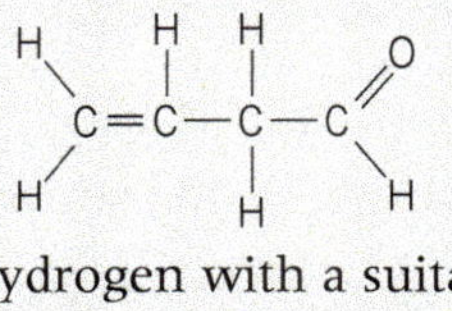

7 Hydrogen with a suitable catalyst will add on to C=C bonds as well as reducing the carbonyl group to an alcohol. Predict the product when hydrogen reacts with the compound in question 6c in the presence of a suitable catalyst.
8 Explain why the reaction of CH_3CHO with HCN forms a racemic mixture, whilst that with CH_3COCH_3 forms a single compound.
9 Give the displayed formula of butanone. State if this compound give a positive result with the tri-iodomethane reaction.

Carboxylic acids

The carboxylic acid functional group is

This is sometimes written as –COOH or as –CO$_2$H. This group can only be at the end of a carbon chain.

Carboxylic acids have two functional groups that you have seen before:

- the carbonyl group, $C=O$, found in aldehydes and ketones
- the hydroxy group, –OH, found in alcohols.

Having two groups on the same carbon atom changes the properties of each group. The most obvious difference is that the –OH group in carboxylic acids is much more acidic than the –OH group in alcohols.

The most familiar carboxylic acid is ethanoic acid (acetic acid), which is the acid in vinegar.

How to name carboxylic acids

Carboxylic acids are named using the suffix *-oic* acid. The carbon atom of the functional group is counted as part of the carbon chain of the root. So, HCOOH is methanoic acid, CH$_3$COOH is ethanoic acid, and so on.

methanoic acid   ethanoic acid

Where there are substituents or side chains on the carbon chain, they are numbered, counting from the carbon of the functional group as carbon number one. So, CH$_3$CHBrCOOH is 2-bromopropanoic acid and CH$_3$CH(CH$_3$)CH$_2$COOH is 3-methylbutanoic acid.

2-bromopropanoic acid   3-methylbutanoic acid

Physical properties of carboxylic acids

The carboxylic acid group can form hydrogen bonds with water molecules (Figure 2). For this reason carboxylic acids up to, and including, four carbons (butanoic acid) are completely soluble in water.

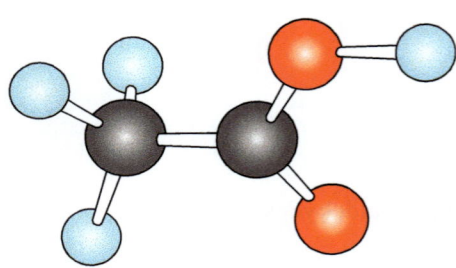

Figure 1 *3D representation of ethanoic acid, a carboxylic acid. Ethanoic acid is also known as acetic acid, and is the main component of vinegar aside from water*

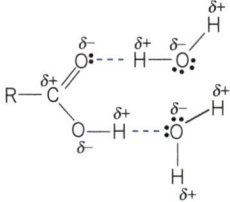

Figure 2 *A molecule of a carboxylic acid forming hydrogen bonds with water*

The acids also form hydrogen bonds with one another in the solid state (Figure 3) to form **dimers**. They therefore have much higher melting points than the alkanes of similar relative molecular mass. Ethanoic acid ($M_r = 60$) melts at 290 K whilst butane ($M_r = 58$) melts at 135 K.

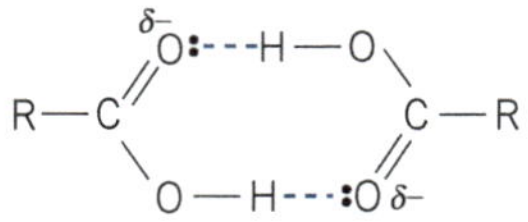

Figure 3 *Two carboxylic acid molecules can hydrogen bond together to form a pair called a dimer*

One way of identifying a carboxylic acid is to measure its melting point and compare it with tables of known melting points. A Thiele tube may be used (Figure 4), or the melting point can be found electrically (Figure 5).

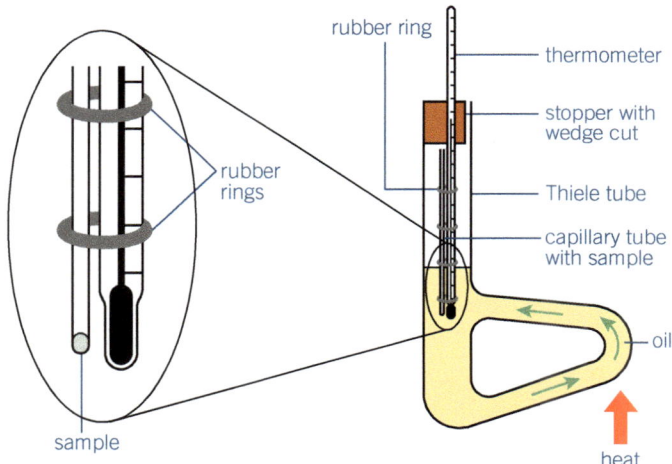

Figure 4 *A Thiele tube may be used to measure melting point*

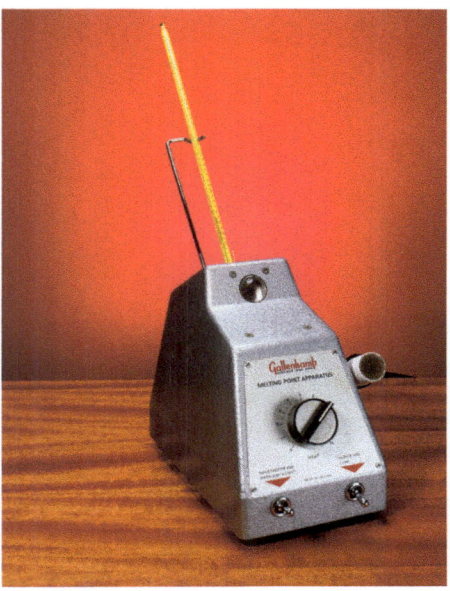

Figure 5 *Modern electric melting point apparatus*

Pure ethanoic acid is sometimes called *glacial* ethanoic acid because it may freeze on a cold day—its freezing point is 13°C (260 K).

The acids have characteristic smells. You will recognise the smell of ethanoic acid as vinegar, whilst butanoic acid has the smell of rancid butter.

The non-systematic names of hexanoic and octanoic acids are *caproic* and *capryllic* acid respectively, from the same derivation as Capricorn the goat. They are present in goat fat and cause its characteristic smell.

Synthesis of carboxylic acids

The formation of carboxylic acids by the oxidation of primary alcohols and aldehydes has been described in Section 16.1. An alternative method is to reflux with acidified potassium manganate(VII) (potassium permanganate, $KMnO_4$) as the oxidising agent. The basic reactions, using ethanol as an example, are shown at the top of the following page.

Exam tip

If you use [O] to represent an oxidising agent, you must still balance the equation.

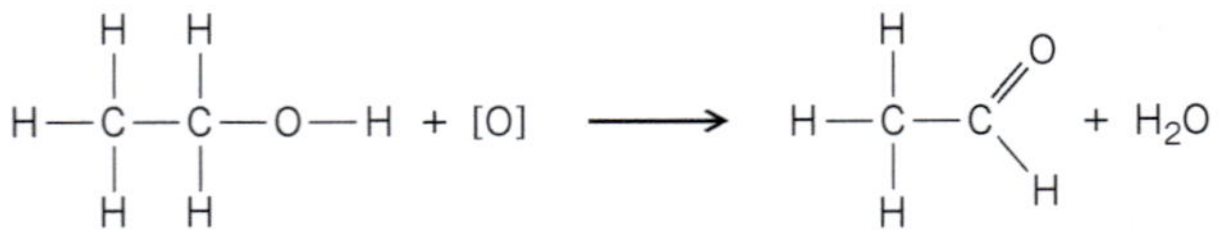

ethanol — ethanal

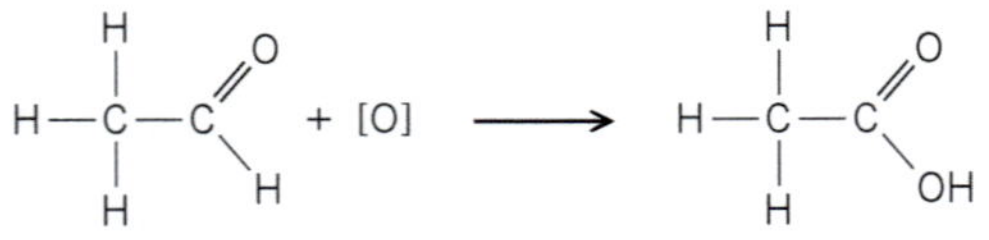

ethanal — ethanoic acid

Carboxylic acids are also formed by:

a the acid hydrolysis of compounds called **nitriles** which contain the $-C\equiv N$ group (see Section 17.1). The basic reaction, using ethanenitrile as an example, is:

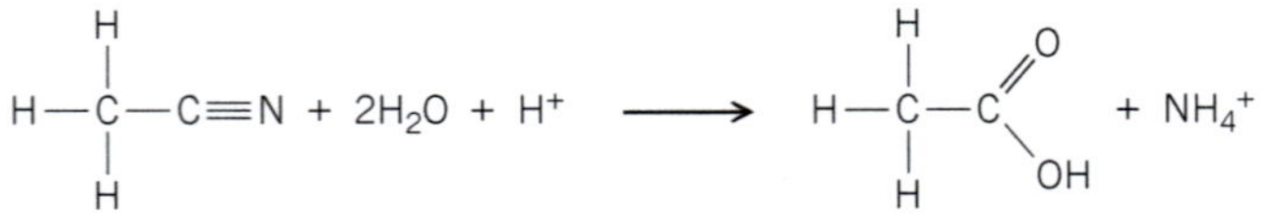

ethanenitrile — ethanoic acid

Alkaline hydrolysis with dilute sodium hydroxide would produce the sodium salt of the acid. Adding acid to this then gives the carboxylic acid.

b the hydrolysis of compounds called **esters** (Section 18.2). This can be done using a dilute acid or with a dilute alkali followed by acidification. The overall reaction is:

$$CH_3COOCH_2CH_3(aq) + H_2O(aq) \rightleftharpoons CH_3COOH(aq) + CH_3CH_2OH(aq)$$

Hot, alkaline hydrolysis forms the salt of the carboxylic acid. This is followed by acidification to produce the carboxylic acid.

Reactivity of carboxylic acids

The carboxylic acid group is polarised as shown:

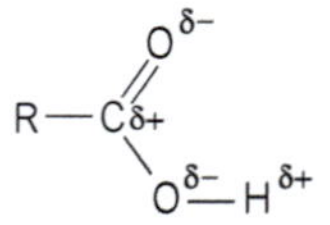

- The $C^{\delta+}$ is open to attack from nucleophiles.
- The $O^{\delta-}$ of the C=O may be attacked by positively charged species (like H^+, in which case we say it has been protonated).
- The $H^{\delta+}$ may be lost as H^+, in which case the compound is behaving as an acid.

Loss of a proton

If the hydrogen of the –OH group is lost, a negative ion—a **carboxylate** ion—is left.

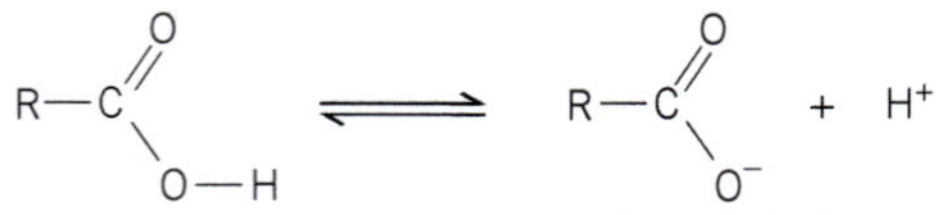

a carboxylate ion

The negative charge is shared over the whole of the carboxylate group.

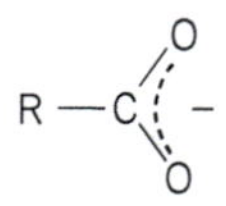

This **delocalisation** makes the resulting ion more stable.

Carboxylic acids are weak acids, so the equilibrium below is well over to the left:

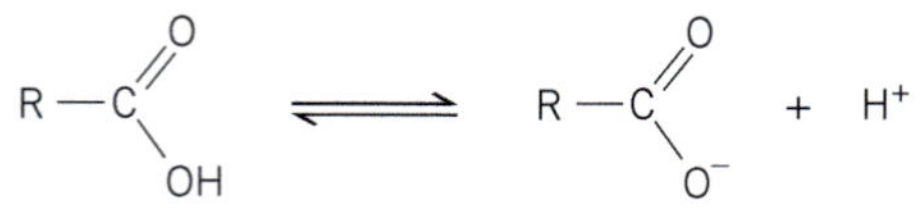

Exam tip

The carboxylic acid group contains both the carbonyl group and the alcohol group. However, the two groups react differently when they are next to each other in a molecule.

Exam tip

The stability of the carboxylate ion is what allows the H^+ ion to be released and makes the molecules acidic.

Even so, they are strong enough to react with sodium hydrogencarbonate, $NaHCO_3$, to release carbon dioxide. This distinguishes them from other organic compounds that contain the –OH group, such as alcohols.

$CH_3COOH(aq) + NaHCO_3(aq) \rightarrow CH_3COONa(aq) + H_2O(l) + CO_2(g)$
ethanoic acid   sodium hydrogencarbonate   sodium ethanoate   water   carbon dioxide

Reactions of carboxylic acids

Carboxylic acids are proton donors and show the typical reactions of acids.

They form ionic salts with the more reactive metals, alkalis, metal oxides, or metal carbonates in the usual way. The salts that are formed have the general name carboxylates, and are named from the particular acid. Methanoic acid gives methanoates, ethanoic acid gives ethanoates, propanoic acid gives propanoates, and so on.

For example, ethanoic acid reacts with aqueous sodium hydroxide in a neutralisation reaction:

$CH_3COOH(aq) + NaOH(aq) \rightarrow CH_3COONa(aq) + H_2O(l)$
ethanoic acid   sodium hydroxide   sodium ethanoate   water

Ethanoic acid reacts with aqueous sodium carbonate in an acid-base reaction:

$2CH_3COOH(aq) + Na_2CO_3(aq) \rightarrow 2CH_3COONa(aq) + H_2O(l) + CO_2(g)$
ethanoic acid   sodium carbonate   sodium ethanoate   water   carbon dioxide

Ethanoic acid reacts with sodium:

$2CH_3COOH(aq) + 2Na(s) \rightarrow 2CH_3COONa(aq) + H_2(g)$
ethanoic acid   sodium   sodium ethanoate   hydrogen

This is a redox reaction in which the oxidation number of the metal increases (from 0 to +1) and the oxidation number of the hydrogen in the –COOH group decreases from +1 to 0. There is no change in oxidation number of the hydrogen atoms in the –CH₃ group.

Formation of esters
Esters, general formula RCOOR′, are **acid derivatives**.

Carboxylic acids react with alcohols to form esters. This reaction is speeded up by a concentrated sulfuric acid catalyst. This is a reversible reaction and forms an equilibrium mixture of reactants and products. For example:

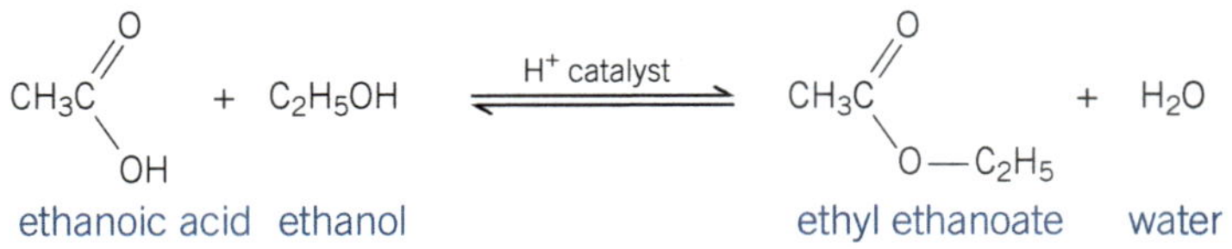

ethanoic acid   ethanol   ethyl ethanoate   water

Reduction of carboxylic acids
Carboxylic acids can be reduced to primary alcohols using lithium tetrahydridoaluminate(III) (lithium aluminium hydride, $LiAlH_4$) in ether solution followed by the addition of water.

$$RCOOH + 4[H] \rightarrow RCH_2OH + H_2O$$

Exam tip
Lithium tetrahydridoaluminate(III) (lithium aluminium hydride, $LiAlH_4$) generates the nucleophile H^-.

Summary test 18.1
1 Carboxylic acids, being acidic, will react with the reactive metals. Give three other reactions that are typical of acids.
2 Identify the acid and the alcohol that would react together to give the ester methyl ethanoate.
3 Identify the acid and the alcohol that would react together to give the ester ethyl methanoate.
4 Methyl ethanoate and ethyl methanoate are a pair of *isomers*. Explain what this means.
5 Give the name of:

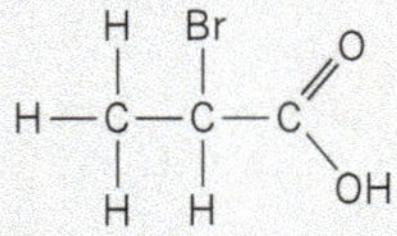

6 Give the displayed formula for 3-chloropropanoic acid.
7 Explain why is it not necessary to call propanoic acid 1-propanoic acid.

18.2 Esters

Learning outcomes

On these pages you will learn to:

- describe the reactions used to make esters
- describe the hydrolysis of esters

Synthesis of esters

Esters are formed by the reaction of a carboxylic acid and an alcohol with a concentrated sulfuric acid catalyst:

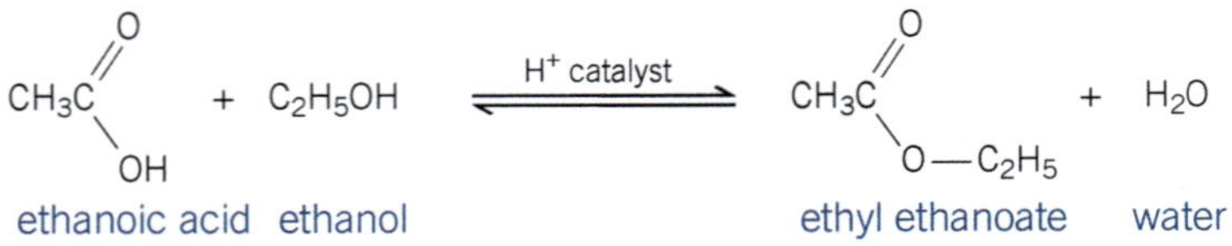

ethanoic acid ethanol   ethyl ethanoate   water

This reaction eliminates a molecule of water and is termed a **condensation reaction**.

Naming esters

Esters are named from the parent acid and the parent alcohol. So $CH_3CH_2CH_2COOCH_3$ is derived from butanoic acid ($CH_3CH_2CH_2COOH$) and methanol (CH_3OH), and is called methyl butanoate. Its isomer $HCOOCH_2CH_2CH_2CH_3$ is derived from methanoic acid ($HCOOH$) and butan-1-ol ($CH_3CH_2CH_2CH_2OH$) and is called butyl methanoate.

It is easy to get ester names the wrong way round unless you are careful. Remember, the name always begins with the alkyl (or aryl) group that has replaced the hydrogen of the acid, rather than the name of the acid.

> **Exam tip**
>
> Take care with the names of esters. It is easy to get them the wrong way round. The part of the name relating to the acid comes last. Also, remember that the acid is named from the number of carbon atoms—including the carbon of the functional group.

Hydrolysis of esters

Esters can be hydrolysed to form a carboxylic acid and an alcohol, for example:

$$CH_3COOCH_2CH_3(aq) + H_2O(aq) \rightleftharpoons CH_3COOH(aq) + CH_3CH_2OH$$

The reaction is reversible and an equilibrium mixture containing all four components is formed.

The reaction is slow. Heating is needed and a base or a strong acid may be used as a catalyst.

If a base such as sodium hydroxide is used, the carboxylic acid produced will react to form a salt ($CH_3COO^-Na^+$ in the example here). This will remove the carboxylic acid from the equilibrium mixture and move the equilibrium to the right by Le Chatelier's principle (see Section 7.1), thus increasing the yield of the reaction.

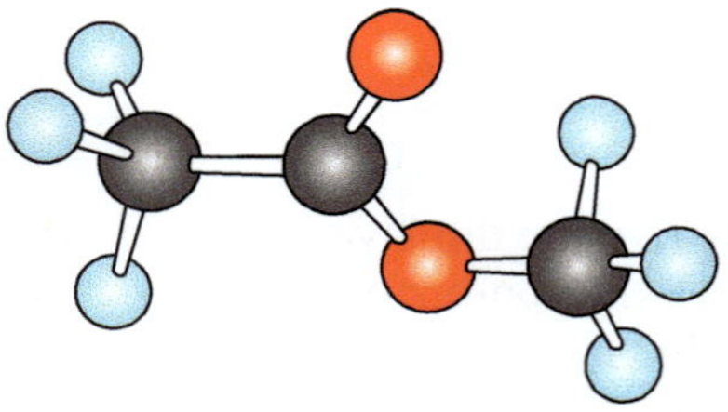

> **Exam tip**
>
> Aryl groups are derived from benzene—see Section 29.1.

Uses of esters

Many esters have pleasant smells and are used as flavourings and in perfumes. Ethyl ethanoate is responsible for the smell of pears in sweets. It is also used as a solvent in industry and in nail polish remover. Esters are also used in the formulation of plastics as **plasticisers**. These are small molecules added to polymers to make them more flexible by allowing the long polymer chains to slide over one another.

Figure 6 *3D representation of ethyl ethanoate, an ester*

Summary test 18.2

1 Give the name of the ester $CH_3CH_2COOCH_2CH_3$. State which carboxylic acid and which alcohol you would react together to produce this compound.

2 Give the formula of ethyl ethanoate and an isomer of ethyl ethanoate that is also an ester. Give the formula of the isomer.

3 Give the names of the following esters:

 and

Figure 7 *Some of the uses of esters (clockwise from top left): nail polish remover, flavouring in sweets, perfume, industrial solvent*

⟮ 📖 **Launch additional digital resources for the chapter** ⟯

1 Ethene, C_2H_4, reacts with cold, dilute acidified potassium manganate(VII) to form a diol with the empirical formula CH_3O.

   a Deduce the molecular formula of the diol produced. *(1 mark)*

   b Give the IUPAC name and draw the displayed formula of the diol produced. *(2 marks)*

2 Consider the following compounds:

A

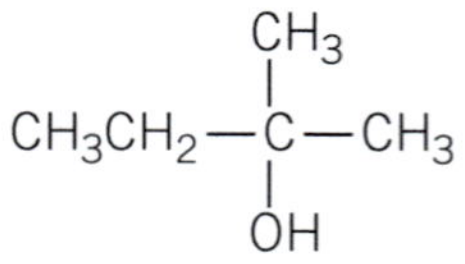

$$CH_3CH_2-\underset{\underset{OH}{|}}{\overset{\overset{CH_3}{|}}{C}}-CH_3$$

B $CH_3CH_2CHOHCH_3$

C CH_3OH

D $CH_3CH_2CH_2CH_2OH$

   a Give the IUPAC name for each compound. *(4 marks)*

   b State the letter(s) for the compound(s) that are a:
     i primary alcohol
     ii tertiary alcohol
     iii secondary alcohol. *(3 marks)*

   c Identify which compound:
     i Reacts with sodium metal.
     ii Could be oxidised to an aldehyde.
     iii Could be oxidised to a ketone.
     iv Forms an alkene when heated with excess concentrated sulfuric acid.
     v Has the lowest boiling point. *(5 marks)*

3 Single-use breathalyser tests are able to detect the presence of ethanol by reaction with acidified potassium dichromate.

   a State the molecular formula for potassium dichromate. *(1 mark)*

   b Describe the role of potassium dichromate in this reaction. *(1 mark)*

   c Describe what would be observed if a person tested positive for ethanol using the breathalyser described above. *(1 mark)*

4 Predict the formulae of the products of the following reactions:

   a $CH_3COCH_2CH_3 + H_2 \rightarrow$

   b $C_6H_5COCH_2CH_3 + H_2 \rightarrow$

   c $CH_3CH_2COCH_2CH_3 + HCN \rightarrow$

   d $CH_3CH_2CHO + Cl_2 \rightarrow$ *(4 marks)*

5 Write structural formulae for all compounds of molecular formula C_4H_8O containing a carbonyl group. How would you distinguish between the different compounds, using simple chemical tests? *(5 marks)*

6 The diagram below summarises some reactions of ethanoic acid:

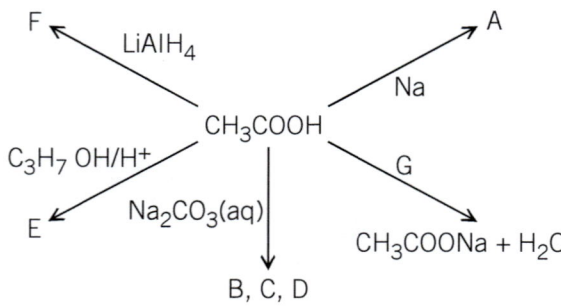

   a Write down the names and formulae of the substances represented by A, B, C, D, E and F. *(6 marks)*

   b Give the identity of the reagent(s) represented by G. *(1 mark)*

7 This question is about an investigation into the mechanism of hydrolysis of ethyl ethanoate:

$$\underset{\text{ethyl ethanoate}}{CH_3COOCH_2CH_3} + H_2O \xrightarrow{\text{HCl}} \underset{\substack{\text{catalyst} \\ \text{acid}}}{\overset{\text{ethanoic}}{CH_3COOH}} + \underset{\text{ethanol}}{CH_3CH_2OH}$$

Two experiments, **A** and **B**, were carried out. Read the accounts of the experiments, then answer the questions which follow.

Experiment A: Ethyl ethanoate was refluxed with deuterated water, D_2O, containing deuterium chloride, DCl (D = ^2_1H). The two hydrolysis products were separated and purified and their relative molecular masses were measured using a mass spectrometer.

The alcohol formed had $M_r = 47$, and the acid had $M_r = 61$.

Experiment B: Ethyl ethanoate was refluxed with enriched water ($H_2^{18}O$) containing HCl. The two hydrolysis products were separated and purified and their relative molecular masses were measured using a mass spectrometer. The alcohol formed had $M_r = 46$, and the acid had $M_r = 62$.

a For **Experiment A**. Compare the M_r values for both the alcohol and the acid formed in this experiment with the values expected from the equation above. Give reasons for any differences. *(2 marks)*

b For **Experiment B**. Compare the M_r values for both the alcohol and the acid formed in this experiment with the values expected from the equation above. Give reasons for any differences. *(2 marks)*

c Using your answers to parts **a** and **b**, what information, if any, do Experiments A and B give about the mechanisms of the hydrolysis reaction? *(3 marks)*

8 Predict the structure of, and name the organic products of the following reactions:

a Butan-2-ol is warmed with acidified potassium manganate(VII). *(2 marks)*

b Propan-2-ol is warmed with excess concentrated sulfuric acid. *(2 marks)*

c Methanol is treated with phosphorous pentachloride. *(2 marks)*

9 The compound with the structural formula below is called but-2-en-1-ol:

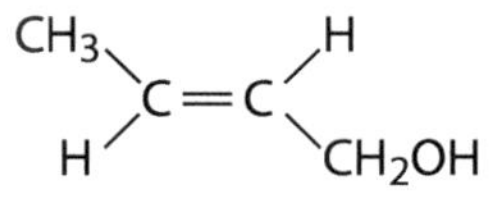

a Name the functional groups in but-2-en-1-ol. *(2 marks)*

b But-2-en-1-ol exists as two *cis–trans* isomers.
  i One of the isomers is shown above. Draw the other isomer. *(1 mark)*
  ii Explain why *cis–trans* isomerism occurs in this compound. *(2 marks)*

c Write an equation for the complete combustion of but-2-en-1-ol. *(2 marks)*

d Give the structural formula of the organic product formed when but-2-en-1-ol is heated under reflux with acidified potassium manganate(VII) solution. Name the type of reaction. *(2 marks)*

e Give the formula of the sodium salt of but-2-en-1-ol, and suggest how it could be made. *(2 marks)*

f 1-bromobut-2-ene can be synthesised from but-2-en-1-ol.
  i Draw the structural formula of 1-bromobut-2-ene. *(1 mark)*
  ii Suggest a suitable reagent and conditions, for this reaction. *(2 marks)*
  iii Name the mechanism for this reaction. *(1 mark)*

g But-2-en-1-ol reacts with bromine. Give the displayed formula of the product formed and name the mechanism for this reaction. *(2 marks)*

10

a Compound **A** has a molecular formula $C_4H_6O_2$.

  A reacts with HCN to form compound **B**, $C_6H_8O_2N_2$.

  A is readily oxidised by acidified potassium dichromate(VI) to an acidic compound **C**, $C_4H_6O_4$.

  When 1.0 g of C is dissolved in water and titrated with 1.0 mol dm^{-3} sodium hydroxide, 16.9 cm^3 of sodium hydroxide is required for neutralisation.

  Suggest structural formulae for **A, B** and **C** and explain the above reactions. *(6 marks)*

b Compound **X** contains 64.3% C, 7.1% H, and 28.6% O by mass. Its relative molecular mass is 56. **X** reduces Fehling's solution to copper(I) oxide. **X** reacts with hydrogen in the presence of a nickel catalyst: 0.1 g of **X** was found to react with 80 cm^3 of hydrogen (measured at stp).

  (1 mol of gas occupies 22 400 cm^3 at stp)

  Suggest a formula for **X** and explain the above reactions. *(8 marks)*

19.1 Primary amines

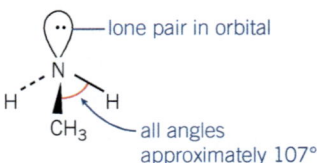

lone pair in orbital

all angles approximately 107°

Figure 1 *The shape of the methylamine molecule*

This section is about a group of compounds called **amines**. Amines can be thought of as derivatives of ammonia in which one or more of the hydrogen atoms in the ammonia molecule have been replaced by alkyl or aryl groups. In this chapter, we shall deal only with primary amines, in which one hydrogen only of an ammonia molecule has been replaced so the general formula is $R-NH_2$.

Amines are very reactive compounds, so they are useful as intermediates in **synthesis**—the making of new molecules.

How to name amines

Amines are named using the suffix *-amine*, for example:

- CH_3-NH_2 is methylamine
- $C_2H_5-NH_2$ is ethylamine.

The properties of primary amines

Shape
Ammonia is a pyramidal molecule with bond angles of approximately 107°. The angles of a perfect tetrahedron are 109.5°. The difference is caused by the lone pair, which repels more than the bonding pairs of electrons in the N–H bonds. Amines keep this basic shape (Figure 1). The lone pair of electrons means that amines can accept a proton (H^+ ion) and are Brønsted–Lowry bases.

Boiling points

Amines are polar:

Primary amines can hydrogen bond to one another using their $-NH_2$ groups (in the same way as alcohols with their $-OH$ groups).

However, as nitrogen is less electronegative than oxygen (electronegativities: O = 3.5, N = 3.0), the hydrogen bonds are not as strong as those in alcohols. The boiling points of amines are lower than those of comparable alcohols:

methylamine, $M_r = 31$, CH_3-NH_2, boiling point = 267 K

methanol, $M_r = 32$, CH_3-OH, boiling point = 338 K

A lower boiling point means the molecules are easier to separate. Shorter chain amines such as methylamine and ethylamine are gases at room temperature. Those with slightly longer chains are volatile liquids. Amines have fishy smells.

Solubility
Primary amines with chain lengths up to about four carbon atoms are very soluble both in water and in alcohols, because they form hydrogen bonds with these solvents. Most amines are also soluble in less polar solvents.

Synthesising amines

Amines may be synthesised by heating ammonia under pressure in ethanol solution with a halogenoalkane, for example:

$$NH_3 + RBr \rightarrow R-NH_2 + HBr$$

An excess of ammonia is used to ensure that the main product is a primary amine. Excess halogenoalkane would result in the formation of secondary and tertiary amines, by the replacement of two or three hydrogen atoms from the ammonia molecule. This reaction is discussed in Section 15.1.

Nitriles and hydroxynitriles

Nitriles

Nitriles are organic compounds containing the $-C\equiv N$ group, for example $CH_3C\equiv N$ called ethanenitrile. When naming nitriles, the carbon of the $-C\equiv N$ group is counted as part of the root of the name.

Synthesising nitriles

Nitriles are produced by the reaction of a halogenoalkane with potassium cyanide dissolved in ethanol and heated. The $C\equiv N^-$ ion is a good nucleophile and this is a nucleophilic substitution reaction. For example:

$$CH_3CH_2Br + KCN \rightarrow CH_3CH_2C\equiv N + KBr$$

This reaction is useful in synthesis since it increases the length of the carbon chain by one. So bromoethane gives propanenitrile. The mechanism of this reaction is considered in Section 15.1.

Reactions of nitriles

Nitriles can be hydrolysed using dilute acid or alkali followed by acidification, to produce carboxylic acids (or their salts if alkaline conditions are used). For example:

$$CH_3CN + 2H_2O + HCl \rightarrow CH_3COOH + NH_4Cl$$

Hydroxynitriles

Hydroxynitriles have the general formula $R-\underset{R'}{\overset{CN}{C}}-OH$, where R and R' are organic groups such as alkyl groups, or a hydrogen atom.

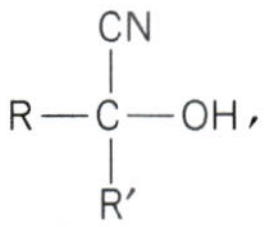

Synthesising hydroxynitriles

Hydroxynitriles are made by the reaction of aldehydes or ketones with hydrogen cyanide, HCN, which contains the nucleophile $C\equiv N^-$. This ion is generated by the reaction of potassium cyanide, KCN, with hydrochloric acid in the reaction vessel.

For example, with an aldehyde such as ethanal:

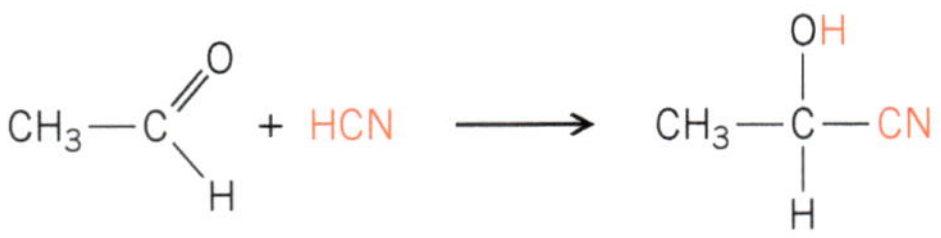

Or with a ketone such as propanone:

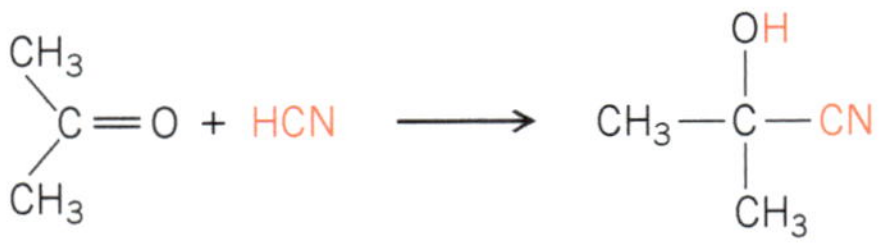

Details of the reaction are discussed in Section 17.1.

There are exam-style questions to test your knowledge of the material in this chapter at the end of Chapter 20.

Learning outcomes

On these pages you will learn to:

- describe the reactions used to make nitriles
- describe the reactions used to make hydroxynitriles
- describe the hydrolysis of nitriles

Summary test 19.2

1 Identify the two functional groups of a hydroxynitrile.
2 a Suggest a method for producing butanoic acid using the reactions in this chapter.
  b State the name of the organic starting material for part **a**.
  c Give the equations for the reactions in part **a**.

20.1 Addition polymerisation

Polymers are very large molecules that are built up from small molecules, called **monomers**. They occur naturally everywhere: starch, proteins, cellulose and DNA are all polymers. The first completely synthetic polymer was Bakelite, which was patented in 1907. Since then, many synthetic polymers have been developed with a range of properties to suit them for very many applications—see Figure 1.

One way of classifying polymers is by the type of reaction by which they are made.

- **Addition polymers** are made from a monomer or monomers with a carbon–carbon double bond (alkenes).
- **Condensation polymers** are made from monomers with two functional groups, which react together with the elimination of a small molecule. Condensation polymers are dealt with in Chapter 35.

The monomers of addition polymers have the general formula:

$$\begin{array}{c} H \qquad\qquad H \\ \diagdown \qquad\qquad \diagup \\ C = C \\ \diagup \qquad\qquad \diagdown \\ H \qquad\qquad R \end{array}$$

When the monomers polymerise, the double bond opens and the monomers bond together to form a backbone of carbon atoms, as shown:

$$\cdots + \begin{array}{c} H \quad H \\ \diagdown \quad \diagup \\ C = C \\ \diagup \quad \diagdown \\ H \quad R \end{array} + \begin{array}{c} H \quad H \\ \diagdown \quad \diagup \\ C = C \\ \diagup \quad \diagdown \\ H \quad R \end{array} + \begin{array}{c} H \quad H \\ \diagdown \quad \diagup \\ C = C \\ \diagup \quad \diagdown \\ H \quad R \end{array} + \cdots$$

$$\downarrow$$

$$\begin{array}{c} \quad H \ H \ H \ H \ H \ H \\ \quad | \ \ | \ \ | \ \ | \ \ | \ \ | \\ -C-C-C-C-C-C- \\ \quad | \ \ | \ \ | \ \ | \ \ | \ \ | \\ \quad H \ R \ H \ R \ H \ R \end{array}$$

This may also be represented by equations such as:

$$n \begin{array}{c} H \ H \\ | \ \ | \\ C = C \\ | \ \ | \\ H \ R \end{array} \longrightarrow \left[\begin{array}{c} H \ H \\ | \ \ | \\ C-C \\ | \ \ | \\ H \ R \end{array} \right]_n$$

R may be a variety of groups—see Table 1.

For example, ethene polymerises to form poly(ethene):

Figure 1 Polymers around us

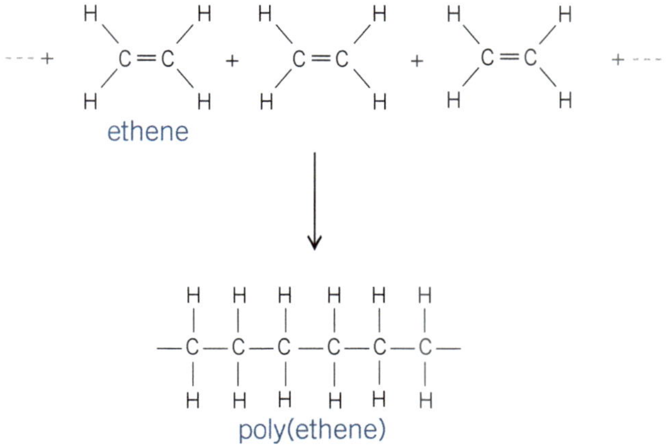

ethene

poly(ethene)

and chloroethene polymerises to form poly(chloroethene), common name polyvinylchloride, PVC:

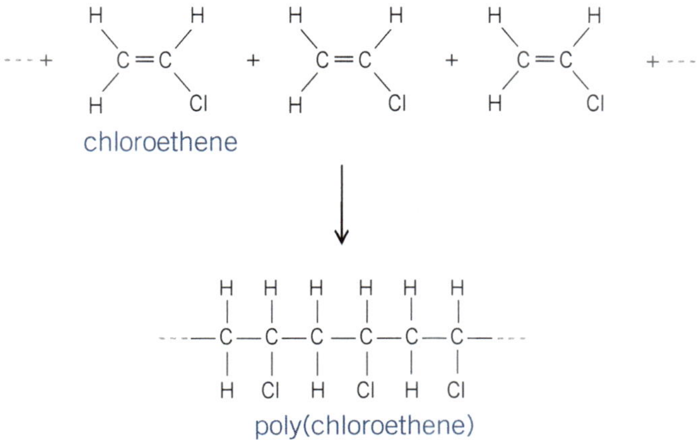

chloroethene

poly(chloroethene)

Table 1 gives some examples of addition polymers based on different substituents.

Table 1 Some addition polymers made from the monomer $H_2C=CHR$

R	Monomer	Polymer	Name of polymer	Common or trade name	Typical uses
—H	$CH_2=CH_2$	$\left[CH_2-CH_2\right]_n$	poly(ethene)	polythene	carrier bags, washing up bowls
—CH_3	CH_3 $CH=CH_2$	CH_3 $\left[CH-CH_2\right]_n$	poly(propene)	polypropylene	yoghurt containers car bumpers
—Cl	Cl $CH=CH_2$	Cl $\left[CH-CH_2\right]_n$	poly(chloroethene)	PVC (polyvinyl chloride)	aprons, 'vinyl' records, drainpipes
—C≡N	CN $CH=CH_2$	Cl $\left[CH-CH_2\right]_n$	poly(propenenitrile)	acrylic (Acrilan, Courtelle)	clothing fabrics
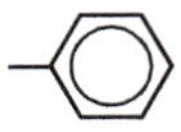 ⬡	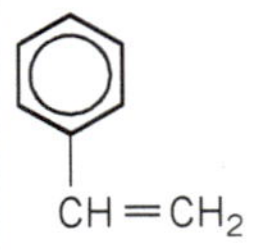 ⬡ $CH=CH_2$	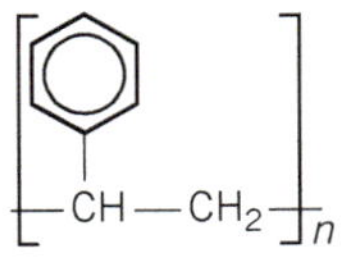 ⬡ $\left[CH-CH_2\right]_n$	poly(phenylethene)	polystyrene	packing materials, electrical insulation

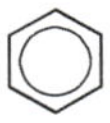

⬡ is the special symbol for a benzene ring, which is discussed in Section 30.1.

Identifying the addition polymer formed from the monomer

The best way to think about identifying the addition polymer formed from the monomer is to remember that an addition polymer is formed from monomers with carbon–carbon double bonds.

There is usually only one monomer (though it is possible to have more), and the double bond opens to form a single bond, see Table 1 on the previous page. This will give the repeat unit for the polymer.

Identifying the monomer(s) used to make an addition polymer

An addition polymer must have a backbone of carbon atoms and the monomer must contain at least two carbons, so that there can be a carbon–carbon double bond. So, in the molecule below the monomer is shown in the red brackets:

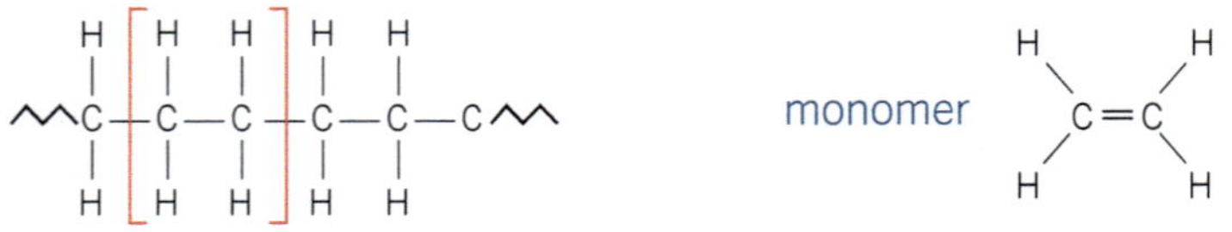

Where some of the carbon atoms have substituents, the monomer must have the substituent, as well as a double bond:

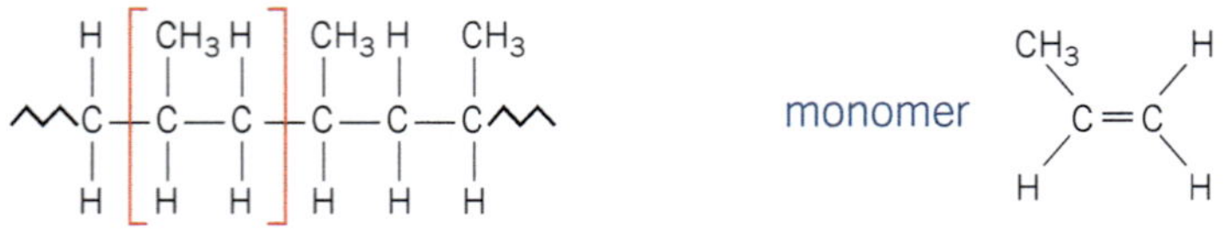

Exam tip

The **repeat unit** of an addition polymer is the smallest group of atoms that produce the polymer when repeated over and over. It normally corresponds to the monomer, shown in brackets in Table 1. In poly(ethene) the repeat unit is just $-CH_2-$, but it is more usual to quote the repeat unit $-CH_2-CH_2-$ based on the monomer.

Modifying the plastics

The properties of polymer materials can be considerably modified by the use of additives such as plasticisers. These are small molecules that get between the polymer chains, forcing them apart and allowing them to slide across each other. For example, PVC is rigid enough for use as drainpipes, but with the addition of a plasticiser it becomes flexible enough for making aprons.

Biodegradability

Poly(alk*enes*) actually have a backbone which is a long chain saturated *alkane* molecule. Alkanes have strong non-polar C–C and C–H bonds so they are very unreactive molecules. This is a useful property, but it also means that they are not attacked by biological agents such as enzymes. So they are not biodegradable. Nor are they susceptible to attack by chemical reagents such as acids, nucleophiles and electrophiles. Non-biodegradable polymers are an increasing problem in today's world, where waste disposal is becoming more and more difficult.

Figure 2 Plastic waste washed up on a beach. These plastics will decompose into tiny nanoparticles that harm the environment and sea life

Some solutions to pollution by plastics

To reduce the amount of plastic it can be reused or recycled.

Mechanical recycling

The simplest form of recycling is called mechanical recycling. The first step is to separate the different types of plastic (see Figure 4). Many recycling facilities provide different containers for different plastics for this purpose. The plastics are then washed and, once they are sorted, they may be ground up into small pellets. These can be melted and remoulded. For example, recycled soft drinks bottles made from PET (polyethylene terephthalate) are used to make fleece clothes. A major difficulty is sorting the polymers by type, as many plastics look very similar and impurities in the remoulded product will lead to degradation of its properties.

Figure 3 Many western nations ship plastic waste to poorer countries to meet recycling targets. Often this plastic is not recyclable due to contamination and has to be sent to landfill. Countries like Malaysia and the Philippines have taken a stand and sent tonnes of contaminated plastic back to their countries of origin

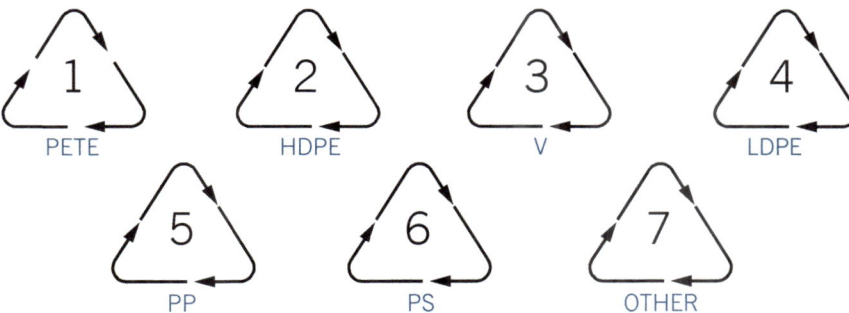

Figure 4 Symbols used for recycling different types of plastic: 1 = polyethyleneterephthalate, 2 = high density polyethylene, 3 = polyvinyl chloride, 4 = low density polyethylene, 5 = polypropylene and 6 = polystyrene. (Note: these are everyday names, rather than systematic names.)

Feedstock recycling

In feedstock recycling, the plastics are heated to a temperature that will break the polymer bonds and produce monomers. These can then be used to make new plastics. Again, impurities due to poor sorting can cause problems.

There are problems with recycling. Poly(propene), for example, is a thermoplastic polymer. This means that it will soften when heated so it can be melted and re-used. However, this can only be done a limited number of times, because at each heating some of the chains break and become shorter, degrading the plastic's properties.

Figure 5 *The properties of plastics can be modified greatly by different additives and manufacturing techniques. Both the model and the packaging here are made from polystyrene with different additives. The bubbles in expanded polystyrene are produced by blowing a gas into the molten polymer during manufacture*

Combustion of polymers

Since addition polymers consist mostly of hydrogen and carbon, they will burn and this is a potential route to disposal. The heat given out can be used as a source of energy for district heating schemes, for example. However, this is not without its problems:

- Complete combustion produces carbon dioxide, a greenhouse gas.
- Incomplete combustion produces toxic carbon monoxide and also soot.
- Combustion of chlorine-containing plastics (such as PVC) produces hydrogen chloride (HCl) and other more-toxic products such as dioxins.

Summary test 20.1

1 State which of the following monomers could form an addition polymer.

A
$$H \quad \quad H$$
$$C=C$$
$$H \quad \quad H$$

B
$$F \quad \quad F$$
$$C=C$$
$$F \quad \quad F$$

C
$$NH_2$$
$$CH_3—C—COOH$$
$$H$$

D
$$H \quad \quad H$$
$$C=C$$
$$H \quad \quad CH_3$$

2 **a** Sketch a section of the polymer formed from the monomer

$$H \quad \quad H$$
$$C=C \quad \quad \text{showing six carbon atoms.}$$
$$H \quad \quad Cl$$

b Give the common name of the monomer.
c Give the systematic name of the polymer.

3 Teflon is a polymer that is used to coat non-stick pans. A section of Teflon is shown below.

$$F \quad F \quad F \quad F \quad F \quad F$$
$$—C—C—C—C—C—C—$$
$$F \quad F \quad F \quad F \quad F \quad F$$

Identify the monomer.

4 This is a section of the polymer that drainpipes are made from, trade name polyvinylchloride (PVC).

$$Cl \quad H \quad Cl \quad H \quad Cl \quad H$$
$$—C—C—C—C—C—C—$$
$$H \quad H \quad H \quad H \quad H \quad H$$

Identify the monomer.

Launch additional digital resources for the chapter

1 Ethylamine, $CH_3CH_2NH_2$, can be synthesised by a nucleophilic substitution reaction using a halogenoalkane, X.

   a Identify suitable reagent(s) and conditions for this reaction. *(1 mark)*

   b Give the displayed formula of halogenoalkane X. *(1 mark)*

   c Suggest a mechanism for the synthesis of ethylamine using the conditions described in your answer to part **a**, give the IUPAC name and draw the displayed formula of ethylamine. *(4 marks)*

2 Answer the questions that follow for each of the following monomers or pairs of monomers:

   A 1,1-dichloroethene,

   B $H_2N(CH_2)_5NH_2$ and $HOOC(CH_2)_5COOH$

   C 3-aminobenzoic acid,

   D ethene and propene

   a Identify the letters from the examples above for the monomers, or pairs of monomers that would undergo addition polymerisation. *(1 mark)*

   b Draw the structures for the repeating units for the polymers produced from your answers to part **a**. *(3 marks)*

3 The following synthetic route shows the formation of compound R. Further analysis showed that compound R was produced as a mixture of two isomers.

   a Explain why HCN should not be used instead of KCN to produce compound R. *(1 mark)*

   b Give the mechanism for the synthesis of compound R. Name this mechanism. *(5 marks)*

   c Give the IUPAC name for Compound R. *(1 mark)*

   d State the type of isomerism exhibited by compound R and explain how the isomers can be distinguished. *(2 marks)*

   e Give diagrams to show the two isomers of compound R. *(2 marks)*

4 1,1-dichloroethane, $Cl_2C=CH_2$ readily undergoes addition polymerisation to give polymer P.

   a What is the systematic name of the polymer P? *(1 mark)*

   b Give the structure of the repeating unit of polymer P. *(1 mark)*

   c Suggest whether or not polymer P is biodegradable. Explain your answer. *(2 marks)*

5 Polychloroethene (polyvinyl chloride, PVC), which is commonly used in water pipes and food wrap, is formed from chloroethene, $H_2C=CHCl$, in an addition polymerisation reaction.

   a Explain the term *polymerisation*. *(2 marks)*

   b Draw a section of a molecule of PVC showing three repeat units. *(3 marks)*

   c What characteristics does chloroethene require as a monomer in order to undergo an addition polymerisation reaction to form PVC. *(1 mark)*

   d Give the displayed formula of the monomer that would form the following polymer: *(1 mark)*

21 Organic synthesis

21.1 Organic synthesis

Learning outcomes

On these pages you will learn to:

- identify the functional groups in an organic molecule
- predict the properties and reactions of organic molecules
- devise and analyse routes for preparing organic molecules

Identifying functional groups

To identify an organic compound, first you need to know the functional groups present.

Some tests for functional groups are very straightforward:

- Is the compound a solid (this suggests a long carbon chain or ionic bonding), a liquid (this suggests a medium length carbon chain or polar or hydrogen bonding), or a gas (this suggests a short carbon chain, or little or no polarity)?
- Does the compound dissolve in water (this suggests polar groups) or not (this suggests no polar groups)?
- Is the compound acidic (this suggests a carboxylic acid)?

Some specific chemical tests are listed in Table 1.

Table 1 *Chemical tests for functional groups*

Functional group	Test	Result
alkene –C=C–	shake with bromine water	red-brown colour disappears
halogenoalkane R–X	**1** add NaOH(aq) and warm **2** acidify with HNO_3 **3** add $AgNO_3$(aq)	precipitate of AgX
alcohol R–OH	add acidified $K_2Cr_2O_7$	orange colour turns green with primary or secondary alcohols (also with aldehydes)
aldehyde R–CHO	warm with Fehling's solution or warm with Tollens' solution or add acidified $K_2Cr_2O_7$	blue colour turns to red-brown precipitate silver mirror forms orange colour turns green
carboxylic acid R–COOH	add $NaHCO_3$(aq)	bubbles observed as carbon dioxide given off

Exam tip

See Sections 11.3 and 15.1 for more detail on how the precipitate AgX can be used to identify the halogen.

Synthetic routes

This section is about working out a series of reactions for making (synthesising) a given molecule, usually called the **target molecule**.

Synthesis of a target molecule is a common challenge in industries like drug or pesticide manufacture. Suppose a molecule is found to have a particular effect, for example, as an antibiotic. Drug companies may synthesise, on a small scale, a number of compounds of similar structures. These will be screened for possible antibiotic properties. Any promising compounds may then be made in larger quantities for thorough investigation of their effectiveness, safety, side effects, and so on, before the final step goes ahead—producing them commercially.

Using the organic reactions you have already met, you can work out a reaction scheme to convert a starting material into a target molecule.

Exam tip

As well as chemical reactions, several instrumental techniques may be used to help identify an organic chemical. These include

- mass spectrometry, see Section 22.2
- nuclear magnetic resonance, see Sections 37.3 and 37.4
- infrared spectroscopy, see Section 22.1.

Working out a scheme

Start by writing down the formula of the starting molecule, A, and that of the target molecule, X.

One way of working out what route to take is to write down all the compounds which can be made from A and all the ways in which X can be prepared (Figure 1).

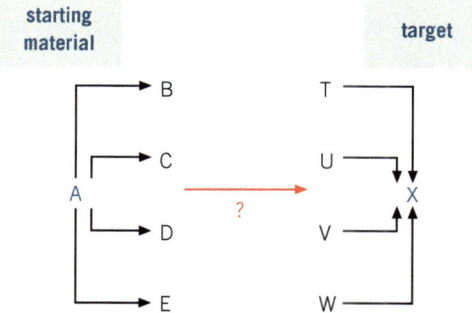

***Figure 1** Devising a synthesis of compound X from compound A*

You may then see how B, C, D, or E can be converted, in one or more steps, to T, U, V, W, or direct to X. It is important to keep the number of steps as small as possible to maximise the yield of the target molecule.

Sometimes you will be able to see straight away that a particular reaction will be needed. For example, if the target molecule has one more carbon atom than the starting material, it is probable that the reaction of cyanide ions with a halogenoalkane will be needed at some stage, as this reaction increases the length of the carbon chain by one. For example:

$$\underset{\text{bromomethane}}{CH_3Br} + CN^- \rightarrow \underset{\text{ethanenitrile}}{CH_3CN} + Br^-$$

Exam tip

A four-step synthesis in which the yield of each step is 90% would give an overall yield of:

$$\frac{90}{100} \times \frac{90}{100} \times \frac{90}{100} \times \frac{90}{100} = 68.6\%$$

How the functional groups are connected

The inter-relationships between the functional groups you should know are shown in Figure 2. Make sure you can recall the reagents and conditions for each conversion.

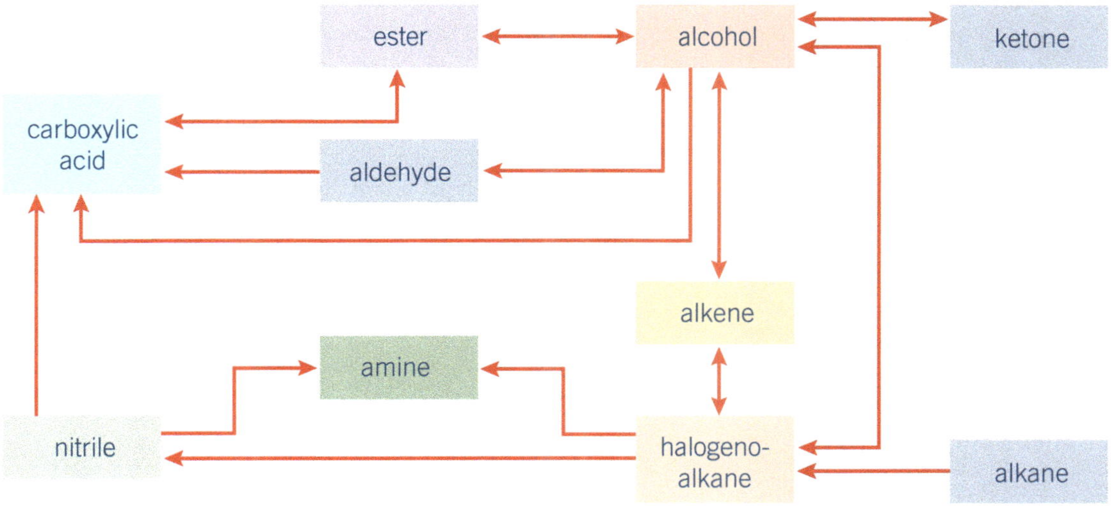

***Figure 2** Interrelationships between functional groups. You can use this chart to revise your knowledge of organic reactions*

Reagents used in organic chemistry

Oxidising agents

Potassium dichromate(VI), $K_2Cr_2O_7$, acidified with dilute sulfuric acid will oxidise primary alcohols to aldehydes, and aldehydes to carboxylic acids. Secondary alcohols are oxidised to ketones.

Reducing agents

Different reducing reagents have different capabilities:

- Lithium tetrahydridoaluminate(III) ($LiAlH_4$) and sodium tetrahydridoborate(III) ($NaBH_4$), will reduce C=O but not C=C.

  These reducing agents will reduce polar unsaturated groups, such as $C^{\delta+}=O^{\delta-}$, but not non-polar ones, such as C=C.

  This is because they generate the nucleophile :H⁻ which attacks $C^{\delta+}$ but is repelled by the electron-rich C=C.
- Hydrogen with a nickel catalyst, H_2/Ni, is used to reduce C=C but not C=O.

Dehydrating agents

Alcohols can be converted to alkenes by passing their vapours over heated aluminium oxide or by acid-catalysed elimination reactions.

Examples of reaction schemes

1 How can propanoic acid be synthesised from 1-bromopropane?

Both the starting material and the target have the same number of carbon atoms, so no alteration to the carbon skeleton is needed.

Write down all the compounds which can be made in one step from 1-bromopropane and all those from which propanoic acid can be made in one step, as shown in Figure 3. You may use Figure 2 to help you.

In this case two of the compounds are the same—the ones in red.

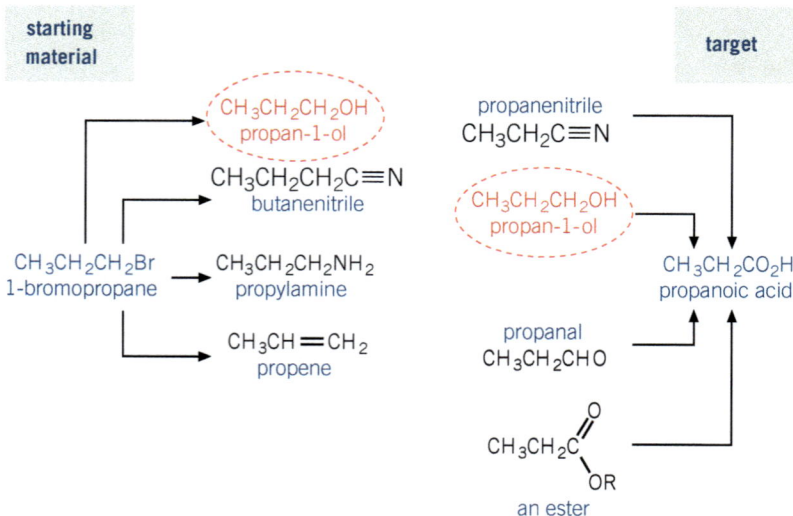

Figure 3 *Devising a synthesis of propanoic acid from 1-bromopropane*

> **Exam tip**
>
> $LiAlH_4$ is a more powerful reducing agent that $NaBH_4$. It must be used in ether solution while $NaBH_4$ works in aqueous solution.

> **Exam tip**
>
> $LiAlH_4$ is also called lithium aluminium hydride. $NaBH_4$ is also called sodium borohydride.

> **Exam tip**
>
> Cover the list of compounds in the centre of Figure 3 and try to remember them. It is also important to recall the reagents and conditions for each reaction.

So, 1-bromopropane can be converted into propan-1-ol which can be converted into propanoic acid. The conversion required can be done in two steps:

Step 1 $CH_3CH_2CH_2Br$ $\xrightarrow{\text{heat with NaOH(aq)}}$ $CH_3CH_2CH_2OH$
          1-bromopropane   propan-1-ol

Step 2 $CH_3CH_2CH_2OH$ $\xrightarrow{\text{reflux with K}_2\text{Cr}_2\text{O}_7/\text{H}^+}$ $CH_3CH_2CO_2H$
          propan-1-ol   propanoic acid

Both of these reactions have a good yield.

2 How can propylamine by synthesised from ethene?

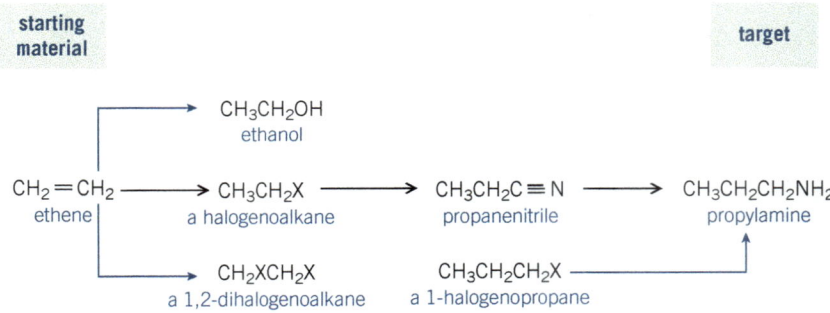

Propylamine has one more carbon atom than ethene. This suggests that the formation of a nitrile is involved at some stage.

Write down all the compounds that can be made from ethene and all the compounds from which propylamine can be made (Figure 4).

starting material   target

CH_3CH_2OH
ethanol

$CH_2{=}CH_2$ → CH_3CH_2X → $CH_3CH_2C{\equiv}N$ → $CH_3CH_2CH_2NH_2$
ethene   a halogenoalkane   propanenitrile   propylamine

CH_2XCH_2X $CH_3CH_2CH_2X$
a 1,2-dihalogenoalkane   a 1-halogenopropane

Figure 4 *Devising a synthesis of propylamine from ethene*

In this instance, there is no compound that can be made in one step from the starting material and then converted into the product. So more than two steps must be required. You already know that the formation of a nitrile is required. A halogenoethane can be converted into propanenitrile so the synthesis can be completed in three steps:

Step 1 CH_2CH_2 $\xrightarrow{\text{HBr}}$ CH_3CH_2Br
               ethene   bromoethane

Step 2 CH_3CH_2Br $\xrightarrow{\text{KCN in ethanol}}$ $CH_3CH_2C{\equiv}N$
               bromoethane   propanenitrile

Step 3 $CH_3CH_2C{\equiv}N$ $\xrightarrow{\text{Ni/H}_2}$ $CH_3CH_2CH_2NH_2$
               propanenitrile   propylamine

Chloroethane or iodoethane could have been chosen instead of bromoethane.

Exam tip

You need to be able to recall the reactions of all the functional groups you have met, including conditions such as heating, refluxing, use of acidic or alkaline conditions, and catalysts.

Synthetic robots

Routine chemical synthesis in the pharmaceutical industry is now often done by robots—not androids but arrays of reaction tubes along with computer-controlled syringes to measure out and mix the reactants. This produces a 'library' of related compounds. For example, you could oxidise several alcohols of different chain lengths to produce a library of aldehydes. The target compound can then be tested to see if any of them have any potential for use as medicines. A chemist is still needed to work out the reaction and program the computer.

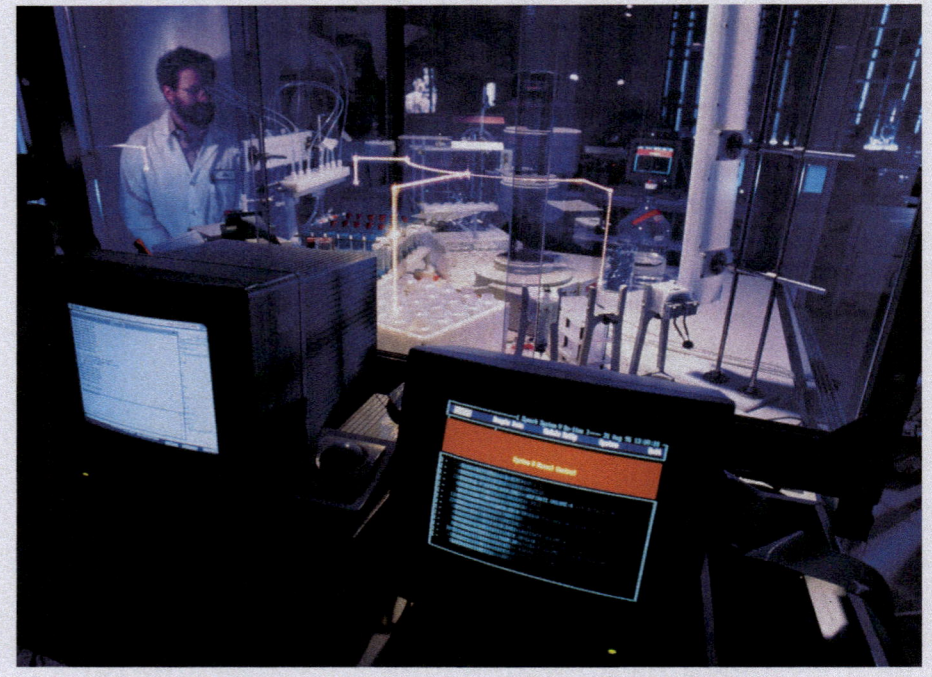

Other considerations

When planning an organic synthesis, there are many considerations to be taken into account as well as just the reaction scheme. These include

- cost of starting materials, reagents and solvents
- any hazards associated with the reaction—toxicity of chemicals involved (including by-products), flammability etc.
- the percentage yield of the reaction; many organic reactions do not go to completion
- competing reactions that may form unwanted by-products. For example, the reaction of a halogenoalkane with sodium hydroxide can result in substitution or elimination, depending on the conditions of the reaction (see Figure 5). Warming with hydroxide ions, dissolved in water (aqueous) favours substitution. Hydroxide ions at high temperature, dissolved in ethanol, favour elimination. So, it is important to choose the conditions of the reaction.

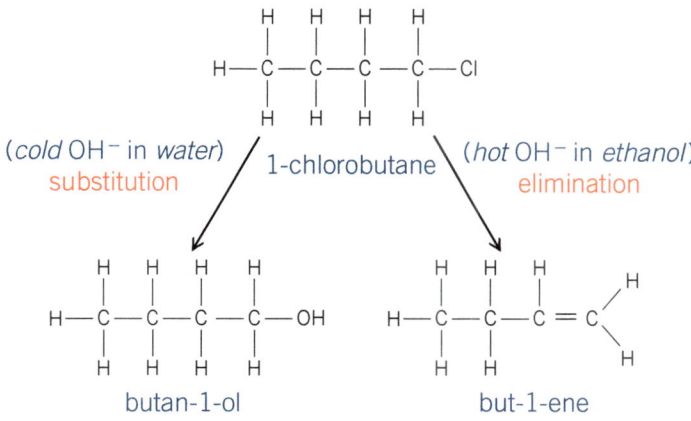

Figure 5 *Reacting 1-chlorobutane with sodium hydroxide will produce two different products depending on the conditions*

Summary test 21.1

1 Explain how you could tell if R–X was a chloroalkane, a bromoalkane, or an iodoalkane.
2 In the test for a halogenoalkane:
   a Explain why it is necessary to acidify with dilute acid before adding silver nitrate.
   b Explain why acidifying with hydrochloric acid would not be suitable.
3 A compound decolourises bromine solution and fizzes when sodium hydrogencarbonate solution is added:
   a Identify its two functional groups.
   b Its relative molecular mass is 72. Give its structural formula.
   c Give equations for the two reactions.
4 Give a one-step reaction to convert:
   a 1-bromobutane to pentanenitrile
   b ethanoic acid to methyl ethanoate
   c but-1-ene to butan-2-ol
   d cyclohexanol to cyclohexene.
5 Give a two-step reaction to convert:
   a ethene to ethanoic acid
   b propanone to 2-bromopropane.
6 For each step in **5a** and **b**, identify the type of reaction taking place and the reagents required.
7 Aspartame, the structure below, is an artificial sweetener.

Identify the functional groups ringed in:
   a blue
   b red
   c brown.

There are exam-style questions to test your knowledge of the material in this chapter at the end of Chapter 22.

22.1 Infrared spectroscopy

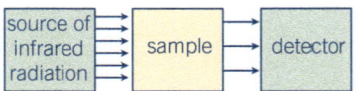

Figure 1 *Schematic diagram of an infrared spectrometer*

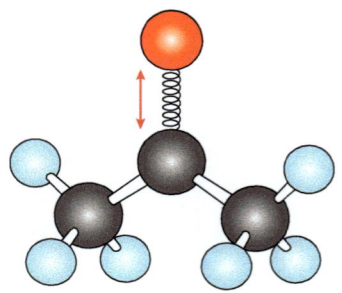

Figure 2 *Vibration of the C=O bond in propanone*

Infrared (IR) spectroscopy is often used by organic chemists to help them identify compounds.

How infrared spectroscopy works

A pair of atoms joined by a chemical bond is always vibrating. The system behaves rather like two balls (the atoms) joined by a spring (the bond). Stronger bonds vibrate faster (at higher frequency) and heavier atoms make the bond vibrate more slowly (at lower frequency). Every bond has its own unique natural frequency that is in the infrared region of the electromagnetic spectrum.

When you shine a beam of infrared radiation (heat energy) through a sample, the bonds in the sample can absorb energy from the radiation and vibrate more. However, any particular bond can only absorb radiation that has the same frequency as the natural frequency of the bond. Therefore, the radiation that emerges from the sample will be missing the frequencies that correspond to the bonds in the sample (see Figure 1).

The infrared spectrometer

This is what happens in an infrared spectrometer:

1 A beam of infrared radiation containing a spread of frequencies is passed through a sample.
2 The radiation that emerges is missing the frequencies that correspond to the types of bonds found in the sample.
3 The instrument plots a graph of the intensity of the radiation emerging from the sample, called the transmittance, against the frequency of radiation.
4 The frequency is expressed as a **wavenumber**, measured in cm^{-1}.

The infrared spectrum

A typical graph, called an **infrared spectrum**, is shown in Figure 3. The dips in the graph (confusingly, they are usually called peaks) represent particular bonds. Figure 4 and Table 1 show the wavenumbers of some bonds commonly found in organic chemistry.

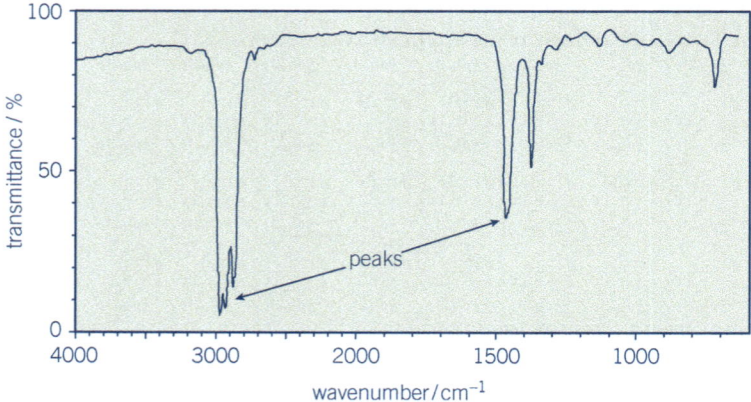

Figure 3 *A typical infrared spectrum. Note that wavenumber gets smaller going from left to right*

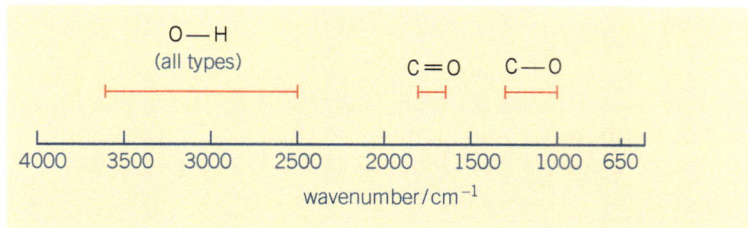

Figure 4 *The ranges of wavenumbers at which some bonds absorb infrared radiation*

Table 1 *Characteristic infrared absorptions in organic molecules*

Bond	Location	Wavenumber / cm⁻¹
C–O	alcohols, esters	1040–1300
C=O	aldehydes, ketones, carboxylic acids, esters	1670–1750
O–H	hydrogen bonded in carboxylic acids	2500–3000 (broad)
N–H	primary amines	3300–3500
O–H	hydrogen bonded in alcohols	3200–3600

These can help us to identify the functional groups present in a compound. For example:

- The O–H bond produces a broad peak at about between 3200 and 3600 cm⁻¹ and this is found in alcohols, ROH; and a very broad O–H peak between 2500 and 3000 cm⁻¹ in carboxylic acids, RCOOH.
- The C=O bond produces a peak between 1670 and 1750 cm⁻¹. This bond is found in aldehydes, RCHO, ketones, R₂CO, and carboxylic acids, RCOOH and esters RCOOR.

Data about the frequencies that correspond to different bonds can be found in the data section at the back of the book.

Figures 5, 6, and 7 show the infrared spectra of ethanal, ethanol, and ethanoic acid with the key peaks marked.

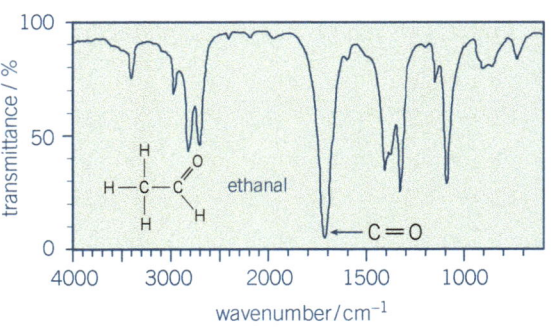

Figure 5 *Infrared spectrum of ethanal*

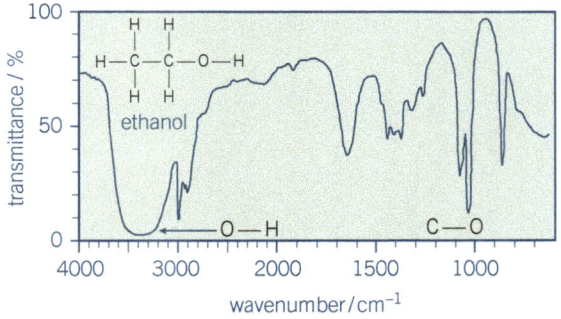

Figure 6 *Infrared spectrum of ethanol*

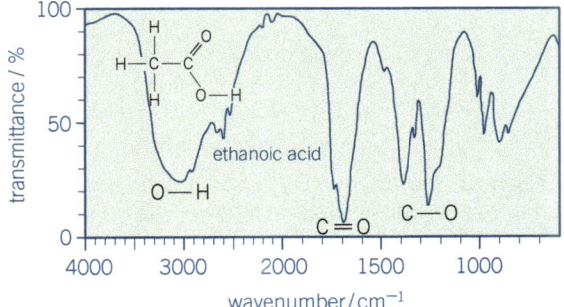

Figure 7 *Infrared spectrum of ethanoic acid*

Exam tip

There is a fuller table of infrared absorptions at the end of this book, and you will be given a copy in your exam.

Exam tip

When analysing peaks in infrared spectra, you should name the bond, not just the functional group.

Exam tip

You should be able to interpret infrared spectra using infrared absorption data.

Exam tip

Notice that the wavenumber decreases from left to right on an infrared spectrum.

The fingerprint region

The area of an infrared spectrum below about 1500 cm⁻¹ usually has many peaks caused by complex vibrations of the whole molecule. This shape is unique for any particular substance. It can be used to identify the chemical, just as people can be identified by their fingerprints. It is therefore called the **fingerprint region**.

Chemists can use a computer to match the fingerprint region of a sample with those on a database of compounds. An exact match confirms the identification of the sample.

Figures 8 and 9 show the IR spectra of two very similar compounds, propan-1-ol and propan-2-ol.

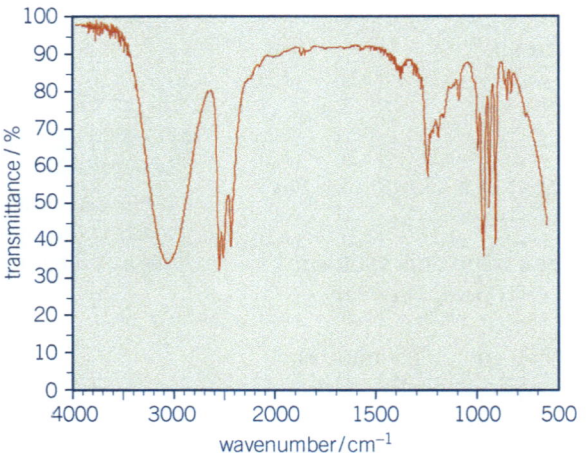

Figure 8 *Infrared spectrum of propan-1-ol*

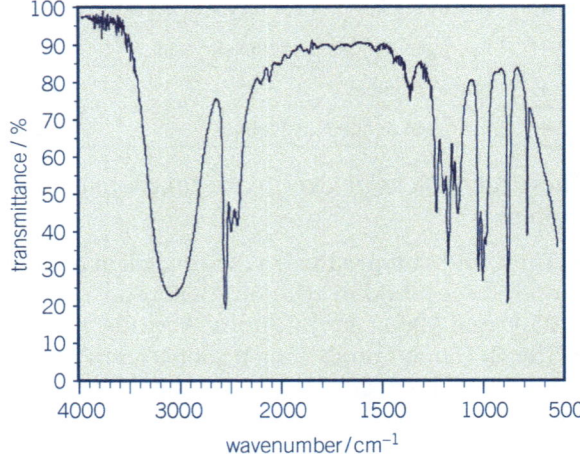

Figure 9 *Infrared spectrum of propan-2-ol*

They are, as expected, very similar overall. However, superimposing the spectra, Figure 10, shows that their fingerprint regions are quite distinct. This is shown more clearly in Figure 11, where the fingerprint region has been enlarged.

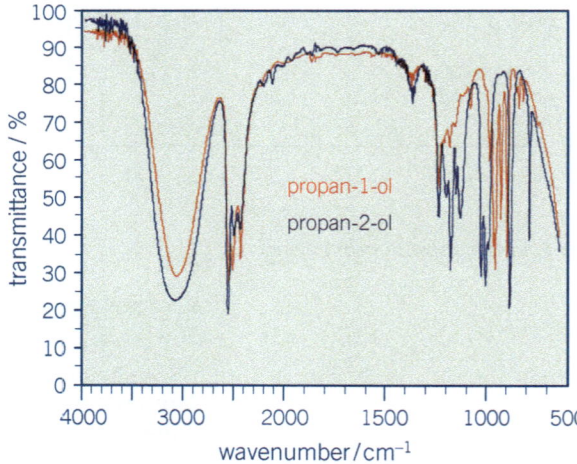

Figure 10 *Infrared spectra of propan-1-ol superimposed on propan-2-ol*

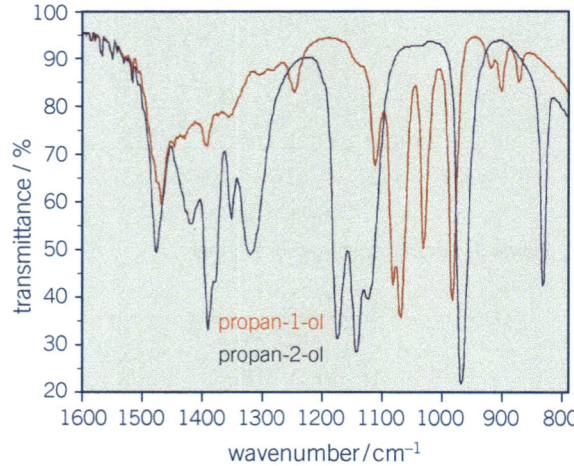

Figure 11 *The fingerprint region of the infrared spectra of propan-1-ol superimposed on propan-2-ol enlarged*

Extension

Identifying impurities

Infrared spectra can also be used to show up the presence of impurities. These may be revealed by peaks that should not be there in the pure compound. Figures 13 and 14 show the spectrum of a sample of pure caffeine and that of caffeine extracted from tea. The broad peak at around 3000 cm^{-1} in the impure sample (Figure 15) is an O–H stretch caused by water in the sample that has not been completely dried. Notice that there are no O–H bonds in caffeine (Figure 15).

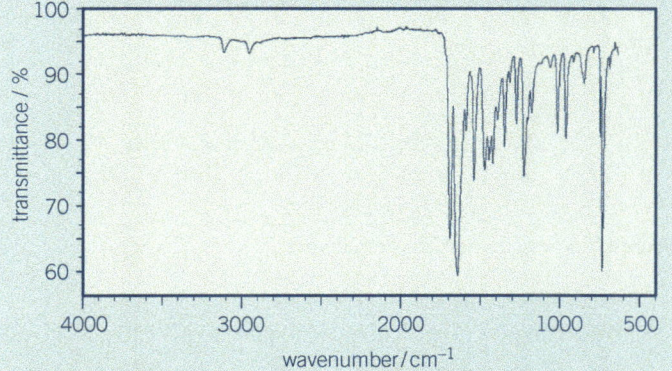

Figure 13 *The infrared spectrum of pure caffeine*

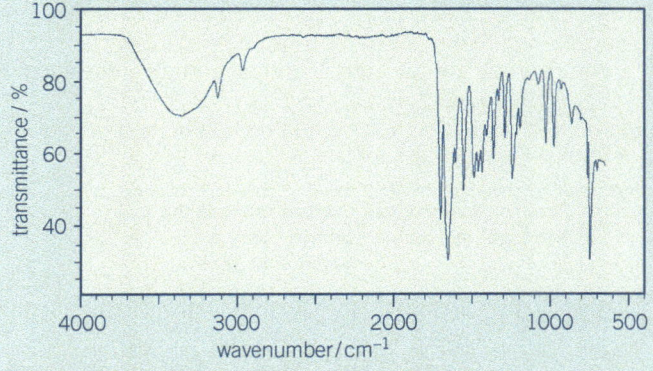

Figure 14 *The infrared spectrum of impure caffeine*

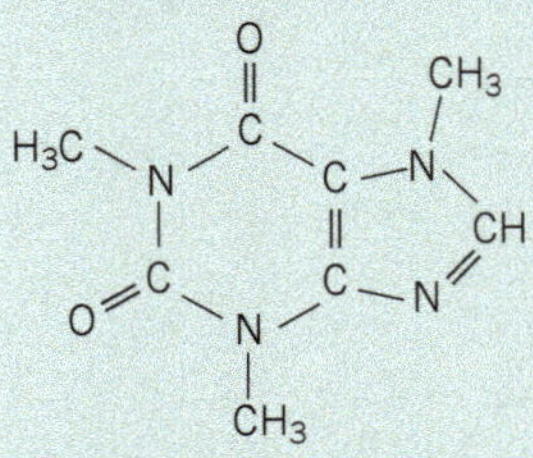

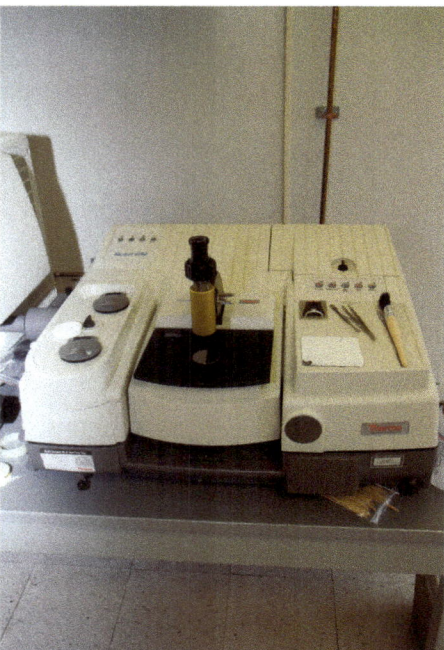

Figure 12 *This IR spectrometer is used in a forensics crime lab*

![Structural formula of caffeine]

Figure 15 *The structural formula of caffeine. It has no O–H bonds*

In practice, analytical chemists will often use a combination of spectroscopic techniques to identify unknown compounds.

Figures 16 and 17 show two more examples of IR spectra.

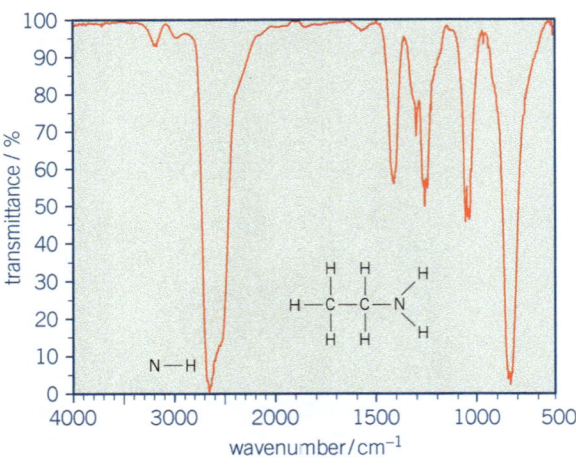

Figure 16 *IR spectrum of ethylamine*

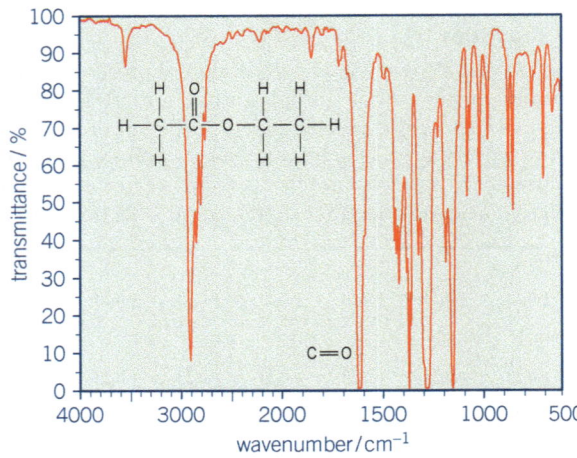

Figure 17 *IR spectrum of ethyl ethanoate*

Summary test 22.1

1 An organic compound has a peak in the IR spectrum at about 1725 cm^{-1}. State which of the following compounds it could be.

a H—C—C—C—H (with H, O, H)

b H—C—C (with H, O, O—H)

c H—C—C—O—H (with H, H, H, H)

2 Explain your answer to question 1.

3 An organic compound has a peak in the IR spectrum at about 3300 cm^{-1}.
State which of the compounds in question 1 it could be.

4 Explain your answer to question 3.

5 An organic compound has a peak in the IR spectrum at 1725 and 3300 cm^{-1}.
State which of the compounds in question 1 it could be.

6 Explain your answer to question 5.

Extension

Greenhouse gases

The greenhouse effect contributes to global warming. It is caused by gases in the atmosphere that absorb the infrared radiation given off from the surface of the Earth. This radiation would otherwise be lost into space. The table gives some data about some of these gases. The infrared radiation is absorbed by bonds in these gases in the same way as in an infrared spectrometer. Carbon dioxide has two C=O bonds which absorb in the infrared region of the spectrum.

Gas	Relative greenhouse effect per molecule	Concentration in the atmosphere / part per million (ppm)
carbon dioxide, CO_2	1	350
methane, CH_4	30	1.7
nitrous oxide (dinitrogen monoxide, N_2O)	160	0.31
ozone, O_3	2000	0.06
trichlorofluoromethane (a CFC)	21 000	0.000 26
dichlorodifluoromethane (a CFC)	25 000	0.000 24

Water vapour is a powerful greenhouse gas, absorbing IR via its O–H bonds. It is not included in the table because its concentration in the atmosphere is very variable.

1 Write the displayed formulae of trichlorofluoromethane and of dichlorodifluoromethane (showing all the atoms and the bonds).
2 What bonds are present in these compounds? Suggest why the relative greenhouse effects of these compounds are so similar.
3 Suggest why the concentration of water vapour is so variable.
4 One way of comparing the overall greenhouse contribution of a gas would be to multiply its concentration by its relative effect. Use this method to compare the contribution of carbon dioxide and methane.

Mass spectrometry

The mass spectrometer is the most useful instrument for the accurate determination of relative atomic masses, A_r. Relative atomic masses are measured on a scale on which the mass of an atom of ^{12}C is defined as *exactly* 12. No other isotope has a relative atomic mass that is exactly a whole number. This is because neither the proton nor the neutron has a relative mass of exactly 1.

$$\text{Relative atomic mass } A_r = \frac{\text{average mass of an element}}{\frac{1}{12} \text{ mass of one atom of } ^{12}C}$$

$$\text{Relative molecular mass } M_r = \frac{\text{average mass of molecule}}{\frac{1}{12} \text{ mass of one atom of } ^{12}C}$$

The mass spectrometer determines the mass of separate atoms (or molecules). Mass spectrometers are an essential part of a chemist's 'toolkit' of equipment. For example, they are used by forensic scientists to help identify substances.

There are several types of mass spectrometer. They all work on the principle of forming ions from the sample and then separating the positive ions according to the ratio of their charge to their mass, m/e.

Mass spectra of elements

The mass spectrometer can be used to identify the different isotopes that make up an element. It detects individual ions, so different isotopes are detected separately because they have different masses. This is how the data for the neon, germanium, and chlorine isotopes in Figures 18, 19, and 20 were obtained. The peak height gives the relative abundance of each isotope and the horizontal scale gives the m/e (mass /charge ratio), which for a singly charged ion is numerically the same as the mass number A.

Learning outcomes

On these pages you will learn to:

- describe the technique of mass spectrometry
- calculate the relative atomic mass of an element from its mass spectrum
- deduce the molecular mass of a compound from its molecular ion peak
- identify molecules from their fragmentation patterns
- calculate the number of carbon atoms in a compound using the M+1 peak
- identify chlorine and bromine in a compound using the M+2 peak

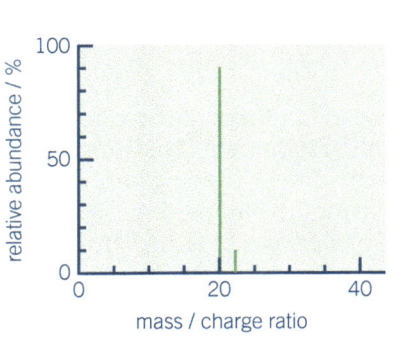

Figure 18 *The mass spectrum of neon. There is no peak at 20.2 because no neon atoms actually have this mass*

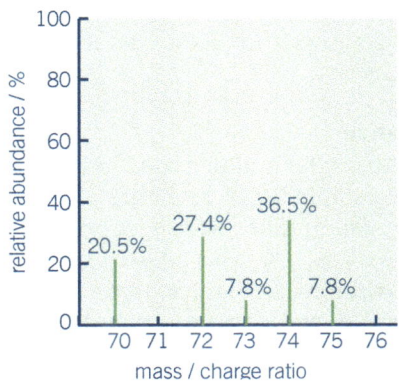

Figure 19 *The mass spectrum of germanium (the percentage abundance of each peak is given)*

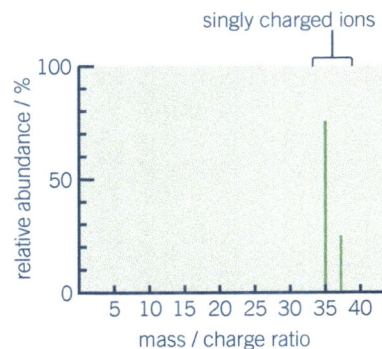

Figure 20 *The mass spectrum of chlorine*

Mass spectrometers can measure relative atomic masses to five decimal places of an atomic mass unit—this is called **high resolution mass spectrometry**. However most work is done to one decimal point—this is called **low resolution mass spectrometry**.

Analysing spectra

Isotopes of neon

The low resolution mass spectrum of neon is shown in Figure 18. This shows that neon has two isotopes, with mass numbers 20 and 22, and abundances to the nearest whole number of 90% and 10%, respectively. From this we can calculate the average relative atomic mass of neon, as shown on the following page.

$$\frac{(90 \times 20) + (10 \times 22)}{100} = 20.2$$

When calculating the relative atomic mass of an element, you must take account of the relative abundances of the isotopes. The relative atomic mass of neon is not 21 because there are far more atoms of the lighter isotope.

Another example is the mass spectrum of the element germanium, which is shown in Figure 19.

Isotopes of chlorine

From Figure 20 we can see that chlorine has two isotopes. They are $^{35}_{17}Cl$, with a mass number of 35, and $^{37}_{17}Cl$, with a mass number of 37. They occur in the ratio of almost exactly 3:1.

$^{35}Cl \quad ^{35}Cl \quad ^{35}Cl$ $\qquad\qquad\qquad ^{37}Cl$

three of these $\qquad\qquad$ to every one of this

So there is 75% ^{35}Cl and 25% ^{37}Cl atoms in naturally occurring chlorine gas.

The average mass of these is 35.5, as shown below:

$$\text{Mass of 100 atoms} = (35 \times 75) + (37 \times 25) = 3550$$

$$\text{Average mass} = \frac{3550}{100} = 35.5$$

This explains why the relative atomic mass of chlorine is approximately 35.5.

Mass spectrometry and relative molecular mass

We have seen how mass spectrometry is used to measure the relative atomic masses of atoms. It is also the main method for finding the relative molecular masses of organic compounds.

The organic molecules are passed through the spectrometer, ionised, and as before the mass to charge ratio, m/e, of the ion is detected and represented as a line on a mass spectrum. However, there may also be other lines associated with **fragmentation**.

Fragmentation

There are many techniques used in mass spectrometry. In some of these the ions of the sample break up as they pass through the instrument. This process is called fragmentation. The spectrum will have many lines, showing ions of smaller mass and not just one. In a simplified spectrum, the ion of the original molecule will have the largest mass and so is the furthest to the right. This is called the **molecular ion peak** and tells us the relative molecular mass.

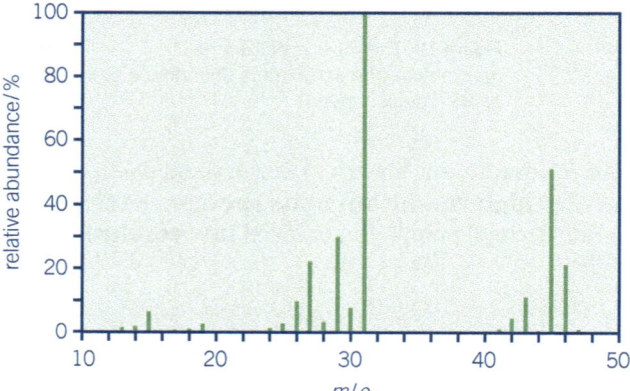

Figure 21 The mass spectrum of ethanol

A mass spectrum of ethanol is shown in Figure 21. When ethanol is ionised it forms the ion $C_2H_5OH^+$ ($CH_3CH_2OH^+$)—the molecular ion with $m/e = 46$. Many of these ions will then break up. Each of the fragments that is an ion produces a line in the mass spectrum. These can provide information that will help to deduce the structure of the compound. They also act as a 'fingerprint' to help identify it.

Some commonly encountered fragments are given in Table 2.

Table 2 *Some fragments commonly encountered in mass spectrometry*

Fragment	Mass
CH_3^+	15
CO^+	28
$CH_3CH_2^+$	29
CH_2OH^+	31
$CH_3CH_2CH_2^+$	43
$C_6H_5^+$	77

Worked example

Butane and methylpropane are isomers and so they will have molecular ions of the same mass. However, the fragmentation patterns will be different. Look at the two mass spectra.

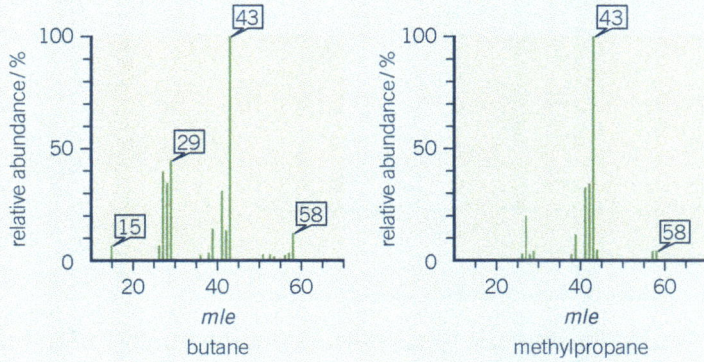

butane   methylpropane

Butane,

$$\begin{array}{ccccc} & H & H & H & H \\ & | & | & | & | \\ H- & C- & C- & C- & C-H \\ & | & | & | & | \\ & H & H & H & H \end{array}$$

shows the following main peaks:

- $\dfrac{m}{e} = 58$, molecular ion $CH_3CH_2CH_2CH_3^+$

- and $\dfrac{m}{e} = 43$, $CH_3CH_2CH_2^+$, formed when the bond between the atoms in red breaks:

- and $\dfrac{m}{e} = 29$, $CH_3CH_2^+$, formed when the bond between the green atoms breaks:

Methylpropane, shows the following main peaks:

- $\dfrac{m}{e} = 58$, molecular ion $CH_3CH(CH_3)CH_3^+$

- $\dfrac{m}{e} = 43$, $CH_3CHCH_3^+$, formed when any of the bonds between the red atoms breaks:

It is not possible to get a peak of $m/e = 29$ from methylpropane by breaking just one bond.

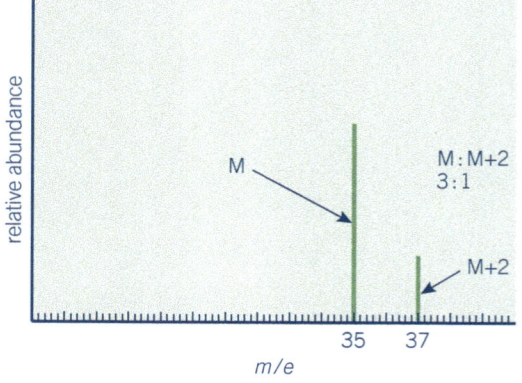

Figure 23 *Mass spectrum showing only chlorine M and M+2 peaks*

The M+1 and M+2 peaks

The M+1 peak

In a more detailed mass spectrum, the molecular ion peak (M) is not the peak of highest mass. There is a smaller peak that is one mass unit to the right of it caused by molecules containing the isotope carbon-13. This is called the M+1 peak (because the mass of this isotope is greater by 1 than carbon-12)—see Figure 22.

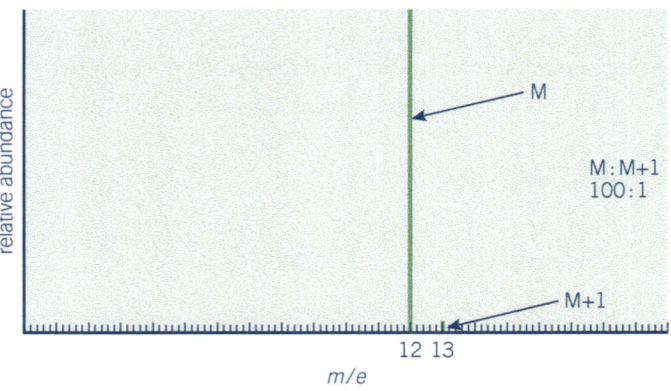

Figure 22 *Mass spectrum showing only the M and M+1 peaks from carbon*

We can use the relative heights of these peaks to work out the number of carbon atoms in the molecule we are investigating:

- The natural abundance of carbon-13 is 1.1%.
- Each carbon atom in a molecule has a 1.1% chance of being a carbon-13 atom.
- So if there is just one carbon atom in the molecule, the height of the M+1 peak will be 1.1 % of the height of the molecular ion peak.
- If there are two carbon atoms, the height of the M+1 peak will be 2.2% of the height of the molecular ion peak, and so on.
- In general, you can find the number of carbon atoms, n, in the molecule by working out the percentage abundance of the M+1 peak compared with the M peak and dividing by 1.1%.

So $n = \dfrac{100 \times \text{abundance of M+1 ion}}{1.1 \times \text{abundance of M}^+ \text{ ion}}$

The M+2 peak

Both chlorine and bromine exist as pairs of isotopes, in ratios much greater than that of the carbon isotopes. This also affects the lines we see on a spectrum.

Chlorine

Chlorine-containing compounds produce two separate molecular ion peaks, M and M+2. This is because chlorine exists as two isotopes ^{35}Cl and ^{37}Cl.

The isotopes are present in the ratio 3:1. So they produce two molecular ion peaks, one of two mass units greater than the other, and in the abundance ratio M : M+2 = 3:1.

So if there are a pair of lines of height in the ratio of 3:1 which are 2 *m/e* units apart, you know that there is one chlorine atom present in the parent ion, Figure 23.

Bromine

Bromine-containing compounds also produce two molecular ion peaks, M and M+2. This is because bromine has two isotopes ^{79}Br and ^{81}Br. These are of almost exactly equal abundance. So bromine-containing compounds produce an M peak and an M+2 peak of almost equal heights that are 2 *m/e* units apart (see Figure 24).

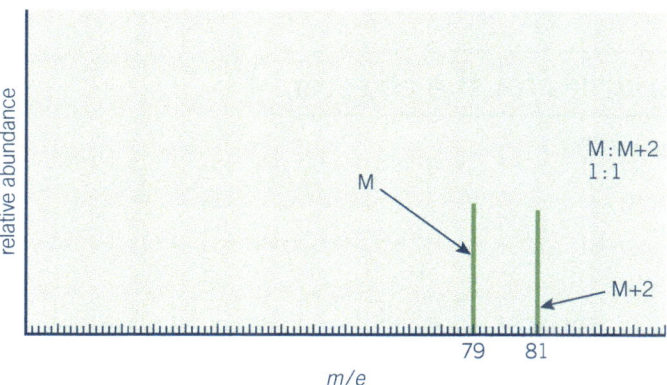

Figure 24 *Mass spectrum showing only bromine M and M+2 peaks*

Summary test 22.2

1 Use the information about germanium in Figure 19, earlier in this section, to calculate its relative atomic mass.
2 Figure 25 shows the mass spectrum of copper. Calculate the relative atomic mass of copper.

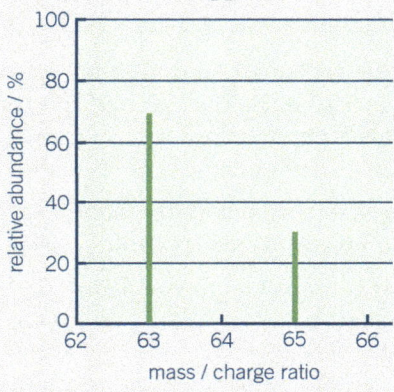

Figure 25 *The mass spectrum of copper*

3 Look at Figure 21 earlier in this section. With the knowledge that the ethanol molecule, CH_3CH_2OH, is breaking up, suggest formulae for the fragments represented by the peaks at *m/e* 46, 45, 31, and 29.
4 Look at the mass spectrum below of compound Z.

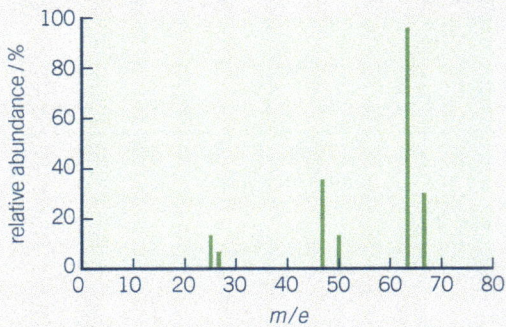

   a Suggest which two peaks could represent the molecular ion.
   b State what the relative abundances of these peaks suggest about the composition of compound Z.
   c There are two fragment peaks at *m/e* 49 and 51. State what their abundance ratios suggest about these fragments.
   d Identify the fragment that has been lost from the molecular ions to produce the peaks at *m/e* 49 and 51.
   e Suggest the formula of compound Z.
5 The relative heights of the M and M+1 peaks in a mass spectrum are in the ratio 23:1. State how many carbon atoms are in the molecule.

Worked example

The relative heights of the M and M+1 peaks in the mass spectrum of a molecule are in the ratio 15:1. How many carbon atoms are in the molecule?

Using

$$n = \frac{100 \times \text{abundance of M+1 ion}}{1.1 \times \text{abundance of M}^+ \text{ ion}}$$

$$= \frac{100 \times 1}{1.1 \times 15} = 6.06$$

So there are 6 carbon atoms, to the nearest whole number.

Launch additional digital resources for the chapter

1 An organic compound was subjected to combustion analysis. 1.0 g of the compound formed 1.37 g carbon dioxide, 1.12 g water and no other products.

a Calculate the percentage by mass of carbon and hydrogen in the compound. *(4 marks)*

b Give the name, and deduce the percentage by mass of the other element present in the organic compound. *(2 marks)*

c Calculate the empirical formula of the organic compound. *(3 marks)*

d The mass spectrum of the compound is shown in the figure below.

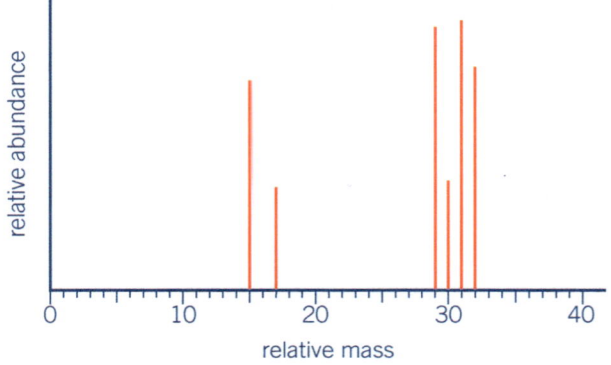

i Use this to deduce the relative molecular mass (M_r) of the organic compound and calculate the molecular formula. *(3 marks)*

ii Using the fragments shown in the mass spectrum, deduce the structural formula of the compound. *(1 mark)*

iii Give the formula of the fragment to which each peak can be attributed. *(6 marks)*

2 Write down the reagents and conditions represented by B, D and F in the reaction scheme at the top of the following column. Give the name and displayed formulae for the compounds represented by A, C and E. *(6 marks)*

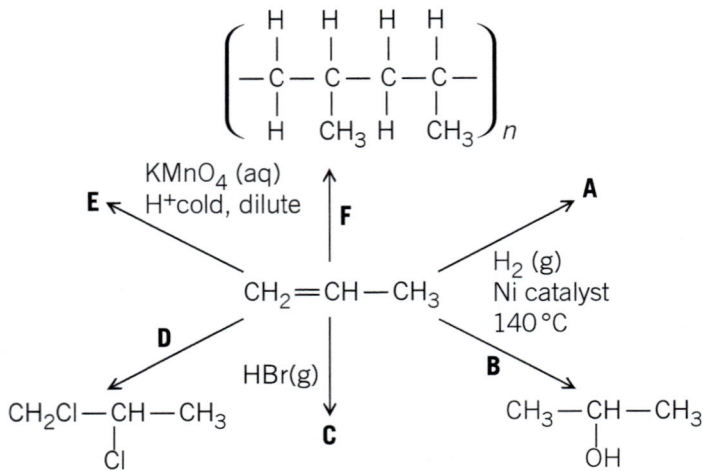

3 Ethoxyethane, $CH_3CH_2OCH_2CH_3$, and butan-2-ol, $CH_3CH_2CH_2CH_2OH$ are isomers. The infrared spectra of these two compounds is given below. The spectra are labelled A and B.

a Use the data section at the back of the book to assign each spectrum to the correct compound. *(1 mark)*

b Identify the bonds responsible for the peaks marked 1, 2, 3 and 4 the spectra. *(4 marks)*

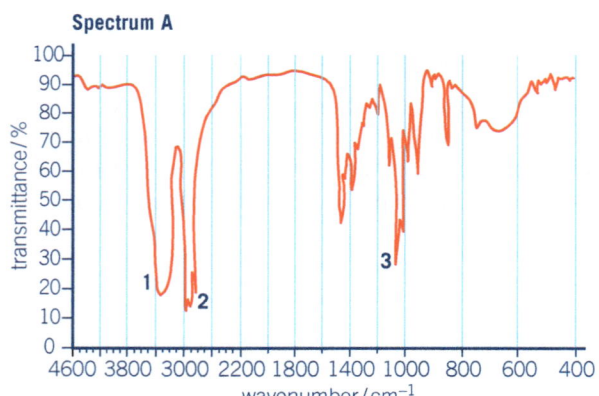

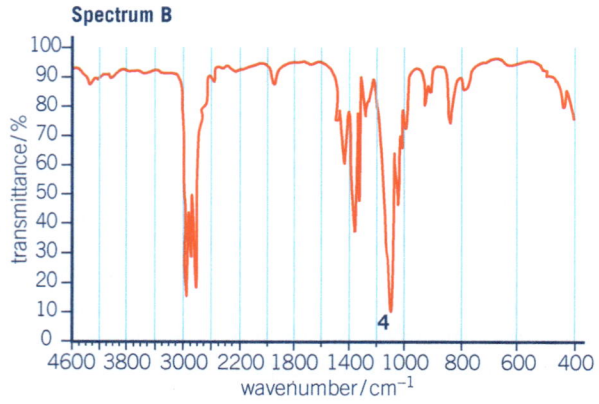

4 The diagram below summarises some reactions of propan-1-ol.

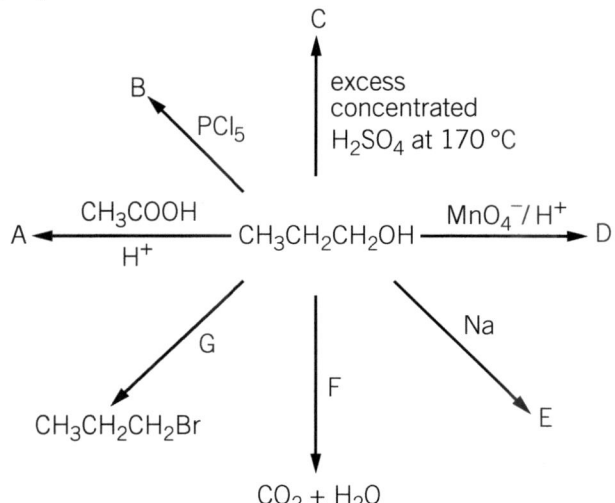

a Give the IUPAC names and structural formulae of the substances represented by A, B, C, D and E. *(5 marks)*

b Specify the reagents and conditions represented by F and G. *(2 marks)*

5 Give the reagents and conditions you would use to carry out the following conversions. One or more steps may be involved in the conversions:

a ethanol to ethyl ethanoate (using ethanol as the only reagent)

b ethanol to 1,2-dibromoethane

c ethene to ethanoic acid

d ethanol to propanenitrile. *(8 marks)*

6 Describe how the following conversions can be carried out in the laboratory:

a CH_3CHO to $CH_3COOCH_2CH_3$

b $CH_2=CH_2$ to CH_3COOH

c CH_3CH_2CN to $CH_3CH_2COO^-Na^+$

d CH_3CH_2OH to $CH_3COO^-Na^+$

(8 marks)

7 State the homologous series to which each of the following compounds belongs:

a $CH_3CH_2CH_2OH$

b $CH_3CH_2COCH_3$

c CH_3CH_2Cl

d CH_3CH_2COOH

e $CH_3CH=CHCH_3$

f $CH_3CH_2OCH_2CH_2CH_3$

8 Write the full displayed formulae of all isomers of the following molecular formulae, stating which type of isomerism is involved:

a C_3H_7Cl

b C_6H_{14}

c $C_2H_3Cl_2Br$ *(6 marks)*

A Level

This section of the book contains the material that you will cover in the second year of the Cambridge International AS & A Level Chemistry course.

The content builds on the chemistry you will have studied earlier at AS Level. If you choose to go on to study chemistry at a higher level, the material in section will provide a foundation for those studies.

The material is divided into three parts:

• Physical chemistry: Chapters 23–26
• Inorganic chemistry: Chapters 27–28
• Organic chemistry: Chapters 29–37

Each chapter is matched to the syllabus and is followed by exercises that will test your understanding and give you practice at tackling Cambridge examination questions.

23 Chemical energetics

23.1 Lattice energy and Born–Haber cycles

Exam tip

An enthalpy change is a heat energy change measured at constant pressure. Often the terms 'energy' and 'enthalpy' are used interchangeably, for example 'lattice energy' and 'bond energy'.

Exam tip

You should know the following definitions:

The standard enthalpy of atomisation ΔH_{at} is the enthalpy change which accompanies the formation of one mole of gaseous atoms from the element in its standard state under standard conditions.

The first ionisation energy ΔH_{i1} (first IE) is the standard enthalpy change when one mole of gaseous atoms is converted into a mole of gaseous ions each with a single positive charge.

The second ionisation energy ΔH_{i2} (second IE) is the standard enthalpy change when one mole of gaseous ions each with a single positive charge is converted into a mole of gaseous ions each with a double positive charge.

The first electron affinity EA_1 is the standard enthalpy change when a mole of gaseous atoms is converted to a mole of gaseous ions, each with a single negative charge.

The second electron affinity EA_2 is the enthalpy change when a mole of gaseous ions each with a single negative charge is converted to a mole of ions each with two negative charges.

The enthalpy of lattice formation (lattice energy ΔH_{latt}) is the standard enthalpy change when one mole of solid ionic compound is formed from its gaseous ions.

It is important to have an equation to refer to for enthalpy changes.

This chapter looks at the enthalpy changes that are involved in the formation of an ionic compound from its elements.

Enthalpy changes on forming ionic compounds

If a cleaned piece of solid sodium is placed in a gas jar containing chlorine gas, an exothermic reaction takes place, forming solid sodium chloride:

$$Na(s) + \tfrac{1}{2}Cl_2(g) \rightarrow NaCl(s) \qquad \Delta H_f = -411\,kJ\,mol^{-1}$$

You can think of it as taking place in several steps:

- The reaction involves solid sodium, not gaseous, and chlorine molecules, not separate atoms. So you must start with the **enthalpy changes for atomisation, ΔH_{at}**:

$$Na(s) \rightarrow Na(g) \qquad \Delta H_{at} = +108\ kJ\,mol^{-1}$$

$$\tfrac{1}{2}Cl_2(g) \rightarrow Cl(g) \qquad \Delta H_{at} = +122\,kJ\,mol^{-1}$$

Energy has to be put in to pull apart the atoms (ΔH_{at} is positive in both cases).

- The gaseous sodium atoms must each give up an electron to form gaseous Na$^+$ ions:

$$Na(g) \rightarrow Na^+(g) + e^-$$

The enthalpy change for this process is the **enthalpy change of first ionisation, ΔH_{i1}**, of sodium and is +494 kJ mol^{-1} (Ionisation energy also has the symbol *IE*).

- The chlorine atoms must gain an electron to form gaseous Cl$^-$ ions:

$$Cl(g) + e^- \rightarrow Cl^-(g)$$

The enthalpy change for this process of electron gain is the **first electron affinity, *EA***. The first electron affinity for the chlorine atom is −349 kJ mol^{-1} (i.e., energy is given out when this process occurs).

There is a further energy change. At room temperature, sodium chloride exists as a solid lattice of alternating positive and negative ions, and not as separate gaseous ions. If positively charged ions come together with negatively charged ions, they form a solid lattice and energy is given out due to the attraction between the oppositely charged ions. This is called the **lattice formation energy, ΔH_{latt}**, and it refers to the process:

$$Na^+(g) + Cl^-(g) \rightarrow NaCl(s) \qquad \Delta H_{latt} = -786\,kJ\,mol^{-1}$$

The following five processes lead to the formation of NaCl(s) from its elements.

- Atomisation of Na:

$$Na(s) \rightarrow Na(g) \qquad \Delta H_{at} = +108\,kJ\,mol^{-1}$$

- Atomisation of Cl:

$$\frac{1}{2}Cl_2(g) \rightarrow Cl(g) \qquad \Delta H_{at} = +122\,\text{kJ mol}^{-1}$$

- Ionisation (e^- loss) of Na:

$$Na(g) \rightarrow Na^+(g) + e^- \qquad \Delta H_{i1} = +494\,\text{kJ mol}^{-1}$$

- Electron affinity of Cl:

$$Cl(g) + e^- \rightarrow Cl^-(g) \qquad EA = -349\,\text{kJ mol}^{-1}$$

- Formation of lattice:

$$Na^+(g) + Cl^-(g) \rightarrow NaCl(s) \qquad \Delta H_{latt} = -786\,\text{kJ mol}^{-1}$$

Hess's law tells us that the total energy (or enthalpy) change for a chemical reaction is the same whatever route is taken, provided that the initial and final conditions are the same. It does not matter whether the reaction actually takes place via these steps or not.

So the sum of the first five energy changes (taking the signs into account) is equal to the enthalpy change of formation of sodium chloride. You can calculate any one of the quantities, provided all the others are known. You do this by using a thermochemical cycle, called a Born–Haber cycle.

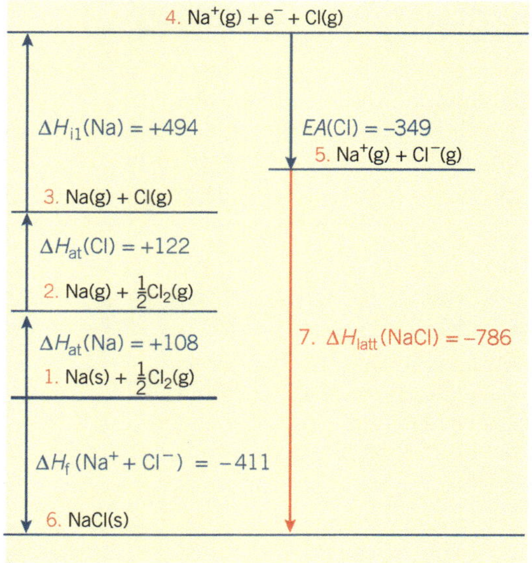

Born–Haber cycles

A **Born–Haber cycle** is a thermochemical cycle that includes all the enthalpy changes involved in the formation of an ionic compound. Born–Haber cycles are constructed by starting with the elements in their standard states. All elements in their standard states have zero enthalpy by definition.

Figure 1 *Stages in the construction of the Born–Haber cycle for sodium chloride, NaCl, to find the lattice enthalpy. All enthalpies are in kJ mol^{-1}*

The Born–Haber cycle for sodium chloride

There are six steps in the Born–Haber cycle for the formation of sodium chloride. Here you will use the cycle to calculate the lattice energy (ΔH_{latt}). The other five steps are shown in Figure 1. (Remember that if you know any five, you can calculate the other). Figure 1 shows you how each step is added to the one before, starting from the elements in their standard state. Positive (endothermic changes) are shown upwards, and negative (exothermic changes) are shown downwards.

$$Na(s) \rightarrow Na(g) \qquad \Delta H_{at}\,Na = +108\,\text{kJ mol}^{-1}$$

$$\frac{1}{2}Cl_2(g) \rightarrow Cl(g) \qquad \Delta H_{at}\,Cl = +122\,\text{kJ mol}^{-1}$$

$$Na(g) \rightarrow Na^+(g) + e^- \qquad \Delta H_{i1} = +494\,\text{kJ mol}^{-1}$$

$$Cl(g) + e^- \rightarrow Cl^-(g) \qquad EA = -349\,\text{kJ mol}^{-1}$$

$$Na(s) + \frac{1}{2}Cl_2(g) \rightarrow NaCl(s) \qquad \Delta H_f = -411\,\text{kJ mol}^{-1}$$

Using a Born–Haber cycle you can see why the formation of an ionic compound from its elements is an *exothermic* process. This is mainly due to the large amount of energy given out when the lattice forms.

To construct a Born–Haber cycle:

1 Start with elements in their standard states. This is the energy zero of the diagram.
2 Add in the atomisation of sodium. This is positive, so it is drawn 'uphill'.
3 Add in the atomisation of chlorine. This too is positive, so drawn 'uphill'.
4 Add in the ionisation of sodium, also positive and so drawn 'uphill'.
5 Add in the electron affinity of chlorine. This is a negative energy change and so it is drawn 'downhill'.

> **Exam tip**
>
> The symbol $\ominus$ used with $\Delta H^{\ominus}$ and other quantities means that the value has been measured at or corrected to standard conditions (a temperature of 298 K and a pressure of 100 000 Pa, which are close to normal room conditions).

6 Add in the enthalpy of formation of sodium chloride, also negative and drawn 'downhill'.

7 The final unknown quantity is the lattice energy of sodium chloride. The size of this is 786 kJ mol^{-1} from the diagram. Lattice energy is the change from separate ions to solid lattice and you must therefore go 'downhill', so:

$$\Delta H_{latt}(Na^+ + Cl^-)(s) = -786\,kJ\,mol^{-1}$$

When drawing Born–Haber cycles:
- Make up a rough scale, for example, one line of lined paper to 100 kJ mol^{-1}.
- Plan out roughly first to avoid going off the top or bottom of the paper. (The zero line representing elements in their standard state will need to be in the middle of the paper.).
- Remember to put in the sign of each enthalpy change and an arrow to show its direction. Positive enthalpy changes go up, negative enthalpy changes go down.

Exam tip

Remember, the standard enthalpy of atomisation is the enthalpy change which accompanies the formation of one mole of gaseous atoms. To form one mole of Cl(g) you need $\frac{1}{2}Cl_2$(g).

Exam tip

Most errors in Born–Haber cycle calculations result either from lack of knowledge of the enthalpy change definitions, or lack of care with + and – signs.

Worked example

The Born–Haber cycle for magnesium chloride

Figure 2 shows the complete Born–Haber cycle for the formation of magnesium chloride, MgCl$_2$, from its elements, together with notes on how it is constructed.

Since two chlorine atoms are involved all the quantities related to chlorine are doubled: ΔH_{at}(Cl) and the first electron affinity are both *multiplied* by two.

Also notice that the ionisation of magnesium, Mg → Mg^{2+}, is the first ionisation enthalpy plus the second ionisation enthalpy. The second ionisation enthalpy is larger because it is more difficult to lose an electron from a positively charged ion than from a neutral atom.

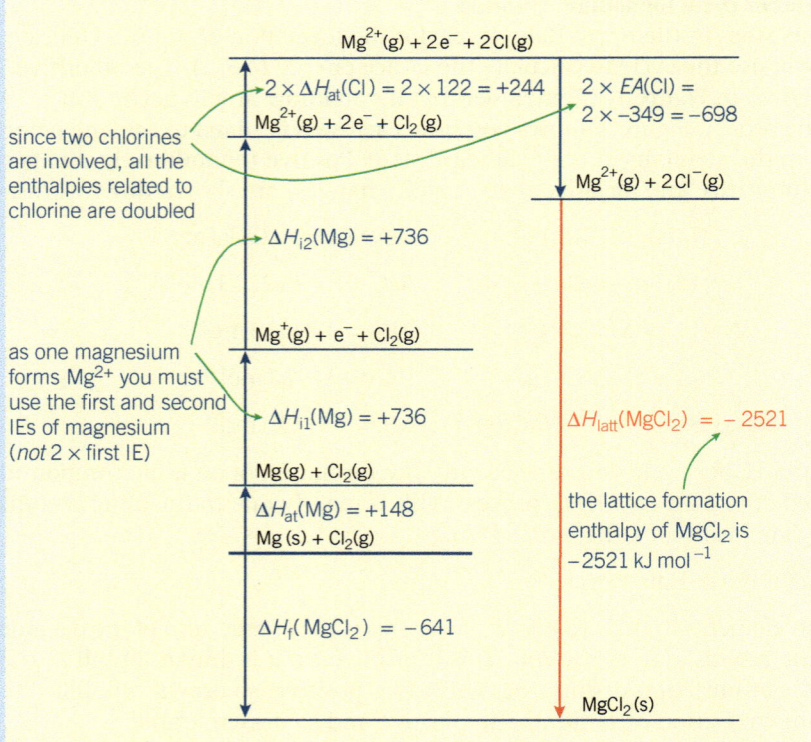

Figure 2 The Born–Haber cycle for magnesium chloride, MgCl$_2$. All enthalpies are in kJ mol^{-1}

Exam tip

For chlorine, you multiply the first electron affinity and ΔH_{at} by two. For magnesium, you add together the first and second ionisation energy. This is because there are two chlorine atoms each gaining one electron (so two first electron affinities), but one magnesium atom losing two electrons (so the first and second ionisation energy).

Here are some more examples of Born–Haber cycles.

1 The Born–Haber cycle for magnesium oxide

All enthalpies are in $kJ\,mol^{-1}$.

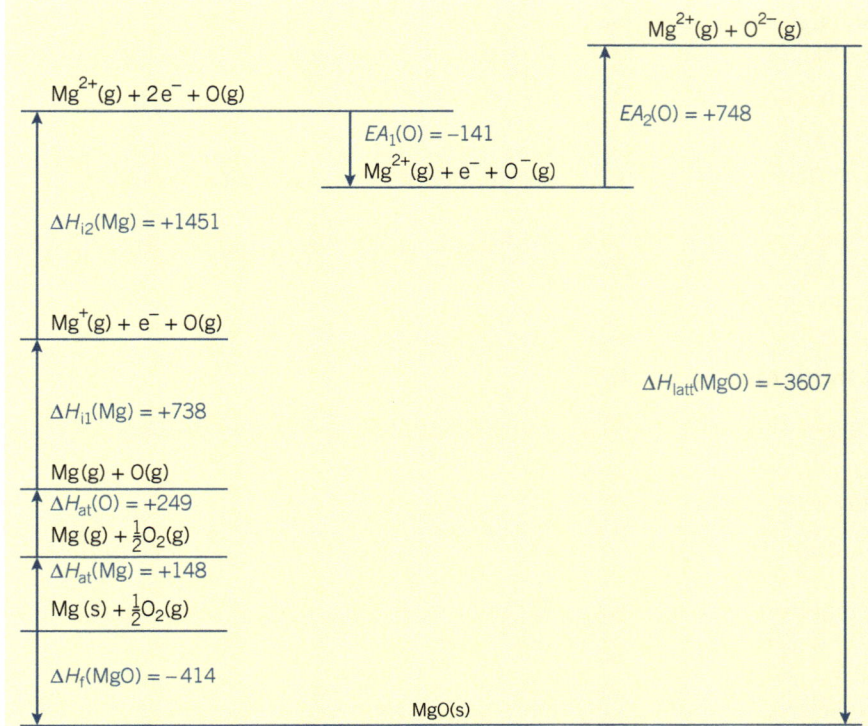

We use the first and second ionisation energies of magnesium, and the first and second electron affinities of oxygen.

2 The Born–Haber cycle for sodium oxide

All enthalpies are in $kJ\,mol^{-1}$.

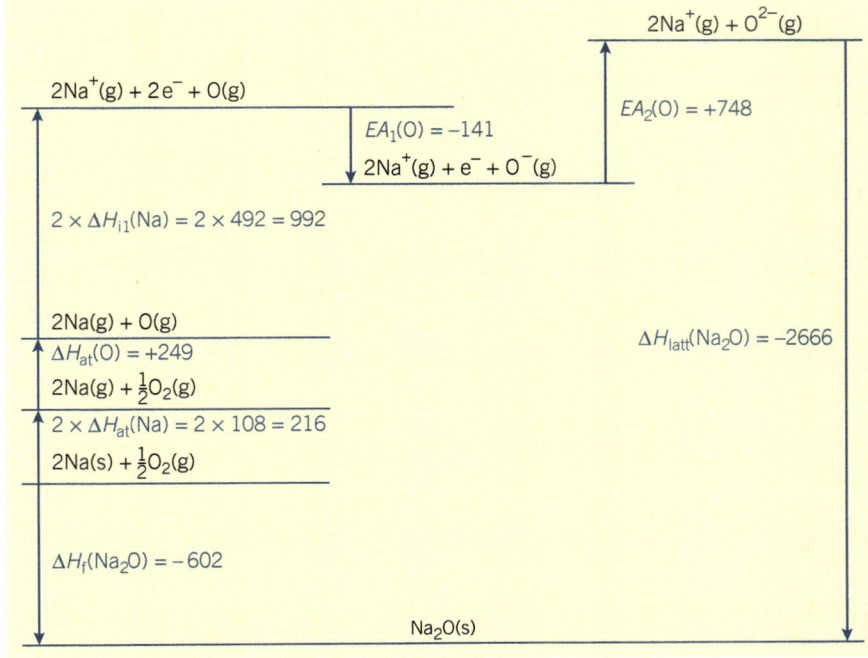

We use the first and second electron affinities of oxygen. The values of atomisation energy and ionisation energy for sodium are doubled as there are two atoms of sodium.

Table 1 *Some values of lattice energies in kJ mol⁻¹ for compounds M⁺X⁻*

		Larger negative ions (anions)			
		F⁻	Cl⁻	Br⁻	I⁻
Larger positive ions (cations)	Li⁺	−1031	−848	−803	−759
	Na⁺	−918	−786	−742	−705
	K⁺	−817	−711	−679	−651
	Rb⁺	−783	−685	−656	−628
	Cs⁺	−747	−661	−635	−613

Table 2 *Some values of lattice energies for compounds M²⁺X²⁻*

		Larger anions	
		O²⁻	S²⁻
Larger cations	Be²⁺	−4443	−3832
	Mg²⁺	−3791	−3299
	Ca²⁺	−3401	−3013
	Sr²⁺	−3223	−2848
	Ba²⁺	−3054	−2725

Table 3 *The electron affinities for the halogen elements (Group 17)*

Element	First electron affinity / kJ mol⁻¹
fluorine	−328
chlorine	−349
bromine	−324
iodine	−295

Table 4 *The first and second electron affinities some Group 16 elements*

Element	First electron affinity / kJ mol⁻¹	Second electron affinity / kJ mol⁻¹
oxygen	−141	+798
sulfur	−200	+640

Trends in lattice energies

The lattice energies of some simple ionic compounds of formula M^+X^- are given in Table 1.

Ions with larger ionic radii lead to smaller lattice energies. This is because the opposite charges do not approach each other as closely when the ions are larger.

Table 2 shows lattice energies for some compounds $M^{2+}X^{2-}$. You can see the same trend related to size of ions as before in Table 1.

Comparing Table 1 with Table 2 shows that for ions of approximately similar size (i.e., formed from elements in the same period of the Periodic Table, such as Na^+ and Mg^{2+} or F^- and O^{2-}) the lattice enthalpy increases with the size of the charge. This is because ions with double the charge give out roughly twice as much energy when they come together.

Trends in electron affinities

Table 3 shows the first electron affinities for the halogen elements (Group 17).

The general pattern is for electron affinities to become less negative (less energy given out) as we descend the group.

In forming a negative ion, an electron is being added to the outer shell of the atom and being attracted by the nucleus. The electron affinity is a measure of the attraction between this electron and the nucleus. The stronger the attraction, the more energy is released. As we go down the group:

- The nuclear charge increases but it is shielded by the inner electrons, so the added electron feels essentially the same positive charge (+7) in each case.
- However, the added electron is being added to a shell *further from the nucleus*. It is this factor that causes less energy to be given out.

Fluorine does not fit the pattern. Its small size means that the added electron feels the repulsion of the electrons already in the outer shell (shell 2) and this partly cancels the charge of the nucleus.

Table 4 shows the first and second electron affinities some Group 16 elements.

The most important thing to note here is that the *first* electron affinity is the enthalpy change for the process:

$$X(g) + e^- \rightarrow X^-(g)$$

Energy is given out in this process because the added electron is attracted by the shielded nuclear charge.

The second electron affinity is the enthalpy change for:

$$X^-(g) + e^- \rightarrow X^{2-}(g)$$

The added electron is now being repelled by the negatively charged X^- ion, so energy has to be put in for this process to occur.

The small size of the oxygen atom explains why the first electron affinity of oxygen is less negative than that of sulfur, in the same way as that of fluorine in Group 17.

Extension

The first noble gas compound

The noble gases are often called the **inert gases** and, until 1962, they seemed to be just that—inert. There were no known compounds of them at all. This was explained on the basis of their stable electron arrangements. It had been predicted that there might be compounds of krypton and xenon with fluorine, but no one took much notice. However, in 1962 British chemist Neil Bartlett created a chemical sensation when he announced that he had prepared the first noble gas compound, xenon hexafluoroplatinate(V). Although the name may seem exotic, Bartlett predicted that the compound had a good chance of existing by using a very simple piece of chemical theory.

He had previously found that the powerful oxidising agent platinum(VI) fluoride gas, PtF_6, would oxidise oxygen molecules to form the compound dioxygenyl hexafluoroplatinate(V), $O_2^+PtF_6^-$, in which the oxidising agent has removed an electron from an oxygen molecule.

He then realised that the first ionisation energy of xenon (the energy required to remove an electron from an atom of it) was a little less positive than that of the oxygen molecule, so that if platinum(VI) fluoride could remove an electron from oxygen, it should also be able to remove one from xenon. The values are:

$Xe(g) \rightarrow Xe^+(g) + e^-$ $\quad \Delta H_{i1} = +1170\,kJ\,mol^{-1}$ (first IE of xenon)

$O_2(g) \rightarrow O_2^+(g) + e^-$ $\quad \Delta H_{i1} = +1183\,kJ\,mol^{-1}$ (first IE of an oxygen *molecule*)

This is not the same as the first ionisation energy of an oxygen *atom*.

The experiment itself was surprisingly simple. As soon as the two gases came into contact, the compound was formed immediately—no heat or catalyst was required. In Bartlett's own words: "When I broke the seal between the red PtF_6 gas and the colourless xenon gas, there was an immediate interaction, causing an orange-yellow solid to precipitate. At once I tried to find someone with whom to share the exciting finding, but it appeared that everyone had left for dinner!"

This was one of those moments when all the textbooks had to be re-written.

The reaction can be represented:

$$Xe(g) + PtF_6(g) \rightarrow Xe^+PtF_6^-(s)$$

More recently it has been realised that the formula of the product may be a little more complex than this.

There are now over 100 noble gas compounds known, although most are highly unstable.

1 Write an equation to represent the first ionisation energy of an oxygen *atom*.
2 If you assume that noble gas compounds are formed with positive noble gas ions, suggest why compounds of xenon and krypton were predicted rather than ones of helium or neon.
3 Why might you *not* expect platinum(VI) fluoride to be a gas?
4 Explain the oxidation states of the elements in $Xe^+PtF_6^-$.

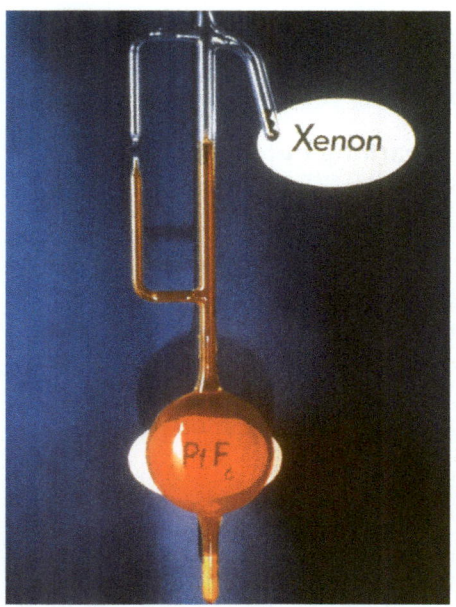

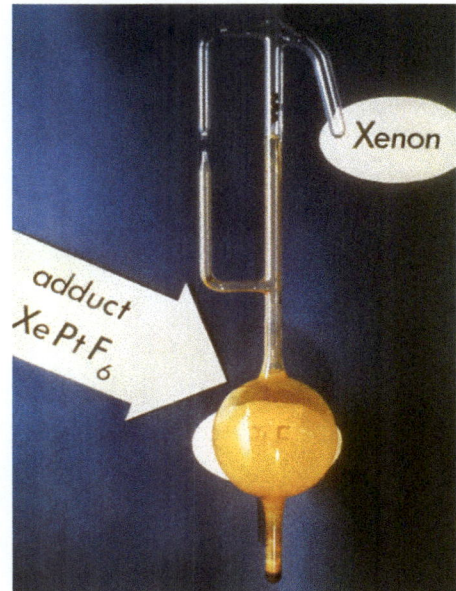

Figure 3 *The formation of XePtF$_6$*

Summary test 23.1

1 a Sketch a Born–Haber cycle to find the lattice energy for sodium fluoride, NaF. The values for the relevant enthalpy terms are given below.
 b Deduce the lattice energy of NaF, given these values.

$Na(s) \rightarrow Na(g)$ $\qquad \Delta H_{at} = +108\,kJ\,mol^{-1}$

$\frac{1}{2}F_2(g) \rightarrow F(g)$ $\qquad \Delta H_{at} = +79\,kJ\,mol^{-1}$

$Na(g) \rightarrow Na^+(g) + e^-$ $\qquad \Delta H_{i1} = +494\,kJ\,mol^{-1}$

$\frac{1}{2}F(g) + e^- \rightarrow F^-(g)$ $\qquad EA = -328\,kJ\,mol^{-1}$

$Na(s) + \frac{1}{2}F_2(g) \rightarrow NaF(s)$ $\qquad \Delta H_f = -574\,kJ\,mol^{-1}$

Learning outcomes

On these pages you will learn to:

- explain and use enthalpy changes of hydration
- explain and use enthalpy changes of solution
- explain how ionic charge and ionic radius affect the enthalpy change of hydration

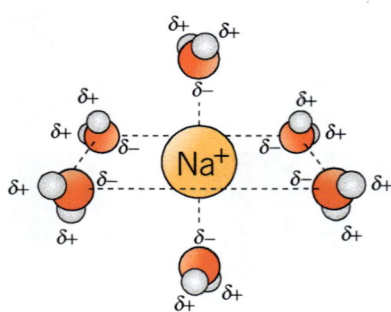

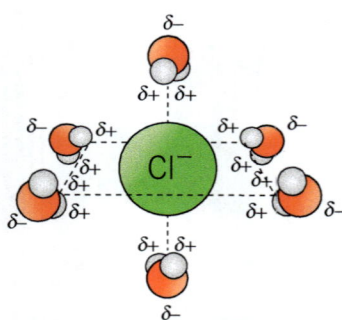

Figure 4 *The hydration of sodium and chloride ions by water molecules*

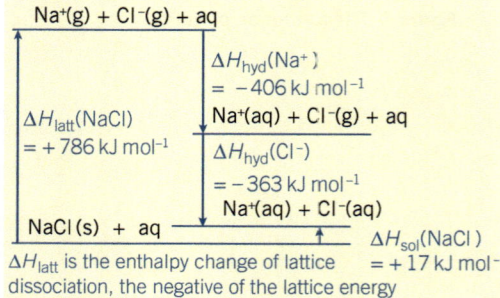

ΔH_{latt} is the enthalpy change of lattice dissociation, the negative of the lattice energy

Figure 5 *Thermochemical cycle for the enthalpy of hydration of sodium chloride*

Ionic solids can only dissolve well in polar solvents. In order to dissolve an ionic compound the lattice must be broken up. This requires an input of energy—the lattice enthalpy. The separate ions are then solvated by the solvent molecules, usually water. These cluster round the ions so that the positive ions are surrounded by the negative ends of the dipole of the water molecules, and the negative ions are surrounded by the positive ends of the dipoles of the water molecules. This is called hydration when the solvent is water (Figure 4).

The **enthalpy change of solution, ΔH_{sol}**, is the enthalpy change when 1 mole of an ionic substance dissolves in water to give a very dilute solution eg $Na^+(g) + (aq) \rightarrow Na^+(aq)$.

The **enthalpy change of hydration, ΔH_{hyd}**, of an ion is the enthalpy change when 1 mole of gaseous ions dissolves in water to give a very dilute solution. Enthalpies of hydration are always negative (ie energy is given out) eg $NaCl(s) + (aq) \rightarrow Na^+(aq) + Cl^-(aq)$.

The enthalpy change of hydration, ΔH_{hyd}, shows the same trends as lattice enthalpy—it is more negative (ie more energy given out) for more highly charged ions and less negative for bigger ions (ie those with greater ionic radius).

You can think of dissolving an ionic compound in water as the sum of three processes:

1 Breaking the ionic lattice to give separate gaseous ions—the lattice energy has to be put in.
2 Hydrating the positive ions (cations)—the enthalpy of hydration is given out.
3 Hydrating the negative ions (anions)—the enthalpy of hydration is given out.

For ionic compounds the enthalpy change of solution $\Delta H^{\ominus}_{sol}$ has rather a small value and may be positive or negative. For example, the enthalpy change of solution (ΔH_{sol}) for sodium chloride is given by the equation:

$$NaCl(s) + aq \rightarrow Na^+(aq) + Cl^-(aq)$$

It may be calculated via a thermochemical cycle as shown below. These are the steps that are needed:

1 $NaCl(s) \rightarrow Na^+(g) + Cl^-(g)$ $\Delta H_{latt} = +786\,kJ\,mol^{-1}$

This is the enthalpy change for lattice dissociation.

2 $Na^+(g) + aq \rightarrow Na^+(aq)$ $\Delta H_{hyd} = -406\,kJ\,mol^{-1}$

This is the enthalpy change for the hydration of the sodium ion.

3 $Cl^-(g) + aq \rightarrow Cl^-(aq)$ $\Delta H_{hyd} = -363\,kJ\,mol^{-1}$

This is the enthalpy change for the hydration of the chloride ion.

4 So =

$$\Delta H_{sol}(NaCl) = \Delta H_{latt}(NaCl) + \Delta H_{hyd}(Na^+) + \Delta H_{hyd}(Cl^-)$$
$$+786 \quad\quad -406 \quad\quad -363 \quad = +17\,kJ\,mol^{-1}$$

The process of dissolving can be represented on an enthalpy diagram (Figure 5) or calculated directly as above. Either method is equally acceptable.

Summary test 23.2

1 Sketch a diagram to calculate the enthalpy change of solution of potassium bromide using the value of lattice enthalpy ΔH_{latt} of $-679\,kJ\,mol^{-1}$.
$$\Delta H_{hyd}(K^+) = -322\,kJ\,mol^{-1}$$
$$\Delta H_{hyd}(Br^-) = -335\,kJ\,mol^{-1}$$
Explain why the value is relatively small.

Entropy change, ΔS

Chemists use the terms 'feasible' or 'spontaneous' to describe reactions which could take place of their own accord. The terms take no account of the rate of the reaction, which could be so slow as to be unmeasurable at room temperature.

You may have noticed that many of the reactions that occur of their own accord are exothermic (ΔH is negative). For example, if you add magnesium to copper sulfate solution, the reaction to form copper and magnesium sulfate takes place and the solution gets hot.

Negative ΔH is a factor in whether a reaction is spontaneous, but it does not explain why a number of endothermic reactions are spontaneous.

For example, both the following reactions, which occur spontaneously, are endothermic (ΔH is positive):

$C_6H_8O_7(aq)$ + $3NaHCO_3(aq)$ → $Na_3C_6H_5O_7(aq)$ + $3H_2O(l)$ + $3CO_2(g)$
citric acid   sodium   sodium citrate   water   carbon
              hydrogencarbonate   dioxide

$NH_4NO_3(s)$ + aq → $NH_4NO_3(aq)$
ammonium nitrate   aqueous ammonium nitrate

Entropy or randomness

Many processes which take place spontaneously involve mixing or spreading out, for example, liquids evaporating, solids dissolving to form solutions, or gases mixing:

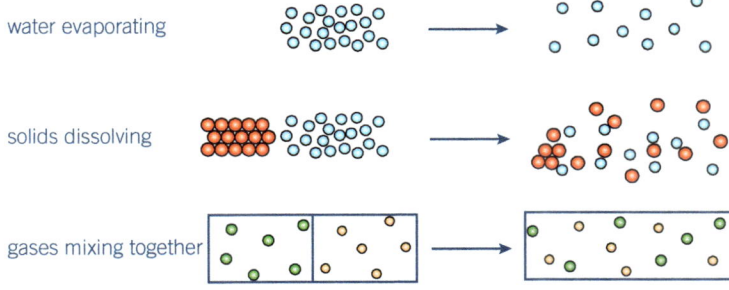

water evaporating

solids dissolving

gases mixing together

Figure 6 *Spontaneous processes*

This is the clue to the second factor which drives chemical processes—a tendency towards randomising or disordering, that is, towards chaos. Gases are more random than liquids, and liquids are more random than solids, because of the arrangement of their particles.

So endothermic reactions may be spontaneous if they involve spreading out, randomising, or disordering. This is true of the two reactions above—the arrangement of the particles in the products is more random than in the reactants.

The randomness of a system, expressed mathematically, is called the **entropy** of the system and is given the symbol S. The entropy is related to the number of possible arrangements of the particles and their energy in a given system. A reaction like the two above, in which the products are more disordered than the reactants, will have a positive value for the entropy change $ΔS$. The units of entropy are $J K^{-1} mol^{-1}$.

Entropies have been determined for a vast range of substances and can be looked up in databases. They are usually quoted for standard conditions: 298 K and 100 kPa pressure. Table 5 on the following page gives some examples.

Table 5 *Some values of entropy*

Substance	State at standard conditions	Entropy S / J K^{-1} mol^{-1}
carbon (diamond)	solid	2.4
carbon (graphite)	solid	5.7
copper	solid	33.0
iron	solid	27.0
ammonium chloride	solid	95.0
calcium carbonate	solid	93.0
calcium oxide	solid	40.0
iron(III) oxide	solid	88.0
water (ice)	solid	48.0
water (liquid)	liquid	70.0
mercury	liquid	76.0
water (steam)	gas	189.0
hydrogen chloride	gas	187.0
ammonia	gas	192.0
carbon dioxide	gas	214.0

Exam tip

A positive entropy change implies an increase in disorder of the system. A negative entropy change implies an increase in order.

In general, gases have larger values than liquids, which have larger values than solids.

Table 5 shows that the entropy increases when water turns to steam. Entropies increase with temperature, partly because at higher temperatures particles spread out and randomness increases, and partly because there are more quanta of energy to be distributed between the particles.

Estimating entropy changes

We can get a feel for the sign of entropy changes by thinking about the randomness of the particles in a system before and after the changes.

- Entropy increases during melting, because the particles in a solid are in clearly defined positions but in a liquid they are free to move.
- Entropy increases when a solid dissolves in a liquid for the same reason as above.
- Entropy increases on boiling, because in a gas the molecules have much more freedom of movement than in a liquid.
- Entropy increases as we increase the temperature of a system for two reasons—the particles are moving faster and there are more quanta of energy distributed through the system.
- Entropy increases in a chemical reaction where there is an increase in the number of gas molecules (or a solid or liquid turns into a gas), because there are far more ways of arranging the molecules in a gas than in a solid or liquid.

Calculating entropy changes

The entropy change for a reaction can be calculated by adding all the entropies of the products and subtracting the sum of the entropies of the reactants.

Expressed mathematically this is:

$$\Delta S = \Sigma S\ (\text{products}) - \Sigma S(\text{reactants})$$

For example:

$$CaCO_3(s) \rightarrow CaO(s) + CO_2(g)$$

Using the values from Table 5:

entropy of products = 40 + 214 = 254 $J K^{-1} mol^{-1}$

entropy of reactant = 93 $J K^{-1} mol^{-1}$

ΔS = 254 − 93 = +161 $J K^{-1} mol^{-1}$

This is a large positive value—a gas is formed from a solid.

Exam tip

The symbol 'Σ' means 'the sum of'.

Summary test 23.3

1 a Without doing a calculation, predict whether the entropy change for the following reactions will be significantly positive, significantly negative, or approximately zero and explain your reasoning.

i $Mg(s) + ZnO(s) \rightarrow MgO(s) + Zn(s)$
ii $2Pb(NO_3)_2(s) \rightarrow 2PbO(s) + 4NO_2(g) + O_2(g)$
iii $MgO(s) + CO_2(g) \rightarrow MgCO_3(s)$
iv $H_2O(l) \rightarrow H_2O(g)$

b Calculate ΔS for each reaction from part **a** using data from the table below. Explain your answers.

Substance	**S / J K^{-1} mol^{-1}**
Mg(s)	33.0
MgO(s)	27.0
$MgCO_3(s)$	66.0
Zn(s)	42.0
ZnO(s)	44.0
$Pb(NO_3)_2(s)$	213.0
PbO(s)	69.0
$NO_2(g)$	240.0
$O_2(g)$	205.0
$CO_2(g)$	214.0
$H_2O(l)$	70.0
$H_2O(g)$	189.0

Learning outcomes

On these pages you will learn to:

- use the Gibbs equation
- determine the feasibility of a reaction
- predict the effect of temperature change on the feasibility of a reaction

You have seen above that a combination of two factors determines the feasibility of a chemical reaction:

- the enthalpy change
- the entropy change.

These two factors are combined in a quantity called the **Gibbs free energy** G. If the change in G, ΔG, for a reaction is negative, then this reaction is feasible. If ΔG is positive, the reaction is not feasible.

ΔG combines the enthalpy change ΔH and entropy change ΔS as follows:

$$\Delta G = \Delta H - T\Delta S$$

ΔG depends on temperature, because of the term $T\Delta S$. This means that some reactions may be feasible at one temperature and not at another. So an endothermic reaction can become feasible when temperature is increased if there is a large enough positive entropy change. (A positive value for ΔS will make ΔG more negative because of the negative sign in the $T\Delta S$ term.)

Here are some examples of how this works.

Take the reaction that we used in Section 23.3:

$$CaCO_3(s) \rightarrow CaO(s) + CO_2(g) \quad \Delta H = +178\,kJ\,mol^{-1}$$

You have seen that $\Delta S = +161\,J\,K^{-1}\,mol^{-1} = 0.161\,kJ\,K^{-1}\,mol^{-1}$

So at room temperature (298 K):

$$\Delta G = \Delta H - T\Delta S$$

$$\Delta G = 178 - (298 \times 0.161) = +130\,kJ\,mol^{-1}$$

This positive value means that the reaction is not feasible at room temperature. The reverse reaction will have $\Delta G = -130\,kJ\,mol^{-1}$ and will be feasible:

$$CaO(s) + CO_2(g) \rightarrow CaCO_3(s)$$

This is the reaction that occurs in desiccators to absorb carbon dioxide.

However, if you do the calculation for a temperature of 1500 K, you get a different result:

At 1500 K:

$$\Delta G = \Delta H - T\Delta S$$
$$\Delta G = -178 - (1500 \times 0.161)$$
$$\Delta G = -178 - 242$$
$$\Delta G = -64\,kJ\,mol^{-1}$$

ΔG is negative and the reaction is feasible at this temperature. This is the reaction that occurs in a lime kiln to make lime (calcium oxide) from limestone (calcium carbonate).

What happens when ΔG = 0?

There is a temperature at which $\Delta G = 0$ for this reaction. This is the point at which the reaction is just feasible. You can calculate this temperature for the reaction above:

$$\Delta G = \Delta H - T\Delta S$$
$$0 = \Delta H - T\Delta S$$
$$\Delta H = T\Delta S \text{ where } \Delta H = +178\,kJ\,mol^{-1} \text{ and } \Delta S = 0.161\,kJ\,K^{-1}$$

So $T = \dfrac{178}{0.161} = 1105.6\,K$

Exam tip

Remember to convert the entropy units by dividing by 1000, because enthalpy is measured in $kJ\,mol^{-1}$ and entropy in $J\,K^{-1}\,mol^{-1}$.

Figure 7 *A traditional lime kiln, where limestone would have been burnt to make lime*

In fact, the reaction does not suddenly flip from feasible to non-feasible. In a closed system an equilibrium exists around this temperature in which both products and reactants are present.

Calculating an entropy change

You can use the temperature at which $\Delta G = 0$ to calculate an entropy change. For example, a solid at its melting point is equally likely to exist as a solid or a liquid—an equilibrium exists between solid and liquid. So ΔG for the melting process must be zero and:

$$0 = \Delta H - T\Delta S$$

For example, the melting point for water is 273 K and the enthalpy change for melting is $6.0 \, \text{kJ} \, \text{mol}^{-1}$. Putting these values into the equation:

$$0 = 6.0 - 273 \times \Delta S$$

$$\Delta S = \frac{6.0}{273} = 0.022 \, \text{kJ} \, \text{K}^{-1} \, \text{mol}^{-1} = +22 \, \text{J} \, \text{K}^{-1} \, \text{mol}^{-1}$$

This is the entropy change that occurs when ice changes to water. It is positive, which you would expect as the molecules in water are more disordered than those in ice.

Extension

Measuring an entropy change

Surprisingly, you can measure an entropy change using kitchen equipment.

A chemistry teacher set out to find the entropy change for the vaporisation of water at home using the household kettle and a top pan balance.

At its boiling point, water is equally likely to exist as liquid or vapour (water or steam), so for vaporisation, $\Delta G = 0$.

Inserting the ΔG value into $\Delta G = \Delta H - T\Delta S$ gives:

$$0 = \Delta H - T\Delta S$$

$$\text{So } \Delta H = T\Delta S$$

Rearrange to $\Delta S = \frac{\Delta H}{T}$

The boiling point of water (at atmospheric pressure) is 100 °C (373 K), $T = 373$ K so all she needed to measure was ΔH.

The kettle had a power rating of 2.4 kW, which means it supplies 2.4 kJ of energy per second.

She brought some water to the boil in an ordinary kitchen kettle, and weighed the kettle and its contents on a top pan balance that read to the nearest gram. She switched on the kettle again and allowed it to boil for 100 seconds holding down the automatic switch. She then reweighed the kettle to find how much water had boiled away. She found that 100 g of water had boiled away, that is, turned from water to steam (vaporised).

Calculate ΔS_{vap} by the following steps:

1 Calculate how many kilojoules of heat were supplied to the water in 100 s.
2 Calculate the value of M_r for water and hence find how many moles of water were vaporised.
3 Calculate ΔH_{vap} for the process in $\text{kJ} \, \text{mol}^{-1}$ and convert this into $\text{J} \, \text{mol}^{-1}$.
4 Use $\Delta S_{vap} = \frac{\Delta H_{vap}}{T}$ to calculate the entropy change of vaporisation.
5 To how many significant figures can you quote your answer?
6 What systematic error (experimental design error) is there in this experiment? How could you reduce it?
7 If the top pan balance weighs to the nearest gram, what is the percentage error in the weighing?
8 The value of ΔS_{vap} for water is higher than for most liquids. Suggest why. Hint—you are measuring the increase in disorder between the liquid and vapour phases. Think about what causes order in the liquid state of water.

Extension

Extracting metals

A good way of extracting metals from their oxide ores is to heat them with carbon. This removes the oxygen as carbon dioxide and leaves the metal. It has the advantage that carbon (in the form of coke, an impure type of carbon formed by heating coal in the absence of air) is cheap. The gaseous carbon dioxide simply diffuses away, so there is no problem separating it from the metal (although it does contribute to global warming as it is a greenhouse gas). You can use ΔG to investigate under what conditions the reaction might be feasible for different metals.

One of the most important metals is iron and its ore is largely iron(III) oxide, Fe_2O_3. You can calculate ΔG from a thermochemical cycle.

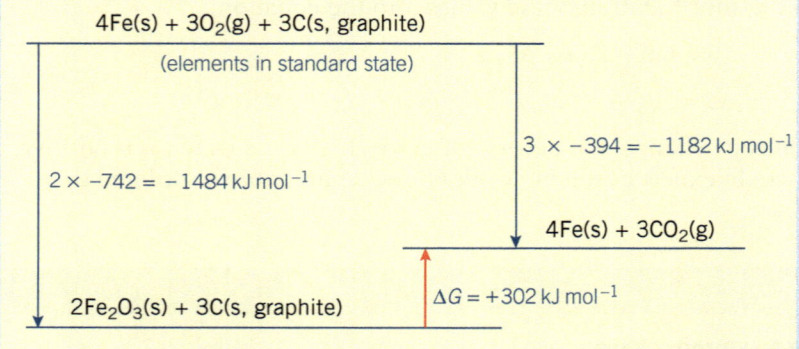

Figure 7 *Free energy diagram for the reduction of iron(III) oxide by graphite*

$$2Fe_2O_3(s) + 3C(s, graphite) \rightarrow 4Fe(s) + 3CO_2(g) \quad \Delta G = +302 \, kJ \, mol^{-1}$$

ΔG is positive, so this reaction is not feasible under standard conditions (298 K).

Will the reaction take place at a higher temperature?

You can work out the temperature at which the reaction just becomes feasible (this is when $\Delta G = 0$) using:

$$\Delta G = \Delta H - T\Delta S$$

Calculating ΔH

ΔH for the reaction can be calculated from the following thermochemical cycle.

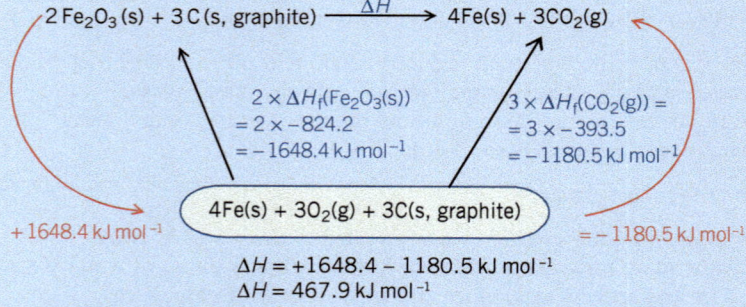

$$\Delta H = +1648.4 - 1180.5 \, kJ \, mol^{-1}$$
$$\Delta H = 467.9 \, kJ \, mol^{-1}$$

Calculating ΔS

You can calculate the entropy change of the reaction by finding the difference between the sum of the entropies of all the products and the sum of the entropies of all the reactants:

$2Fe_2O_3(s)$	+	$3C(s, graphite)$	$\rightarrow$	$4Fe(s)$	+	$3CO_2(g)$
(2×87.4)	+	(3×5.7)		(4×27.3)	+	(3×213.6)
191.9				750.0		

$$\Delta S = +558.1 \, J \, K^{-1} \, mol^{-1}$$

This value is large and positive, as you would expect from a reaction in which two solids produce a gas.

Putting these values into $\Delta G = \Delta H - T\Delta S$:

$$0 = +467.9 - T \times \frac{55.81}{1000}$$
$$T = 467.9 \times \frac{1000}{558.1}$$
$$= 838.4 \text{ K}$$

The reaction is not feasible below this temperature.

In fact, in the blast furnace a higher temperature is used, above the melting point of iron (1808 K), so that the iron is formed as a liquid. Also, the carbon is not pure graphite, but coke.

1 Calculate ΔG at 2000 K using the values above.

Answer: -648.3 kJ mol^{-1}

Kinetic factors

Gibbs Free Energy changes tell us nothing about how quickly or slowly a reaction is likely to go. You might predict that a certain reaction should occur spontaneously because of a negative value of ΔG, but the reaction might take place so slowly that for practical purposes it does not occur at all. In other words, there is a large activation energy barrier for the reaction.

Carbon in the form of graphite gives an interesting example:

$$C(s, \text{graphite}) + O_2(g) \rightarrow CO_2(g) \quad \Delta H = -393.5 \text{ kJ mol}^{-1}$$

The reaction is exothermic and you can calculate the actual value of ΔS and so find ΔG.

Calculating ΔS
ΔS for the reaction is the sum of the entropies of the product minus the sum of the entropies of the reactants.

C(s, graphite)	+	O$_2$(g)	$\rightarrow$	CO$_2$(g)	$\Delta H = -394$ kJ mol^{-1}
5.7		205.0		213.6	

So $\Delta S = 213.6 - (5.7 + 205.0)$

$\Delta S = +2.9$ J K^{-1} mol^{-1}, positive as predicted

Calculating ΔG
$\Delta G = \Delta H - T\Delta S$

So under standard conditions (approximately room temperature and pressure):

$$\Delta G = -394 - \frac{298 \times 2.9}{1000}$$

Remember to divide the entropy value by 1000 to convert from J K^{-1} mol^{-1} to kJ K^{-1} mol^{-1}.

$\Delta G = -394 - 0.86$

$\Delta G = -394.86$ kJ mol^{-1}, negative so the reaction is feasible.

However, experience with graphite (the 'lead' in pencils) tells you that the reaction does not take place at room temperature—although it will take place at higher temperatures. At room temperature, the reaction is so slow that in practice it doesn't take place at all.

Since the branch of chemistry dealing with enthalpy and entropy changes is called **thermodynamics**, and that dealing with rates is called **kinetics**, graphite is said to be thermodynamically unstable but kinetically stable.

1 For the reaction:
$$MgO(s) \rightarrow Mg(s) + O_2(g)$$
$$\Delta H = +602 \text{ kJ mol}^{-1}$$
$$\Delta S = +109 \text{ J K}^{-1} \text{mol}^{-1}$$
a Using the equation $\Delta G = \Delta H - T\Delta S$, calculate ΔG at:
i 1000 K
ii 6000 K
iii At which of these temperatures is the reaction feasible?
b Calculate the temperature when $\Delta G = 0$.
2 Calculate the entropy change for:

$$NH_3(g) + HCl(g) \rightarrow NH_4Cl(s)$$

The entropy values are:
$S(NH_3) = 192$ J K^{-1} mol^{-1}
$S(HCl) = 187$ J K^{-1} mol^{-1}
$S(NH_4Cl) = 95$ J K^{-1} mol^{-1}

Launch additional digital resources for the chapter

1 Explain the following terms with reference to the hypothetical substance X^+Y^-.

   a i Lattice energy, ΔH_{latt}. *(1 mark)*
     ii Enthalpy change of hydration, ΔH_{hyd}. *(1 mark)*
     iii Enthalpy change of solution, ΔH_{soln}. *(1 mark)*

   b Construct an energy cycle linking the terms given in part **a**. *(2 marks)*

   c Calculate the enthalpy change of hydration, ΔH_{hyd} of potassium iodide, assuming that its enthalpy change of solution, ΔH_{soln} is +21 kJ mol^{-1} and its lattice energy ΔH_{latt} is –642 kJ mol^{-1}.

2 **a** Define the following terms:
     i ionisation energy
     ii atomisation energy
     iii first electron affinity. *(3 marks)*

   b Construct a fully labelled Born–Haber cycle for the formation of potassium bromide. *(3 marks)*

   c Using the information in the table below, calculate the lattice energy of potassium bromide. *(2 marks)*

Reaction	ΔH / kJ mol^{-1}
K(s) + ½Br$_2$(l) → K$^+$Br$^-$(s)	–392
K(s) → K(g)	+90
K(s) → K$^+$(g) + e$^-$	+420
½Br$_2$(l) → Br(g)	+112
Br(g) + e$^-$ → Br$^-$(g)	–342

   d Use the table below to explain the trend in the lattice energy for the given potassium halides. *(3 marks)*

Compound	KF	KCl	KI
Lattice energy/kJ mol^{-1}	–392	–710	–643

3 Consider the entropy changes involved in the burning of magnesium:

$$2Mg(s) + O_2(g) \rightarrow 2MgO(s); \quad \Delta H^\ominus = -1204 \text{ kJ mol}^{-1}$$

$$S^\ominus(Mg) = 32.7 \text{ J K}^{-1} \text{mol}^{-1}$$

$$S^\ominus(O_2) = 204.9 \text{ J K}^{-1} \text{mol}^{-1}$$

$$S^\ominus(MgO) = 26.8 \text{ J K}^{-1} \text{mol}^{-1}$$

   a Calculate the standard entropy change, ΔS, for the reaction as written in the equation above. *(3 marks)*

   b Use your answer to part **a** to deduce whether this is a favourable entropy change. Give a reason for your answer. *(2 marks)*

4 The diagram below shows the synthesis of calcium oxide. The numerical values given are in kJ mol^{-1}.

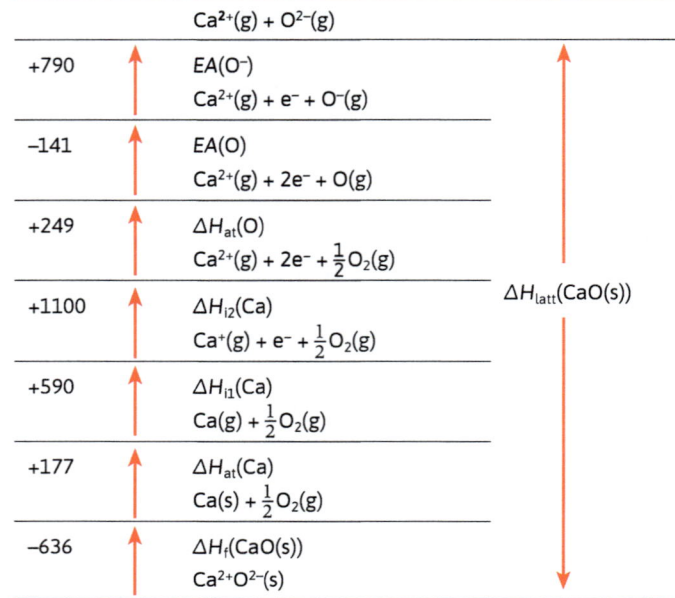

   a Calculate the lattice energy, ΔH_{latt}, for calcium oxide, CaO(s). *(2 marks)*

   b Explain why the electron affinity of (O) is negative whereas the electron affinity of (O$^-$) is positive. *(2 marks)*

   c Compare the first ionisation energies of magnesium with calcium. Give reasons for any differences. *(3 marks)*

5 **a** Define the term 'entropy'. *(1 mark)*

   b Describe what happens to the stability of a system as its entropy increases. *(1 mark)* c The table below gives standard entropy values for the halogens:

Substance	Standard entropy at 298 K / J K⁻¹ mol⁻¹		
	in gas state	in liquid state	in solid state
fluorine	203		
chlorine	223		
bromine	245	152	
iodine	261		117

i Explain the difference in entropy values for bromine in the liquid and gas states. *(2 marks)*

ii Suggest why the difference in entropy between the two states given for bromine is less than the difference in the entropy between the two states given for iodine. *(2 marks)*

iii Describe the trend in entropy values for the halogens in the gas state. Give reasons for your answer. *(2 marks)*

d Consider the sublimation of iodine:

i Use the data from the table above to calculate the standard entropy change of sublimation of iodine. *(2 marks)*

ii The enthalpy of formation of $I_2(s) = 0$ and the enthalpy of formation of $I_2(g) = 62.2\,kJ\,mol^{-1}$. Use this data, and your answer to **di**, to calculate ΔG for the sublimation of iodine at 298 K. *(3 marks)*

iii State whether the sublimation of iodine is spontaneous at room temperature. Explain your answer. *(2 marks)*

6 State whether for each of the following changes, the entropy change of the substances will be positive, negative, or approximately zero.

a $Cu(s) \rightarrow Cu(l)$

b $I_2(g) \rightarrow I_2(s)$

c Heating water from 20°C to 50°C.

d $NH_4Cl(s) \rightarrow NH_3(g) + HCl(g)$

e $CH_4(g) + 2O_2(g) \rightarrow CO_2(g) + 2H_2O(g)$

f $C_3H_8(g) + 5O_2(g) \rightarrow 3CO_2(g) + 4H_2O(g)$

g $NaCl(s) \xrightarrow{aq} Na^+(aq) + Cl^-(aq)$

h $H^+(aq) + OH^-(aq) \rightarrow H_2O(l)$

7 a Consider the thermal decomposition of calcium carbonate:

$$CaCO_3(s) \rightarrow CaO(s) + CO_2(g);\ \Delta H^\ominus = +178\,kJ\,mol^{-1}$$
$$\Delta S^\ominus = +165\,J\,K^{-1}\,mol^{-1}$$

i Calculate $\Delta G^\ominus$ for this reaction at 298 K. State whether this reaction will occur spontaneously. *(4 marks)*

ii Calculate the temperature to which calcium carbonate must be heated to make it decompose. Give your answer in °C. *(4 marks)*

b Consider the thermal decomposition of magnesium carbonate:

$$MgCO_3(s) \rightarrow MgO(s) + CO_2(g);\ \Delta H^\ominus = +100.3\,kJ\,mol^{-1}$$
$$\Delta S^\ominus = +174.8\,J\,K^{-1}\,mol^{-1}$$

i Calculate $\Delta G^\ominus$ for this reaction at 298 K. State whether this reaction will occur spontaneously. *(4 marks)*

ii Calculate the temperature to which magnesium carbonate must be heated to make it decompose. Give your answer in °C. *(4 marks)*

c Use your answers to parts **a** and **b** to compare the thermal stability of magnesium carbonate with that of calcium carbonate. Explain your answer. *(3 marks)*

24 Electrochemistry

24.1 Electrolysis

Learning outcomes

On these pages you will learn to:

- describe and explain electrolysis
- use the relationship $F = Le$
- calculate the mass of a substance liberated during electrolysis

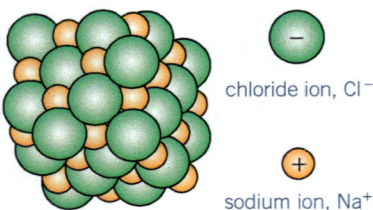

chloride ion, Cl$^-$

sodium ion, Na$^+$

Figure 1 *The sodium chloride structure*

Exam tip

To remember which electrode is negative, 'copper goes to the cathode' helps.

Electrolysis is the process by which ionic compounds are decomposed by passing electricity through them. Ionic compounds conduct electricity only when liquid—either molten or dissolved in water—so that the ions are free to move. In the solid state, the ions are held in place in a lattice arrangement, Figure 1.

A schematic diagram of an experimental setup for electrolysing molten sodium chloride is shown in Figure 2.

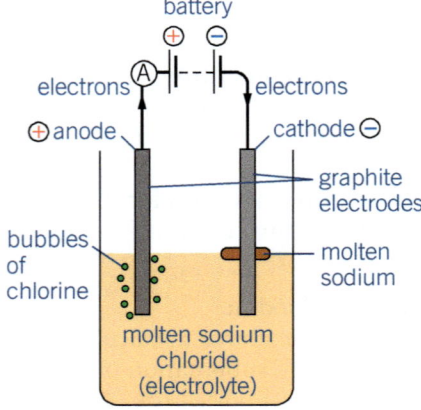

Figure 2 *Electrolysis of molten sodium chloride*

The **electrodes** are the terminals by which the current enters and leaves the ionic liquid which is called the **electrolyte**. The **cathode** is the negative electrode and the **anode** the positive.

Chemical reactions take place at the electrodes.

In the above example these are:

At the cathode: $Na^+(l) + e^- \rightarrow Na(l)$

At the anode: $Cl^-(l) \rightarrow \frac{1}{2}Cl_2(g) + e^-$

At the cathode, electron(s) are gained by positive ions, so this is a reduction reaction (the oxidation number of the sodium has decreased from +1 to 0).

At the anode, electron(s) are lost by negative ions, so this is an oxidation reaction (the oxidation number of the chlorine has increased from −1 to 0).

The overall process can be found by adding the two half equations above and cancelling the electrons:

$Na^+(l) + e^- \rightarrow Na(l)$

$Cl^-(l) \rightarrow \frac{1}{2}Cl_2(g) + e^-$

$Na^+(l) + \cancel{e^-} + Cl^-(l) \rightarrow Na(l) + \frac{1}{2}Cl_2(g) + \cancel{e^-}$

or $NaCl(l) \rightarrow Na(l) + \frac{1}{2}Cl_2(g)$

The Faraday constant

We can see that to produce 1 mole of sodium atoms requires 1 mole of electrons. This number is the Avogadro constant, L, (6.02×10^{23}) of electrons and each electron has a charge e, $1.60 \times 10^{-19}\,C$.

So the amount of electric charge to produce one mole of sodium atoms is therefore 1.60×10^{-19} C $\times 6.02 \times 10^{23}$ mol^{-1} = 96 320 C mol^{-1}.

This number is called the **Faraday constant**, *F*, after Michael Faraday, who developed many of the theories around electrolysis. We usually round this up to 96 500 C mol^{-1}. We can also refer to a quantity of electric charge of 96 500 C as 1 faraday.

The number of coulombs in 1 faraday = $L \times e$

1 faraday will produce 1 mole of atoms of product from singly charged ions. 2 faradays are needed to produce 1 mole of product atoms from doubly charged ions and so on.

Calculating the Avogadro constant *L* (the number of particles in a mole of substance)

This equation allows us to measure a value for the Avogadro constant. If we electrolyse a liquid with a known current (*I* amps), for a known time (*t* seconds), the amount of charge (*Q* coulombs) passed is given by $Q = It$.

So $It = Le$ and $L = \dfrac{It}{e}$

Figure 3 *Michael Faraday, whom the Faraday constant and faraday unit of measurement were named after*

Worked example

Using the electrolysis of silver nitrate to calculate the Avogadro constant

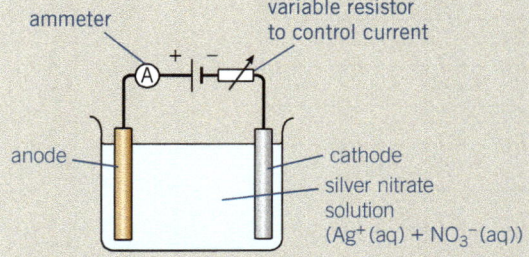

Figure 4 *Apparatus for the electrolysis of silver nitrate solution*

Using the apparatus in Figure 4, a current of 0.1 A (0.1 Cs^{-1}) was passed for 30 minutes (30 × 60 = 1800 s) and the cathode was found to have increased in mass by 0.200 g.

$$A_r(\text{Ag}) = 107.9$$

The reaction at the cathode is:

$$\text{Ag}^+(\text{aq}) + \text{e}^- \rightarrow \text{Ag}(\text{s})$$

The quantity of electricity passed is, $Q = It = 0.1$ C s^{-1} × 1800 s = 180 C

The charge on one electron is 1.60×10^{-19} C

So the number of electrons in 180 C is:

$$\frac{180}{1.60 \times 10^{-19}} = 1.125 \times 10^{21} \text{ electrons}$$

The number of moles of Ag atoms produced was:

$$\frac{0.200}{107.9} = 1.85 \times 10^{-3}$$

Since 1 Ag$^+$ ion requires 1 electron, and 1.85×10^{-3} moles required 1.125×10^{21} electrons

$$1 \text{ mol} = \frac{1.125 \times 10^{21}}{1.85 \times 10^{-3}} = 6.08 \times 10^{23}$$

This is the value of the Avogadro constant obtained in this experiment—it is close to the accepted value of 6.02×10^{23}.

Exam tip

- Remember to convert the time into seconds.
- 1 Amp = 1 Coulomb per second

Figure 5 *A statue of Humphry Davy. Davy discovered the elements potassium, sodium, calcium, strontium, barium, magnesium, and boron using electrolysis*

What are the products of electrolysis?

If we electrolyse a pure molten electrolyte, there can only be one cation discharged at the cathode, and one anion discharged at the anode. In the case of an aqueous solution of an ionic compound there will also be H^+ and OH^- ions present in the electrolyte. These come from the water which is partially dissociated.

$$H_2O(l) \rightleftharpoons H^+(aq) + OH^-(aq)$$

Which ion is discharged can be predicted from the **electrochemical series**, Table 1. This is a list of ions in order of the voltage they produce in an electrochemical cell (it is covered in more detail in Section 24.2).

Table 1 *The electrochemical series*

Reduction half equation	$E^{\ominus}/V$
$Li^+(aq) + e^- \rightarrow Li(s)$	−3.04
$Ca^{2+}(aq) + 2e^- \rightarrow Ca(s)$	−2.87
$Al^{3+}(aq) + 3e^- \rightarrow Al(s)$	−1.66
$Zn^{2+}(aq) + 2e^- \rightarrow Zn(s)$	−0.76
$Cr^{3+}(aq) + e^- \rightarrow Cr^{2+}(aq)$	−0.41
$Pb^{2+}(aq) + 2e^- \rightarrow Pb(s)$	−0.13
$2H^+(aq) + 2e^- \rightarrow H_2(g)$	0.00
$Cu^{2+}(aq) + e^- \rightarrow Cu^+(aq)$	+0.15
$Cu^{2+}(aq) + 2e^- \rightarrow Cu(s)$	+0.34
$I_2(aq) + 2e^- \rightarrow 2I^-(aq)$	+0.54
$Fe^{3+}(aq) + e^- \rightarrow Fe^{2+}(aq)$	+0.77
$Ag^+(aq) + e^- \rightarrow Ag(s)$	+0.80
$Br_2(aq) + 2e^- \rightarrow 2Br^-(aq)$	+1.07
$Cl_2(aq) + 2e^- \rightarrow 2Cl^-(aq)$	+1.36
$MnO_4^- + 8H^+(aq) + 5e^- \rightarrow Mn^{2+}(aq) + 4H_2O(l)$	+1.52
$Ce^{4+}(aq) + e^- \rightarrow Ce^{3+}(aq)$	+1.70

Where there is more than one cation present in the electrolyte, positive ions lower down in the series will be discharged at the cathode in preference to ones higher up. So in a solution of silver nitrate, silver will be discharged in preference to hydrogen. And in a solution of zinc nitrate, the hydrogen will be discharged. (This rule may not apply if the cation higher in the series is present in a significantly higher concentration than the other cation.)

For anions, it is the ion *higher* in the series that is preferred, although the concentration effect also occurs.

Nitrate and sulfate ions are never discharged.

OH^- ions produce oxygen:

$$4OH^-(aq) - 4e^- \rightarrow O_2(g) + 2H_2O(l)$$

Electrodes can also have an effect. For example, the electrolysis of copper sulfate solution with copper electrodes leads to copper being deposited at the cathode, while the copper anode dissolves.

Exam tip

Remember that cations are positive and anions are negative.

The amount of substance discharged

This depends on the current used and the time for which it is passed. Multiplying these gives the number of coulombs used and hence the number of faradays. One faraday will discharge a mole of singly charged ions or half a mole of doubly charged ions etc.

Worked example

The amount of substance discharged

A current of 0.1 amp is passed for 1 hour through two molten electrolytes in separate containers, potassium chloride, KCl, and calcium chloride, $CaCl_2$.

How many moles of **a** potassium, **b** calcium will this discharge from each molten electrolyte?

The steps are:

1 How many coulombs of charge are passed?

$Q = It$

So $Q = 0.1 \times 60 \times 60 = 360\,C$

2 How many faradays is this?

1 faraday = 96 500 C

So

$360 \div 96\,500 = 3.7 \times 10^{-3}$ faradays

3 What mass of potassium is discharged from K^+ ions?

3.7×10^{-3} faradays will discharge 3.7×10^{-3} mol of singly charged ions.

1 mole of potassium has a mass of 39.1 g

So $39.1 \times 3.7 \times 10^{-3}$ g of potassium will be discharged.

= 0.145 g

4 What mass of calcium is discharged from Ca^{2+} ions?

3.7×10^{-3} faradays will discharge $\dfrac{3.7 \times 10^{-3}}{2}$ mol of doubly charged ions = 1.85×10^{-3} mol.

1 mol of calcium has a mass of 40.1 g

So $40.1 \times 1.85 \times 10^{-3}$ g of calcium will be discharged.

= 0.074 g

Exam tip

$Q = It$

and Q tells us the quantity of charge passed.

Summary test 24.1

1 In the worked example above state how many moles of chlorine gas, Cl_2, will be produced in each case?

2 A mixture of zinc sulfate and copper sulfate is electrolysed in aqueous solution using carbon electrodes. Predict the products would you get at the anode and cathode. Explain your answer.

3 A solution containing nickel ions was electrolysed with a current of 0.1 amp for 160 minutes. 0.295 g of nickel was deposited at the cathode. Deduce the charge on the nickel ion.

Standard electrode potentials, $E^\ominus$, standard cell potentials, $E^\ominus_{cell}$, and the Nernst equation

Learning outcomes

On these pages you will learn to:

- explain how standard electrode potential is measured
- calculate standard cell potentials and predict the feasibility of a reaction
- explain the direction of electron flow in a simple cell
- calculate how electrode potential varies with ion concentration using the Nernst equation
- describe the relationship between Gibbs free energy and standard cell potential

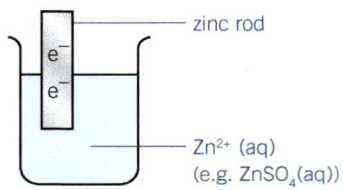

Figure 7 *A zinc electrode*

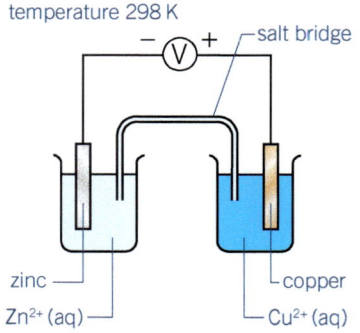

Figure 8 *Two electrodes (here copper and zinc) connected together with a voltmeter to measure the potential difference*

If you place two different metals in a salt solution and connect them together (Figure 6) an electric current flows so that electrons pass from the more reactive metal to the less reactive. This is the basis of batteries that power everything from mobile phones to electric cars.

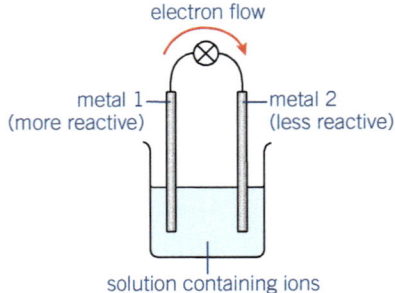

Figure 6 *Flow of electric current*

This section looks at how electricity is produced by electrochemical cells and how this can be used to explain and predict redox reactions (which are all about electron transfer).

Half-cells

When a rod of metal is dipped into a solution of its own ions, an equilibrium is set up.

For example, dipping zinc into zinc sulfate solution sets up the following equilibrium:

$$Zn(s) \rightleftharpoons Zn^{2+}(aq) + 2e^-$$

This arrangement is called an electrode, or a **half-cell**, because two half-cells can be joined together to make an electrical cell (Figure 7).

If you could measure the electrical potential of this cell, it would tell us how readily electrons are released by the metal, that is, how good a reducing agent the metal is. (Remember that reducing agents release electrons.)

However, electrical potential cannot be measured directly, only **potential difference** (often called **voltage**). What you can do is to connect together two different electrodes and measure the potential difference between them with a voltmeter (Figure 8).

The electrical circuit is completed by a **salt bridge**, the simplest form of which is a piece of filter paper soaked in a solution of a salt (usually saturated potassium nitrate). A salt bridge is used rather than a piece of wire, to avoid further metal/ion potentials in the circuit.

If you connect the two electrodes in Figure 8 to the voltmeter (Figure 9) you get a potential difference (voltage) of 1.10 V (if the solutions are 1.00 mol dm^{-3} and the temperature 298 K). The voltmeter shows that the zinc electrode is the more negative.

The fact that the zinc electrode is negative tells you that zinc loses its electrons more readily than does copper—zinc is a better reducing agent. If we remove the voltmeter and electrons are allowed to flow, they will do so from zinc to copper. The following changes will take place:

1 Zinc dissolves to form $Zn^{2+}(aq)$, increasing the concentration of $Zn^{2+}(aq)$.
2 The electrons flow through the wire to the copper rod where they combine with $Cu^{2+}(aq)$ ions (from the copper sulfate solution), depositing fresh copper on the rod and decreasing the concentration of $Cu^{2+}(aq)$.

The following two **half reactions** take place:

$$Zn(s) \rightarrow Zn^{2+}(aq) + 2e^-$$

and $Cu^{2+}(aq) + 2e^- \rightarrow Cu(s)$

adding the two equations: $\quad Zn(s) + Cu^{2+}(aq) + 2\cancel{e^-} \rightarrow Zn^{2+}(aq) + Cu(s) + 2\cancel{e^-}$

When the two half reactions are added together, the electrons cancel out and you get the overall reaction:

$$Zn(s) + Cu^{2+}(aq) \rightarrow Zn^{2+}(aq) + Cu(s)$$

This is the reaction you get on putting zinc directly into a solution of copper ions. It is a redox reaction with zinc being oxidised and copper ions reduced. If the two half-cells are connected they generate electricity. This forms an electrical cell called the Daniell cell (Figure 10).

The hydrogen electrode

To compare the tendency of different metals to release electrons, a standard electrode is needed to which any other half-cell can be connected for comparison. The half-cell chosen is called the **standard hydrogen electrode** (Figure 11).

Hydrogen gas is bubbled into a solution of $H^+(aq)$ ions. Since hydrogen doesn't conduct, electrical contact is made via a piece of unreactive platinum metal (coated with finely divided platinum to increase the surface area and allow any reaction to proceed rapidly). The electrode is used under standard conditions of $[H^+(aq)] = 1.00 \, mol \, dm^{-3}$, pressure 100 kPa, and temperature 298 K.

The potential of the standard hydrogen electrode is *defined* as zero. If it is connected to another electrode (Figure 12), the measured voltage, called the electromotive force E (emf), is the **standard electrode potential** for that half-cell. It is also called the standard reduction potential. If a second cell is at standard conditions ([metal ions] = $1.00 \, mol \, dm^{-3}$, temperature = 298 K), then the emf is given the symbol $E^\ominus$. Electrodes with negative values of $E^\ominus$ are better at releasing electrons (better reducing agents) than hydrogen

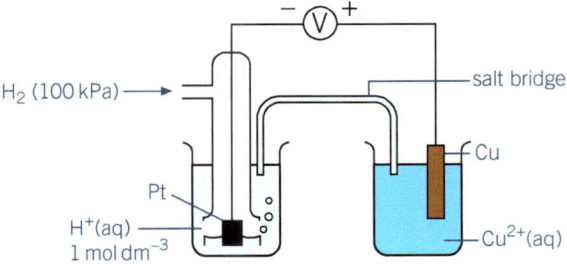

Figure 12 Measuring $E^\ominus$ for a copper electrode

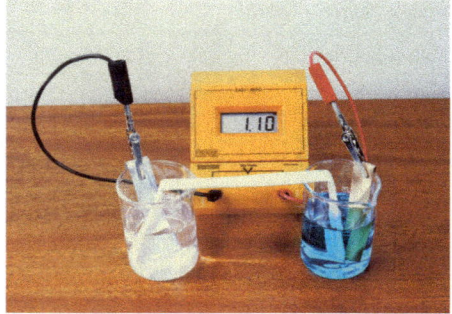

Figure 9 Measuring the potential difference of zinc and copper half-cells

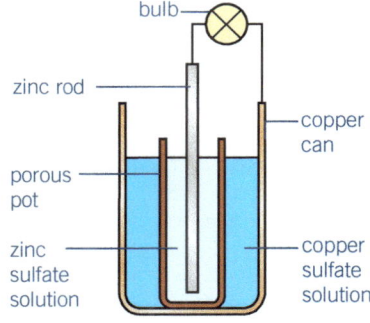

Figure 10 A Daniell cell lighting a bulb. The porous pot acts like a salt bridge

Exam tip

A perfect voltmeter does not allow any current to flow—it merely measures the electrical 'push' or pressure (the potential difference) which tends to make current flow.

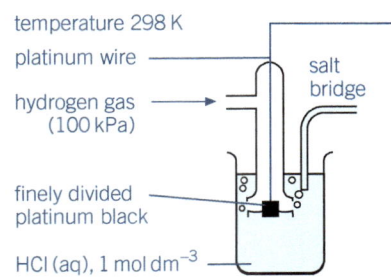

Figure 11 The standard hydrogen electrode

Exam tip

The salt chosen for the salt bridge must not react with either of the solutions in the half-cells.

Table 2 Some $E^{\ominus}$ values

Half reaction	$E^{\ominus}$/V
$Li^+(aq) + e^- \rightleftharpoons Li(s)$	−3.04
$Ca^{2+}(aq) + 2e^- \rightleftharpoons Ca(s)$	−2.87
$Al^{3+}(aq) + 3e^- \rightleftharpoons Al(s)$	−1.66
$Zn^{2+}(aq) + 2e^- \rightleftharpoons Zn(s)$	−0.76
$Pb^{2+}(aq) + 2e^- \rightleftharpoons Pb(s)$	−0.13
$2H^+(aq) + 2e^- \rightleftharpoons H_2(g)$	0.00
$Cu^{2+}(aq) + 2e^- \rightleftharpoons Cu(s)$	+0.34
$Ag^+(aq) + e^- \rightleftharpoons Ag(s)$	+0.80

> **Exam tip**
>
> The standard cell potential, $E^{\ominus}_{cell}$, is the emf (voltage) when two half-cells are connected together under standard conditions (298 K, 100 kPa and 1 mol dm^{-3} concentration of solutions).

> **Exam tip**
>
> $Cu^{2+}(aq)/Cu(s)$ is shorthand for $Cu^{2+}(aq) + 2e^- \rightleftharpoons Cu(s)$. The state symbols may be omitted.

> **Exam tip**
>
> It is worth sketching diagrams like the ones shown in Figure 13 and Figure 14. It will prevent you getting confused with + and − signs. Remember that more negative values are drawn to the left on the diagrams.

Changing the conditions of an electrode, such as the temperature or the concentration of ions, will change its electrical potential.

> **Exam tip**
>
> The standard electrode potential, $E^{\ominus}$, of a metal / metal ion half-cell is the emf (voltage) when that half-cell is connected to a hydrogen electrode under standard conditions (298 K, 100 kPa and 1 mol dm^{-3} concentration of solutions).

The electrochemical series

A list of some $E^{\ominus}$ values for metal/metal ion standard electrodes is given in Table 2.

In Table 2, the equilibria are written with the electrons on the left of the arrow. These are called reduction potentials. Arranged in this order with the most negative values at the top, this list is called the electrochemical series. The number of electrons involved in the reaction has no effect on the value of $E^{\ominus}$.

The voltage obtained by connecting two standard electrodes together is found by the difference between the two $E^{\ominus}$ values and is called the **standard cell potential**, $E^{\ominus}_{cell}$. So connecting an $Al^{3+}(aq)/Al(s)$ standard electrode to a $Cu^{2+}(aq)/Cu(s)$ standard electrode would give a voltage of 2.00 V (Figure 13) and the $Al^{3+}(aq)/Al(s)$ electrode will be negative.

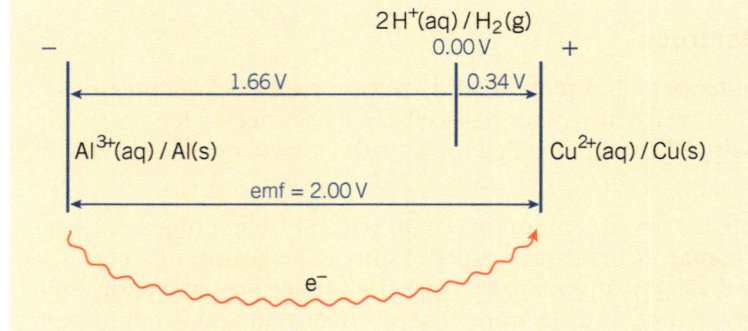

Figure 13 Calculating the value of the voltage when two electrodes are connected

If you connect an $Al^{3+}(aq)/Al(s)$ standard electrode to a $Pb^{2+}(aq)/Pb(s)$ standard electrode $E^{\ominus}_{cell}$ will be 1.53 V and the $Al^{3+}(aq)/Al(s)$ electrode will be the negative electrode of the cell (Figure 14).

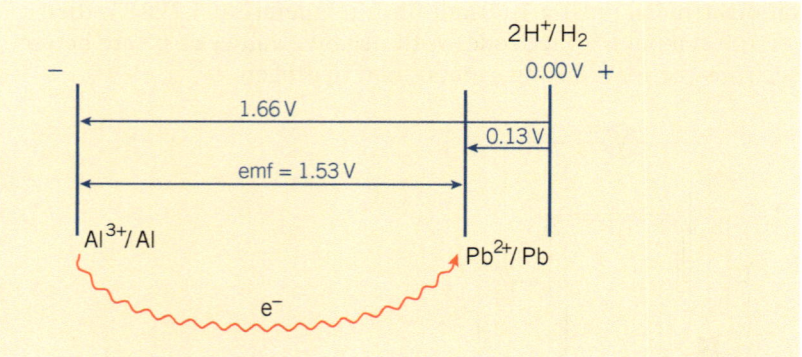

Figure 14 Calculating the value of $E^{\ominus}_{cell}$ for an Al^{3+}/Al electrode connected to a Pb^{2+}/Pb electrode connected

If you connect an $Al^{3+}(aq)/Al(s)$ standard electrode to a $Zn^{2+}(aq)/Zn(s)$ standard electrode as shown in Figure 15, the voltmeter will read 0.90 V. The $Al^{3+}(aq)/Al(s)$ electrode will be the negative electrode of the cell. To get a reading on the voltmeter, the voltmeter must be connected as shown with its negative terminal to the aluminium.

Predicting the direction of redox reactions

It is possible to use standard electrode potentials to decide on the feasibility of a redox (ie electron transfer) reaction. When you connect a pair of electrodes, the electrons will flow from the more negative to the more positive electrode and not in the opposite direction. So the signs of the electrodes tell us the direction of a redox reaction.

Think of the following electrodes (Figure 16):

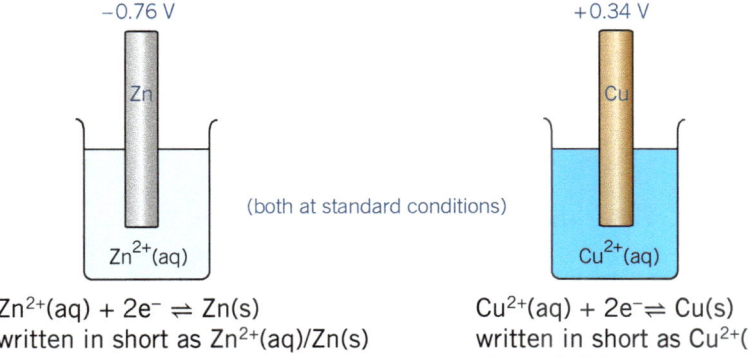

$Zn^{2+}(aq) + 2e^- \rightleftharpoons Zn(s)$
written in short as $Zn^{2+}(aq)/Zn(s)$
$E^\ominus = -0.76\,V$

$Cu^{2+}(aq) + 2e^- \rightleftharpoons Cu(s)$
written in short as $Cu^{2+}(aq)/Cu(s)$
$E^\ominus = +0.34\,V$

Figure 16 *The two electrodes. Their potentials are measured with respect to a standard hydrogen electrode*

Figure 17 shows these two electrodes connected together. Electrons will tend to flow from zinc (the more negative) to copper (the more positive).

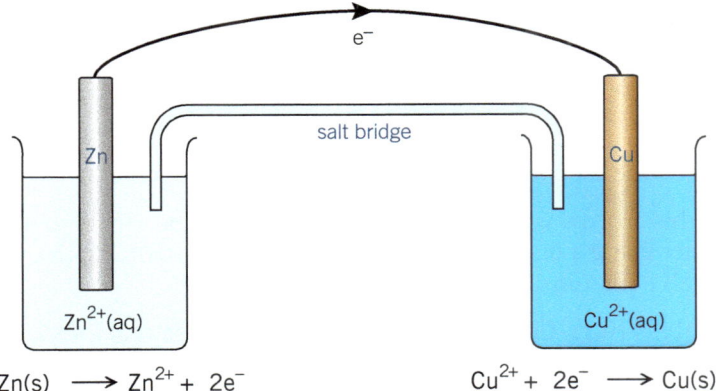

$Zn(s) \longrightarrow Zn^{2+} + 2e^-$

$Cu^{2+} + 2e^- \longrightarrow Cu(s)$

Figure 17 *Connecting $Zn^{2+}(aq)/Zn(s)$ and $Cu^{2+}(aq)/Cu(s)$ electrodes*

So you know which way the two half reactions must go. You can use a diagram to represent the cell. If you then include $E^\ominus$ values for the two electrodes you can find $E^\ominus_{cell}$. Figure 19 overleaf is the diagram for the zinc/copper cell. You can see

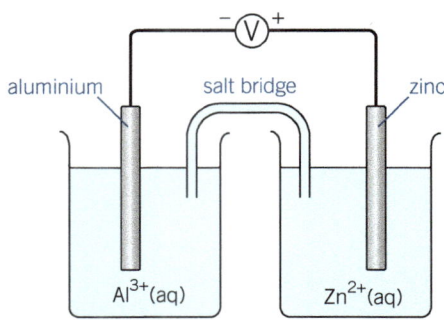

Figure 15 *Connecting a pair of electrodes*

Figure 18 *All of these batteries work using the principles described in this chapter*

how it is related to the apparatus.

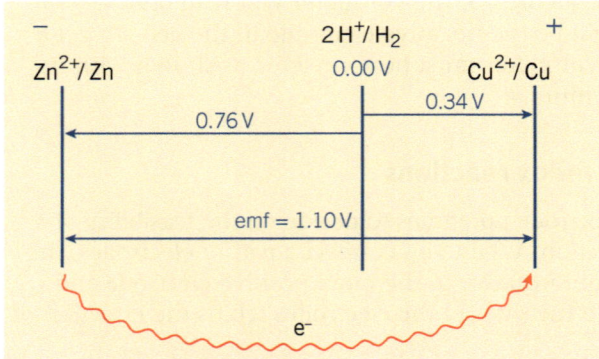

Figure 19 *Predicting the direction of electron flow when a Zn^{2+}/Zn electrode is connected to a Cu^{2+}/Cu electrode*

Electrons will flow from more negative to more positive, ie from the Zn^{2+}/Zn to Cu^{2+}/Cu as shown in Figure 19.

To calculate $E^{\ominus}_{cell}$:

the two equations are:

$Zn(s) \rightarrow Zn^{2+}(aq) + 2e^-$ This is the reverse of the half reaction in Table 2, so the sign of $E^{\ominus}$ must be reversed and becomes +0.76 V.

$Cu^{2+}(aq) + 2e^- \rightarrow Cu(s)$ $E^{\ominus}$ from Table 2 is +0.34 V

The overall effect is:

$$Zn(s) \rightarrow Zn^{2+}(aq) + 2e^- \quad E^{\ominus} = +0.76\,V$$

$$Cu^{2+}(aq) + 2e^- \rightarrow Cu(s) \quad E^{\ominus} = +0.34\,V$$

Adding the half equations and the values of $E^{\ominus}$ and cancelling the electrons gives

$$Cu^{2+}(aq) + Zn(s) + 2e^- \rightarrow Cu(s) + Zn^{2+}(aq) + 2e^- \quad E^{\ominus}_{cell} = +1.1\,V$$

So this reaction is feasible and is the reaction that actually happens, either by connecting the two electrodes or more directly by adding Zn to $Cu^{2+}(aq)$ ions in a test tube. The reverse reaction (which has $E^{\ominus}_{cell}$ of −1.1 V) is not feasible and does not occur.

$$Cu(s) + Zn^{2+}(aq) \rightarrow Zn(s) + Cu^{2+}(aq) \quad E^{\ominus}_{cell} = -1.1\,V$$

This reaction would require electrons to flow from positive to negative.

You can go through this process whenever you want to predict the outcome of a redox reaction, either using a diagram like Figure 19 or by constructing the equation from two half equations.

The general rule is that if $E^{\ominus}_{cell}$ is positive, the reaction is feasible.

With redox systems that only involve metal ions but no metal (e.g., Fe^{3+}/Fe^{2+}), a beaker containing all the relevant ions and a platinum electrode to make electrical contact is used in order to measure $E^{\ominus}$ by connecting to a hydrogen electrode (Figure 20).

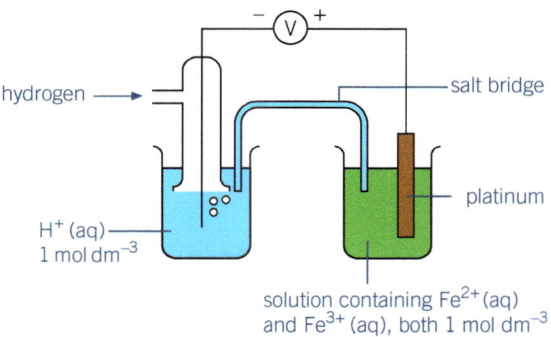

Figure 20 *A Fe^{3+}/Fe^{2+} cell connected to a hydrogen electrode to measure $E^{\ominus}$ for $Fe^{3+}(aq) + e^- \rightleftharpoons Fe^{2+}(aq)$*

Further examples

You can extend the electrochemical series to systems other than simple metal / metal ion ones (Table 3). Good reducing agents (eg Li) are found top right and good oxidising agents such as MnO_4^- / H^+ bottom left.

So Ca(s) is also a good reducing agent, but not as good as Li(s). The best oxidising agent listed in Table 3 is $Ce^{4+}(aq)$.

Exam tip

$E^{\ominus}_{cell}$ values tell you whether a reaction is feasible or not. They do not give any information about the speed of the reaction. A feasible reaction may be so slow that in practice it does not take place at all at room temperature.

Table 3 *$E^{\ominus}$ values for more half equations*

Reduction half equation	$E^{\ominus}/V$
$Li^+(aq) + e^- \rightarrow Li(s)$	−3.04
$Ca^{2+}(aq) + 2e^- \rightarrow Ca(s)$	−2.87
$Al^{3+}(aq) + 3e^- \rightarrow Al(s)$	−1.66
$Zn^{2+}(aq) + 2e^- \rightarrow Zn(s)$	−0.76
$2CO_2(g) + 2H^+(aq) + 2e^- \rightarrow C_2O_4H_2(aq)$	−0.43
$Cr^{3+}(aq) + e^- \rightarrow Cr^{2+}(aq)$	−0.41
$Pb^{2+}(aq) + 2e^- \rightarrow Pb(s)$	−0.13
$2H^+(aq) + 2e^- \rightarrow H_2(g)$	0.00
$Cu^{2+}(aq) + e^- \rightarrow Cu^+(aq)$	+0.15
$Cu^{2+}(aq) + 2e^- \rightarrow Cu(s)$	+0.34
$I_2(aq) + 2e^- \rightarrow 2I^-(aq)$	+0.54
$Fe^{3+}(aq) + e^- \rightarrow Fe^{2+}(aq)$	+0.77
$Ag^+(aq) + e^- \rightarrow Ag(s)$	+0.80
$Br_2(aq) + 2e^- \rightarrow 2Br^-(aq)$	+1.07
$Cl_2(aq) + 2e^- \rightarrow 2Cl^-(aq)$	+1.36
$MnO_4^- + 8H^+(aq) + 5e^- \rightarrow Mn^{2+}(aq) + 4H_2O(l)$	+1.52
$Ce^{4+}(aq) + e^- \rightarrow Ce^{3+}(aq)$	+1.70

Worked example

Finding $E^{\ominus}_{cell}$ for an iron-chlorine electrochemical cell

Will the following reaction occur or not?

$$Fe^{3+}(aq) + Cl^{-}(aq) \rightarrow Fe^{2+}(aq) + \tfrac{1}{2}Cl_2(aq)$$

Figure 21 shows that the $E^{\ominus}_{cell}$ is 0.59 V, with iron the more negative. So, electrons will flow from the Fe^{3+}/Fe^{2+} standard electrode to the $\tfrac{1}{2}Cl_2/Cl^{-}$ standard electrode.

$Fe^{2+} \rightarrow Fe^{3+} + e^{-}$ $E^{\ominus} = -0.77\,V$ (the sign is reversed as the half reaction is the reverse of that in Table 1)

$\tfrac{1}{2}Cl_2 + e^{-} \rightarrow Cl^{-}$ $E^{\ominus} = +1.36\,V$

$Fe^{2+} + \tfrac{1}{2}Cl_2 + \cancel{e^{-}} \rightarrow Fe^{3+} + Cl^{-} + \cancel{e^{-}}$ $E^{\ominus}_{cell} = +0.59\,V$

This is the reaction that will occur. So, chlorine will oxidise iron(II) ions to iron(III) ions.

The original reaction is *not* feasible.

$$Fe^{3+}(aq) + Cl^{-}(aq) \rightarrow Fe^{2+}(aq) + \tfrac{1}{2}Cl_2(aq)\quad E^{\ominus}_{cell} = -0.59\,V$$

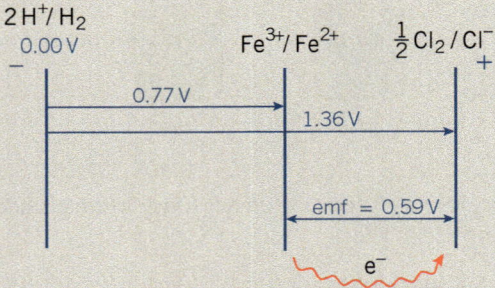

Figure 21 Working out the $E^{\ominus}_{cell}$ for $Fe^{3+}(aq) + Cl^{-}(aq) \rightarrow Fe^{2+}(aq) + \tfrac{1}{2}Cl_2(aq)$

Worked example

Finding $E^{\ominus}_{cell}$ for an iron-iodine electrochemical cell

Will the following reaction occur or not?

$$Fe^{3+}(aq) + I^{-}(aq) \rightarrow Fe^{2+}(aq) + \tfrac{1}{2}I_2(aq)$$

Figure 22 shows that the $E^{\ominus}_{cell}$ is 0.23 V with iodine the more negative. So, electrons will flow from the I_2/I^{-} electrode to the Fe^{3+}/Fe^{2+} electrode.

$I^{-} \rightarrow \tfrac{1}{2}I_2 + e^{-}$ $E^{\ominus} = -0.54\,V$ (the sign is reversed as the half reaction is the reverse of that in Table 1)

$Fe^{3+} + e^{-} \rightarrow Fe^{2+}$ $E^{\ominus} = +0.77\,V$

$I^{-} + Fe^{3+} + \cancel{e^{-}} \rightarrow \tfrac{1}{2}I_2 + Fe^{2+} + \cancel{e^{-}}$ $E^{\ominus}_{cell} = +0.23\,V$

This is the reaction that will occur. So, iron(III) ions will oxidise iodide ions to iodine.

The following reaction is *not* feasible.

$$Fe^{2+}(aq) + \tfrac{1}{2}I_2(aq) \rightarrow Fe^{3+}(aq) + I^{-}(aq)\quad E^{\ominus}_{cell} = -0.23\,V$$

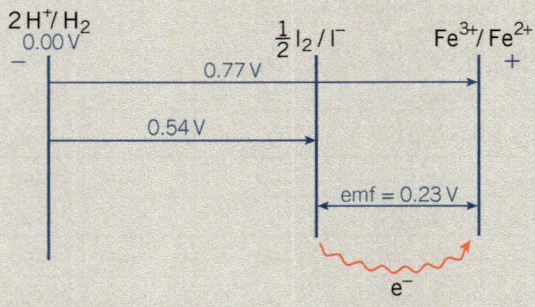

Figure 22 Working out the $E^{\ominus}_{cell}$ for $Fe^{3+}(aq) + I^{-}(aq) \rightarrow Fe^{2+}(aq) + \tfrac{1}{2}I_2(aq)$

To decide on the feasibility of a redox reaction

1 Split the reaction into two half reactions.
2 Look up the values of $E^\ominus$ for each half reaction.
3 One of the half reactions will have to be reversed.
4 Reverse the sign of $E^\ominus$ for this reaction.
5 Add the half reactions and the values of $E^\ominus$ to give the value of $E^\ominus_{cell}$.
6 The reaction is feasible if the value of $E^\ominus_{cell}$ is positive.

 The 'anticlockwise rule'—a handy shortcut

Table 4 *The anti-clockwise rule*

Reduction half equation	$E^\ominus$/ V
$Zn^{2+}(aq) + 2e^- \longrightarrow Zn(s)$	−0.76
$2CO_2(g) + 2H^+(aq) + 2e^- \longrightarrow C_2O_4H_2(aq)$	−0.43
$Cr^{3+}(aq) + e^- \longrightarrow Cr^{2+}(aq)$	−0.41
$Pb^{2+}(aq) + 2e^- \longrightarrow Pb(s)$	−0.13
$2H^+(aq) + 2e^- \longrightarrow H_2(g)$	0.00
$Cu^{2+}(aq) + e^- \longrightarrow Cu^+(aq)$	+0.15
$Cu^{2+}(aq) + 2e^- \longrightarrow Cu(s)$	+0.34

Look at the table of $E^\ominus$ values above. They are listed in order with most negative at the top. This means that electrons will move downwards from negative to positive. So electrons will flow from the $Zn^{2+}(aq)/Zn(s)$ cell to a $Cu^{2+}(aq)/Cu(s)$ half-cell.

For this to happen the two equilibria must move in the directions shown by the red arrows:

• $Zn^{2+}(aq)/Zn(s)$ will release electrons
• $Cu^{2+}(aq)/Cu(s)$ will accept electrons

This means that the upper equilibrium reaction moves from right to left, and the lower equilibrium reaction moves from left to right. So we call it the **anticlockwise rule**.

$$Zn(s) \rightarrow Zn^{2+}(aq) + 2e^-$$

$$Cu^{2+}(aq) + 2e^- \rightarrow Cu(s)$$

Adding:

$$Zn(s) + Cu^{2+}(aq) + 2e^- \rightarrow Zn^{2+}(aq) + 2e^- + Cu(s)$$

$$Zn(s) + Cu^{2+}(aq) \rightarrow Zn^{2+}(aq) + Cu(s)$$

which correctly predicts the reaction that actually happens.

$E^\ominus_{cell}$ is the difference in $E^\ominus$ between the two half-cells—in this case −0.76 V and +0.34 V = 1.10 V with the $Zn^{2+}(aq)/Zn(s)$ negative and the $Cu^{2+}(aq)/Cu(s)$ positive.

Note: you may have to multiply one or both of the equations to get the electrons to cancel and give a balanced equation. This does not affect the value of $E^\ominus$ for either half equation.

You may see half equations listed in alphabetical order rather than in $E^\ominus$ order. In this case you should select the two half equations you need, and write them down in $E^\ominus$ order, with the most negative at the top. Then apply the anticlockwise rule.

It is always the case that on connecting two half-cells the upper (more negative) one in the table moves from right to left and the lower one from left to right.

The Nernst equation

Values of $E^{\ominus}$ and $E^{\ominus}_{cell}$ refer, by definition, to standard conditions—a temperature of 298 K and concentrations of $1\,mol\,dm^{-3}$. The **Nernst equation** allows us to calculate the effect of changing the concentration of one or more species involved.

The Nernst equation is:

$$E = E^{\ominus} + \left(\frac{0.059}{z}\right)\log_{10}\frac{[\text{oxidised species}]}{[\text{reduced species}]}$$

For a simple half-cell such as $Cu^{2+}(aq)/Cu(s)$ the equation can be written:

$$E = E^{\ominus} + \left(\frac{0.059}{z}\right)\log_{10}[Cu^{2+}(aq)]$$

where z represents the number of electrons involved in the reaction at the electrode (ie the difference in oxidation state between Cu and Cu^{2+}). In this example, we cannot change the concentration of the Cu(s) because it is a solid so it is not included in the equation. This is similar to the way in which concentrations of solids and pure liquids do not appear in equilibrium constant expressions. This is discussed in more detail in Section 25.1.

This means that if we increase the concentration of Cu^{2+} ions, the value of E for the half-cell will increase (ie become more positive). This makes sense, because if $[Cu^{2+}(aq)]$ increases the equilibrium $Cu^{2+}(aq) + 2e^{-} \rightleftharpoons Cu(s)$ will move to the right, removing electrons from the Cu(s) cathode and making it more positive.

So if we double the $[Cu^{2+}(aq)]$ from $1\,mol\,dm^{-3}$ to $2\,mol\,dm^{-3}$.

$$E = E^{\ominus} + \left(\frac{0.059}{z}\right)\log_{10}[Cu^{2+}(aq)]$$

$$E = 0.34 + \left(\frac{0.059}{2}\right)\log_{10}(2)$$

$$E = 0.34 + 0.0295 \times 0.301$$

$$E = 0.34 + 0.0088$$

$$E = 0.3488\,V$$

Now consider a half-cell where two species can be changed. In this case both species must appear in the equation. So for a $Fe^{3+}(aq)/Fe^{2+}(aq)$ the equation becomes:

$$E = E^{\ominus} + \left(\frac{0.059}{z}\right)\log_{10}\frac{[Fe^{3+}(aq)]}{[Fe^{2+}(aq)]}$$

Here $z = 1$, so if we had a half-cell in which the concentration of Fe^{3+} was ten times that of Fe^{2+} then

$$E = 0.77 + 0.059\,\log_{10}(10)$$

$$E = 0.77 + 0.059 \times 1$$

$$E = 0.829\,V$$

$E^\ominus$ and ΔG

We have seen in Section 23.4 that the sign of the Gibbs Free Energy change, ΔG, is the factor that determines the feasibility of chemical reactions. The value of $E^\ominus_{cell}$ determines the feasibility of redox reactions, so it is no surprise that the two are linked mathematically.

The relationship is:

$$\Delta G = -n\, E^\ominus_{cell}\, F$$

where n is the number of electrons transferred in the redox reaction, and F is the value of the Faraday constant. The $-$ sign appears because reactions are feasible when $E^\ominus_{cell}$ is positive but when ΔG is negative.

So for the reaction:

$Cu^{2+}(aq) + Zn(s) + 2e^- \rightarrow Cu(s) + Zn^{2+}(aq) + 2e^- \quad E^\ominus_{cell} = +1.1V$

$\Delta G = -n\, E^\ominus_{cell}\, F$

$\Delta G = -2 \times 1.1 \times 96500 = -212300\,J\,mol^{-1}$, or $-212.30\,kJ\,mol^{-1}$

Summary test 24.2

1 Calculate the value of the $E^\ominus_{cell}$ for an $Al^{3+}(aq)/Al(s)$ standard electrode connected to a $Zn^{2+}(aq)/Zn(s)$ standard electrode? Sketch a diagram like Figure 13 earlier in this section to illustrate your answer.

2 Use the values of $E^\ominus$ in Table 3 to calculate $E^\ominus_{cell}$ for the following:
   a $Ce^{4+}(aq) + Fe^{2+}(aq) \rightarrow Ce^{3+}(aq) + Fe^{3+}(aq)$
   b $I_2(aq) + 2Br^-(aq) \rightarrow Br_2(aq) + 2I^-(aq)$
   c $MnO_4^-(aq) + 8H^+(aq) + 5I^-(aq) \rightarrow Mn^{2+}(aq) + 4H_2O(l) + 2I_2(aq)$
   d $2H^+(aq) + Pb(s) \rightarrow Pb^{2+}(aq) + H_2(g)$

3 Predict which of the halogens could oxidise $Ag(s)$ to $Ag^+(aq)$ ions.

4 a Predict whether the reaction $Br_2(aq) + 2Cl^-(aq) \rightarrow Cl_2(aq) + 2Br^-(aq)$ is feasible.
   b Predict whether the reaction $Fe^{3+}(aq) + Br^-(aq) \rightarrow Fe^{2+}(aq) + Br_2(aq)$ is feasible.

5 Use the Nernst equation to calculate the value of $E^\ominus$ for the Zn/Zn^{2+} half-cell in which the concentration of $Zn^{2+}(aq)$ is $5\ mol\ dm^{-3}$.

6 Calculate ΔG for the following reactions:
   a $Fe^{3+}(aq) + Cl^-(aq) \rightarrow Fe^{2+}(aq) + \frac{1}{2}Cl_2(aq) \qquad E^\ominus_{cell} = -0.59\,V$
   b $I^- + Fe^{3+} \rightarrow \frac{1}{2}I_2 + Fe^{2+} \qquad E^\ominus_{cell} = +0.23\,V$

Launch additional digital resources for the chapter

1 This question is about the electrolysis of molten lead bromide.

a Identify the formula of the products formed at the cathode and anode in the electrolysis of molten lead bromide. *(2 marks)*

b Construct half equations for the reactions that occur at the cathode and anode in the electrolysis cell. *(2 marks)*

c Use your answer to part **b** to construct a full redox equation for the electrolysis of molten lead bromide. *(1 mark)*

d A current of 0.20 A is passed through the electrolysis cells for 30 minutes. Calculate the quantity of electric charge used. *(2 marks)*

e 193 000 C is required to produce 1 mole of lead by electrolysis.
  i Calculate the mass of lead produced in the electrolysis cell for the quantity of charge you calculated in **c**. *(2 marks)*
  ii Calculate the mass of liquid bromine, Br_2, produced in the electrolysis cell in the same time. *(2 marks)*

f Calculate the time required to collect 10 g of lead in the electrolysis cell, assuming a constant current of 0.20 A. *(2 marks)*

2 a Explain the terms oxidation and reduction. *(2 marks)*

b Using your answer to part **a**, construct separate half-equations involving electrons for the oxidation and reduction processes when the following substances react: *(4 marks)*
  i Mg with Cl_2
  ii MnO_2 with conc. HCl
  iii H_2O_2 with I^-
  iv Fe^{3+} with I^-

c Draw a fully labelled diagram showing how a student would set up apparatus to determine the standard electrode potential of a $Zn^{2+}(aq) \mid Zn(s)$ half-cell. *(3 marks)*

3 a Draw a fully labelled diagram of a cell incorporating the following two electrode reactions.

$$H_2O_2(aq) + 2H^+(aq) + 2e^- \rightarrow 2H_2O(l) \quad E^\ominus = +1.77 \text{ V}$$

$$Sn^{4+}(aq) + 2e^- \rightarrow Sn^{2+}(aq) \quad E^\ominus = +0.15 \text{ V}$$

b Show the direction in which electrons flow in the external circuit when the cell is used to generate electricity. *(1 mark)*

c Construct equations for the reactions at each electrode when the cell produces an electric current. Give the cell potential when the cell operates under standard conditions. *(3 marks)*

4 The following table gives the standard electrode potentials for a number of half-reactions:

	$E^\ominus$/ V
$Zn^{2+}(aq) + 2e^- \rightarrow Zn(s)$	−0.76
$Fe^{2+}(aq) + 2e^- \rightarrow Fe(s)$	−0.44
$I_2(aq) + 2e^- \rightarrow 2I^-(aq)$	+0.54
$Fe^{3+}(aq) + e^- \rightarrow Fe^{2+}(aq)$	+0.77
$Ce^{4+}(aq) + e^- \rightarrow Ce^{3+}(aq)$	+1.61

a Give the half equation that is used as the standard for these electrode potentials *(1 mark)*

b Identify which of the substances in the table above is:
  i The strongest oxidising agent *(1 mark)*
  ii The strongest reducing agent *(1 mark)*

c Give the identity of the substance(s) in the table that could be used to convert iodide ions to iodine? Construct balanced equations for any possible conversions. *(4 marks)*

d A half-cell is constructed by putting a platinum electrode in a solution which is 1.0 mol dm^{-3} with respect to both Fe^{2+} and Fe^{3+} ions. This half-cell is then connected by means of a 'salt bridge' to another half-cell containing an iron electrode in a 1.0 mol dm^{-3} solution of Fe^{2+} ions.
  i Calculate the potential of this cell. *(1 mark)*
  ii If the two electrodes are connected externally, what reactions take place in each half-cell. *(2 marks)*

e Construct an equation for the reaction you would expect to occur when an iron nail is placed in a solution of iron(II) sulfate. *(2 marks)*

5 Nerve cells maintain a small voltage across their membranes. This voltage, or potential, is the result of the presence of solutions containing different concentrations of K^+ and Na^+ inside and outside the cell.

The version of the Nernst equation given below can be used to calculate the potential across the membrane:

$$E = \frac{0.059}{Z} \log_{10} \frac{[\text{ion inside cell}]}{[\text{ion outside cell}]}$$

a Assuming that the cell potential is the result of differences in the concentration of potassium ions only, calculate the potential (in mV) across a nerve cell membrane when:
[K^+] inside the cell = 150 mmol dm^{-3}
[K^+] outside the cell = 4 mmol dm^{-3} *(2 marks)*

b Assuming that the cell potential is the result of differences in the concentration of calcium ions only, calculate the potential (in mV) across a nerve cell membrane when:

[Ca^{2+}] inside the cell = 70 nmol dm^{-3}
[Ca^{2+}] outside the cell = 4 nmol dm^{-3} *(2 marks)*

6 Iodide ions are able to reduce aqueous iron(III) ions to iron(II) ions. This reaction can produce a current in an electrochemical cell.

a Construct a labelled diagram to show part of an electrochemical cell that can be connected to a standard hydrogen electrode to measure the standard electrode potential for the reduction of iron(III) ions. *(3 marks)*

b State the role of the salt bridge in an electrochemical cell. *(1 mark)*

7 The table below gives some standard electrode potential data:

Electrode reaction	$E^{\ominus}/V$
$F_2 + 2e^- \rightleftharpoons 2F^-$	+2.87
$2HOCl + 2H^+ + 2e^- \rightleftharpoons Cl_2 + 2H_2O$	+1.64
$Cl_2 + 2e^- \rightleftharpoons 2Cl^-$	+1.36
$O_2 + 4H^+ + 4e^- \rightleftharpoons 2H_2O$	+1.23
$Ag^+ + e^- \rightleftharpoons Ag$	+0.80
$Fe^{3+} + e^- \rightleftharpoons Fe^{2+}$	+0.77
$2H^+ + 2e^- \rightleftharpoons H_2$	0.00
$Fe^{2+} + 2e^- \rightleftharpoons Fe$	−0.44

a i Define the term oxidising agent, in terms of electrons. *(1 mark)*
   ii From the data in the table above, identify the species that is the weakest oxidising agent. Give reasons for your answer. *(2 marks)*

b The standard electrode potential for the Ag^+/Ag electrode is +0.80. Write the conventional representation of this cell. *(2 marks)*

c A simple cell is constructed by connecting two half cells consisting of Fe^{2+}/Fe and Ag^+/Ag using a salt bridge.

   i Explain why saturated potassium chloride solution is used. *(1 mark)*

   ii Calculate the emf of this cell. *(1 mark)*

   iii Explain, using the data in the table above, why potassium chloride would not be a suitable material to use for the salt bridge in this cell. *(2 marks)*

d Using the data in the table above, predict what would happen to a solution of iron(II) chloride when left in a beaker with no lid. *(2 marks)*

25.1 Acids and bases

In Section 7.2, we used the Brønsted–Lowry theory to define acids as proton (H^+) donors, and bases as proton acceptors. This means that acids and bases occur in pairs:

$$HCl(g) + NH_3(g) \rightleftharpoons NH_4Cl(s)$$

In the forward reaction (left to right) hydrogen chloride is acting as an acid and ammonia is acting as as a base. In the reverse reaction (right to left) NH_4^+ is acting as an acid, donating a proton to Cl^- which is acting as a base. We say that there are two **conjugate acid–base pairs**:

- HCl is the **conjugate acid** and Cl^- is its **conjugate base**
- NH_4^+ is the conjugate acid and NH_3 is its conjugate base.

Water as an acid and a base

Whether a species is acting as an acid or a base depends on the reactants. Water is a good example of this.

Hydrogen chloride can donate a proton to water, so that water acts as a base:

$$HCl + H_2O \rightarrow H_3O^+ + Cl^-$$

In this case H_2O and H_3O^+ are a conjugate pair: H_3O^+ is the conjugate acid and H_2O is the conjugate base.

H_3O^+ is called the **oxonium ion**, but the names hydronium ion and hydroxonium ion are also used.

Water may also act as an acid. For example:

$$H_2O + NH_3 \rightarrow OH^- + NH_4^+$$

Here water is donating a proton to ammonia. So water is a conjugate acid and OH^- is its conjugate base. NH_4^+ is a conjugate acid and NH_3 is its conjugate base.

Another example of a conjugate acid–base pair is a mixture of concentrated sulfuric acid, H_2SO_4, and concentrated nitric acid, HNO_3.

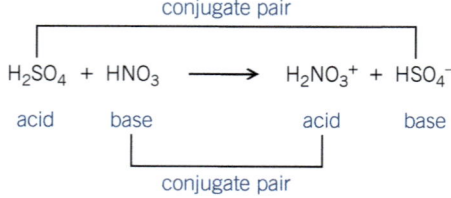

Sulfuric acid donates a proton to nitric acid, so is acting as the acid, while in this example nitric acid is acting as a base.

The proton in aqueous solution

It is important to realise that the H^+ ion is just a proton. The hydrogen atom has only one electron and if this is lost all that remains is a proton (the hydrogen nucleus). This is about 10^{-15} m in diameter, compared to 10^{-10} m or more for any other chemical entity. This extremely small size and consequent intense electric field cause it to have unusual properties compared with other positive ions. It is never found isolated. In aqueous solutions it is always bonded to at least one water molecule to form the ion H_3O^+. For simplicity, protons are represented in an aqueous solution by $H^+(aq)$ rather than $H_3O^+(aq)$.

Since the H^+ ion has no electrons of its own, it can only form a bond with another species that has a lone pair of electrons.

The ionisation of water

Water is slightly ionised:

$$H_2O(l) \rightleftharpoons H^+(aq) + OH^-(aq)$$

This may also be written:

$$H_2O(l) + H_2O(l) \rightleftharpoons H_3O^+(aq) + OH^-(aq)$$

This emphasises that this is an acid–base reaction in which one water molecule donates a proton to another.

This equilibrium is established in water and all aqueous solutions:

$$H_2O(l) \rightleftharpoons H^+(aq) + OH^-(aq)$$

You can write an equilibrium expression:

$$K_c = \frac{[H^+(aq)][OH^-(aq)]}{[H_2O(l)]}$$

The concentration of water $[H_2O(l)]$ is constant and is incorporated into a modified equilibrium constant K_w, where $K_w = K_c \times [H_2O(l)]$.

So, $\quad K_w = [H^+(aq)][OH^-(aq)]$

K_w is called the **ionic product of water** and at 298 K it is equal to $1.0 \times 10^{-14}\,mol^2\,dm^{-6}$. Each H_2O that dissociates (splits up) gives rise to one H^+ and one OH^- so, in pure water, at 298 K:

$$[OH^-(aq)] = [H^+(aq)]$$

So, $\quad 1.0 \times 10^{-14} = [H^+(aq)]^2$

$$[H^+(aq)] = 1.0 \times 10^{-7}\,mol\,dm^{-3} = [OH^-(aq)] \text{ at } 298\,K\ (25\,°C)$$

The pH scale

The acidity or alkalinity of a solution depends on the concentration of $H^+(aq)$ and is measured on the pH scale. Acids have pH values of less than 7 and alkalis values of greater than 7. A pH of 7 is neutral.

$$pH = -\log_{10}[H^+(aq)]$$

This expression is more complicated than simply stating the concentration of $H^+(aq)$. However, using the logarithm of the concentration does away with awkward numbers like 10^{-13}, etc. These would occur because the concentration of $H^+(aq)$ in most aqueous solutions is so small. The minus sign makes almost all pH values positive (because the logs of numbers less than 1 are negative).

On the pH scale:

- The smaller the pH, the greater the concentration of $H^+(aq)$.
- A difference of one pH number means a tenfold difference in $[H^+]$. So, for example, pH 2 has ten times the H^+ concentration of pH 3.

Remember that, at 298 K, $K_w = [H^+(aq)][OH^-(aq)] = 1.00 \times 10^{-14}\,mol^2\,dm^{-6}$.

This means that in neutral aqueous solutions:

$$[H^+(aq)] = [OH^-(aq)] = 1.0 \times 10^{-7}\,mol\,dm^{-3}$$

$$pH = -\log_{10}[H^+(aq)] = -\log_{10}[1.0 \times 10^{-7}] = 7.00$$

so the pH is 7.00.

Exam tip

In any aqueous solution if $[H^+(aq)]$ increases then $[OH^-(aq)]$ decreases in proportion, so their product remains constant.

Extension

The concentration of water

What is the concentration of water in $mol\,dm^{-3}$? This question often catches out even experienced chemists. $1\,dm^3$ of water weighs $1000\,g$. The M_r of water is 18.0.

1 How many moles of water is $1000\,g$?
2 What does this make the concentration of water in $mol\,dm^{-3}$?

Answers:
1 55.5 moles   2 55.5 $mol\,dm^{-3}$

Extension

Like all equilibrium constants, K_w depends on temperature. So the pH of pure water is only 7.00 at 298 K.

At 373 K (the boiling point of water), $K_w = 5.13 \times 10^{-13}\,mol^2\,dm^{-6}$. Have a go at calculating the pH of water at this temperature and determine whether boiling water is acidic or basic.

Answer: The pH is 6.14 but it is still neutral as there are still the same number of OH^- and H^+ ions present

---

Exam tip

Remember that square brackets [ ] mean concentration in $mol\,dm^{-3}$.

Exam tip

$-\log_{10} 1.0 \times 10^{-7}\,\text{mol dm}^{-3} = 7$. This is why the pH of pure water is 7 at 298 K.

pH measures alkalinity as well

pH measures alkalinity as well as acidity, because as $[H^+(aq)]$ goes up, $[OH^-(aq)]$ goes down. At 298 K, if a solution contains more $H^+(aq)$ than $OH^-(aq)$, its pH will be less than 7 and it is called **acidic**. If a solution contains more $OH^-(aq)$ than $H^+(aq)$, its pH will be greater than 7 and it is called **alkaline** (Figure 1).

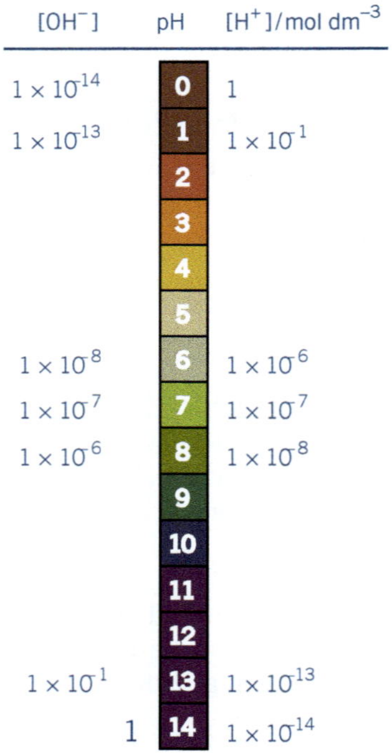

[OH⁻]
pH
[H⁺]/mol dm⁻³
1 × 10⁻¹⁴
0
1
1 × 10⁻¹³
1
1 × 10⁻¹
2
3
4
5
1 × 10⁻⁸
6
1 × 10⁻⁶
1 × 10⁻⁷
7
1 × 10⁻⁷
1 × 10⁻⁶
8
1 × 10⁻⁸
9
10
11
12
1 × 10⁻¹
13
1 × 10⁻¹³
1
14
1 × 10⁻¹⁴

***Figure 1** The pH scale. The colours are those of universal indicator*

Working with the pH scale

You can work out the concentration of hydrogen ions $[H^+]$ in an aqueous solution if you know the pH. It is the antilogarithm of the pH value.

Finding $[H^+(aq)]$ from pH

For example, an acid has a pH of 3.00:

$$\text{pH} = -\log_{10}[H^+(aq)]$$

$$3.00 = -\log_{10}[H^+(aq)]$$

$$-3.00 = \log_{10}[H^+(aq)]$$

Take the antilog of both sides:

$$[H^+(aq)] = 1.0 \times 10^{-3}\,\text{mol dm}^{-3}$$

Finding $[OH^-(aq)]$ from pH

With bases, you need two steps. Suppose the pH of a solution is 10.00:

$$\text{pH} = -\log_{10}[H^+(aq)]$$

$$10.00 = -\log_{10}[H^+(aq)]$$

$$-10.00 = \log_{10}[H^+(aq)]$$

Take the antilog of both sides:

$$[H^+(aq)] = 1.0 \times 10^{-10}$$

You know $[H^+(aq)]\,[OH^-(aq)] = 1.0 \times 10^{-14}\,\text{mol}^2\,\text{dm}^{-6}$

Substituting your value for $[H^+(aq)] = 1.0 \times 10^{-10}$ into the equation:

$$[1.0 \times 10^{-10}] \, [OH^-(aq)] = 1.0 \times 10^{-14} \, mol^2 \, dm^{-6}$$

$$[OH^-(aq)] = 1.0 \times 10^{-4} \, mol \, dm^{-3}$$

Worked example

The pH of strong acid solutions

HCl dissociates completely in dilute aqueous solution to $H^+(aq)$ ions and $Cl^-(aq)$ ions, that is, the reaction goes to completion:

$$HCl(aq) \rightarrow H^+(aq) + Cl^-(aq)$$

Acids that dissociate completely like this are called strong acids.

Example 1

In $1.00 \, mol \, dm^{-3}$ HCl:

$$[H^+(aq)] = 1.00 \, mol \, dm^{-3}$$

$$\log_{10}[H^+(aq)] = \log_{10} 1.00 = 0.00$$

$$-\log_{10}[H^+(aq)] = 0.00$$

So the pH of $1 \, mol \, dm^{-3}$ HCl = 0.00

Example 2

In $0.16 \, mol \, dm^{-3}$ solution of HCl:

$$[H^+(aq)] = 0.16 \, mol \, dm^{-3}$$

$$\log_{10}[H^+(aq)] = \log 0.160 = -0.796$$

$$-\log_{10}[H^+(aq)] = 0.796$$

So the pH of $0.16 \, mol \, dm^{-3}$ HCl = 0.80 to 2 decimal places

---

Worked example

The pH of strong alkaline solutions

The relationship

$$K_w = [H^+(aq)][OH^-(aq)] = 1.00 \times 10^{-14} \, mol^2 \, dm^{-6}$$

is true for all aqueous solutions—acid, alkaline or neutral. So to find the pH of an alkaline solution we must first calculate $[OH^-(aq)]$, then use it in the above equation to calculate $[H^+(aq)]$.

For example, to find the pH of $1 \, mol \, dm^{-3}$ NaOH:

NaOH is fully dissociated so we know:

$$NaOH(aq) \rightarrow Na^+(aq) + OH^-(aq)$$

$$[OH^-(aq)] = 1 \, mol \, dm^{-3}$$

Substituting this into the equation: $[H^+(aq)][OH^-(aq)] = 1.00 \times 10^{-14} \, mol^2 \, dm^{-6}$

$$[H^+(aq)] \times 1 = 1.00 \times 10^{-14} \, mol \, dm^{-3}$$

$$\text{and } [H^+(aq)] = 1.00 \times 10^{-14} \, mol^2 \, dm^{-6}$$

Taking logs $\log[H^+(aq)] = -14$

pH = 14

The pH of barium hydroxide solution

The concentration of a saturated solution of barium hydroxide is $0.15 \, mol \, dm^{-3}$. What is the pH of a saturated solution of barium hydroxide?

$$Ba(OH)_2(aq) \rightleftharpoons Ba^{2+}(aq) + 2OH^-(aq)$$

Each $Ba(OH)_2$ that dissolves produces two OH^- ions.

So $[OH^-] = 0.3 \, mol \, dm^{-3}$

At 298 K, $K_w = [H^+][OH^-] = 1 \times 10^{-14} \, mol^2 \, dm^{-6}$

$1 \times 10^{-14} = [H^+] \times 0.3$

$[H^+] = \dfrac{1 \times 10^{-14}}{0.3} = 3.33 \times 10^{-14}$

$pH = -\log_{10}[H^+] = -\log(3.33 \times 10^{-14})$

$pH = 13.48$

Weak acids and bases

Strong acids and bases are those which fully dissociate into ions in aqueous solutions, for example:

$$HCl(g) + aq \rightarrow H^+(aq) + Cl^-(aq)$$

$$NaOH(aq) + aq \rightarrow Na^+(aq) + OH^-(aq)$$

Weak acids and bases are those which dissociate only partially into ions in aqueous solutions, for example:

$$CH_3COOH(aq) \rightleftharpoons H^+(aq) + CH_3COO^-(aq)$$

$$NH_3(aq) + H_2O(l) \rightleftharpoons NH_4^+(aq) + OH^-(aq)$$

Many acids and bases are only slightly ionised (not fully dissociated) when dissolved in water. Ethanoic acid (the acid in vinegar, also known as acetic acid) is a typical example. In a $1 \, mol \, dm^{-3}$ solution of ethanoic acid, only about four in every thousand ethanoic acid molecules are dissociated into ions (so the degree of dissociation is 4/1000) – the rest remain dissolved as wholly covalently bonded molecules. In fact an equilibrium is set up:

	$CH_3COOH(aq)$ ethanoic acid		$H^+(aq)$ hydrogen ions		$CH_3COO^-(aq)$ ethanoate ions
		$\rightleftharpoons$		+	
Before dissociation:	1000		0		0
At equilibrium:	996		4		4

Acids like this are called **weak acids**. Weak refers only to the degree of dissociation. In a $5 \, mol \, dm^{-3}$ solution, ethanoic acid is still a weak acid, while in a $10^{-4} \, mol \, dm^{-3}$ solution, hydrochloric acid is still a strong acid.

When ammonia dissolves in water, it forms an alkaline solution. The equilibrium lies well to the left and ammonia is weakly basic:

$$NH_3(aq) + H_2O(l) \rightleftharpoons NH_4^+(aq) + OH^-(aq)$$

The dissociation of weak acids

Imagine a weak acid HA which dissociates:

$$HA(aq) \rightleftharpoons H^+(aq) + A^-(aq)$$

The equilibrium constant is given by:

$$K_c = \frac{[H^+(aq)]_{eqm}[A^-(aq)]_{eqm}}{[HA(aq)]_{eqm}}$$

For a weak acid, this is usually given the symbol K_a and called the **acid dissociation constant**.

$$K_a = \frac{[H^+(aq)]_{eqm}[A^-(aq)]_{eqm}}{[HA(aq)]_{eqm}}$$

The larger the value of K_a, the further the equilibrium is to the right, the more the acid is dissociated, and the stronger it is. Acid dissociation constants for some acids are given in Table 1.

Table 1 Values of K_a for some weak acids

Acid	K_a / mol dm^{-3}
chloroethanoic	1.30×10^{-3}
benzoic	6.30×10^{-5}
ethanoic	1.70×10^{-5}
hydrocyanic	4.90×10^{-10}

K_a has units and it is important to state these. They are found by multiplying and cancelling the units in the expression for K_a.

$$K_a = \frac{\text{mol dm}^{-3} \times \text{mol dm}^{-3}}{\text{mol dm}^{-3}} = \text{mol dm}^{-3}$$

Calculating the pH of weak acids

We can calculate the pH of solutions of strong acids, by assuming that they are fully dissociated. For example, in a 1.00 mol dm^{-3} solution of nitric acid, $[H^+] = 1.00 \text{ mol dm}^{-3}$. In weak acids this is no longer true, and you must use the acid dissociation expression to calculate $[H^+]$.

Calculating the pH of 1.00 mol dm^{-3} ethanoic acid

The concentrations in mol dm^{-3} are:

	$CH_3COOH(aq)$	$\rightleftharpoons$	$CH_3COO^-(aq)$	+	$H^+(aq)$
Before dissociation:	1.00		0		0
At equilibrium:	$1.00 - [CH_3COO^-(aq)]$		$[CH_3COO^-(aq)]$		$[H^+(aq)]$

But as each CH_3COOH molecule that dissociates produces one CH_3COO^- ion and one H^+ ion:

$$[CH_3COO^-(aq)] = [H^+(aq)]$$

Since the degree of dissociation of ethanoic acid is so small (it is a weak acid), $[H^+(aq)]_{eqm}$ is very small and, to a good approximation, $1.00 - [H^+(aq)] \approx 1.00$.

$$K_a = \frac{[H^+(aq)]^2}{1.00}$$

From Table 1,

$$K_a = 1.70 \times 10^{-5} \text{ mol dm}^{-3}$$

$$1.70 \times 10^{-5} = [H^+(aq)]^2$$

$$[H^+(aq)] = \sqrt{1.75 \times 10^{-5}}$$

$$[H^+(aq)] = 4.18 \times 10^{-3} \text{ mol dm}^{-3}$$

Taking logs: $\log_{10}[H^+(aq)] = -2.378$

pH $= 2.378 = 2.38$ to 2 decimal places

Figure 2 Vinegar is an aqueous solution of ethanoic acid. It contains about 5% ethanoic acid by volume—approximately 0.1 mol dm^{-3}

Calculating the pH of 0.100 mol dm^{-3} ethanoic acid

Using the same method, you get:

$$K_a = \frac{[H^+(aq)]^2}{0.10 - [H^+(aq)]}$$

Again, $0.100 - [H^+(aq)] \approx 0.10$

so: $\quad 1.70 \times 10^{-5} = \dfrac{[H^+(aq)]^2}{0.10}$

$$1.70 \times 10^{-6} = [H^+(aq)]^2$$

$$[H^+(aq)] = 1.30 \times 10^{-3}\,mol\,dm^{-3}$$

$$pH = 2.89 \text{ to 2 decimal places}$$

pK_a

pK_a is often used for a weak acid. This is defined as:

$$pK_a = -\log_{10} K_a$$

Think of p as meaning '$-\log_{10}$ of '.

pK_a can be useful in calculations (Table 2). It gives a measure of how strong a weak acid is—the smaller the value of pK_a, the stronger the acid.

Table 2 Values of K_a and pK_a for some weak acids

Acid	K_a / mol dm^{-3}	pK_a
chloroethanoic	1.30×10^{-3}	2.88
benzoic	6.30×10^{-5}	4.20
ethanoic	1.70×10^{-5}	4.77
hydrocyanic	4.90×10^{-10}	9.31

Buffer solutions

Buffers are solutions that can resist changes of pH when small amounts of acid or alkali are added to them.

How buffers work

Buffers are designed to keep the concentration of hydrogen ions and hydroxide ions in a solution almost unchanged. They are based on an equilibrium reaction which will move in the direction to remove either additional hydrogen ions or hydroxide ions if these are added.

Acidic buffers

Acidic buffers are made from weak acids. They work because the dissociation of a weak acid is an equilibrium reaction.

Consider a weak acid, HA. It will dissociate in solution:

$$HA(aq) \rightleftharpoons H^+(aq) + A^-(aq)$$

From the equation, $[H^+(aq)] = [A^-(aq)]$. As it is a weak acid, $[H^+(aq)]$ and $[A^-(aq)]$ are both very small because most of the HA is undissociated.

Adding alkali

If a little alkali is added, the OH^- ions from the alkali will react with HA to produce water molecules and A^-:

$$HA(aq) + OH^-(aq) \rightarrow H_2O(aq) + A^-(aq)$$

This removes the added OH^- so the pH tends to remain almost the same.

Adding acid

If H^+ is added, the equilibrium shifts to the left. H^+ ions combine with A^- ions to produce undissociated HA. But, since $[A^-]$ is small, the supply of A^- soon runs out and there is no A^- left to combine the added H^+. So the solution is not a buffer.

However, you can add to the solution a supply of extra A^- by adding a soluble salt of HA, which fully ionises, such as Na^+A^-. This increases the supply of A^- so that more H^+ can be used up. So, there is a way in which both added H^+ and OH^- can be removed.

> **An acidic buffer is made from a mixture of a weak acid and a soluble salt of that acid. It will maintain a pH of below 7 (acidic).**

The function of the weak acid component of a buffer is to act as a source of HA which can remove any added OH^-:

$$HA(aq) + OH^-(aq) \rightarrow A^-(aq) + H_2O(l)$$

The function of the salt component of a buffer is to act as a source of A^- ions which can remove any added H^+ ions:

$$A^-(aq) + H^+(aq) \rightarrow HA(aq)$$

Buffers don't ensure that *no* change in pH occurs. The addition of acid or alkali will still change the pH, but only slightly. The change will be far less than adding the same amount of acid or alkali to a non-buffer. It is also possible to saturate a buffer—to add so much acid or alkali that all of the available HA or A^- is used up.

Another way of achieving a mixture of weak acid and its salt is by neutralising some of the weak acid with an alkali such as sodium hydroxide. If you neutralise half the acid, you end up with a buffer whose pH is equal to the pK_a of the acid, as it has an equal supply of HA and A^-.

> **At half-neutralisation: pH = pK_a**

This is a very useful buffer because it is equally efficient at resisting a change in pH whether acid or alkali is added.

Basic buffers

Basic buffers also resist change but maintain a pH at above 7.

> **Basic buffers are made from a mixture of a weak base and a salt of that base.**

A mixture of aqueous ammonia and ammonium chloride, $NH_4^+Cl^-$, acts as a basic buffer. In this case:

- The aqueous ammonia removes added H^+:

$$NH_3(aq) + H^+(aq) \rightarrow NH_4^+(aq)$$

- the ammonium ion, NH_4^+, removes added OH^-:

$$NH_4^+(aq) + OH^-(aq) \rightarrow NH_3(aq) + H_2O(l)$$

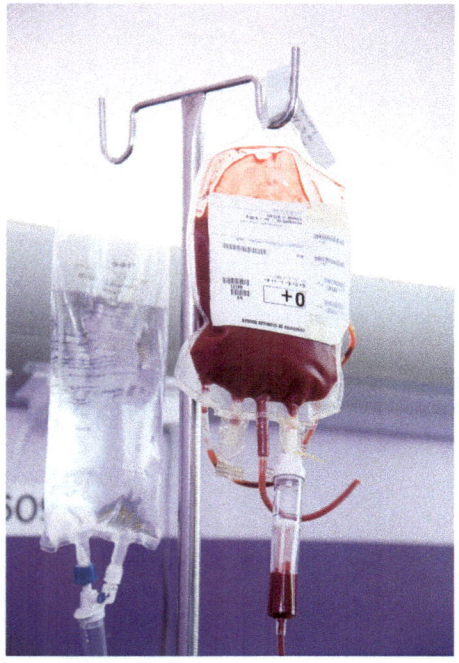

Figure 3 *Blood is buffered to a pH of 7.40*

Figure 4 *Most shampoos are buffered so that they are slightly alkaline*

Examples of buffers

An important example of a system involving a buffer is blood, the pH of which is maintained at approximately 7.4. A change of as little as 0.5 of a pH unit may be fatal.

Blood is buffered to a pH of 7.4 by a number of mechanisms. The most important is:

$$H^+(aq) + HCO_3^-(aq) \rightleftharpoons CO_2(aq) + H_2O(l)$$

Addition of extra H^+ ions moves this equilibrium to the right, removing the added H^+. Addition of extra OH^- ions removes H^+ by reacting to form water. The equilibrium above moves to the left releasing more H^+ ions. (The same equilibrium reaction acts to buffer the acidity of soils.)

Another equilibrium that contributes to the buffering of blood is:

$$HPO_4^{2-}(aq) + H^+(aq) \rightleftharpoons H_2PO_4^-$$

There are many examples of buffers in everyday products, such as detergents and shampoos. If either of these substances were too acidic or too alkaline, they could damage fabric or skin and hair.

Calculations on buffers

Different buffers can be made which will maintain different pHs. When a weak acid dissociates:

$$HA(aq) \rightleftharpoons H^+(aq) + A^-(aq)$$

You can write the expression:

$$K_a = \frac{[H^+(aq)]\ [A^-(aq)]}{[HA(aq)]}$$

You can use this expression to calculate the pH of buffers.

Worked example

Calculating pH of a buffer solution 1

A buffer consists of $0.100\,mol\,dm^{-3}$ ethanoic acid and $0.100\,mol\,dm^{-3}$ sodium ethanoate. What is the pH of the buffer?
(K_a for ethanoic acid is 1.7×10^{-5}, $pK_a = 4.77$.)

Calculate $[H^+(aq)]$ from the equation.

$$K_a = \frac{[H^+(aq)][A^-(aq)]}{[HA(aq)]}$$

Sodium ethanoate is fully dissociated, so $[A^-(aq)] = 0.100\,mol\,dm^{-3}$

Ethanoic acid is almost undissociated, so $[HA(aq)] \approx 0.100\,mol\,dm^{-3}$

$$1.7 \times 10^{-5} = [H^+(aq)] \times \frac{0.100}{0.100}$$

$$1.7 \times 10^{-5} = [H^+(aq)] \text{ and pH} = -\log_{10} [H^+(aq)]$$

$$pH = 4.77$$

When you have equal concentrations of acid and salt, pH of the buffer is equal to pK_a of acid used, and this is exactly the same situation as the half-neutralisation point.

Changing the concentration of HA or A^- will affect the pH of the buffer. If you use $0.200\,mol\,dm^{-3}$ ethanoic acid and $0.100\,mol\,dm^{-3}$ sodium ethanoate, the pH will be 4.50. Check that you can do this by doing a calculation like the one above.

Worked example

Calculating pH of a buffer solution 2

Calculate the pH of the buffer formed when $500 \, cm^3$ of $0.400 \, mol \, dm^{-3}$ NaOH is added to $500 \, cm^3$ $1.00 \, mol \, dm^{-3}$ HA.

$$K_a = 6.25 \times 10^{-5}$$

Some of the weak acid is neutralised by the sodium hydroxide leaving a solution containing A^- and HA, which will act as a buffer.

$$\text{moles HA} = c \times \frac{V}{1000} = 1.00 \times \frac{500}{1000} = 0.500 \, mol$$

$$\text{moles NaOH} = \text{moles OH}^- = c \times \frac{V}{1000} = 0.400 \times \frac{500}{1000} = 0.200 \, mol$$

Equation:	HA	+	NaOH	→	H_2O	+	NaA
Initially:	0.500 mol		0.200 mol				0
Finally:	0.300 mol		0 mol				0.200 mol

This leaves $1000 \, cm^3$ of a solution containing $0.300 \, mol$ HA and $0.200 \, mol$ A^- since all the NaA is dissociated to give A^-.

The concentrations are $[HA] = 0.300 \, mol \, dm^{-3}$ and $[A-] = 0.200 \, mol \, dm^{-3}$

$$K_a = \frac{[H^+(aq)][A^-(aq)]}{[HA(aq)]}$$

$$6.25 \times 10^{-5} = [H^+(aq)] \times \frac{0.200}{0.300}$$

$$[H^+(aq)] = 6.25 \times 10^{-5} \times \frac{3}{2} = 9.375 \times 10^{-5}$$

$$\text{So pH} = -\log [H^+(aq)] = 4.03$$

Exam tip

In a buffer solution $[H^+] \neq [A^-]$ so do *not* use the simplified expression

$$K_a = \frac{[H^+]^2}{[HA]}$$

The pH change when an acid or a base is added to a buffer

Adding acid

You can calculate how the pH changes when acid is added to a buffer. Suppose you start with $1.00 \, dm^3$ of a buffer solution of ethanoic acid at concentration $0.10 \, mol \, dm^{-3}$ and sodium ethanoate at concentration $0.10 \, mol \, dm^{-3}$. K_a is 1.7×10^{-5}. This has a pH of 4.77, as shown in the calculation previously.

Now add $10.0 \, cm^3$ of hydrochloric acid of concentration $1.00 \, mol \, dm^{-3}$ to this buffer. Virtually all the added H^+ ions will react with the ethanoate ions, $[A^-]$, to form molecules of ethanoic acid, $[HA]$.

Before adding the acid:

- Number of moles of ethanoic acid = 0.10
- Number of moles of sodium ethanoate = 0.10

number of moles of hydrochloric acid added is $c \times \frac{V}{1000}$

$$1.00 \times \frac{10.0}{1000} = 0.010$$

After adding the acid, this means:

- The amount of acid is increased by 0.010 mol to 0.110 mol.
- The amount of salt is decreased by 0.010 mol to 0.090 mol.

So, the concentration of acid $[HA]$ is now $0.110 \, mol \, dm^{-3}$.

And the concentration of salt $[A^-]$ is now $0.090 \, mol \, dm^{-3}$.

303

In calculating these concentrations you have ignored the volume of the added hydrochloric acid, $10 \, cm^3$ in $1000 \, cm^3$, only a 1% change.

$$K_a = \frac{[H^+(aq)] \, [A^-(aq)]}{[HA(aq)]}$$

So $\quad 1.7 \times 10^{-5} = [H^+(aq)] \dfrac{0.090}{0.110}$

$\quad\quad [H^+(aq)] = 1.7 \times 10^{-5} \times \dfrac{0.110}{0.090}$

$\quad\quad = 2.08 \times 10^{-5}$

$\quad\quad pH = 4.68$

Note how small the pH change is from 4.77 to 4.68.

Adding base

If you add $10 \, cm^3$ of $1.00 \, mol \, dm^{-3}$ sodium hydroxide to the original buffer, it will react with the H^+ ions and more HA will ionise. So this time you decrease the concentration of the acid [HA] by $0.01 \, mol \, dm^{-3}$ and increase the concentration of ethanoate ions by $0.010 \, mol \, dm^{-3}$.

Using similar steps to those above gives the new pH as 4.89. Check that you agree with this answer.

Note how small the pH change is from 4.77 to 4.89.

📖 Making a buffer solution

You may need to make up buffer solutions of specified pHs, to calibrate a pH meter, for example.

To find suitable concentrations of weak acid and its salt, you use the equation:

$$K_a = \frac{[H^+(aq)] \, [A^-(aq)]}{[HA(aq)]}$$

You can rearrange this equation to make [H^+] the subject and then taking logs of both sides. This results in an equation that is easier to use for calculations on buffers. It is called the **Henderson–Hasselbalch equation**:

$$pH = pK_a - \log\left(\frac{[HA]}{[A^-]}\right)$$

It tells you that the pH of a buffer solution depends on the pK_a of the weak acid on which it is based and on the ratio of the concentration of the acid to that of its salt.

For example, to make a buffer of pH = 4.50 you first select a weak acid whose pK_a is close to the required pH. Benzoic acid has a pK_a of 4.20, so you could make a buffer from benzoic acid and sodium benzoate.

Then substituting into the Henderson–Hasselbalch equation:

$$4.50 = 4.20 - \log\left(\frac{[HA]}{[A^-]}\right)$$

$$\log\left(\frac{[HA]}{[A^-]}\right) = -0.30$$

Taking antilogs:

$$\left(\frac{[HA]}{[A^-]}\right) = 0.50$$

This means that the concentration of acid is half the concentration of the salt.

So, for example, a solution that is $0.05 \, mol \, dm^{-3}$ in benzoic acid and $0.10 \, mol \, dm^{-3}$ in sodium benzoate would be a suitable buffer.

Exam tip

Always quote pH values to two decimal places.

Solubility product

There is a large variation in the extent to which ionic solids dissolve in water.

> **Solubility products are equilibrium constants used for sparingly soluble solids in a saturated solution at a given temperature.**

For example, if the slightly soluble salt, silver chloride, is added to water until no more will dissolve, the following equilibrium is set up which is well over to the left.

$$AgCl(s) \rightleftharpoons Ag^+(aq) + Cl^-(aq)$$

The equilibrium expression is:

$$K_c = \frac{[Ag^+(aq)]_{eqm}\,[Cl^-(aq)]_{eqm}}{[AgCl(s)]_{eqm}}$$

As we cannot change the concentration of a solid (by squeezing more of it into the same volume), the term $[AgCl(s)]_{eqm}$ is constant. It is therefore incorporated into the equilibrium constant, to form a modified equilibrium constant. This is the **solubility product** of silver chloride, K_{sp}.

So

$$K_{sp} = [Ag^+(aq)]_{eqm}\,[Cl^-(aq)]_{eqm}$$

and in this case the units of K_{sp} are $mol^2\,dm^{-6}$.

Finding concentrations from K_{sp}

If we know the value of K_{sp} we can find the concentrations of ions present in a saturated solution.

For example, for silver chloride, AgCl, in the example above, $K_{sp} = 1.8 \times 10^{-10}\,mol^2\,dm^{-6}$.

Let y be the concentration of Ag^+ (in $mol\,dm^{-3}$), which is the same as the concentration of Cl^-.

$$K_{sp} = [Ag^+(aq)]_{eqm}\,[Cl^-(aq)]_{eqm}$$
$$1.8 \times 10^{-10} = y^2$$
$$So\ y = 1.34 \times 10^{-5}\,mol\,dm^{-3}$$

To find the concentration in $g\,dm^{-3}$, multiply the number of mol by M_r for AgCl (which is 143.4):

$$143.4 \times 1.34 \times 10^{-5}\,g\,dm^{-3}$$
$$= 192.16 \times 10^{-5} = 1.9 \times 10^{-3}\,g\,dm^{-3}$$

Finding K_{sp} from the concentration

At a particular temperature the solubility of lead(II) fluoride is $2.61 \times 10^{-3}\,mol\,dm^{-3}$. What is the value of K_{sp}?

$$PbF_2(s) \rightleftharpoons Pb^{2+}(aq) + 2F^-(aq)$$

So the concentration of $Pb^{2+}(aq) = 2.61 \times 10^{-3}\,mol\,dm^{-3}$

And the concentration of $F^-(aq) = 2 \times 2.61 \times 10^{-3}\,mol\,dm^{-3}$, because there are 2 fluoride ions for every lead ion.

Now $K_{sp} = [Pb^{2+}(aq)]\,[F^-(aq)]^2$

Putting the values for concentrations into this equation:

$$K_{sp} = 2.61 \times 10^{-3}\,mol\,dm^{-3} \times (2 \times 2.61 \times 10^{-3}\,mol\,dm^{-3})^2$$
$$= 7.11 \times 10^{-8}\,mol^3\,dm^{-9}$$

Notice how the units are multiplied together.

For lead iodide, PbI_2, (solubility 1.65×10^{-3} mol dm⁻³) the expression for the equilibrium is:

$$PbI_2(s) \rightleftharpoons Pb^{2+}(aq) + 2I^-(aq) \text{ so:}$$

$$K_{sp} = [Pb^{2+}(aq)]_{eqm} [I^-(aq)]_{eqm}^2 \text{ (units mol}^3 \text{dm}^{-9})$$

$$= (1.65 \times 10^{-3})(2 \times 1.65 \times 10^{-3})^2$$

$$= 1.8 \times 10^{-8} \text{ mol}^3 \text{dm}^{-9}$$

The common ion effect

If we were to add soluble lead nitrate to a solution of lead iodide to increase $[Pb^{2+}(aq)]$, Le Chatelier's principle tells us that the equilibrium would move to the left and precipitate solid lead iodide. Adding soluble potassium iodide to increase $[I^-(aq)]$ would have the same effect.

This is called the **common ion effect** and the process of precipitation is sometimes called 'salting out'. It is a useful technique for removing a partially soluble salt from a solution. The rule is that if the value of the solubility product is exceeded, solid will precipitate.

Worked example

The common ion effect

Lead iodide (PbI_2) is relatively insoluble, $K_{sp} = 1.8 \times 10^{-8}$ mol³ dm⁻⁹

$$PbI_2(s) \rightleftharpoons Pb^{2+}(aq) + 2I^-(aq)$$

So if we take a saturated solution of lead iodide in pure water:

$$[I^-(aq)] = 2 \times [Pb^{2+}(aq)]$$

$$K_{sp} = 1.8 \times 10^{-8} \text{ mol}^3 \text{dm}^{-9} = [Pb^{2+}(aq)] [I^-(aq)]^2$$

$$\text{So } 1.8 \times 10^{-8} = [Pb^{2+}(aq)] \times (2 \times [Pb^{2+}(aq)])^2 = 4 [Pb^{2+}(aq)]^3$$

$$[Pb^{2+}(aq)]^3 = \frac{1.8 \times 10^{-8}}{4}$$

$$= 4.5 \times 10^{-9} \text{ mol}^3 \text{dm}^{-9}$$

$$[Pb^{2+}(aq)] = \sqrt[3]{4.5 \times 10^{-9}} = 1.7 \times 10^{-3} \text{ mol dm}^{-3}$$

$$\mathbf{[Pb^{2+}(aq)] = 1.7 \times 10^{-3} \text{ mol dm}^{-3}}$$

If we now try to dissolve some solid lead iodide in 1 mol dm⁻³ potassium iodide solution:

then $[I^-(aq)] = 1$ mol dm⁻³ (we can ignore the tiny extra concentration that comes from the relatively insoluble lead iodide).

What will be $[Pb^{2+}(aq)]$ now?

$$K_{sp} = 1.8 \times 10^{-8} \text{ mol}^3 \text{dm}^{-9} = [Pb^{2+}(aq)] \times 1^2$$

So $\mathbf{[Pb^{2+}(aq)] = 1.8 \times 10^{-8} \text{ mol dm}^{-3}}$

Notice that this is much less than the previous value. This has come about by trying to dissolve lead iodide in a solution already containing a comparatively large concentration I^- ions. The common ion has reduced $[Pb^{2+}(aq)]$ dramatically.

We would get the same effect if we added a concentrated solution of potassium iodide to a solution of lead iodide. The common ion would cause lead iodide to precipitate out.

So a precipitate will form if the concentration of lead ions in the mixture is greater than or equal to 1.8×10^{-8} mol dm⁻³.

Extension

The sensitivity of the test for lead ions

This equilibrium is often used as the basis of a test for lead ions—we add potassium iodide solution to the solution we are testing, and look for a yellow precipitate of lead iodide. We can use the idea of solubility product to work out how sensitive the test is. The value of K_{sp} for lead iodide is $1.8 \times 10^{-8} \, mol^3 \, dm^{-9}$, so:

$$[Pb^{2+}(aq)]_{eqm} \, [I^-(aq)]_{eqm}{}^2 = 1.8 \times 10^{-8} \, mol^3 \, dm^{-9}$$

If we add an equal volume of $2 \, mol \, dm^{-3}$ potassium iodide to the lead solution:
$[I^-]$ in the mixture = $1 \, mol \, dm^{-3}$

A precipitate will form if the solubility product is just exceeded. The limit is when:

$$[Pb^{2+}(aq)]_{eqm} \times 1^2 = 1.8 \times 10^{-8} \, mol \, dm^{-3}$$

$$\text{So } [Pb^{2+}(aq)]_{eqm} = \frac{1.8 \times 10^{-8}}{1} = 1.8 \times 10^{-8} \, mol \, dm^{-3}$$

So the test will detect Pb^{2+} ions if the concentration is greater than $1.8 \times 10^{-8} \, mol \, dm^{-3}$.

Summary test 25.1

1 Identify which reactant is an acid and which a base in the following:

   a $HNO_3 + OH^- \rightarrow NO_3{}^- + H_2O$
   b $CH_3COOH + H_2O \rightarrow CH_3COO^- + H_3O^+$

2 At $298 \, K$ in an acidic solution, $[H^+]$ is $1 \times 10^{-4} \, mol \, dm^{-3}$. Deduce $[OH^-(aq)]$.

3 Identify the species formed when the following bases accept a proton:

   a OH^-
   b NH_3
   c H_2O
   d Cl^-

4 Calculate the pH of $0.1 \, mol \, dm^{-3}$ sodium hydroxide solution.

5 **a** A solution of hydrochloric acid has a pH of 2. Deduce its concentration.
   b A solution of hydrochloric acid has a pH of 3. Without doing a calculation state the concentration of this acid.
   c Explain your answer to part **b**.

6 Identify the strongest acid in Table 2 earlier in this section.

7 State what you can say about the concentration of H^+ ions compared with the concentration of ethanoate ions in all solutions of pure ethanoic acid.

8 Calculate the pH of the following solutions:
   a $0.100 \, mol \, dm^{-3}$ chloroethanoic acid
   b $0.0100 \, mol \, dm^{-3}$ benzoic acid

9 Calculate the pH of the following buffers.
   a Using [ethanoic acid] = $0.10 \, mol \, dm^{-3}$, [sodium ethanoate] = $0.20 \, mol \, dm^{-3}$ (K_a of ethanoic acid = $1.7 \times 10^{-5} \, mol \, dm^{-3}$).
   b Using [benzoic acid] = $0.10 \, mol \, dm^{-3}$, [sodium benzoate] = $0.10 \, mol \, dm^{-3}$ (K_a of benzoic acid = $6.3 \times 10^{-5} \, mol \, dm^{-3}$).

10 Silver sulfide (Ag_2S) is sparingly soluble in water.
   a Give the equation for the equilibrium set up when solid silver sulfide dissolves in water.
   b Give the expression for the solubility product of silver sulfide.
   c State the units of this solubility product.

11 K_{sp} for Ag_2S is $8 \times 10^{-5} \, mol^3 \, dm^{-9}$. State the concentration of ions in a saturated solution.

Learning outcomes

On these pages you will learn to:

- explain and perform calculations using partition coefficients
- describe the effect of polarity on solubility

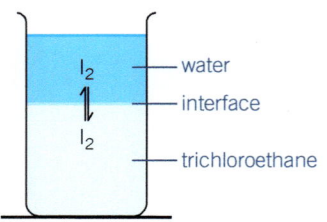

Figure 5 *Partition of iodine between water and trichloroethane*

If we add together two solvents that do not mix, such as trichloroethane and water, they will form two layers with the denser liquid on the bottom (see Figure 5). If we now add a solute and shake well, the solute will divide itself between the two solvents according to its solubility in each. At the interface some solute will move 'up' and some will move 'down', but at the same rate. A dynamic equilibrium will have been set up:

$$\text{solute in solvent A} \rightleftharpoons \text{solute in solvent B}$$

The equilibrium expression is:

$$\frac{[\text{solute in solvent A}]}{[\text{solute in solvent B}]} = \text{constant}$$

The constant is called the **partition coefficient**. It has the symbol K_{pc} and has no units.

This forms the basis of a technique called **solvent extraction** for extracting a solute. For example, iodine is around 80 times more soluble in tetrachloromethane than it is in an aqueous solution of potassium iodide (ie the partition coefficient is 80). Shaking an aqueous iodine solution with tetrachloromethane will extract the majority of the iodine into the tetrachloromethane layer. We can then evaporate off the solvent leaving the iodine behind.

The reason for the different solubilities of a solute in different solvents is explained by the polarity of the solute and solvents. Polar (and ionic) solutes dissolve well in polar solvents such as water or ethanol, while non-polar solutes dissolve well in non-polar solvents.

For example, sodium chloride (ionic) will dissolve in water (polar) but not in hexane (non-polar). Iodine (non-polar) will dissolve much better in trichloromethane (relatively non-polar) than in water.

Figure 6 *A solute is partitioned between two immiscible solvents in a separating funnel in order to separate it from other products of a reaction*

Worked example

Partition of iodine between trichloroethane and an aqueous solution

The partition coefficient for iodine between an aqueous solution and an organic solvent trichloroethane (abbreviated as TCE) as shown in Figure 5 is 80.

So $\dfrac{[I_2(TCE)]_{eqm}}{[I_2(aq)]_{eqm}} = 80$

If we start with a 0.1 mol solution of aqueous iodine, and shake with an equal volume of TCE, what will be the concentration of iodine in the TCE layer?

Imagine the iodine divided into 81 units. 80 of them will be dissolved in TCE and 1 in the aqueous solution. So $\dfrac{80}{81}$ of the original iodine will end up dissolved in TCE,

so $[I_2(TCE)]_{eqm}$ will be $\left(\dfrac{80}{81}\right) \times 0.1\,\text{mol}$

$$[I_2(TCE)]_{eq} = 0.099\,\text{mol}$$

0.001 mol of iodine remains in the aqueous solution. If this is extracted again with an equal volume of trichloroethane, $\dfrac{80}{81} \times 0.001$ will be the amount in the trichloroethane layer, ie 0.000 99 mol. This can be added to the amount extracted by the first extraction giving 0.099 + 0.000 99 = 0.099 99 mol iodine in total.

This technique is useful for extracting a non-polar organic compound, for example, from an aqueous reaction mixture. This is because non-polar compounds will dissolve to a much greater extent in non-polar solvents than they do in water.

Summary test 25.2

1 Water and trichloromethane do not mix and trichloromethane is the denser liquid. Iodine dissolved in aqueous solution is brown and it is purple in trichloromethane. Describe what you would observe if an aqueous solution of iodine is shaken with an equal volume of trichloromethane. Explain how you would know when equilibrium had been reached.

2 The partition coefficient of compound X between ether and water is 10.1 g of X is dissolved in 100 cm³ of water. Calculate how much of X can be extracted using 100 cm³ of ether. Calculate how much more a second extraction with 100 cm³ of ether would extract.

(🖺 **Launch additional digital resources for the chapter**)

1 a Give the formula for the conjugate acid of:
   i CH_3COO^- **ii** NH_3 **iii** HCO_3^- *(3 marks)*

   b Give the formula for the conjugate base of:
   i H_3O^+ **ii** HSO_4^- **iii** NH_3 *(3 marks)*

2 a Construct an expression for the ionic product of water, K_w.   *(1 mark)*

   b For a 0.01 mol dm^{-3} solution of sulfuric acid at 298 K, calculate:
   i The sulfate ion concentration   *(1 mark)*
   ii The hydroxide ion concentration.
   $(K_w = 1 \times 10^{-14}$ mol^2 dm$^{-6})$ *(2 marks)*

3 The concentration of a sample of sodium hydroxide solution is 2.0 mol dm^{-3}. Sodium hydroxide is a strong alkali.

   a Use the value of K_w in question 2 to calculate [H$^+$] for the solution.   *(2 marks)*

   b Write an expression for pH.   *(1 mark)*

   c Use your answer to part **b** to calculate the pH of the solution.   *(1 mark)*

4 The solubility product of lead(II) sulfate, PbSO$_4$, in water is 1.6×10^{-8} mol^2 dm^{-6}.

   a Calculate the solubility of lead(II) sulfate in:
   i Pure water
   ii 0.1 mol dm^{-3} Pb(NO$_3$)$_2$
   iii 0.01 mol dm^{-3} Na$_2$SO$_4$ *(3 marks)*

   b Explain why lead(II) sulfate is more soluble in water than in any solution containing either Pb^{2+}(aq) or SO$_4^{2-}$(aq)?   *(2 marks)*

   c Use your understanding of solubility product to explain the 'common ion' effect. Give an example to support your answer.   *(3 marks)*

5 0.6 g of a solute, X, was shaken with an immiscible mixture of 20 cm^3 trichloromethane and 100 cm^3 water at 298 K. At equilibrium, analysis showed that the trichloromethane contained 0.5 g of X and the water contained 0.1 g of X.

   a Calculate the partition coefficient of X between trichloromethane and water at 298 K.   *(2 marks)*

   b Partition coefficients have no units. Give a reason for this.   *(1 mark)*

   c Explain why temperature should always be given when measuring a partition coefficient.   *(1 mark)*

6 A buffer solution is made by making a solution of a weak acid with its salt.

   a Explain the role of a buffer solution.   *(1 mark)*

   b Construct two expressions to show what happens when the two substances are added to water.   *(2 marks)*

   c Use your answer to part **b** to help you to explain how the pH of a buffer solution remains stable when an acid is added.   *(2 marks)*

   d Use your answer to part **b** to help you explain how the pH of a buffer solution remains stable on addition of an alkali.   *(2 marks)*

   e Butanoic acid is a weak monobasic acid. $(K_a = 6.3 \times 10^{-5})$.
   i Calculate the mass of butanoic acid needed to make 1 dm^3 of a 0.02 mol dm^{-3} solution of butanoic acid.   *(2 marks)*
   ii Calculate the pH of a solution of the acid of concentration 0.02 mol dm^{-3}.   *(4 marks)*
   iii Calculate the pH of a buffer solution made from 0.02 mol dm^{-3} butanoic acid and 0.04 mol dm^{-3} sodium butanoate.   *(3 marks)*
   iv Calculate the pH change of the buffer solution on addition of 1 cm^3 of 1.0 mol dm^{-3} of NaOH solution. Assume the total volume remains unchanged.   *(6 marks)*

7 The table below gives he pK_a values for three substances:

Substance	pK_a
ethanoic acid	4.8
ethanol	16.0
phenol	9.9

  a Write the names of the substances in order of increasing acidity, starting with the least acidic. *(1 mark)*

  b Give the structural formula for all substances in the table. *(3 marks)*

  c Explain in terms of Brønsted–Lowry theory, the relative acidities of the three compounds in the table. *(3 marks)*

8 An oil company is worried about an impurity, **M**, in its petrol. 1 dm³ of petrol contains 5 g of **M**. In an effort to reduce the concentration of **M** in its petrol, the company use a solvent, **S**. The partition coefficient of **M** between petrol and S is 0.01 at 298 K.

  a Explain what is meant by the term partition coefficient. *(2 marks)*

  b Explain the principles of solvent extraction. *(2 marks)*

  c Calculate the total mass of *M* removed from 1 dm³ of petrol by shaking it with 100 cm³ of **S** at 298 K. *(3 marks)*

9 5 mol of ethanol, 6 mol of ethanoic acid, 6 mol of ethyl ethanoate, and 4 mol of water were mixed together in a stoppered bottle at 15°C. After equilibrium had been attained, the bottle was found to contain only 4 mol of ethanoic acid.

  a Write an equation for the reaction between ethanol and ethanoic acid to form ethyl ethanoate and water. *(1 mark)*

  b Construct an expression for the equilibrium constant, K_c, for this reaction at 15°C. *(1 mark)*

  c Calculate how many moles of ethanol, ethyl ethanoate, and water are present in the equilibrium mixture. *(3 marks)*

  d Calculate the value of K_c for this reaction. Give the units for K_c. *(2 marks)*

  e In a separate container, 1 mol of ethanol, 1 mol of ethanoic acid, 3 mol of ethyl ethanoate, and 3 mol of water are mixed together in a stoppered flask at 15°C. Calculate the number of moles of ethanol and ethyl ethanoate present at equilibrium. *(2 marks)*

10 A buffer solution with a pH of 3.95 was prepared using ethanoic acid and sodium ethanoate. In the buffer solution, the concentration of ethanoate ions was 0.125 mol dm⁻³.

K_a for ethanoic acid = 1.74×10^{-5} mol dm⁻³ at 25°C.

  a Construct an expression for the acid dissociation constant, K_a, for ethanoic acid. *(1 mark)*

  b Use your answer to part a to calculate the concentration of ethanoic acid in the buffer solution. Give your answer to 3 significant figures. *(3 marks)*

  c A different buffer solution was prepared. The concentration of ethanoic acid was 0.25 mol dm⁻³ and the concentration of sodium ethanoate was 0.125 mol dm⁻³.

    A 5×10^{-3} mol sample of sodium hydroxide was added to 500 cm³ of this buffer solution. Calculate the pH of the buffer solution after the addition of sodium hydroxide. Give your answer to two decimal places. *(5 marks)*

11 Butanoic acid is an example of a weak acid.

  a Define the term *weak acid*. *(1 mark)*

  b Construct a balanced equation for the reaction of butanoic acid with sodium carbonate. *(1 mark)*

  c Calcium hydroxide was dissolved in water to give a solution with a concentration of 0.02 mol dm⁻³. Calculate the pH of this solution. ($K_W = 1.00 \times 10^{-14}$ mol² dm⁻⁶).

Give your answer to two decimal places. *(3 marks)*

26 Reaction kinetics

26.1 Simple rate equations, orders of reaction and rate constants

Learning outcomes

On these pages you will learn to:

- use the rate equation to determine the order of a reaction
- calculate the initial rate of a reaction
- describe the relationship between half-life and order of reaction
- calculate the value of the rate constant, k
- interpret reaction mechanisms to identify intermediates and the rate-determining step
- explain how temperature change affects the rate of a reaction

Exam tip

Species in chemistry is a general term that includes molecules, ions, and atoms that might be involved in a chemical reaction.

Exam tip

When $X \propto Y$, if X is doubled then Y also doubles.

The rate of chemical reactions

The rate of a chemical reaction depends on the concentrations of some or all of the species in the reaction vessel—reactants and catalysts. But these do not necessarily all make the same contribution to how fast the reaction goes.

For example, in the reaction $X + Y \rightarrow Z$, the concentration of X, [X], may have more effect than the concentration of Y, [Y]. Or, it may be that [X] has no effect on the rate and only [Y] matters. The detail of how each species contributes to the rate of the reaction can only be found out by experiment. A species that does not appear in the chemical equation may also affect the rate, for example, a catalyst.

The **rate equation**, also called the rate expression, tells us about the contributions of the species that do affect the rate of a reaction.

The rate equation

The rate equation is the result of experimental investigation. It is an equation that describes how the rate of the reaction at a particular temperature depends on the concentration of species involved in the reaction. It is quite possible that one (or more) of the species that appear in the chemical equation will not appear in the rate equation. This means that they do not affect the rate. For example, the reaction:

$$X + Y \rightarrow Z$$

This reaction might have the rate equation:

$$\text{rate} \propto [X][Y]$$

The symbol $\propto$ means proportional to.

This would mean that both [X] and [Y] have an equal effect on the rate. Doubling either [X] or [Y] would double the rate of the reaction. Doubling the concentration of both would quadruple the rate.

But it might be that the rate equation for the reaction is:

$$\text{rate} \propto [X][Y]^2$$

This would mean that doubling [X] would double the rate of the reaction, but doubling [Y] would quadruple the rate.

A species, such as a catalyst, that is not in the chemical equation may also appear in the rate equation.

The rate constant *k*

By introducing a constant into the equation you can get rid of the proportionality sign. For example, suppose the rate equation were:

$$\text{rate} \propto [X][Y]^2$$

This can be written:

$$\text{rate} = k[X][Y]^2$$

k is called the **rate constant** for the reaction. k is different for every reaction and varies with temperature, so the temperature at which it was measured needs to be stated. If the concentrations of all the species in the rate equation are 1 mol dm^{-3}, then the rate of reaction is equal to the value of k.

The order of a reaction

Suppose the rate equation for a reaction is:

$$\text{rate} = k[X][Y]^2$$

This means that $[Y]$, which is raised to the power of 2, has double the effect on the rate than that of $[X]$. The **order of reaction**, with respect to one of the species, is the power to which the concentration of that species is raised in the rate equation. It tells us how the rate depends on the concentration of that species.

So, for rate = $k[X][Y]^2$ the order with respect to X is one ($[X]$ and $[X]^1$ are the same thing), and the order with respect to Y is two.

The **overall order of the reaction** is the sum of the orders of all the species that appear in the rate equation. In this case, the overall order is three. So this reaction is said to be **first-order** with respect to X, **second-order** with respect to Y, and **third-order** overall.

So if the rate equation for a reaction is rate = $k[A]^m[B]^n$, where m and n are the orders of the reaction with respect to A and B, the overall order of the reaction is $m + n$.

The chemical equation and the rate equation

The rate equation tells us about the species that affect the rate. Species that appear in the chemical equation do not necessarily appear in the rate equation. Also, the coefficient of a species in the chemical equation—the number in front of it—has no relevance to the rate equation. But catalysts, which do not appear in the chemical equation, may appear in the rate equation.

For example, in the reaction:

$$CH_3COCH_3(aq) + I_2(aq) \xrightarrow{\text{H}^+ \text{ catalyst}} CH_2ICOCH_3(aq) + HI(aq)$$

propanone · · · · · iodine · · · · · · · · · · · · · · · iodopropanone · · · · · hydrogen iodide

The rate equation has been found by experiment to be:

$$\text{rate} = k[CH_3COCH_3(aq)][H^+(aq)]$$

So the reaction is first-order with respect to propanone, first-order with respect to H$^+$ ions, and second-order overall. The rate does not depend on $[I_2(aq)]$, so you can say the reaction is zero order with respect to iodine. The H$^+$ ions act as a catalyst in this reaction.

Units of the rate constant

The units of the rate constant vary depending on the overall order of reaction.

For a **zero-order** reaction:

$$\text{rate} = k$$

The units of rate are mol dm^{-3} s^{-1}.

For a first-order reaction where:

$$\text{rate} = k[A]$$

$$k = \frac{\text{rate}}{[A]}$$

The units of rate are $mol\,dm^{-3}\,s^{-1}$ and the units of [A] are $mol\,dm^{-3}$, so the units of k are s^{-1} obtained by cancelling:

$$k = \frac{\cancel{mol\,dm^{-3}}\,s^{-1}}{\cancel{mol\,dm^{-3}}}$$

Therefore, the units of k for a first-order rate constant are s^{-1}.

For a second-order reaction where:

$$rate = k[B][C]$$

$$k = \frac{rate}{[B][C]}$$

The units of rate are $mol\,dm^{-3}\,s^{-1}$ and the units of both [B] and [C] are $mol\,dm^{-3}$, so the units of k are s^{-1} obtained by cancelling:

$$k = \frac{\cancel{mol\,dm^{-3}}\,s^{-1}}{\cancel{mol\,dm^{-3}}\,mol\,dm^{-3}}$$

Therefore the units of k for a second-order rate constant are $dm^3\,mol^{-1}\,s^{-1}$.

For a third-order reaction:

$$rate = k[D][E]^2$$

$$k = \frac{rate}{[D][E]^2}$$

The unit of rate is $mol\,dm^{-3}\,s^{-1}$, the unit of [D] is $mol\,dm^{-3}$, and the unit of $[E]^2$ is $(mol\,dm^{-3})^2$.

$$k = \frac{\cancel{mol\,dm^{-3}}\,s^{-1}}{\cancel{mol\,dm^{-3}}\,(mol\,dm^{-3})^2}$$

Therefore, the units of k for a third-order rate constant are $dm^6\,mol^{-2}\,s^{-1}$.

Determining the rate equation

- If the rate is not affected by the concentration of a species, the reaction is *zero order* with respect to that species. The species is not included in the rate equation.
- If the rate is directly proportional to the concentration of the species, the reaction is *first-order* with respect to that species.
- If the rate is proportional to the square of the concentration of the species, the reaction is *second-order* with respect to that species, and so on.

Finding the order of a reaction by using rate–concentration graphs

One method of finding the order of a reaction with respect to a particular species, A, is by plotting a graph of rate against concentration.

First plot a graph of [A] against time, and draw tangents at different values of [A]. The gradients of these tangents are the reaction rates (the changes in concentration over time) at different concentrations (Figure 1). The values for these rates can then be used to construct a second graph of rate against concentration (Figure 2).

- If the graph is a horizontal straight line (Figure 2a), this means that the rate is unaffected by [A] so **the order is zero**.
- If the graph is a sloping straight line through the origin (Figure 2b) then rate $\propto [A]^1$ so **the order is 1**.
- If the graph is not a straight line (Figure 2c), the order cannot be found directly—it could be two. Try plotting rate against $[A]^2$. If this is a straight line, then **the order is two**.

> **Exam tip**
>
> Remember that the rate equation is found entirely from experimental evidence. It cannot be predicted from the chemical equation for the reaction. It is therefore very different to the equilibrium law expression with the equilibrium constant K_c (although it looks similar).

> **Exam tip**
>
> It is always better to work out the units of k rather than try to remember them.

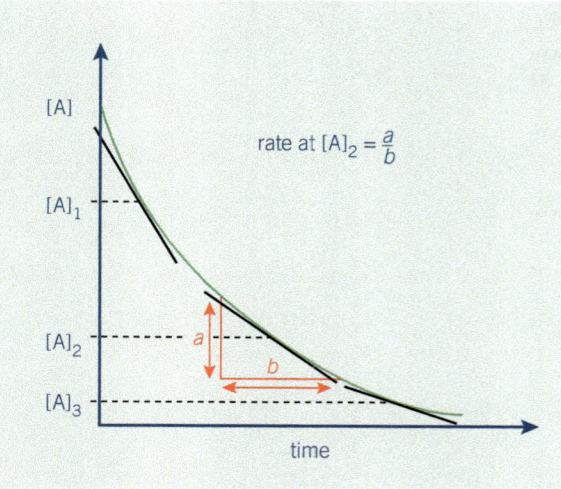

rate at $[A]_2 = \dfrac{a}{b}$

Figure 1 *Finding the rates of reaction for different values of [A]*

The initial rate method

With the initial rate method, a series of experiments is carried out at constant temperature. Each experiment starts with a different combination of initial concentrations of reactants, catalyst, and so on. The experiments are planned so that, between any pair of experiments, the concentration of only one species varies—the rest stay the same. Then, for each experiment, the concentration of one reactant is followed and a concentration–time graph plotted (Figure 3). The tangent to the graph at time = 0 is drawn. The gradient of this tangent is the **initial rate**. By measuring the initial rate, the concentrations of all substances in the reaction mixture are known *exactly* at this time.

Comparing the initial concentration and the initial rates for pairs of experiments allows the order with respect to each reactant to be found. For example, for the reaction:

$$2NO(g) \quad + \quad O_2(g) \quad \rightarrow \quad 2NO_2(g)$$

nitrogen monoxide   oxygen   nitrogen dioxide

The initial rates are shown in Table 1.

Table 1 *Results obtained for the reaction $2NO(g) + O_2(g) \rightarrow 2NO_2(g)$*

Experiment number	Initial [NO] / mol dm^{-3}	Initial [O$_2$] / mol dm^{-3}	Initial rate / mol dm^{-3} s^{-1}
1	1.0×10^{-3}	1.0×10^{-3}	7.0×10^{-4}
2	2.0×10^{-3}	1.0×10^{-3}	28.0×10^{-4}
3	3.0×10^{-3}	1.0×10^{-3}	63.0×10^{-4}
4	2.0×10^{-3}	2.0×10^{-3}	56.0×10^{-4}
5	3.0×10^{-3}	3.0×10^{-3}	189.0×10^{-4}

Comparing Experiment 1 with Experiment 2, [NO] is doubled whilst [O$_2$] stays the same. The rate quadruples (from 7.0×10^{-4} mol dm^{-3} s^{-1} to 28.0×10^{-4} mol dm^{-3} s^{-1}) which suggests rate $\propto$ [NO]^2. This is confirmed by considering Experiments 1 and 3 where [NO] is trebled whilst [O$_2$] stays the same. Here the rate is increased ninefold, as would be expected if rate $\propto$ [NO]^2. So the order with respect to nitrogen monoxide is two.

Now compare Experiment 2 with Experiment 4. Here [NO] is constant but [O$_2$] doubles. The rate doubles (from 28.0×10^{-4} mol dm^{-3} s^{-1} to 56.0×10^{-4} mol dm^{-3} s^{-1}) so it looks as if rate $\propto$ [O$_2$]. This is confirmed by considering Experiments 3 and 5. Again [NO] is constant, but [O$_2$] triples. The rate triples too, confirming that the order with respect to oxygen is one.

So   rate $\propto$ [NO]^2 and rate $\propto$ [O$_2$]^1

That is, rate $\propto$ [NO]^2[O$_2$]^1

Provided that no other species affect the reaction rate, the overall order is three and the rate equation is: rate $= k$ [NO]^2[O$_2$]^1

Finding the rate constant k

To find k in the reaction of NO and O$_2$ substitute any set of values of rate, [NO], and [O$_2$] in the equation. Taking the values for Experiment 2:

$$28.0 \times 10^{-4} = k \times (2 \times 10^{-3})^2 \times 1 \times 10^{-3}$$

$$28.0 \times 10^{-4} = k \times 4 \times 10^{-9}$$

$$k = \frac{28.0}{4} \times 10^{5}$$

$$k = 7.0 \times 10^{5}$$

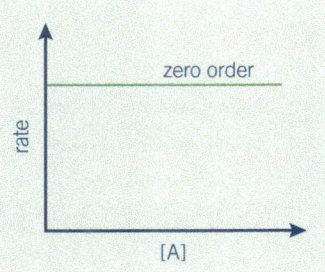

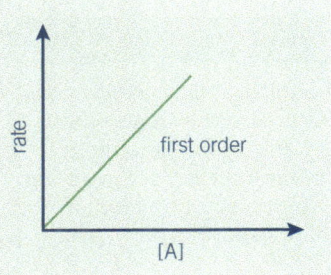

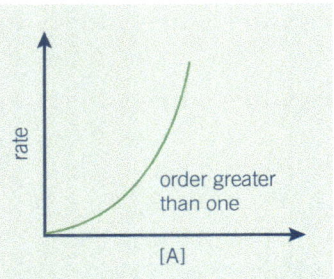

Figure 2 *Graphs of rate against concentration*

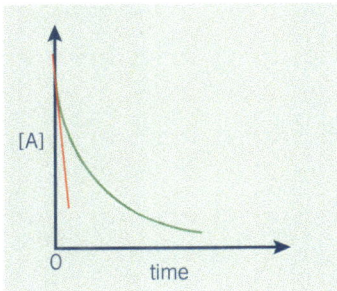

Figure 3 *Finding the initial rate of a reaction. The initial rate is the gradient at time = 0*

> **Exam tip**
>
> It is easier to apply the initial rate technique to problems than to read about it. You can practise using other pairs of experiments in Table 1.

But you need to work out the units for k, as these are different for reactions of different overall order. Putting in the units gives:

$$28.0 \times 10^{-4}\,\text{mol}\,\text{dm}^{-3}\,\text{s}^{-1} = k \times (2 \times 10^{-3})^2\,(\text{mol}\,\text{dm}^{-3})\,(\text{mol}\,\text{dm}^{-3}) \times 1 \times 10^{-3}\,\text{mol}\,\text{dm}^{-3}$$

Units can be cancelled in the same way as numbers, so cancelling the units gives:

$$28.0 \times 10^{-4}\,\cancel{\text{mol}\,\text{dm}^{-3}}\,\text{s}^{-1} = k \times (4 \times 10^{-6})\,(\cancel{\text{mol}\,\text{dm}^{-3}})\,(\text{mol}\,\text{dm}^{-3}) \times 1 \times 10^{-3}\,\text{mol}\,\text{dm}^{-3}$$

$$28.0 \times 10^{-4} = k \times 4 \times 10^{-9}\,\text{mol}^2\,\text{dm}^{-6}\,\text{s}^{-1}$$

$$k = \frac{28.0}{4} \times 10^{5}\,\text{mol}^{-2}\,\text{dm}^{6}\,\text{s}^{-1}$$

$$k = 7.0 \times 10^{5}\,\text{dm}^{6}\,\text{mol}^{-2}\,\text{s}^{-1}$$

Exam tip

You should get the same value of k using the figures for any of the experiments.

Exam tip

Since the units of k vary for reactions of different orders, it is important to put the units for rate and the concentrations in first. Then, you can cancel them to make sure you have the correct units for k.

The effect of temperature on k

Small changes in temperature produce large changes in reaction rates. A rough rule is that for every 10 K rise in temperature, the rate of a reaction doubles. Suppose the rate equation for a reaction is rate = k[A][B]. You know that [A] and [B] do not change with temperature, so the rate constant k must increase with temperature.

In fact, the rate constant k allows you to compare the speeds of different reactions at a given temperature. It is a property specific to a particular reaction.

> **k is the rate of the reaction at a particular temperature when the concentrations of all the species in the rate equation are 1 mol dm^{-3}.**

The larger the value of k, the faster the reaction. Look at Table 2. You can see that the value of k increases with temperature. This is true for all reactions.

Table 2 The values of the rate constant, k, at different temperatures for the reaction $2HI(g) \rightarrow I_2(g) + H_2(g)$

Temperature / K	k / mol^{-1} dm^3 s^{-1}
633	0.0178×10^{-3}
666	0.107×10^{-3}
697	0.501×10^{-3}
715	1.05×10^{-3}
781	15.1×10^{-3}

Exam tip

Remember, increasing the temperature always increases the rate of reaction and the value of the rate constant k.

Why the rate constant depends on temperature

Temperature is a measure of the average kinetic energy of particles. Particles will only react together if their collisions have enough energy to start bond breaking. This energy is called the **activation energy** E_A. Figure 4 shows how the energies of the particles in a gas (or in a solution) are distributed at three different temperatures. Only molecules with energy greater than E_A can react (see Section 5.1).

The shape of the graph changes with temperature. As the temperature increases, a greater proportion of molecules have enough energy to react. This is the main reason for the increase in reaction rate with temperature (see Figure 4).

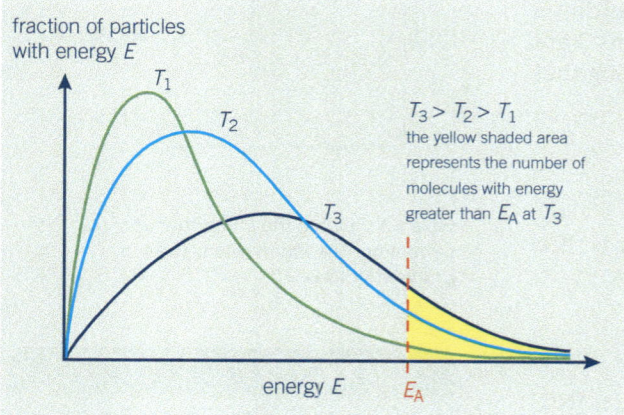

$T_3 > T_2 > T_1$

the yellow shaded area represents the number of molecules with energy greater than E_A at T_3

Figure 4 The distribution of molecular energies at three temperatures

The half-life of a reaction

> **The half-life of a reaction, $t_{1/2}$, is the time taken for the concentration of a reactant to fall to half its original value.**

The **half-life** may be measured from the start of the reaction or from any chosen point after this. It gives us an alternative method of determining the order of a reaction from a graph of concentration of reactant (or product) against time.

Figure 5 shows a graph of the concentration of a reactant, R, against time for a first-order reaction. The red lines show how to calculate the first half-life (starting from the initial concentration) and then successive half-lives (starting from the end of the previous half-life). *For a first-order reaction, successive half-lives are exactly the same.* In other words, the half-life of a first-order reaction is independent of concentration.

We can find the rate constant for a first-order reaction from the half-life.

$$\text{Rate constant, } k = \frac{0.693}{t_{1/2}}$$

Worked example

Finding the order and rate constant using half-lives

The graph on the right shows the concentration of reactant against time for a chemical reaction:

- determine the first three half-lives
- show that the reaction is first-order with respect to this reactant
- calculate the rate constant, k.

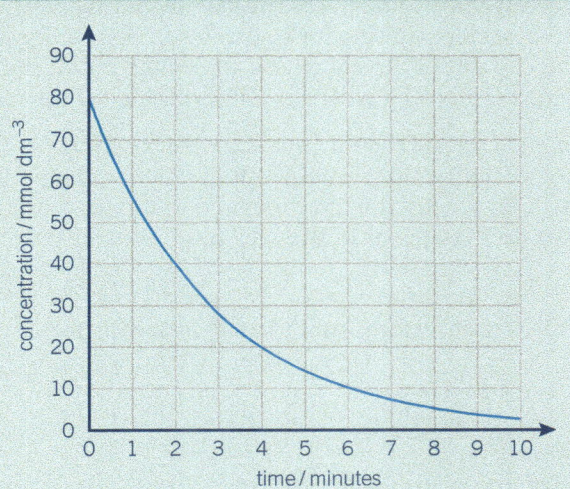

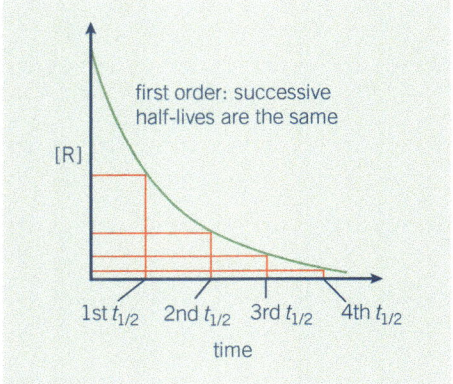

Figure 5 Successive half-lives for a first-order reaction

The second graph shows the first three half-lives:

- The first half-life is the time for the concentration to drop from 80 mmol dm⁻³ to 40 mmol dm⁻³. It is 2 minutes.

- The second half-life is the time for the concentration to drop from 40 mmol dm⁻³ to 20 mmol dm⁻³. It is 2 minutes.

- The third half-life is the time for the concentration to drop from 20 mmol dm⁻³ to 10 mmol dm⁻³. It is also 2 minutes.

So successive half-lives are constant and the reaction is first-order with respect to the reactant whose concentration was measured.

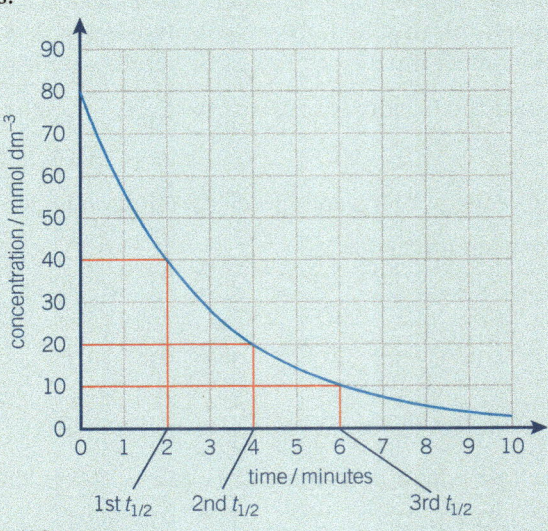

Using rate constant, $k = \dfrac{0.693}{t_{1/2}} = \dfrac{0.693}{2\,\text{min}} = 0.347\ \text{min}^{-1}$

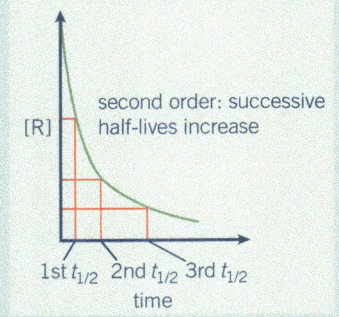

Reaction mechanisms

Most reactions take place in more than one step. The separate steps that lead from reactants to products are together called the **reaction mechanism**. For example, the reaction below involves 12 ions:

$$BrO_3^-(aq) + 6H^+(aq) + 5Br^-(aq) \rightarrow 3Br_2(aq) + 3H_2O(l)$$

This reaction *must* take place in several steps—it is very unlikely indeed that the 12 ions of the reactants will all collide at the same time. The steps in-between will involve very short-lived **intermediates**. These intermediate species, which give information about the mechanism of the reaction, are usually difficult or impossible to isolate and therefore identify. So, other ways of working out the mechanism of the reaction are used.

The rate-determining step

In a multi-step reaction, the steps nearly always follow after each other, so that the product of one step is the starting material for the next. Therefore the rate of the slowest step will govern the rate of the whole process. The slowest step may form a 'bottleneck', called the **rate-determining step**, or rate-limiting step. Suppose you had everything you needed to make a cup of instant coffee, starting with cold water. The rate of getting your drink will be governed by the slowest step: waiting for the water to boil. It doesn't matter how quickly you get the cup out of the cupboard, the coffee out of the jar, or add the milk—the rate-determining step will always be waiting for the water to boil.

In a chemical reaction, any step that occurs *after* the rate-determining step will not affect the rate, provided that it is fast compared with the rate-determining step. So species that are involved in the mechanism after the rate-determining step do not appear in the rate equation. For example, the reaction:

$$A + B + C \rightarrow Y + Z$$

This reaction might occur in the following steps:

1 $A + B \xrightarrow{\text{fast}} D$

2 $D \xrightarrow{\text{slow}} E$

3 $E + C \xrightarrow{\text{fast}} Y + Z$

Step 2 is the slowest step and so determines the rate. Then, as soon as some E is produced, it rapidly reacts with C to produce Y and Z.

But the rate of Step 1 might affect the overall rate—the concentration of D depends on this. So, any species involved in or *before* the rate-determining step could affect the overall rate and therefore appear in the rate equation.

So, for the reaction $A + B + C \rightarrow Y + Z$, the rate equation will be:

$$\text{rate} \propto [A][B][D]$$

The reaction between iodine and propanone demonstrates this.

The overall reaction is:

| propanone | iodine | | iodopropane |

The rate equation is found to be rate = $k[CH_3COCH_3][H^+]$

The mechanism is:

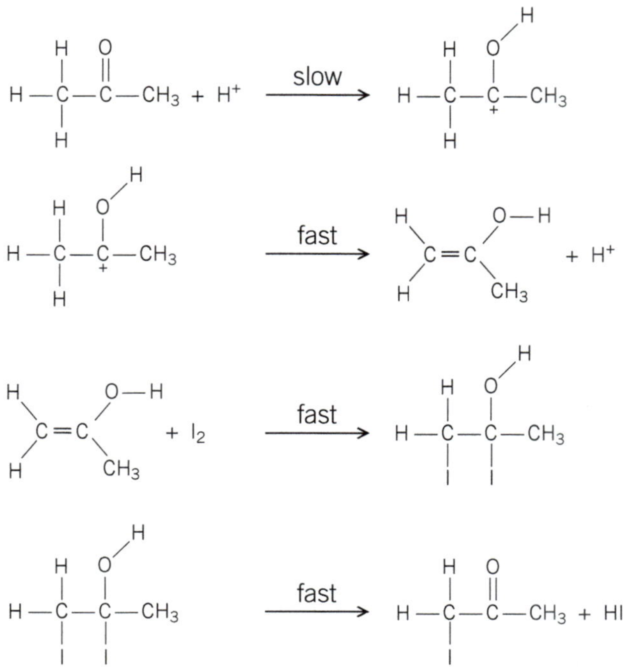

The rate-determining step is the first one, which explains why $[I_2]$ does not appear in the rate equation. Notice how a H^+ ion is used up in the first step and regenerated in the second step. It is therefore acting as a catalyst.

Using the order of a reaction to find the rate-determining step

Here is a simple example. The three structural isomers with formula C_4H_9Br all react with alkali. The overall reaction is represented by the following equation:

$$C_4H_9Br + OH^- \rightarrow C_4H_9OH + Br^-$$

Two mechanisms are possible.

a A two-step mechanism:

$$\textbf{Step 1} \qquad C_4H_9Br \xrightarrow{\text{slow}} C_4H_9^+ + Br^-$$
$$\textbf{Step 2} \qquad C_4H_9^+ + OH^- \xrightarrow{\text{fast}} C_4H_9OH$$

The slow step involves breaking the C–Br bond while the second (fast) step is a reaction between oppositely charged ions.

b A one-step mechanism:

$$C_4H_9Br + OH^- \xrightarrow{\text{slow}} C_4H_9OH + Br^-$$

The C–Br bond breaks at the same time as the C–OH bond is forming.

The three isomers of formula C_4H_9Br are:

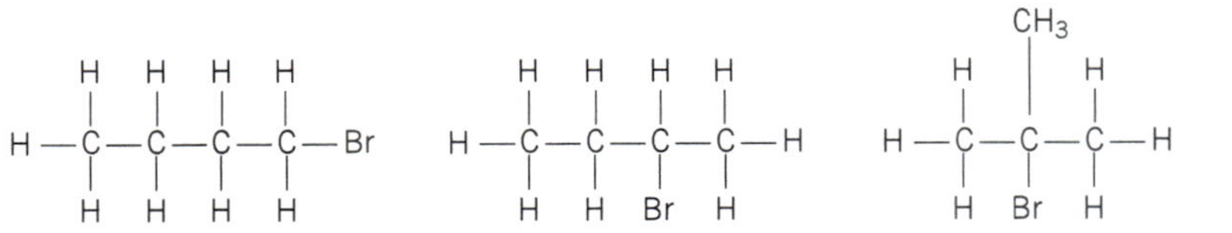

1-bromobutane   2-bromobutane   2-bromo-2-methylpropane

Experiments show that 1-bromobutane reacts by a second-order mechanism:

$$\text{rate} = k[C_4H_9Br][OH^-]$$

The rate depends on the concentration of *both* the bromobutane and the OH⁻ ions, suggesting mechanism **b**, a one-step reaction.

Experiments show that 2-bromo-2-methylpropane reacts by a first-order mechanism:

$$\text{rate} = k[\text{C}_4\text{H}_9\text{Br}]$$

This suggests mechanism **a** in which a slow step, breaking the C–Br bond, is followed by a rapid step in which two oppositely charged ions react together. So, the breaking of the C–Br bond is the rate-determining step.

The compound 2-bromobutane reacts by a mixture of both mechanisms and has a more complex rate equation.

Exam tip

The species in the rate equation are the reactants involved in reactions occurring before the rate-determining step.

Summary test 26.1

1 Give the rate equation for a reaction that is first-order with respect to [A], first-order with respect to [B], and second-order with respect to [C].

2 Consider the reaction:

$$\text{BrO}_3^-(\text{aq}) + 5\text{Br}^-(\text{aq}) + 6\text{H}^+(\text{aq}) \rightarrow 3\text{Br}_2(\text{aq}) + 3\text{H}_2\text{O(l)}$$

| bromate ions | bromide ions | hydrogen ions | bromine | water |

The rate equation is:

$$\text{rate} = k[\text{BrO}_3^-(\text{aq})][\text{Br}^-(\text{aq})][\text{H}^+(\text{aq})]^2$$

a State the order with respect to:
  i $\text{BrO}_3^-(\text{aq})$
  ii $\text{Br}^-(\text{aq})$
  iii $\text{H}^+(\text{aq})$

b Deduce the units for the rate constant.

c State what would happen to the rate if you doubled the concentration of:
  i $\text{BrO}_3^-(\text{aq})$
  ii $\text{Br}^-(\text{aq})$
  iii $\text{H}^+(\text{aq})$

d Give the coefficients of the following in the chemical equation above.
  i $\text{BrO}_3^-(\text{aq})$
  ii $\text{Br}^-(\text{aq})$
  iii $\text{H}^+(\text{aq})$
  iv $\text{Br}_2(\text{aq})$
  v $\text{H}_2\text{O(l)}$

3 In the reaction L + M → N the rate equation is found to be:
$$\text{rate} = k[\text{L}]^2[\text{H}^+]$$

a Identify *k*.

b State the order of the reaction with respect to:
  i L
  ii M
  iii N
  iv H⁺

c State the overall order of the reaction.

d The rate is measured in $\text{mol dm}^{-3}\text{s}^{-1}$. Give the units of *k*.

e Suggest the function of H⁺ in the reaction.

4 In the reaction $G + 2H \rightarrow I + J$, state which is the correct rate equation.

A rate = $k[G][H]^2$

B rate = $k \dfrac{[G][H]}{[I][J]}$

C rate = $k[G][H]$

D It is impossible to tell without experimental data.

5 For the reaction $A + B \rightarrow C$, the following data were obtained:

Initial [A] / mol dm^{-3}	Initial [B] / mol dm^{-3}	Initial rate / mol dm^{-3} s^{-1}
1	1	3
1	2	12
2	2	24

  a Give the order of reaction with respect to:

    i A

    ii B

  b Give the overall order.

  c Give the initial rate if the initial [A] were $1 \, mol \, dm^{-3}$ and [B] were $3 \, mol \, dm^{-3}$.

  d Explain what these results suggest is the rate equation for this reaction.

  e State if we can be certain that this is the full rate equation. Explain your answer.

6 A particular rate constant for the decomposition of N_2O_5 is $5 \times 10^{-5} \, s^{-1}$.

  a Give the order of this reaction. Explain your answer.

  b Give the first half-life of this reaction.

  c Explain what you can say about subsequent half-lives.

7 The following reaction schemes show possible mechanisms for the overall reaction:

$$A + E \xrightarrow{\text{catalyst}} G$$

Scheme 1	Scheme 2	Scheme 3
slow (i) A + B ⟶ C	fast (i) A + B ⟶ C	fast (i) A + B ⟶ C
fast (ii) C ⟶ D + B	fast (ii) C ⟶ D + B	slow (ii) C ⟶ D + B
fast (iii) D + E ⟶ F	slow (iii) D + E ⟶ F	fast (iii) D + E ⟶ F
fast (iv) F ⟶ G	slow (iv) F ⟶ G	fast (iv) F ⟶ G

  a In Scheme 2, state which species is the catalyst.

  b State which species *cannot* appear in the rate equation for Scheme 1.

  c State the rate-determining step in Scheme 3.

Homogeneous and heterogeneous catalysts

We have seen in Section 8.3 that catalysts reduce the activation energy of a reaction. So at a given temperature more of the collisions between reactants are effective. Catalysts work by providing an alternative reaction mechanism of lower activation energy. They are divided into **homogeneous catalysts** where the catalyst is in the same phase (solid, liquid or gas) as the reactants, and **heterogeneous catalysts** where the catalyst is in a different phase from the reactants (usually solid catalyst and gaseous or liquid reactants).

Homogeneous catalysis

Example 1—the oxidation of iodine by peroxodisulfate ions

Peroxodisulfate ions, $S_2O_8^{2-}$, oxidise iodide ions to iodine. This reaction is catalysed by Fe^{2+} ions. The overall reaction is:

$$S_2O_8^{2-}(aq) + 2I^-(aq) \rightarrow 2SO_4^{2-}(aq) + I_2(aq)$$

The catalysed reaction takes place in two steps. First the peroxodisulfate ions oxidise iron(II) to iron(III):

$$S_2O_8^{2-}(aq) + 2Fe^{2+}(aq) \rightarrow 2SO_4^{2-}(aq) + 2Fe^{3+}(aq)$$

The Fe^{3+} then oxidises the I^- to I_2, regenerating the Fe^{2+} ions so that none are used up in the reaction:

$$2Fe^{3+}(aq) + 2I^-(aq) \rightarrow 2Fe^{2+}(aq) + I_2(aq)$$

So iron first gives an electron to the peroxodisulfate and later takes one back from the iodide ions.

The uncatalysed reaction takes place between two ions of the same charge (both negative), which repel, therefore giving a high activation energy. Both steps of the catalysed reaction involve reaction between pairs of oppositely charged ions. This helps to explain the increase in rate.

Figure 7 shows a possible reaction profile. Although there are two steps in the catalysed reaction, the overall activation energy is lower than that for the uncatalysed reaction.

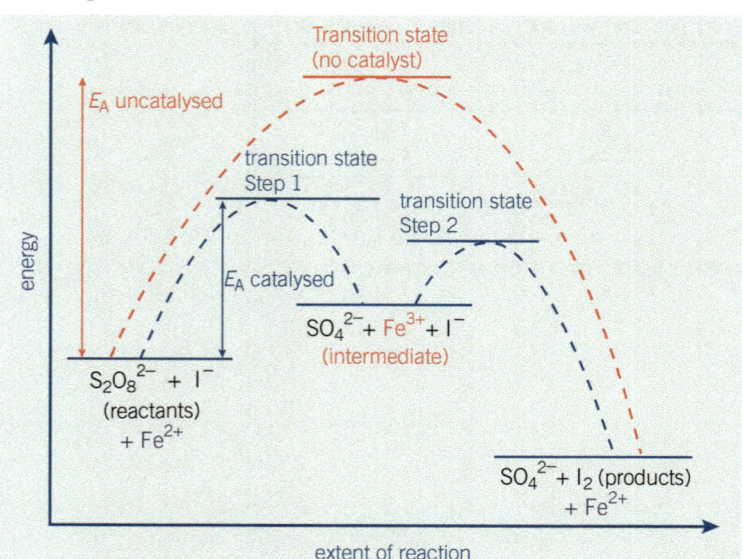

Figure 7 *Possible reaction profile for the iodine/peroxodisulfate reaction. E_A for the catalysed reaction is the energy gap between the reactants and the higher of the two transition states (transition state Step 1)*

Example 2—the oxidation of sulfur dioxide in the atmosphere

Sulfur dioxide gas is present in the atmosphere due to the burning of coal and other fossil fuels which contain sulfur and sulfur-containing impurities:

$$S(s) + O_2(g) \rightarrow SO_2(g)$$

In the atmosphere sulfur dioxide is oxidised to sulfur trioxide which is in turn converted to sulfuric acid:

$$2SO_2(g) + O_2(g) \rightarrow 2SO_3(g)$$

and

$$SO_3(g) + H_2O(l) \rightarrow H_2SO_4(l)$$

which is a significant cause of acid rain.

The oxidation of sulfur dioxide is catalysed by nitrogen oxides (NO and NO_2, sometimes called NO_x) in the atmosphere. These are produced from atmospheric nitrogen and oxygen by internal combustion engines and also lightning strikes. This happens as follows:

$$2NO(g) + O_2(g) \rightarrow 2NO_2(g) \qquad \text{reaction 1}$$
$$SO_2(g) + NO_2(g) \rightarrow SO_3(g) + NO(g) \qquad \text{reaction 2}$$

Reaction 2 converts sulfur dioxide to sulfur trioxide, then reaction 1 converts nitrogen monoxide back to nitrogen dioxide. So that the nitrogen oxides effectively form a catalyst.

Adding the two reactions shows that no nitrogen oxide is used up in the process:

$$2NO(g) + O_2(g) \rightarrow 2NO_2(g) \qquad \text{reaction 1}$$
$$2SO_2(g) + 2NO_2(g) \rightarrow 2SO_3(g) + 2NO(g) \qquad \text{reaction 2}$$

$$\cancel{2NO(g)} + O_2(g) + 2SO_2(g) + \cancel{2NO_2(g)} \rightarrow \cancel{2NO_2(g)} + 2SO_3(g) + \cancel{2NO(g)}$$

Heterogeneous catalysis

This frequently takes place on the surface of a transition metal. Molecules of gas interact with atoms on the surface of the metal and form weak temporary bonds, often part way between covalent bonds and van der Waals forces. This process, called **adsorption**, holds the reactant molecules close together and therefore more likely to react. Once reacted, molecules of the product escape from the surface (a process called **desorption**). This leaves room on the catalyst surface for more reactant molecules. The strength of the bonds is critical—they must be strong enough to hold the reactants but weak enough to allow the products to escape.

While adsorbed on the catalyst, bonds *within* the reactant molecule may weaken, as electrons in these bonds are used to form weak bonds with atoms of the catalyst, see Figure 8.

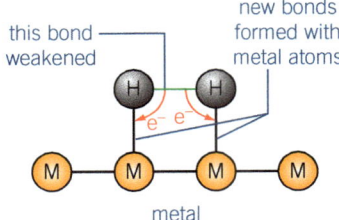

Figure 8 *Hydrogen adsorbed on a metal catalyst*

> **Exam tip**
>
> Transition metals have partly full d-orbitals which can be used to form weak chemical bonds with the reactants. This has two effects: weakening bonds within the reactant, and holding the reactants close together on the metal surface in the correct orientation for reaction.

323

Example 1—the Haber process

Ammonia is made in the Haber process by direct combination of nitrogen and hydrogen using an iron-based catalyst.

$$N_2(g) + 3H_2(g) \rightarrow 2NH_3(g)$$

During Haber's research, many transition metals were tried as catalysts and iron was found to have a suitable strength of adsorption. (The fact that it is cheap helped, too!) Because the catalyst is a solid and the products and reactants are gases, the products can be pumped away continuously.

It is thought that N_2 molecules and H_2 molecules adsorbed on the catalyst surface split into separate atoms. These then react to form ammonia, shown schematically in Figure 9.

$$N_2(g) \rightarrow 2N$$

$$H_2(g) \rightarrow 2H$$

Then $N + H \rightarrow NH + H \rightarrow NH_2 + H \rightarrow NH_3$

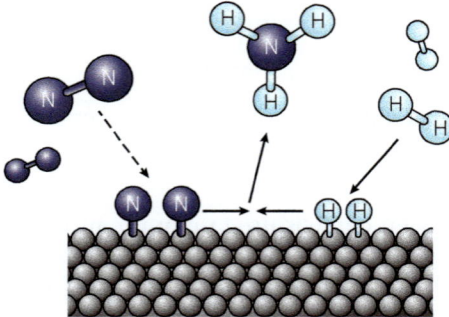

Figure 9 *Formation of ammonia on the surface of a catalyst*

Example 2—catalytic converters

A similar situation occurs in the reactions on catalytic converters in motor vehicles. The catalyst is a mixture of the transition metals palladium, platinum, and rhodium, deposited on a ceramic honeycomb to increase the surface area of these expensive metals. Two of the reactions that occur are:

carbon monoxide + nitrogen oxides → nitrogen + carbon dioxide

hydrocarbons + nitrogen oxides → nitrogen + carbon dioxide + water

What happens on the surface of the converter is shown schematically in Figure 10.

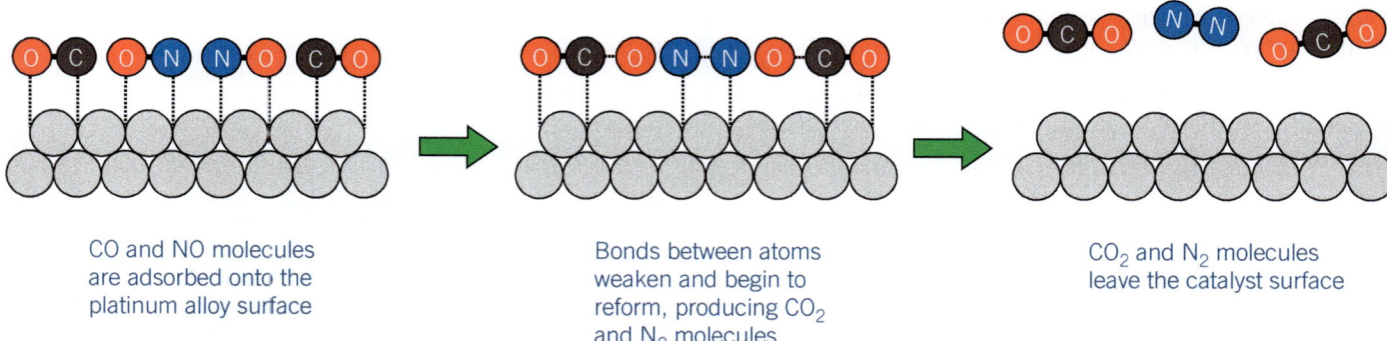

CO and NO molecules are adsorbed onto the platinum alloy surface

Bonds between atoms weaken and begin to reform, producing CO_2 and N_2 molecules

CO_2 and N_2 molecules leave the catalyst surface

Figure 10 *The reaction between nitrogen monoxide and carbon monoxide on the surface of a catalyst*

Summary test 26.2

1 **a** State the difference between a homogeneous and a heterogeneous catalyst.

   b Identify each of the examples below as homogeneous or heterogeneous:

     i A gauze of platinum and rhodium catalyses the oxidation of ammonia gas to nitrogen monoxide during the manufacture of nitric acid.

     ii Nickel catalyses the hydrogenation of vegetable oils.

     iii The enzymes in yeast catalyse the production of ethanol from sugar.

2 Explain why a catalyst makes a reaction go faster. Explain why this is particularly important for industry.

3 The peroxodisulfate / iodide reaction above is catalysed by Fe^{3+} ions (as well as by Fe^{2+} ions):

   a $S_2O_8^{2-}(aq) + 2I^-(aq) \rightarrow 2SO_4^{2-}(aq) + I_2(aq)$

   b Give the two equations that explain this and explain why it is slow in the absence of the catalyst.

1 In aqueous solution, ammonium ions react with nitrite ions according to following equation:

$$NH_4^+(aq) \ + \ NO_2^-(aq) \ \rightarrow \ N_2(g) \ + \ 2H_2O(l)$$

The table below shows how the initial rate of the reaction depends on the concentration of the reactants:

$[NH_4^+]$ / mol dm^{-3}	$[NO_2^-]$ / mol dm^{-3}	Initial rate / mol dm^{-3} dm^{-3}
0.0100	0.200	5.4×10^{-7}
0.0200	0.200	10.8×10^{-7}
0.0300	0.200	16.2×10^{-7}
0.200	0.0202	10.8×10^{-7}
0.200	0.0204	21.6×10^{-7}
0.200	0.0606	32.4×10^{-7}

a Deduce the order of reaction with respect to NH_4^+. *(2 marks)*

b Deduce the order of reaction with respect to NO_2^-. *(2 marks)*

c State the rate equation for this reaction. *(1 mark)*

d Give the overall order of reaction. *(1 mark)*

2 Hydrogen peroxide reacts with iodide ions in acid solution according to the following equation:

$$H_2O_2(aq) \ + \ 2I^-(aq) \ + \ 2H^+(aq) \ \rightarrow \ I_2(aq) \ + \ 2H_2O(l)$$

The rate of the reaction can be calculated by measuring the time for the first appearance of I_2 in the solution. When iodine first appears the concentration of I_2 is 10^{-5} mol dm^{-3}.

a For a particular experiment, the initial concentrations are $[H_2O_2]$ = 0.010 mol dm^{-3}, $[I^-]$ = 0.010 mol dm^{-3} and $[H^+]$ = 0.1 mol dm^{-3}. Calculate the rate if I_2 appears after 6 seconds. *(1 mark)*

b In a second experiment, the initial concentrations are $[H_2O_2]$ = 0.005 mol dm^{-3}, $[I^-]$ = 0.010 mol dm^{-3} and $[H^+]$ = 0.1 mol dm^{-3}. Calculate the rate if I_2 appears after 12 seconds. *(1 mark)*

c Use your answers to parts **a** and **b** to explain why the reaction is first order with respect to H_2O_2. *(2 marks)*

d Use the rate equation below to calculate the value of the rate constant, k. Give its units. *(2 marks)*

$$Rate = k[H_2O_2][H^+][I^-]$$

e Calculate the rate of reaction when $[H_2O_2]$ = 0.05 mol dm^{-3}, $[H^+]$ = 0.1 mol dm^{-3} and $[I^-]$ = 0.02 mol dm^{-3}. *(2 marks)*

3 In an experiment to study the acid-catalysed reaction of propanone (CH_3COCH_3) with iodine, 50 cm^3 of 0.02 mol dm^{-3} I_2 was mixed with 50 cm^3 of acidified 0.25 mol dm^{-3} propanone solution.

10 cm^3 portions of the reaction mixture were removed at 5 minute intervals and added rapidly to excess $NaHCO_3(aq)$. The remaining iodine was then titrated against $Na_2S_2O_3(aq)$. The graph below shows the volume of $Na_2S_2O_4(aq)$ required to react with the remaining iodine at different times from the start of the reaction.

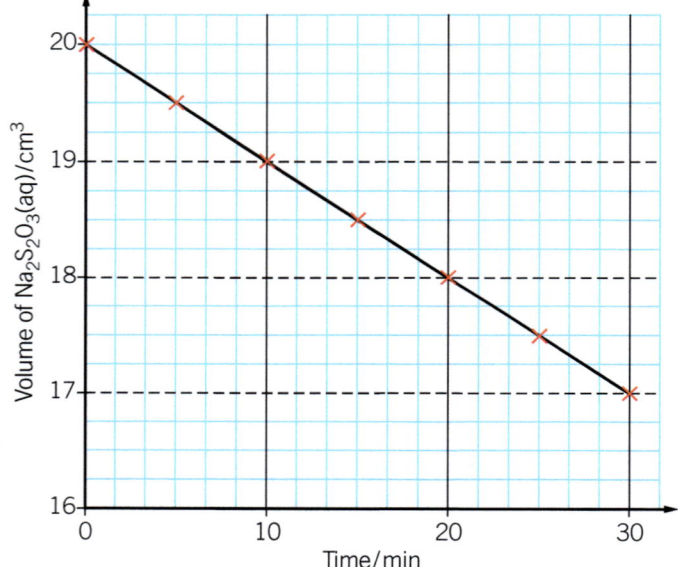

a Explain why the reaction mixture is added rapidly to excess $NaHCO_{3(aq)}$ before titration with aqueous $Na_2S_2O_3$? *(1 mark)*

b Calculate the rate of reaction for $Na_2S_2O_3$ in terms of cm^3 min^{-1}. *(2 marks)*

c Describe how the rate of change of iodine concentration varies during this experiment. *(1 mark)*

d State whether the reaction rate is dependent on the iodine concentration and give the order of reaction with respect to iodine. *(2 marks)*

e Calculate the concentration of I_2 in the 100 cm^3 of the reaction mixture at t = 0 min? *(1 mark)*

f Construct an equation for the reaction between $S_2O_3^{2-}$ and I_2. Use this to calculate the concentration of $Na_2S_2O_3$ used in the titrations. *(4 marks)*

g Use your answer to part **b** to calculate the rate of reaction (in cm^3 of $Na_2S_2O_3$ min^{-1}) if the reaction is first order with respect to propanone and 0.5 mol dm^{-3} propanone were used in place of 0.25 mol dm^{-3}. *(2 marks)*

4 This question is about the reactions of three compounds, A, B and C. Their formulae are shown below:

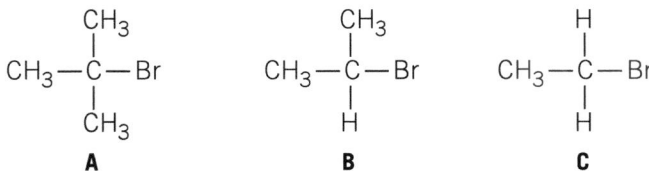

a Give the IUPAC name for compound B. *(1 mark)*

b Give the IUPAC name for the compound formed when compound B is heated with aqueous sodium hydroxide under reflux. *(1 mark)*

c Compound A also reacts with aqueous sodium hydroxide when heated under reflux.
  i Give the IUPAC name of the product formed. *(1 mark)*
  ii Explain using your knowledge of the inductive effect, why the mechanism for this reaction is S_N1. *(3 marks)*

d Compound C reacts with aqueous sodium hydroxide mainly by the S_N2 mechanism. Describe the S_N2 mechanism, including a diagram of the transition state. Use your answer to explain why the mechanism for this reaction is mainly S_N2. *(4 marks)*

e Compound C also reacts with sodium hydroxide dissolved in ethanol.
  i State the type of reaction that occurs under these conditions. *(1 mark)*
  ii Name the organic product of this reaction. *(1 mark)*

5 Classify the following catalysts as heterogeneous or homogeneous:

a H^+ ions in the reaction of I_2 with propanone.

b The enzyme catalase in the decomposition of hydrogen peroxide in plant cells.

c Nickel in the hydrogenation of unsaturated fats.

d Salivary amylase in the breakdown of starch. *(4 marks)*

6 Two gases, X and Y, react according to the equation:

$$X(g) + 2Y(g) \rightarrow XY_2(g)$$

An experiment was performed at 400 K in order to determine the order of this reaction and the following results were obtained.

Experiment number	Initial concentration of X / mol dm^{-3}	Initial concentration of Y / mol dm^{-3}	Initial rate of formation of XY_2 / mol dm^{-3} s^{-1}
1	0.10	0.10	0.0001
2	0.10	0.20	0.0004
3	0.10	0.30	0.0009
4	0.20	0.10	0.0001
6	0.30	0.10	0.0001

a Give the order of reaction with respect to:
  i X   **ii** Y *(2 marks)*

b Write a rate equation for the reaction of X with Y. *(1 mark)*

c Using the rate equation, suggest a possible mechanism for this reaction. *(2 marks)*

d Use the results from the experiment to calculate the value of the rate constant, k, and state its units. *(3 marks)*

e Suggest why chemists are interested in obtaining orders of reaction and determining rate equations. *(2 marks)*

7 a Draw an energy profile curve for the reaction:

$$H_2(g) + I_2(g) \rightarrow 2HI(g) \qquad \Delta H^{\ominus} = -10 \text{ kJ mol}^{-1}$$

Annotate your diagram to indicate the activation energy and the enthalpy change for the reaction. *(3 marks)*

b The rate of this reaction is given by:

$$\text{rate} = k[H_2][I_2]$$

  i Identify the order of reaction with respect to iodine *(1 mark)*
  ii Deduce the overall order of reaction. *(1 mark)*

c When 0.1 mol of H_2 and 0.2 mol of I_2 were mixed at 400°C in a 1 dm^3 vessel, the initial rate of formation of hydrogen iodide was 2.3×10^{-5} mol dm^{-3} s^{-1}. Calculate the value of k at 400°C. *(4 marks)*

27.1 Similarities and trends in the properties of the Group 2 metals, magnesium to barium, and their compounds

The thermal decomposition of the salts of Group 2 metals

We have seen in Section 10.1 that there is a trend in the ease of thermal decomposition of both the carbonates and nitrates of the Group 2 metals.

- The carbonates decompose to the oxide and carbon dioxide, for example:

$$MgCO_3(s) \rightarrow MgO(s) + CO_2(g)$$

As we descend the group, higher temperatures are needed to decompose the carbonate.

- The nitrates all decompose to brown nitrogen dioxide gas and oxygen, eg

$$Ca(NO_3)_2(s) \rightarrow CaO(s) + 2NO_2(g) + \frac{1}{2}O_2(g)$$

Again higher temperatures are needed to decompose the nitrate as we descend the group.

These trends can be explained in terms of the *size* of the M^{2+} ion. All the ions have two positive charges, so the smaller the ion, the greater **charge density** it has (think of it as the ability to attract electrons), Figure 1. The charge density decreases as we go down the group because the volume of the ions get larger and the charge is more spread out.

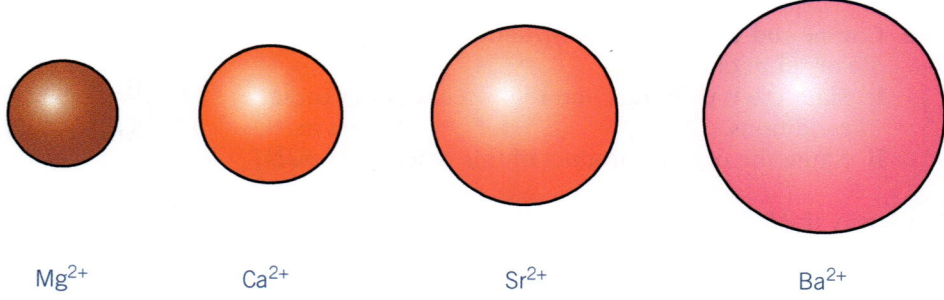

Mg²⁺   Ca²⁺   Sr²⁺   Ba²⁺

Figure 1 *All the ions have the same charge, so as they get larger their charge density (represented by the intensity of the colour) gets smaller*

A small M^{2+} ion, with its high charge density, distorts the electron distribution in the larger carbonate ion by pulling electrons towards it (we say the carbonate ion is polarised). This is shown in Figure 2, where the arrows ('curly arrows') show the movement of pairs of electrons.

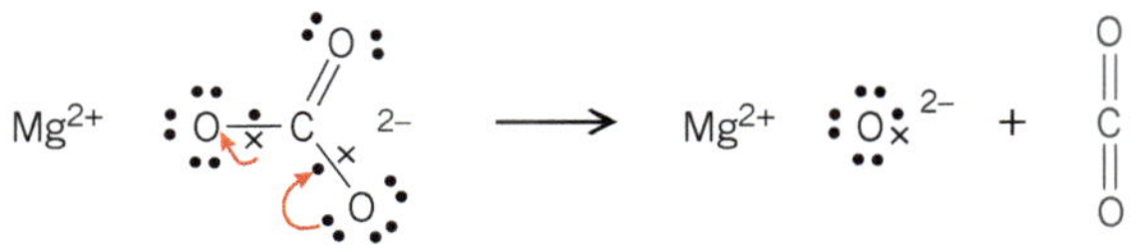

$$Mg^{2+} \quad :\overset{..}{\underset{..}{O}} \overset{x}{-} \overset{\overset{..}{\overset{O}{\parallel}}}{C} \Big)^{2-} \longrightarrow \quad Mg^{2+} \quad :\overset{..}{\underset{..}{O}}x^{2-} + \quad \overset{O}{\underset{O}{\overset{\parallel}{C}}}$$

Figure 2 *The decomposition of a carbonate ion. Curly arrows show electron movement towards the Mg^{2+} ion*

This allows one of the C–O bonds to break forming an O^{2-} ion and leaving behind a molecule of carbon dioxide.

The larger the metal ion, the lower its charge density. This means that there is less polarisation of the carbonate ion, so it is more stable when heated. This effect is explained schematically in Figure 3.

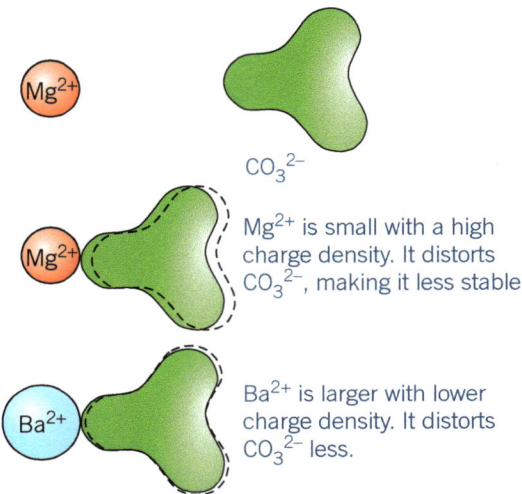

CO_3^{2-}

Mg^{2+} is small with a high charge density. It distorts CO_3^{2-}, making it less stable.

Ba^{2+} is larger with lower charge density. It distorts CO_3^{2-} less.

Figure 3 *The distorting effect of Group 2 ions on a CO_3^{2-} ion*

A similar polarisation effect explains the trend of decomposition of the nitrates. So again the thermal stability of the nitrates increases as we go down the group.

The solubilities of the Group 2 metal hydroxides and sulfates

Hydroxides
As we have seen in Section 10.1, there is a clear trend in the solubilities of the hydroxides—going down the group they become more soluble. So barium hydroxide is the most soluble:

$$Ba(OH)_2(s) + aq \rightarrow Ba^{2+}(aq) + 2OH^- (aq)$$

Sulfates
The solubility trend in the sulfates is exactly the opposite—the sulfates become less soluble going down the group. So, barium sulfate is virtually insoluble.

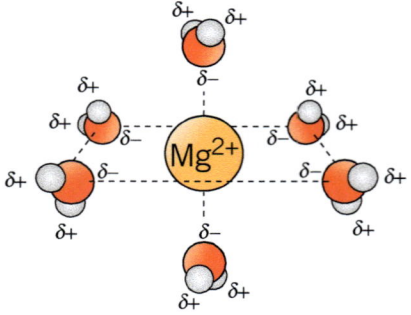

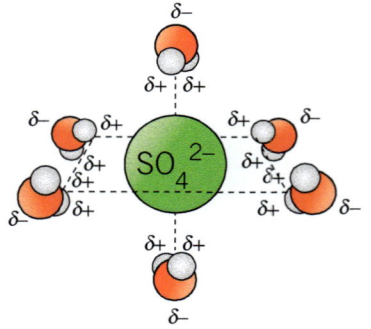

Figure 4 *The dissolving of magnesium sulfate by water molecules*

An enthalpy diagram

These trends in solubility can be explained by using an enthalpy diagram to look at the process of dissolving (see Section 23.2).

We can think of dissolving an ionic solid in water as happening in two stages.

1 **Endothermic.** The ionic lattice breaks to give separate ions in the gas phase. This requires the lattice energy to be *put in.*

2 **Exothermic.** The water molecules cluster around the ions. The positive ions are surrounded by the negative ends of the dipoles of the water molecules. The negative ions are surrounded by the positive ends of the dipoles of the water molecules (see Figure 4). These are both exothermic processes so energy is *given out*—the enthalpy of hydration of the positive ions (cations) and the enthalpy of hydration of the negative ions (anions) respectively. The process of dissolving can be shown on enthalpy (energy) diagrams like that in Figure 5.

It is the difference between these two processes that tell us the size of the $\Delta H_{sol.}$ In general, the more exothermic this is, the more soluble the compound is.

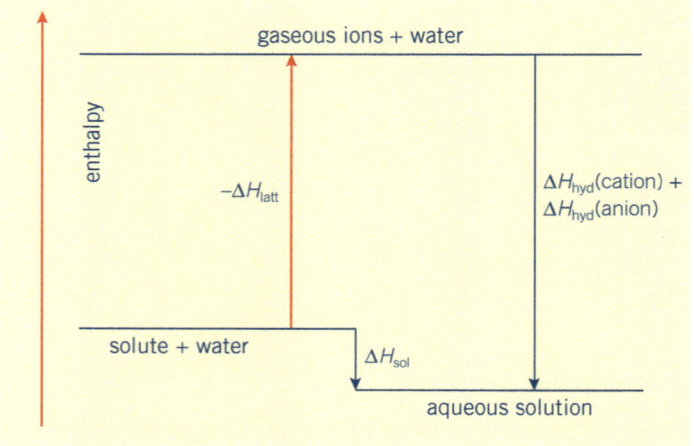

Figure 5 *Thermochemical cycle for a negative enthalpy of solution*

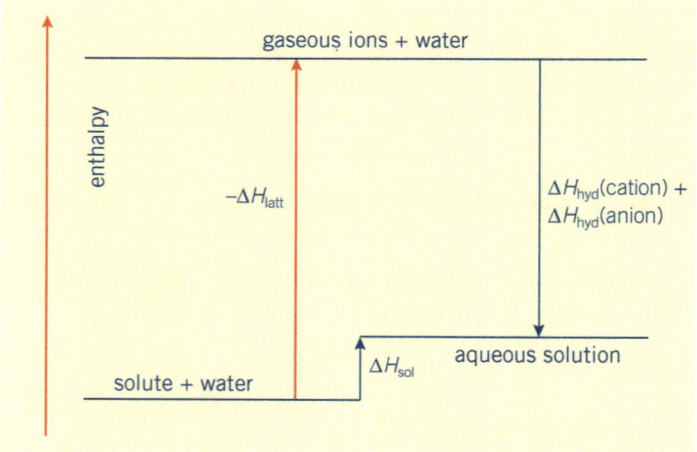

Figure 6 *Thermochemical cycle for a positive enthalpy of solution*

The lattice energy and the enthalpy change of hydration affect solubility, and they are both affected by the size of the M^{2+} ion. As we go down the group, the M^{2+} ion gets larger. This means that the lattice energy and the enthalpy of hydration of the ion both become smaller. The solubility of the compounds is related to the enthalpy of solution, ΔH_{sol}, which is related to the *difference* in these two values.

However the balance between these two factors is different for hydroxides and sulfates.

- For hydroxides, as we go down the group the lattice energy decreases faster than the enthalpy of hydration. So ΔH_{sol} becomes more exothermic and the solubility of Group 2 hydroxides increases as we descend the group.
- For sulfates, as we go down the group the enthalpy of hydration decreases faster than the lattice energy. So ΔH_{sol} becomes less exothermic and the solubility of Group 2 sulfates decreases as we descend the group.

Summary test 27.1

1 Explain what is meant by the term *charge density*.
2 State which has the larger charge density: Na^+ or Mg^{2+}. Explain your answer.
3 State which has the larger charge density: Ca^{2+} or Sr^{2+}. Explain your answer.
4 Give the equation for the decomposition of strontium carbonate with heat. State if this would be more or less easily decomposed than calcium carbonate. Explain your answer.
5 Magnesium sulfate will dissolve in bath water and is used in many bath products. Barium sulfate can be taken by mouth as a 'barium meal' to outline the gut in medical X-rays, even though $Ba^{2+}(aq)$ ions are poisonous. Explain both of these.
6 Using the data given below, draw an enthalpy diagram to calculate ΔH_{sol}.
ΔH_{latt} for NaCl is -786 kJ mol^{-1}
ΔH_{hyd} for Na^+ is -406 kJ mol^{-1}
ΔH_{hyd} for Cl^- is -364 kJ mol^{-1}

There are exam-style questions to test your knowledge of the material in this chapter at the end of Chapter 28.

28.1 General physical and chemical properties of the first row of transition elements, titanium to copper

Learning outcomes

On these pages you will learn to:

- define the transition elements and describe their characteristic properties
- sketch the shapes of the d-orbitals
- explain why transition elements have variable oxidation states
- explain why transition elements behave as catalysts
- explain why transition elements form complex ions

The elements from titanium to copper lie within the d-block elements (see Figure 1).

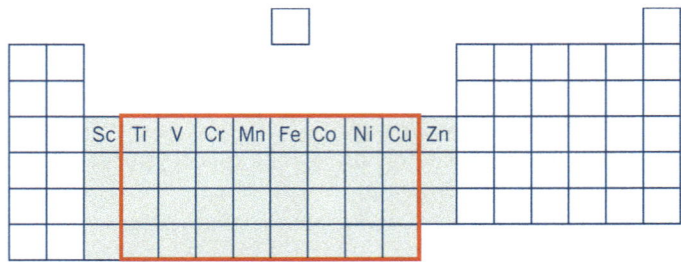

Figure 1 *The d-block elements (shaded) and the first row transition metals (outlined)*

Across a period, electrons are being added to a d-sub-level (3d in the case of titanium to copper). The elements from titanium to copper are metals. They are good conductors of heat and electricity. They are hard, strong, and shiny, and have high melting and boiling points compared with a typical s-block metal such as calcium.

These physical properties, together with fairly low chemical reactivity, make these metals extremely useful. Examples include iron (and its alloy steel) for vehicle bodies and to reinforce concrete, copper for water pipes, and titanium for jet engine parts that must withstand high temperatures.

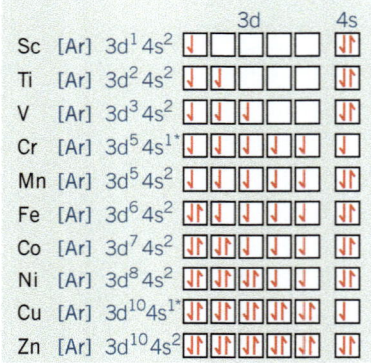

Figure 2 *Electronic arrangements of the elements in the first d-series. [Ar] represents the electron arrangement of argon—$1s^2 2s^2 2p^6 3s^2 3p^6$*

Electronic configurations in the d-block elements and ions

Figure 2 shows the electron arrangements for the elements in the first row of the d-block.

In general there are two outer 4s electrons and as you go across the period, electrons are added to the inner 3d sub-level. This explains the overall similarity of these elements.

The arrangements of chromium, Cr, and copper, Cu, do not quite fit the pattern. The d-sub-level is full ($3d^{10}$) in Cu and half full ($3d^5$) in Cr and there is only one electron in the 4s outer level. It is believed that a half-full d-level makes the atoms more stable, in the same way as a full outer shell makes the noble gas atoms stable. The shapes of the five d-orbitals are shown in Figure 3.

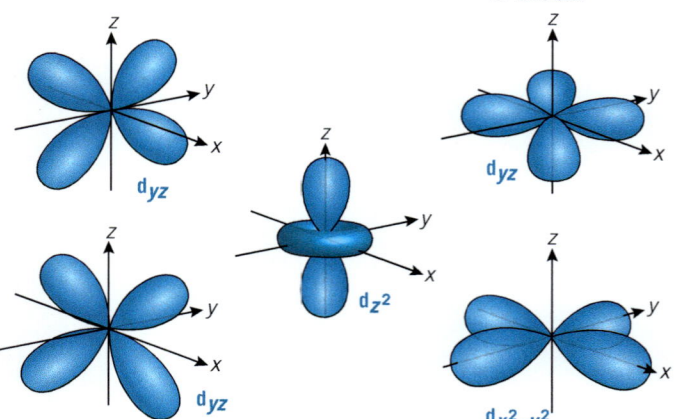

Figure 3 *The shapes of the five d-orbitals*

Electron configurations

To work out the configuration of the ion of an element, first write down the configuration of the element from its proton number, using the Periodic Table.

With all transition elements, the 4s electrons are lost first when ions are formed.

Worked example

Electron configuration of V²⁺

Vanadium, V, has a proton number of 23. Its electron configuration is:

$$1s^2 \ 2s^2 \ 2p^6 \ 3s^2 \ 3p^6 \ 3d^3 \ 4s^2$$

The vanadium ion V²⁺ has lost the two 4s² electrons and has the electron configuration:

$$1s^2 \ 2s^2 \ 2p^6 \ 3s^2 \ 3p^6 \ 3d^3$$

Worked example

Electron configuration of the Cu²⁺ ion

The proton number of copper is 29. The electron configuration is therefore:

$$1s^2 \ 2s^2 \ 2p^6 \ 3s^2 \ 3p^6 \ 3d^{10} \ 4s^1$$

The Cu²⁺ ion has lost two electrons, so it has the electron configuration:

$$1s^2 \ 2s^2 \ 2p^6 \ 3s^2 \ 3p^6 \ 3d^9$$

The definition of a transition element

The formal definition of a transition element is that it forms at least one stable ion with a *part* full d sub-shell of electrons. Scandium only forms Sc^{3+} ($3d^0$) in all its compounds, and zinc only forms Zn^{2+} ($3d^{10}$) in all its compounds. They are therefore d-block elements but not transition elements. The transition elements are outlined in red in Figure 1.

Chemical properties of transition metals

The chemistry of transition metals has four main features which are common to all the elements:

- Variable oxidation states: Transition metals have more than one oxidation state in their compounds, for example, Cu(I) and Cu(II). They can therefore take part in many redox reactions.
- Colour: The majority of transition metal ions are coloured, for example, $Cu^{2+}(aq)$ is blue.
- Catalysis: Catalysts affect the rate of reaction without being used up or chemically changed themselves. Many transition metals, and their compounds, show catalytic activity. For example, iron is the catalyst in the Haber process, vanadium(V) oxide is the catalyst in the Contact process and manganese(IV) oxide is the catalyst in the decomposition of hydrogen peroxide.
- Complex formation: Transition elements form complex ions. A **complex ion** is formed when a transition metal ion is surrounded by ions or molecules, collectively called ligands, which are bonded to it by coordinate bonds. These bonds are formed by lone pairs of electrons on the ligands being donated into empty d-orbitals (called vacant orbitals) of suitable energy on the transition metal ion. For example, $[Cu(H_2O)_6]^{2+}$ is a complex ion that is formed when copper sulfate dissolves in water.

Variable oxidation states of transition elements

Group 1 metals lose their outer electron to form only +1 ions and Group 2 lose their outer two electrons to form only +2 ions in their compounds.

Exam tip

When transition metals form ions, they lose the 4s electrons first.

Figure 4 *Some transition metals in use*

Exam tip

The terms oxidation number and oxidation state mean exactly the same.

A typical transition metal can use its 3d-electrons as well as its 4s-electrons in bonding. This is because there is little difference in energy between the 3d and 4s orbitals. This means that it can have a greater variety of oxidation states in different compounds. Table 1 shows this for the first d-series. Zinc and scandium are shown as part of the d-series although they are not transition metals.

Table 1 *Oxidation states shown by the elements of the first d-series in their compounds*

Sc	Ti	V	Cr	Mn	Fe	Co	Ni	Cu	Zn
	+I	+I	+I	+I	+I	+I	+I	+I	
	+II	+II	+II	+II	+II	+II	+II	+II	+II
+III	+III	+III	+III	+III	+III	+III	+III	+III	
	+IV	+IV	+IV	+IV	+IV	+IV	+IV		
		+V	+V	+V	+V	+V			
			+VI	+VI	+VI				
				+VII					

The most common oxidation states are shown in red, though they are not all stable.

Except for scandium and zinc all the elements show both the +1 and +2 oxidation states. These are formed by the loss of 4s electrons.

For example, nickel has the electron configuration $1s^2 2s^2 2p^6 3s^2 3p^6 3d^8 4s^2$ and Ni^{2+} is $1s^2 2s^2 2p^6 3s^2 3p^6 3d^8$.

Iron has the electron configuration $1s^2 2s^2 2p^6 3s^2 3p^6 3d^6 4s^2$ and Fe^{2+} is $1s^2 2s^2 2p^6 3s^2 3p^6 3d^6$.

Only the lower oxidation states of transition metals actually exist as simple ions. For example, Mn^{2+} ions exist but Mn^{7+} ions do not. In all Mn(VII) compounds, the manganese is covalently bonded to oxygen in a compound ion as in MnO_4^- (Figure 5).

The ability of transition elements to adopt more than one oxidation state is one of the reasons that they can act as catalysts. For example, Fe^{2+} ions act as a catalyst in the oxidation of iodine by peroxodisulfate ions (see Section 26.2).

The part-empty d-orbitals of transition metals also have a role in their catalytic effect. They allow the metals to adsorb reactants onto their surfaces and dative bonds to form with ligands using the vacant d-orbitals. For example, iron acts as a catalyst in the Haber process for making ammonia (see Section 26.2).

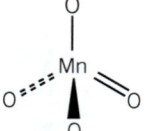

Figure 5 *Bonding in the MnO_4^- ion*

Summary test 28.1

1 The electron arrangement of manganese is:
$1s^2 2s^2 2p^6 3s^2 3p^6 3d^5 4s^2$
Give the electron arrangement of:
a the Mn^{2+} ion
b the Mn^{3+} ion
2 The electron arrangement of iron can be written [Ar] $3d^6 4s^2$.
a State what the [Ar] represents.
b State which two electrons are lost to form Fe^{2+} from Fe.
c State which further electron is lost in forming Fe^{3+}.

General characteristic chemical properties of the first set of transition elements, titanium to copper

All transition metal ions can form coordinate bonds by accepting electron pairs from other ions or molecules. This is because they have empty or part-empty d-orbitals that are energetically available.

The bonds that are formed are **coordinate (dative covalent) bonds**. An ion or molecule with a lone pair of electrons that forms a coordinate bond with a transition metal is called a **ligand**. Some examples of ligands are $H_2O:$, $:NH_3$, $:Cl^-$, $:CN^-$. These ions are **monodentate** ligands as they have just one lone pair.

The formation of complex ions

In some cases, two, four, or six monodentate ligands bond to a single central transition metal ion. The resulting species is called a **complex ion**. The number of coordinate bonds to ligands that surround the d-block metal ion is called the **coordination number**.

- Ions with coordination number six are usually octahedral, for example, $[Co(NH_3)_6]^{3+}$.
- Ions with coordination number four are usually tetrahedral, for example, $[CoCl_4]^{2-}$.
- Some ions with coordination number four are square planar, for example, $[NiCN_4]^{2-}$.

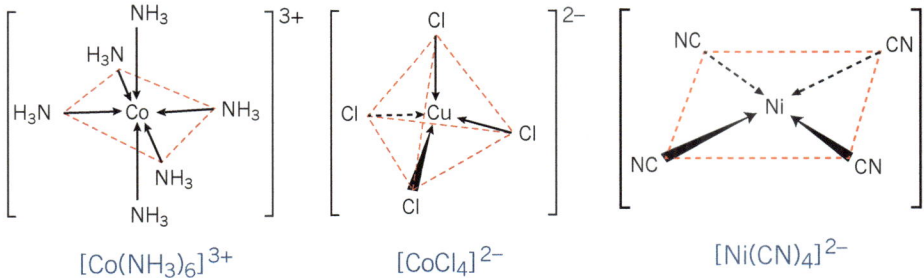

$[Co(NH_3)_6]^{3+}$ $[CoCl_4]^{2-}$ $[Ni(CN)_4]^{2-}$

Aqua ions

If you dissolve the salt of a transition metal, for example, copper sulfate, in water the positively charged metal ions become surrounded by water molecules acting as ligands (Figure 6). Normally there are six water molecules in an octahedral arrangement. Such species are called **aqua ions**.

Multidentate ligands—chelation

Some molecules or ions, called multidentate ligands, have more than one atom with a lone pair of electrons which can bond to a transition metal ion.

Bidentate ligands have two lone pairs which can bond to a transition metal ion. They include:

- 1,2-diaminoethane (also called ethane-1,2-diamine, Figure 7). Each nitrogen has a lone pair which can form a coordinate bond to the metal ion. The name of this ligand is often abbreviated to *en*, for example, $[Cr(en)_3]^{3+}$. It is a neutral ligand and the chromium ion has a 3+ charge, so the complex ion also has a 3+ charge.

Learning outcomes

On these pages you will learn to:

- describe monodentate, bidentate and polydentate ligands
- explain the reactions of transition elements with ligands to form complexes, and predict their formulae, charge and coordination number
- describe the geometry of linear, square planar, tetrahedral and octahedral transition element complexes
- describe ligand exchange reactions
- predict the feasibility of redox reactions using $E^\ominus$ values
- perform calculations involving redox systems

> **Exam tip**
>
> All ligands *must* have a lone pair of electrons.

> **Exam tip**
>
> Coordinate (dative covalent) bonds are ones where both the electrons in the bond come from one of the atoms forming the bond.

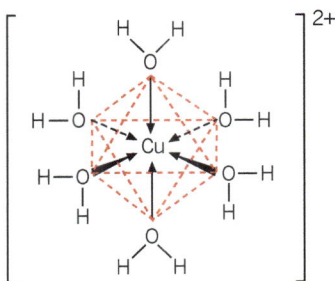

Figure 6 A copper(II) ion surrounded by water molecules

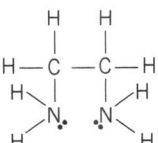

Figure 7 1,2-diaminoethane

335

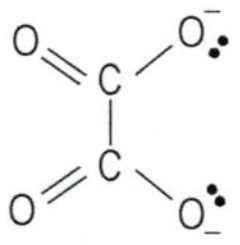

Figure 8 *Ethanedioate*

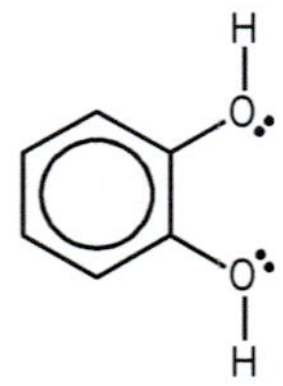

Figure 9 *Benzene-1,2-diol*

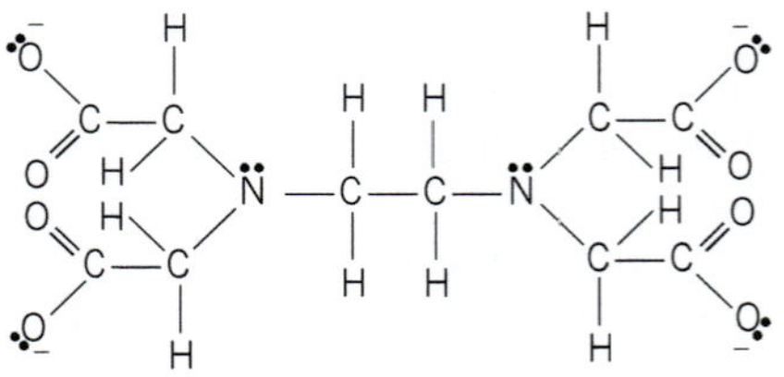

Figure 10 *EDTA*

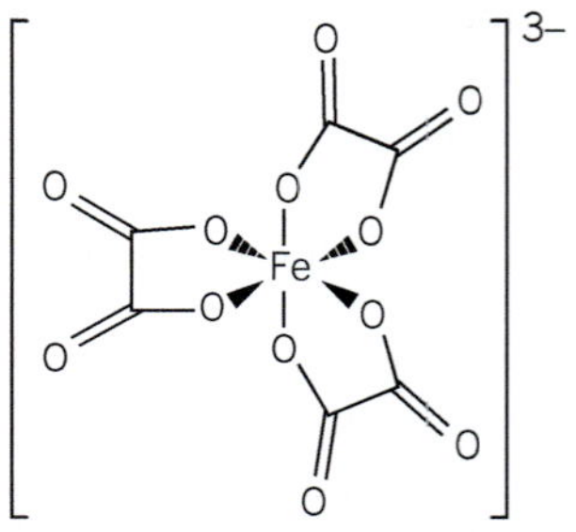

Figure 11 *A complex of Fe^{3+} with three ethanedioate ligands*

Exam tip

Both ligand sites of bidentate ligands usually bond to the same metal, forming a ring. However, they can act as bridges between two metal ions.

Extension

Naming ligands

When molecules or ions are acting as ligands, they are given special names as follows:

Ligand	Name
Water, H_2O	aqua
Ammonia, NH_3	ammine
Hydroxide, OH^-	hydroxo
Chloride, Cl^-	chloro
Cyanide, CN^-	cyano

Note the spelling of ammine with two m's. Don't confuse it with amines in organic chemistry.

- The ethanedioate (oxalate) ion, $C_2O_4^{2-}$ (Figure 8). which has a charge of −2
- Benzene-1,2-diol, sometimes called 1,2-dihydroxybenzene, is also a neutral ligand (Figure 9).

EDTA

An important multidentate ligand is the ion ethylenediaminetetracetate, called $EDTA^{4-}$ (Figure 10).

This can act as a **hexadentate** ligand using lone pairs on four oxygen and both nitrogen atoms. Complex ions with multidentate ligands are called **chelates**. Chelates can be used to effectively remove d-block metal ions from solution and can be used as antidotes for lead or mercury poisoning.

Extension

Haemoglobin

Haemoglobin (Figure 12) is the red pigment in blood. It is responsible for carrying oxygen from the lungs to the cells of the body. The molecule consists of an Fe^{2+} ion with a coordination number of six. Four of the coordination sites are taken up by a ring system called a porphyrin which acts as a tetradentate ligand. This complex is called *haem*.

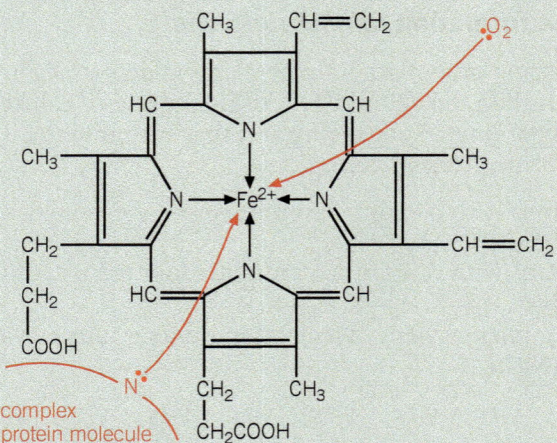

Figure 12 *Haemoglobin*

Below the plane of this ring is a fifth nitrogen atom acting as a ligand. This atom is part of a complex protein called *globin*. The sixth site can accept an oxygen molecule as a ligand. The Fe^{2+} to O_2 bond is weak, as $:O_2$ is not a very good ligand. This allows the oxygen molecule to be easily given up to cells.

Better ligands than oxygen can bond irreversibly to the iron and so destroy haemoglobin's oxygen-carrying capacity. This explains the poisonous effect of carbon monoxide, which is a better ligand than oxygen. Carbon monoxide is often formed by incomplete combustion in faulty gas heaters. Because it binds more strongly to the iron than oxygen, it is possible to suffocate in a room with plentiful oxygen. The cyanide ion behaves in the same way as carbon monoxide.

Anemia is a condition which may be caused by a shortage of haemoglobin. The body suffers from a lack of oxygen and the symptoms include fatigue and breathlessness. The causes may be loss of blood or deficiency of iron in the diet. The latter may be treated by taking 'iron' tablets which contain iron(II) sulfate.

Shapes and charges of complex ions

Complex ions may have a positive charge or a negative charge.

The aqueous copper(II) ion, with six ligands, is an octahedral shape (with six points, but eight faces). The metal ion, Cu^{2+}, has a charge of +2 and as the ligands are all neutral. The complex ion has an overall charge of +2. All the bond angles are 90°.

The $[CoCl_4]^{2-}$ ion, with four ligands, is tetrahedral. The metal ion, Co^{2+}, has a charge of +2 and each of the four ligands $:Cl^-$, has a charge of −1, so the complex ion has an overall charge of −2. Here the bond angles are 109.5°.

A few complexes of coordination number four adopt a square planar geometry, for example Ni(CN)$_4$$^{2-}$ (Figure 13). The metal ion Ni^{2+} has a charge of +2 and each of the four ligands :CN$^-$, has a charge of −1, so the complex ion has an overall charge of −2. The bond angles are all 90°.

Some complexes are linear. One example being [Ag(NH$_3$)$_2$]$^+$: Ag$^+$ has a charge of +1 and each of the NH$_3$ ligands is neutral, so the complex ion has an overall charge of +1. The bond angle is 180°.

$$[H_3N \rightarrow Ag \leftarrow NH_3]^+$$

A solution containing the complex ion [Ag(NH$_3$)$_2$]$^+$ is called **Tollens' reagent**. It is used in organic chemistry to distinguish aldehydes from ketones. Aldehydes reduce the [Ag(NH$_3$)$_2$]$^+$ to Ag (metallic silver), while ketones do not. The silver forms a mirror on the surface of the test tube, giving the name of the test—the silver mirror test.

The shapes of complex ions can be represented using wedge and dotted bonds. Wedge bonds come out of the paper and dotted bonds go in, Figure 14.

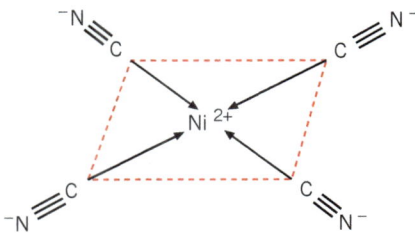

Figure 13 A square planar complex ion

Ligand exchange (substitution) reactions

When a transition metal compound is dissolved in water, the metal ions are surrounded by water molecules acting as ligands. The water molecules that act as ligands in metal aqua ions can be replaced by other ligands—either because the other ligands form stronger coordinate bonds or because they are present in higher concentration and an equilibrium is displaced.

Replacing water as a ligand
There are a number of possibilities:

- The water molecules may be replaced by other neutral ligands, such as ammonia.
- The water molecules may be replaced by negatively charged ligands, such as chloride ions.
- The water molecules may be replaced by bi- or multidentate ligands—this is called **chelation**.
- Replacement of the water ligands may be complete or partial.

Replacement by neutral ligands – no change in coordination number
In general for an M^{2+} ion, water molecules may be replaced one at a time by ammonia. Both ligands are uncharged and are of similar size, so there is no change in coordination number or charge on the ion:

$$[M(H_2O)_6]^{2+} + NH_3 \rightleftharpoons [M(NH_3)(H_2O)_5]^{2+} + H_2O$$

$$[M(NH_3)(H_2O)_5]^{2+} + NH_3 \rightleftharpoons [M(NH_3)_2(H_2O)_4]^{2+} + H_2O$$

$$[M(NH_3)_2(H_2O)_4]^{2+} + NH_3 \rightleftharpoons [M(NH_3)_3(H_2O)_3]^{2+} + H_2O$$

$$[M(NH_3)_3(H_2O)_3]^{2+} + NH_3 \rightleftharpoons [M(NH_3)_4(H_2O)_2]^{2+} + H_2O$$

$$[M(NH_3)_4(H_2O)_2]^{2+} + NH_3 \rightleftharpoons [M(NH_3)_5(H_2O)]^{2+} + H_2O$$

$$[M(NH_3)_5(H_2O)]^{2+} + NH_3 \rightleftharpoons [M(NH_3)_6]^{2+} + H_2O$$

Overall:

$$[M(H_2O)_6]^{2+} + 6NH_3 \rightleftharpoons [M(NH_3)_6]^{2+} + 6H_2O$$

There is a complication, because ammonia is a base as well as a ligand. Its solution contains OH$^-$ ions, so a precipitate may form and then re-dissolve

$$[M(H_2O)_6]^{2+} + 2OH^-(aq) \rightarrow M(H_2O)_4(OH)_2(s) + 2H_2O(l)$$

$$M(H_2O)_4(OH)_2(s) + 6NH_3(aq) \rightleftharpoons [M(NH_3)_6]^{2+}(aq) + 4H_2O(l) + 2OH^-(aq)$$

Exam tip

Remember that an octahedral complex has six ligands surrounding the metal ion, not eight.

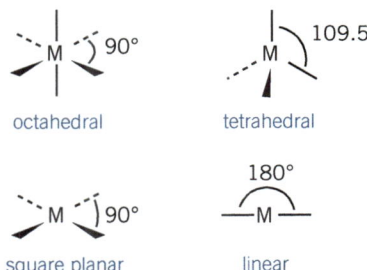

octahedral   tetrahedral

square planar   linear

Figure 14 The four main shapes of transition metal complexes using wedge and dotted bonds

Exam tip

Insoluble hydrated hydroxides are formed when OH$^-$ ions are added to aqueous solutions of transition metal ions.

Cobalt(II)

When M is cobalt, Co, the first step is the formation of a blue precipitate of hydrated cobalt(II) hydroxide when ammonia, NH_3, is added. This is produced by the loss of a proton from each of two of the six water molecules coordinated to the Co^{2+} ion:

$$[Co(H_2O)_6]^{2+} (aq) + 2NH_3(aq) \rightarrow \quad [Co(H_2O)_4(OH)_2](s) \quad + 2NH_4^+(aq)$$
$$\text{hydrated cobalt(II) hydroxide}$$

Here ammonia is acting as a base, by removing a H^+ ion from two of the water molecules.

If you add more of the concentrated ammonia, then both OH^- and all four water ligands are replaced by ammonia. This is for two reasons:

1 Ammonia is a better ligand than water.
2 The high concentration of ammonia displaces equilibria (like those on the previous page) to the right, thus displacing water and OH^-.

Overall:

$$[Co(H_2O)_4 (OH)_2](s) + 6NH_3(aq) \rightleftharpoons [Co(NH_3)_6]^{2+}(aq) + 4H_2O(l) + 2OH^-(aq)$$

The blue precipitate dissolves to form a pale yellow solution (which is oxidised by oxygen in air to a brown mixture containing Co(III)).

Replacement by chloride ions—change in coordination number

Addition of concentrated hydrochloric acid to a pink solution of $[Co(H_2O)_6]^{2+}(aq)$ results in a blue solution containing $CoCl_4^{2-}$ ions.

$$[Co(H_2O)_6]^{2+}(aq) + 4Cl^- (aq) \rightleftharpoons [CoCl_4]^{2-} (aq) + 6H_2O(l)$$

Adding water to the blue solution reverses the reaction.

Copper(II)

When aqueous copper ions react with ammonia in aqueous solution, ligand replacement is only partial—only four of the water ligands are replaced. The overall reaction is:

$$[Cu(H_2O)_6]^{2+}(aq) + 4NH_3(aq) \rightleftharpoons [Cu(NH_3)_4(H_2O)_2]^{2+}(s) + 4H_2O(l)$$

$[Cu(H_2O)_6]^{2+}$ is pale blue while $[Cu(NH_3)_4(H_2O)_2]^{2+}$ is a very deep blue.

The steps are similar to those above for Co^{2+}. The ammonia first acts as a base removing protons from two of the water molecules in $[Cu(H_2O)_6]^{2+}$ to form $[Cu(OH)_2(H_2O)_4](s)$. The first thing we see is a pale blue precipitate of copper hydroxide. When more of the concentrated ammonia is added, the precipitate dissolves to form a deep blue solution containing $[Cu(NH_3)_4(H_2O)_2]^{2+}$ (Figure 15). The ammonia has replaced both OH^- ligands, and two of the H_2O ligands:

$$[Cu(OH)_2(H_2O)_4](s) + 4NH_3(aq) \rightleftharpoons [Cu(NH_3)_4(H_2O)_2]^{2+}(aq) + 2H_2O(l) + 2OH^-(aq)$$

Figure 15 *Pale blue solution of $[Cu(H_2O)_6]^{2+}$, the pale blue precipitate of $[Cu(OH)_2(H_2O)_4]$, and the deep blue solution of $[Cu(NH_3)_4(H_2O)_2]^{2+}$*

The shape of the $[Cu(NH_3)_4(H_2O)_2]^{2+}$ ion

The $[Cu(NH_3)_4(H_2O)_2]^{2+}$ is octahedral, as expected for a six coordinate ion. The four ammonia molecules exist in a square-planar arrangement around the metal ion with the two water molecules above and below the plane (Figure 16).

The Cu–O bonds are longer (and therefore weaker) than the Cu–N bonds, as would be expected because water is a poorer ligand than ammonia. The octahedron is slightly distorted.

Replacement by chloride ions—change in coordination number

When aqueous copper ions react with concentrated hydrochloric acid there is a change in both charge and coordination number. Concentrated hydrochloric acid provides a high concentration of Cl^- ligands:

$$[Cu(H_2O)_6]^{2+} + 4Cl^- \rightleftharpoons [CuCl_4]^{2-} + 6H_2O$$

The pale blue colour of the $[Cu(H_2O)_6]^{2+}$ ion is replaced by the yellow $[CuCl_4]^{2-}$ ion. (Although the solution may look green as some $[Cu(H_2O)_6]^{2+}$ will remain.) Again, the actual replacement takes place in steps. The coordination number of the ion is four and the ion is tetrahedral.

$[Cu(H_2O)_6]^{2+}$ is six coordinate and $[CuCl_4]^{2-}$ is four coordinate (Figure 17), because Cl^- is larger than H_2O and fewer ligands can physically fit around the central copper ion.

Redox reactions in transition metal chemistry

Many of the reactions of transition metal compounds are redox reactions, in which the metals are either oxidised or reduced. For example, iron shows two stable oxidation states, Fe^{3+} and Fe^{2+}.

Fe^{2+} is the less stable state—it can be oxidised to Fe^{3+} by the oxygen in the air and also by chlorine. For example:

$$\begin{array}{cccc} +2 & 0 & +3 & -1 \\ 2Fe^{2+}(aq) & + \quad Cl_2(g) & \rightarrow \quad 2Fe^{3+}(aq) & + \quad 2Cl^-(aq) \end{array}$$

In this reaction, chlorine is the oxidising agent—its oxidation number drops from 0 to −1 (as it gains an electron), whilst the oxidation number of the iron increases from +2 to +3 (as it loses an electron). Remember the phrase OIL RIG—oxidation is loss, reduction is gain (of electrons).

Using half equations to predict redox reactions

Table 2 lists some $E^\ominus$ values for half equations.

Table 2 Some standard electrode potentials

Reduction half equation	$E^\ominus$ / V
$Li^+(aq) + e^- \rightleftharpoons Li(s)$	−3.04
$Ca^{2+}(aq) + 2e^- \rightleftharpoons Ca(s)$	−2.87
$Al^{3+}(aq) + 3e^- \rightleftharpoons Al(s)$	−1.66
$Zn^{2+}(aq) + 2e^- \rightleftharpoons Zn(s)$	−0.76
$2CO_2(g) + 2H^+(aq) + 2e^- \rightleftharpoons C_2O_4H_2(aq)$	−0.43
$Cr^{3+}(aq) + e^- \rightleftharpoons Cr^{2+}(aq)$	−0.41
$Pb^{2+}(aq) + 2e^- \rightleftharpoons Pb(s)$	−0.13
$2H^+(aq) + 2e^- \rightleftharpoons H_2(g)$	0.00
$Cu^{2+}(aq) + e^- \rightleftharpoons Cu^+(aq)$	+0.15
$Cu^{2+}(aq) + 2e^- \rightleftharpoons Cu(s)$	+0.34
$I_2(aq) + 2e^- \rightleftharpoons 2I^-(aq)$	+0.54
$Fe^{3+}(aq) + e^- \rightleftharpoons Fe^{2+}(aq)$	+0.77
$Ag^+(aq) + e^- \rightleftharpoons Ag(s)$	+0.80
$Br_2(aq) + 2e^- \rightleftharpoons 2Br^-(aq)$	+1.07
$Cl_2(aq) + 2e^- \rightleftharpoons 2Cl^-(aq)$	+1.36
$MnO_4^- + 8H^+(aq) + 5e^- \rightleftharpoons Mn^{2+}(aq) + 4H_2O(l)$	+1.52
$Ce_4^+(aq) + e^- \rightleftharpoons Ce^{3+}(aq)$	+1.70

The reaction of potassium manganate(VII) with Fe^{2+}

The technique of using half equations is useful for constructing balanced equations in more complex reactions. Potassium manganate(VII) can act as an oxidising agent in acidic solution (one containing $H^+(aq)$ ions). For example, potassium manganate oxidises Fe^{2+} to Fe^{3+}.

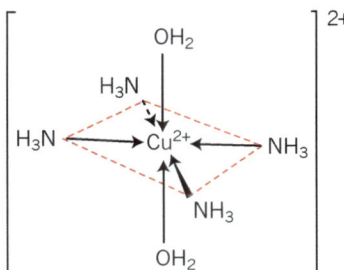

Figure 16 The shape of the $[Cu(NH_3)_4(H_2O)_2]^{2+}$ ion. The dotted lines are not bonds; they are construction lines to show the square-planar arrangement of the NH_3 ligands

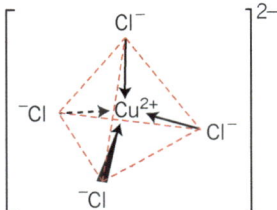

Figure 17 The shape of the $[CuCl_4]^{2-}$ ion

Exam tip

The chloride ions form larger ligands than ammonia or water, because chlorine is in Period 3 and has one more shell of electrons than nitrogen and oxygen.

Exam tip

Because NH_3 and H_2O ligands are similar in size and both are uncharged, ligand exchange occurs without a change in charge or coordination number.

Exam tip

Ammonia is a better ligand than water because the lone pair on the nitrogen atom of ammonia is less strongly held than that on the more electronegative oxygen atom of water. It is therefore more readily donated to positively charged metal ions.

During the reaction the oxidation number of the manganese falls from +7 to +2. We can use $E^{\ominus}$ values (see Table 2 on previous page) to predict the feasibility of redox reactions.

First construct the half equation for the reduction of Mn(VII) to Mn(II):

$$MnO_4^-(aq) \rightarrow Mn^{2+}(aq)$$

The oxygen atoms must be balanced using H^+ ions and H_2O molecules:

$$MnO_4^- + 8H^+(aq) \rightarrow Mn^{2+}(aq) + 4H_2O(l)$$

Then balance for charge using electrons:

$$MnO_4^-(aq) + 5e^- + 8H^+(aq) \rightarrow Mn^{2+}(aq) + 4H_2O(l) \qquad E^{\ominus} = +1.52 \text{ V}$$

The half equation for the oxidation of iron(II) to iron(III) is straightforward:

$$Fe^{2+}(aq) \rightarrow Fe^{3+}(aq) + e^- \qquad E^{\ominus} = -0.77 \text{ V}$$

This is the reverse of the reaction in the table so the sign of $E^{\ominus}$ is reversed.

To construct a balanced symbol equation for the reaction of acidified potassium manganate(VII) with $Fe^{2+}(aq)$, first multiply the Fe^{2+}/Fe^{3+} half reaction by five (so that the numbers of electrons in each half reaction are the same) and then add the two half equations and the values of $E^{\ominus}$ to give $E^{\ominus}_{cell}$:

$$5Fe^{2+}(aq) \rightarrow 5Fe^{3+}(aq) + 5e^- \qquad\qquad E^{\ominus} = -0.77 \text{ V}$$

$$MnO_4^-(aq) + 5e^- + 8H^+(aq) \rightarrow Mn^{2+}(aq) + 4H_2O(l) \qquad E^{\ominus} = +1.52 \text{ V}$$

$$5Fe^{2+}(aq) + MnO_4^-(aq) + \cancel{5e^-} + 8H^+(aq) \rightarrow 5Fe^{3+}(aq) + \cancel{5e^-} + Mn^{2+}(aq) + 4H_2O(l)$$

$$5Fe^{2+}(aq) + MnO_4^-(aq) + 8H^-(aq) \rightarrow 5Fe^{3+}(aq) + Mn^{2+}(aq) + 4H_2O(l) \quad E^{\ominus}_{cell} = +0.75 \text{ V}$$

As $E^{\ominus}_{cell}$ is positive, the reaction is feasible.

This technique makes balancing complex redox reactions much easier.

The reaction of potassium manganate(VII) with ethanedioic acid
The half reactions are:

$$MnO_4^-(aq) + 5e^- + 8H^+(aq) \rightarrow Mn^{2+}(aq) + 4H_2O(l) \qquad E^{\ominus} = +1.52 \text{ V}$$

and

$$2CO_2(g) + 2H^+(aq) + 2e^- \rightarrow C_2O_4H_2(aq) \qquad E^{\ominus} = -0.43 \text{ V}$$

To equalise the number of electrons transferred, we must multiply to top half equation by 2 and the lower one by 5. (Remember this has no effect on the values of $E^{\ominus}$.)

$$2MnO_4^-(aq) + 10e^- + 16H^+(aq) \rightarrow 2Mn^{2+}(aq) + 8H_2O(l) \qquad E^{\ominus} = +1.52 \text{ V}$$

$$10CO_2(g) + 10H^+(aq) + 10e^- \rightarrow 5C_2O_4H_2(aq) \qquad E^{\ominus} = -0.43 \text{ V}$$

Reversing the second equation (and changing the sign of $E^{\ominus}$).

$$5C_2O_4H_2(aq) \rightarrow 10CO_2(g) + 10H^+(aq) + 10e^- \qquad E^{\ominus} = +0.43 \text{ V}$$

We now add the two half equations, cancel the electrons and add the values of $E^{\ominus}$ to give $E^{\ominus}_{cell}$.

$$2MnO_4^-(aq) + \cancel{10e^-} + 16H^+(aq) + 5C_2O_4H_2(aq) \rightarrow 2Mn^{2+}(aq) + 8H_2O(l) + 10CO_2(g) + 10H^+(aq) + \cancel{10e^-} \qquad E^{\ominus}_{cell} = +1.95 \text{ V}$$

$$2MnO_4^-(aq) + 16H^+(aq) + 5C_2O_4H_2(aq) \rightarrow 2Mn^{2+}(aq) + 8H_2O(l) + 10CO_2(g) + 10H^+(aq) \qquad E^{\ominus}_{cell} = +1.95 \text{ V}$$

$E^{\ominus}_{cell}$ is positive, so the reaction is feasible.

Exam tip

Multiplying the half-cell reaction by five has no effect on the value of $E^{\ominus}$.

Exam tip

Potassium manganate(VII) is the systematic name of potassium permanganate, $KMnO_4$.

Analysis of transition metal compounds by titration

Worked example

Iron tablets

A brand of iron tablets has this stated on the pack: 'Each tablet contains 0.200 g of iron(II) sulfate.' The following experiment was done to check this.

One tablet was dissolved in excess sulfuric acid and made up to 250 cm³ in a volumetric flask. 25.00 cm³ of this solution was pipetted into a flask and titrated with 0.00100 mol dm⁻³ potassium manganate(VII) solution until the solution just became purple. Taking an average of several titrations, 26.30 cm³ of potassium manganate(VII) solution was needed.

Number of moles potassium manganate(VII) solution $= c \times \dfrac{V}{1000}$

where c is the concentration of the solution in mol dm⁻³ and V is the volume of solution used in cm³.

No. of moles potassium manganate(VII) solution $= 0.00100 \times \dfrac{26.30}{1000} = 2.63 \times 10^{-5}$ mol

$$5Fe^{2+}(aq) + MnO_4^-(aq) + 8H^+(aq) \rightarrow 5Fe^{3+}(aq) + Mn^{2+}(aq) + 4H_2O(l)$$

From the equation, 5 mol of Fe^{2+} reacts with 1 mol of MnO_4^-:

Number of moles of $Fe^{2+} = 5 \times 2.63 \times 10^{-5}$ mol $= 1.315 \times 10^{-4}$ mol

25.00 cm³ of solution contained $\dfrac{1}{10}$ tablet.

So one tablet contains $1.315 \times 10^{-4} \times 10 = 1.315 \times 10^{-3}$ mol Fe^{2+}

Since 1 mol iron(II) sulfate contains 1 mol Fe^{2+}, each tablet contains 1.315×10^{-3} mol $FeSO_4$.

The relative formula mass of $FeSO_4$ is 151.9.

So, each tablet contains $1.315 \times 10^{-3} \times 151.9 = 0.200$ g of iron(II) sulfate as stated on the bottle.

Measuring the concentration of Cu²⁺ ions

To measure the concentration of copper(II) ions in solution we can use the following reaction with iodide ions in which Cu^{2+} ions oxidise iodide ions to iodine:

$$2Cu^{2+}(aq) + 4I^- \rightarrow 2CuI(s) + I_2(aq)$$

If we add excess sodium iodide to our solution of copper(II)ions, the iodine produced can then be titrated with sodium thiosulfate, using starch as an indicator near the end point:

$$2Na_2S_2O_3(aq) + I_2(aq) \rightarrow Na_2S_4O_6(aq) + 2NaI(aq)$$

The outcome of these two equations is as follows:

2 mol of Cu^{2+} ions produces 1 mol of iodine, which reacts with 2 mol of thiosulfate ions. So 1 mol of sodium thiosulfate is equivalent to 1 mol of Cu^{2+} ions.

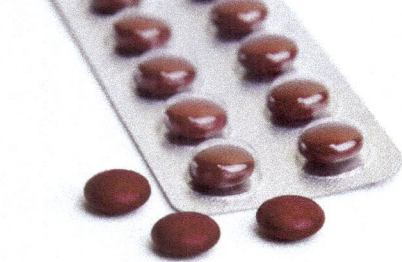

Figure 18 *Tablets of iron(II) sulfate are taken by people with iron deficiency*

Worked example

Finding the concentration of a copper(II) sulfate solution

25 cm³ of a solution of a copper(II) sulfate solution were pipetted into excess potassium iodide solution. Iodine was formed along with a white precipitate of copper(I) iodide. The iodine was titrated with 0.100 mol dm⁻³ sodium thiosulfate solution using starch as an indicator as the end point was reached. The average titre was 20.00 cm³.

20.00 cm³ sodium thiosulfate solution contains $\dfrac{20 \times 0.100}{1000} = 2 \times 10^{-3}$ mol

As shown above, this means that there were 2×10^{-3} mol Cu^{2+} ions in the original 25 cm³ of copper sulfate solution.

So there were $2 \times 10^{-3} \times \dfrac{1000}{25} = 0.080$ mol Cu^{2+} in 1 dm³

So the concentration of the copper sulfate solution was 0.080 mol dm⁻³

Summary test 28.2

1 **a** State the shapes of the following:
    i $[Cu(H_2O)_6]^{2+}$
    ii $[Cu(NH_3)_6]^{2+}$
    iii $[CuCl_4]^{2-}$
  b Give the coordination number of the transition metal in each complex in part **a**.
  c Explain why the coordination numbers are different.

2 Benzene-1,2-dicarboxylate is shown below.

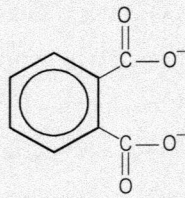

  a Suggest which atoms are likely to form coordinate bonds with a metal ion.
  b Identify the lone pairs.
  c Predict if it is likely to be a mono-, bi- or hexa-dentate ligand.

3 In the stepwise conversion of $[Cu(H_2O)_6]^{2+}$ to $[CuCl_4]^{2-}$, one of the species formed is neutral. Suggest two possible formulae that it could have.

4 **a** Sketch the shape of $[Cu(H_2O)_6]^{2+}$ and **b** predict the shape of $[CuBr_4]^{2-}$. Explain your answer.

5 When concentrated hydrochloric acid is added to an aqueous solution containing Co(II) ions, the following change takes place:
    $[Co(H_2O)_6]^{2+} \rightarrow [CoCl_4]^{2-}$ and the colour changes from pink to blue.
  a State whether there is any change in the oxidation state of the cobalt.
  b Give the shapes of the two ions concerned.

6 Zinc will reduce VO_2^+ ions to VO_2^+, VO^{2+} to V^{3+}, and V^{3+} ions to V^{2+} ions. The relevant half equations are:
  • $Zn(s) \rightarrow Zn^{2+}(aq) + 2e^-$
  • $VO_2^+(aq) + 2H^+(aq) + e^- \rightarrow H_2O(l) + VO^{2+}(aq)$
  • $VO^{2+}(aq) + 2H^+(aq) + e^- \rightarrow H_2O(l) + V^{3+}(aq)$
  • $V^{3+}(aq) + e^- \rightarrow V^{2+}(aq)$
  a Give the balanced equation for each of the reduction steps.
  b V^{2+} has to be protected from air. Suggest a reason for this.

7 A titration to determine the amount of iron(II) sulfate in an iron tablet was carried out. The tablet was dissolved in excess sulfuric acid and made up to $250\,cm^3$ in a volumetric flask. $25.00\,cm^3$ of this solution was pipetted into a flask and titrated with $0.00100\,mol\,dm^{-3}$ potassium manganate(VII) solution until the solution just became purple. Taking an average of several titrations, $25.00\,cm^3$ of potassium manganate(VII) solution was needed. Calculate the number of grams of iron in this tablet. $A_r(Fe) = 55.8$.

8 The $E^\ominus$ value for $Cr_2O_7^{2-}(aq) + 14H^+(aq) + 6e^- \rightleftharpoons 2Cr^{3+}(aq) + 7H_2O(l)$ is $+1.33\,V$.
  Use $E^\ominus$ values to demonstrate that acidified $Cr_2O_7^{2-}$ ions can be used in a redox titration with Fe^{2+} when Cl^- ions are present, that is, that $Cr_2O_7^{2-}$ ions will not oxidise Cl^-.

Colour of complexes

Most transition metal compounds are coloured. The colour is caused by the compounds absorbing energy that corresponds to light in the visible region of the spectrum. If a solution of a substance looks purple, for example, it is because it absorbs all the light from a beam of white light shone at it except red and blue. The red and blue light passes through and the solution appears purple (Figure 19).

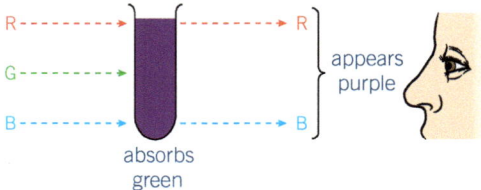

Figure 19 *Solutions look coloured because they absorb some colours and let others pass through*

Why are transition metal complexes coloured?

This is a simplified explanation, but the general principle is as follows:

- Transition metal compounds are coloured because they have part-filled d-orbitals.
- It is therefore possible for electrons to move from one d-orbital to another.
- In an isolated transition metal atom, all the d-orbitals are at exactly the same energy level. They are described as **degenerate orbitals**. But in a complex, the ligands nearby raise the energy levels of the d-orbitals and split them into two groups at slightly different energy levels. The pattern of splitting is different depending on the geometry of the complex, see Figure 20. These are described as **non-degenerate orbitals**.
- When white light is shone through a solution containing transition metal ions, electrons may move from one d-orbital to another at a higher energy level (called an **excited state**). If this occurs they absorb energy, in the visible region of the spectrum equal to the difference in energy between levels.
- This colour is therefore missing from the spectrum and you see the combination of the colours that are not absorbed.

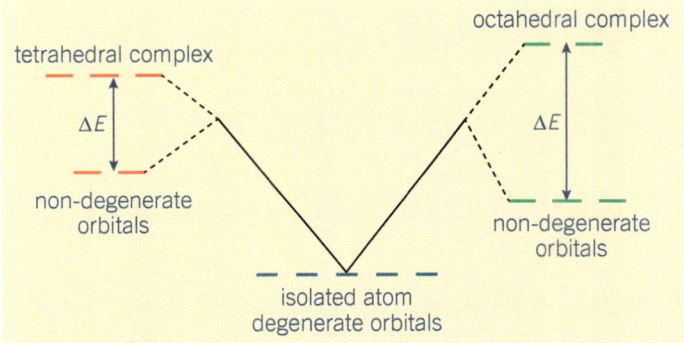

Figure 20 *The effect of ligands on the energies of d-orbitals in transition metal complexes with tetrahedral and octahedral geometries*

The frequency of the light is related to the energy difference by the expression $\Delta E = h\nu$, where E is the energy, ν the frequency, and h a constant called Planck's constant. This is the energy absorbed when an electron jumps from a lower energy non-degenerate orbital to a higher energy non-degenerate orbital, as shown by ΔE in Figure 20. The frequency is related to the colour of light. Violet is of high energy and therefore high frequency, and red is of low energy and low frequency.

Exam tip

Complexes of Cu^+ and Ag^+ are not coloured. Both these ions have full shells of d electrons, so electrons cannot move from one d-orbital to another.

The colour of a transition metal complex depends on the energy gap ΔE. This in turn depends on

- the oxidation state of the metal
- the nature of the ligands
- the shape of the complex ion.

So different compounds of the same metal will have different colours. For example, Table 3 and Figure 21 show the colours of four vanadium species each with a different oxidation state.

Table 3 Colours of four vanadium species

Oxidation number	Species	Colour
5	$VO_2^+(aq)$	yellow
4	$VO^{2+}(aq)$	blue
3	$V^{3+}(aq)$	green
2	$V^{2+}(aq)$	violet

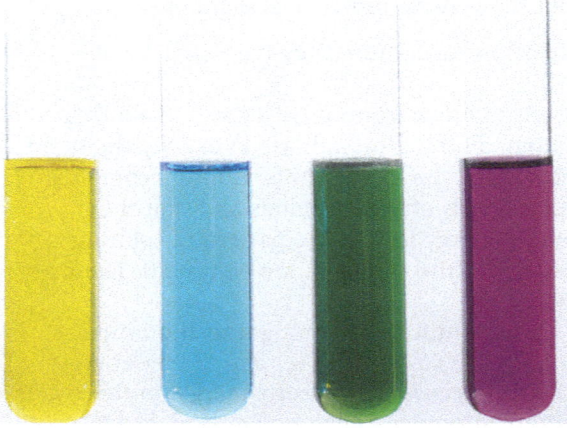

Figure 21 Zinc ions in acid solution will reduce vanadium(V) through oxidation numbers V(IV) and V(III) to V(II). The final test-tube is normally stoppered because oxygen in the air will rapidly oxidise V(II)

Some more examples of how changing the oxidation state of the metal affects the colour of the complex are given in Table 4.

Table 4 The effect of oxidation state on the colour of some metal complexes

Oxidation state of metal	+2	+3
iron complexes	$[Fe(H_2O)6]^{2+}$ green	$[Fe(H_2O)_6]^{3+}$ pale brown
chromium complexes	$[Cr(H_2O)_6]^{2+}$ blue	$[Cr(H_2O)_6]^{3+}$ red-violet
cobalt complexes	$[Co(NH_3)_6]^{2+}$ brown	$[Co(NH_3)_6]^{3+}$ yellow

Different ligands around the same transition metal ion also affect the energy gap between the d-orbitals and hence the colour of the complex. The shape of the complex also affects the colour, as it affects the difference in energy levels of the d-orbitals. Table 5 (and Figures 22 and 23) shows some examples of this for copper(II) complexes.

Table 5 The colours of some copper(II) complexes (*this complex is tetrahedral and the others are octahedral)

Formula of complex	Colour
$[Cu(H_2O)_6]^{2+}$	pale blue
$[Cu(H_2O)_4(OH)_2]$	pale blue (precipitate)
$[Cu(NH_3)_4(H_2O)_2]^{2+}$	dark blue
$CuCl_4^{2-}$ *	yellow

Figure 22 Some octahedral copper(II) complexes; from left to right, $[Cu(H_2O)_6]^{2+}$, $[Cu(H_2O)_4(OH)_2]$, $[Cu(NH_3)_4(H_2O)_2]^{2+}$

The ligand exchange reactions of Co(II) complexes described in Section 28.2 also illustrate the effect of different ligands and coordination number on the colour of complex ions (see Table 6).

*Table 6 The colours of some cobalt (II) complexes (*this complex is tetrahedral and the others are octahedral)*

Formula of complex	Colour
$[Co(H_2O)_6]^{2+}$	pink
$[Cu(H_2O)_4(OH)_2]$	blue (precipitate)
$[Co(NH_3)_6]^{2+}$	red-brown
$CoCl_4^{2-}$ *	blue

Figure 23 $CuCl_4^{2-}$ is a tetrahedral copper(II) complex; it has a distinctive yellow colour

Extension

The colour of gemstones

Transition metal ions are responsible for the colours of most gemstones. Rubies are made of aluminium oxide, Al_2O_3, which is colourless. The red colour is caused by trace amounts of Cr^{3+} ions which replace some of the Al^{3+} ions in the crystal lattice. The oxide ions, O^{2-}, are the ligands surrounding the Cr^{3+} ions.

The green colour of emeralds is also caused by Cr^{3+} ions. In this case, the material from which the gemstone is made is beryllium aluminium silicate, $Be_3Al_2(SiO_3)_6$. The ligand surrounding the Cr^{3+} is silicate, SiO_3^{2-}. This shows the effect of changing the ligand on the colour of the transition metal ions.

The red of garnet and the yellow-green colour of peridot are both caused by Fe^{2+} ions—surrounded by eight silicate ligands in garnet and six in peridot.

Other examples include Cu^{2+} which is responsible for the blue-green of turquoise and Mn^{2+} which is responsible for the pink of tourmaline.

Summary test 28.3

1 a Explain why copper(II) sulfate is coloured, but zinc sulfate is colourless.
  b A solution of copper sulfate is blue. State what colour light passes through this solution.
  c State what happens to the other colours of the visible spectrum.

28.4 Stereoisomerism in transition element complexes

Learning outcomes

On these pages you will learn to:

- describe geometrical isomerism in transition element complexes
- describe optical isomerism in transition element complexes
- explain the overall polarity of transition element complexes

Isomers are compounds with the same molecular formula but with different arrangements of their atoms in space. Transition metal complexes can form both geometrical isomers (*cis–trans*, or *E–Z* isomers) and optical isomers.

Geometrical isomerism

Here ligands differ in their position in space relative to one another. This type of isomerism occurs in octahedral and square planar complexes. Take the octahedral complex ion $[CrCl_2(H_2O)_4]^+$. The Cl^- ligands may be next to each other (the *cis*- or *Z*-form) or on opposite sides of the central chromium ion (the *trans*- or *E*-form) (Figure 24).

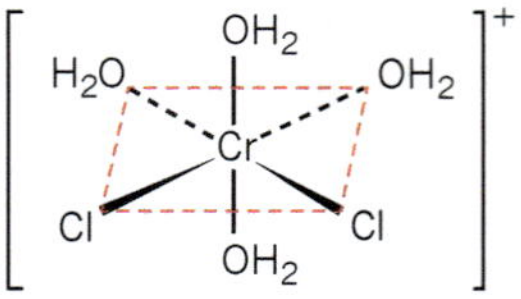

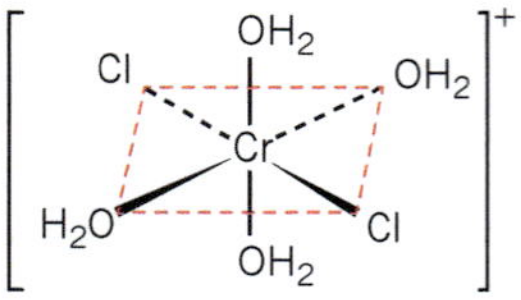

Figure 24 *The cis-isomer (top) and the trans-isomer (bottom) of the $[CrCl_2(H_2O)_4]^+$ ion*

In the square planar complex platin, $[Pt(NH_3)_2Cl_2]$, the Cl^- ligands may be next to each other (the *cis*- or *Z*-form) or on opposite sides of the central chromium ion (the *trans*- or *E*-form). A pair of geometrical isomers will have different chemical properties. For example, cisplatin is one of the most successful anti-cancer drugs whilst the *trans*-isomer has no therapeutic effect.

Exam tip

Geometrical isomerism does not occur in tetrahedral complexes.

Extension

Cisplatin, an anti-cancer drug

Cancer is not one disease but many. What cancers have in common is 'rogue' cells which have lost control over their growth and replication and grow much faster than normal cells. Cisplatin was discovered in 1965 and is one of the most successful cancer treatments, for example, giving survival rates of up to 90% in some cancers.

Cisplatin is square planar and has the formula:

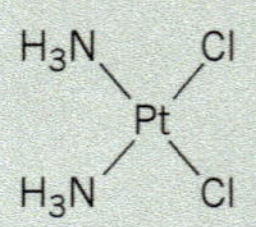

It works by bonding to strands of DNA, distorting their shape and preventing replication of the cells. The molecule bonds to nitrogen atoms on two adjacent guanine bases (see Figure 25) on a strand of DNA.

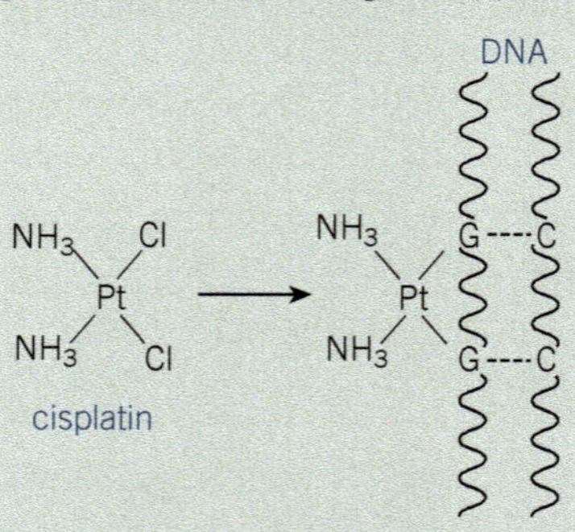

This works because the nitrogen atoms of the guanine molecules have lone pairs of electrons which form dative covalent bonds with the platinum. In aqueous solution, the chloride ions in cisplatin are first displaced by water because of the high concentration of water molecules. The water ligands are then displaced by nitrogen on guanine because the nitrogen is a better ligand. This is an example of a ligand substitution reaction.

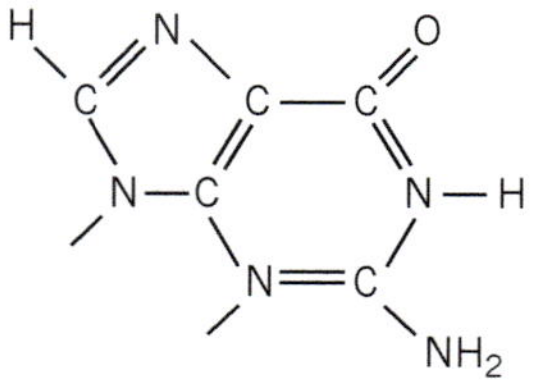

Figure 25 *The structure of the guanine base*

Like all drugs, cisplatin has side effects—it will bond to DNA in healthy cells as well as in cancerous ones. As cancer cells are replicating faster than healthy cells, the effect of the drug is greater on cancer cells than on normal cells. However, healthy cells that replicate quickly, such as hair follicles, are significantly affected and this is why patients undergoing chemotherapy (drug treatment for cancer) often lose their hair. Work is underway to find drugs and delivery systems that can better discriminate between healthy and cancerous cells.

Geometrical (*cis–trans*) isomerism can also occur in an octahedral complex with two bidentate ligands such as 1,2-diaminoethane (en). This is shown in Figure 26 for the complex $[Co(en)_2Cl_2]^+$. The two chloride ligands may be either on opposite sides of the metal ion (*trans*-) or adjacent (*cis*-).

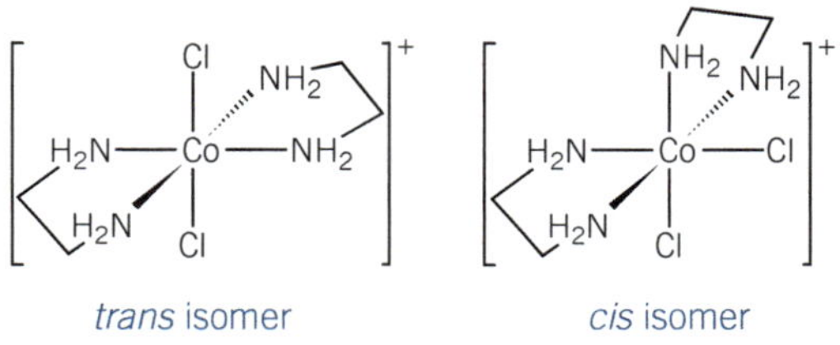

trans isomer   *cis* isomer

Figure 26 Geometrical isomers of $[Co(en)_2Cl_2]^+$

Optical isomerism

Here the two isomers are non-superimposable mirror images of each other. In transition metal complexes this occurs when there are two or more bidentate ligands in a complex (Figure 27).

Look at Figure 27 and imagine rotating one of the complexes around a vertical axis until the two chlorine atoms are in the same position as in the other one. You should be able to see that the positions of the en ligands no longer match. The best way to be sure about this is to use molecular models.

Optical isomers are said to be **chiral**. They have identical chemical properties but can be distinguished by their effect on polarised light. One isomer will rotate the plane of polarisation of polarised light clockwise, and the other will rotate it anticlockwise. Optical isomerism would also occur if the two Cl^- ligands were replaced by another en ligand.

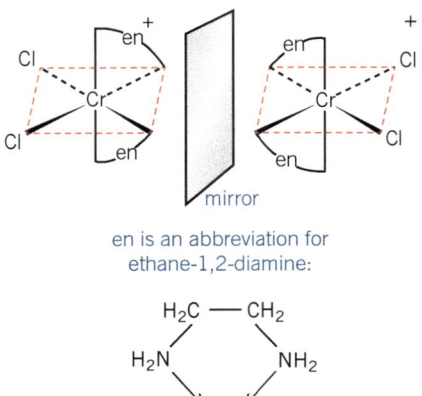

en is an abbreviation for ethane-1,2-diamine:

Figure 27 Transition metal complexes that are non-identical mirror images of each other (top) and the structure of the 1,2-diaminoethane (en) ligand (bottom)

Polarity of transition metal complexes

Transition metal complexes may be polar, depending on their geometry. Look at the isomers cisplatin and transplatin as examples in Figure 28.

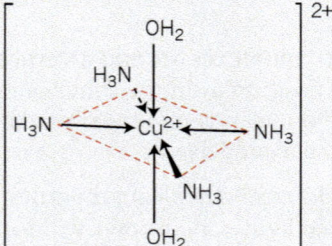

cisplatin   transplatin

Figure 28 *The shapes of the isomers cisplatin and transplatin*

The compounds are both neutral overall as the 2+ charge of the platinum ion is balanced by that of the two Cl⁻ ligands. However. the Pt–Cl bonds and the Pt–NH₃ bonds are polarised as shown in Figure 28. This is because both chlorine and nitrogen are more electronegative than platinum. In transplatin, the dipole moments of the two Pt–Cl and Pt–NH₃ bonds cancel out because they are in opposite directions, so the complex is non-polar. In cisplatin, they will not cancel out because chlorine and nitrogen do not have exactly the same electronegativity so the complex will be polar overall.

Summary test 28.4

1 Sketch the formula of transplatin and suggest why it is not an effective anti-cancer drug.

2 The $[Cu(NH_3)_4(H_2O)_2]^{2+}$ ion is shown below.

$$\left[\begin{array}{c} OH_2 \\ H_3N \\ H_3N \longrightarrow Cu^{2+} \longleftarrow NH_3 \\ NH_3 \\ OH_2 \end{array} \right]^{2+}$$

Sketch a possible geometrical isomer of this ion.

Stability constants, K_{stab}

The water molecules that act as ligands in metal aqua ions can be replaced by other ligands —either because the other ligands form stronger coordinate bonds or because they are present in higher concentration and an equilibrium is displaced.

Ligand substitution reactions—replacing water as a ligand

There are a number of possibilities:

- The water molecules may be replaced by other neutral ligands, such as ammonia.
- The water molecules may be replaced by negatively charged ligands, such as chloride ions.
- The water molecules may be replaced by bi- or multidentate ligands—this is called **chelation**.
- Replacement of the water ligands may be complete or partial.

Replacement by neutral ligands – no change in coordination number

In general for an M^{2+} ion, water molecules may be replaced one at a time by ammonia. Both ligands are uncharged and are of similar size, so there is no change in coordination number or charge on the ion:

In the complex $Cu(H_2O)_6{}^{2+}$, four of the water ligands can be replaced by ammonia molecules, which are better ligands. This takes place in four steps (state symbols are omitted here for clarity):

$$[Cu(H_2O)_6]^{2+} + NH_3 \rightleftharpoons [Cu(H_2O)_5NH_3]^{2+} + H_2O \qquad \text{step 1}$$

$$[Cu(H_2O)_5NH_3]^{2+} + NH_3 \rightleftharpoons [Cu(H_2O)_4(NH_3)_2]^{2+} + H_2O \qquad \text{step 2}$$

$$[Cu(H_2O)_4(NH_3)_2]^{2+} + NH_3 \rightleftharpoons [Cu(H_2O)_3(NH_3)_3]^{2+} + H_2O \qquad \text{step 3}$$

$$[Cu(H_2O)_3(NH_3)_3]^{2+} + NH_3 \rightleftharpoons [Cu(H_2O)_2(NH_3)_4]^{2+} + H_2O \qquad \text{step 4}$$

Overall:

$$[Cu(H_2O)_6]^{2+} + 4NH_3 \rightleftharpoons [Cu(H_2O)_2(NH_3)_4]^{2+} + 4H_2O$$

Each step of the replacement is an equilibrium and we can write an equilibrium constant expression for it (see Section 7.1). The equilibrium constants for ligand substitution reactions in a solvent are called **stability constants, K_{stab}**.

So for the first step:

$$K_{stab} = \frac{[[Cu(H_2O)_5NH_3]^{2+}]}{[[Cu(H_2O)_6]^{2+}] \, [NH_3]} \text{ mol}^{-1} \text{ dm}^3$$

The concentration of water cannot change, so has been omitted from the equilibrium expression.

For the first step, $K_{stab} = 1.78 \times 10^4$ mol^{-1} dm^3. This is a large value and indicates that the equilibrium is well over to the right.

For the second step:

$$K_{stab} = \frac{[[Cu(H_2O)_4(NH_3)_2]^{2+}]}{[[Cu(H_2O)_5NH_3]^{2+}] \, [NH_3]} = 4.07 \times 10^3 \text{ mol}^{-1} \text{ dm}^3$$

For the third step:

$$K_{stab} = \frac{[[Cu(H_2O)_3(NH_3)_3]^{2+}]}{[[Cu(H_2O)_4(NH_3)_2]^{2+}] \, [NH_3]} = 9.50 \times 10^2 \text{ mol}^{-1} \text{ dm}^3$$

Learning outcomes

On these pages you will learn to:

- explain and write an expression for the stability constant K_{stab}
- perform calculations using K_{stab}
- explain ligand exchange reactions and the stability of complexes using K_{stab}

Exam tip

Note the units of K_{stab}. These are obtained by cancelling the units of concentration in the equilibrium expression.

For the fourth step:

$$K_{stab} = \frac{[[Cu(H_2O_2)_2(NH_3)_4]^{2+}]}{[[Cu(H_2O)_3(NH_3)_3]^{2+}][NH_3]} = 1.74 \times 10^2 \text{ mol}^{-1} \text{ dm}^3$$

Notice that the stability constant decreases for each successive step.

The value of the overall stability constant is the product of the stability constants for each of the individual steps. It is given by:

$$K_{stab} = \frac{[[Cu(H_2O)_2(NH_3)_4]^{2+}]}{[[Cu(H_2O)_6]^{2+}][NH_3]^4} = 1.18 \times 10^{13} \text{ mol}^{-4} \text{dm}^{12}$$

This very large value indicates that the overall equilibrium is well to the right and that a stable complex has been formed. The larger the value of K_{stab}, the more stable the complex.

Worked example

Cisplatin

Cisplatin has the formula $PtCl_2(NH_3)_2$. In water the two Cl^- ligands are replaced by H_2O in two steps.

The first step is:

$$PtCl_2(NH_3)_2(aq) + H_2O(l) \rightleftharpoons [PtCl(NH_3)_2(H_2O)]^+ (aq) + Cl^-(aq)$$

Write an expression for the stability constant for this process and give the units:

$$K_{stab} = \frac{[[PtCl(NH_3)_2(H_2O)]^+](aq) \times [Cl^-(aq)]}{[PtCl_2(NH_3)_2(aq)]} \text{ mol dm}^{-3}$$

The concentration of water does not appear as it cannot change.

The second step is:

$$[PtCl(NH_3)_2(H_2O)]^+(aq) + H_2O(l) \rightleftharpoons [Pt(NH_3)_2(H_2O)_2]^{2+}(aq) + Cl^-(aq)$$

Write an expression for the stability constant for this process and give the units:

$$K_{stab} = \frac{[[Pt(NH_3)_2(H_2O)_2]^{2+}](aq) \times [Cl^-(aq)]}{[[PtCl(NH_3)_2(H_2O)(aq)]^+]} \text{ mol dm}^{-3}$$

The overall reaction is:

$$PtCl_2(NH_3)_2(aq) + 2H_2O(l) \rightleftharpoons [Pt(NH_3)_2(H_2O)_2]^{2+}(aq) + 2Cl^-(aq)$$

Write an expression for the stability constant for the overall process and give the units:

$$K_{stab} = \frac{[[Pt(NH_3)_2(H_2O)_2]^{2+}](aq) \times [Cl^-(aq)]^2}{[PtCl_2(NH_3)_2(aq)]} \text{ mol}^2 \text{ dm}^{-6}$$

1 Cl^- is a better ligand than water. Suggest why this replacement takes place.

Answer: Water is present in a much greater concentration than the chloride ion.

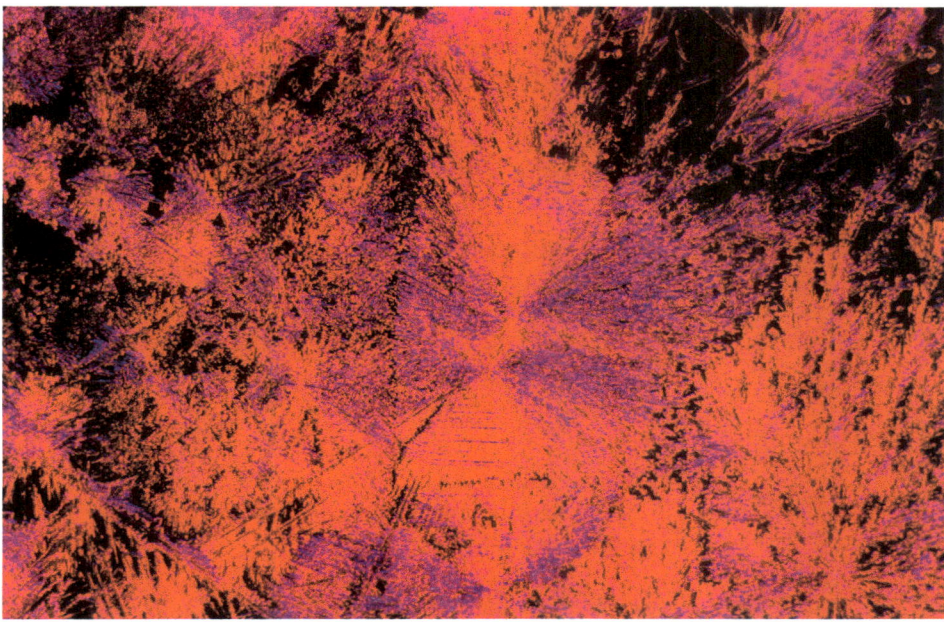

Figure 29 *This image shows crystals of cisplatin. Cisplatin is used as an anti-cancer drug in chemotherapy*

Summary test 28.5

1 a Demonstrate that the overall stability constant for:
$$[Cu(H_2O)_6]^{2+} + 4NH_3 \rightleftharpoons [Cu(H_2O)_2(NH_3)_4]^{2+} + 4H_2O$$
   is the product of the stability constants of the individual steps.
  b Demonstrate that the units of the overall stability constant are found by multiplying the units for each separate stability constant and cancelling.

2 Chloride ions can replace water molecules as ligand for Cu^{2+} ions. The overall reaction is:
$$Cu(H_2O)_6^{2+} + 4Cl^- \rightleftharpoons CuCl_4^{2-} + 6H_2O$$
  a Give an equation for the first step of this process. Give the expression for the stability constant with units.
  b Give the expression for the overall stability constant.

Launch additional digital resources for the chapter

1 The carbonates and nitrates of Group 2 decompose on heating.

a Construct balanced equations for the thermal decomposition reactions of:
  i Calcium carbonate *(2 marks)*
  ii Strontium nitrate *(2 marks)*

b State what observations would be made on heating barium carbonate at 900°C. *(2 marks)*

c Identify and explain the trend in thermal stability down Group 2. *(3 marks)*

2 The table below shows the solubilities of Group 2 hydroxides:

Table A

Substance	Solubility/mol per 100 g water at 298 K
magnesium hydroxide	0.2×10^{-4}
calcium hydroxide	15.3×10^{-4}
strontium hydroxide	33.7×10^{-4}
barium hydroxide	150×10^{-4}

a Describe the trend shown in solubility shown in the table. *(1 mark)*

b The table below shows the hydration enthalpies of some common ions:

Table A

Ion	Hydration enthalpy / kJ mol^{-1}
Mg^{2+}	−1891
Ca^{2+}	−1562
Sr^{2+}	−1480
Ba^{2+}	−1360
OH^-	−460

  i Describe and explain the trend in hydration enthalpies shown in the table above. *(2 marks)*
  ii Suggest the expected trend in solubility of Group 2 hydroxides if enthalpy of hydration were the only factor to affect solubility. Explain your answer. *(3 marks)*
  iii Identify the other factor that affects solubility. Explain how this factor, in combination with hydration enthalpy values, gives rise to the trend you described in part **a**. *(3 marks)*

3 a Define the term 'transition metal'. *(1 mark)*

b Which of the following elements are regarded as transition metals?
  i scandium **ii** iron **iii** zinc *(1 mark)*

c Although the salts of transition elements are usually coloured, there are several copper(I) compounds which are white. Suggest an explanation for this. *(2 marks)*

d The densities of transition elements in the same period gradually increased with relative atomic mass. Explain this trend. *(3 marks)*

4 In a catalytic converter, transition metals are used to catalyse the conversion of oxides of nitrogen and carbon monoxide present in exhaust gases.

a State the type of catalysis that that occurs between the solid transition metal and gaseous reactants. *(1 mark)*

b Identify and explain the three stages in the conversion of exhaust gases in a catalytic converter. *(5 marks)*

5 This question is about the nickel complexes $[Ni(H_2O)_6]^{2+}$ and $[Ni(NH_3)_6]^{2+}$.

a State the coordination number of nickel in both complexes. *(1 mark)*

b Explain why both the nickel complex ions above have an octahedral structure. *(2 marks)*

c A solution containing $[Ni(H_2O)_6]^{2+}$(aq) is green and $[Ni(NH_3)_6]^{2+}$(aq) is blue.
  i State the electronic configuration of a nickel atom. *(1 mark)*
  ii State the electronic configuration of a nickel ion. *(1 mark)*
  iii Use the diagram below to explain why the nickel complexes above form coloured solutions and why these are different.

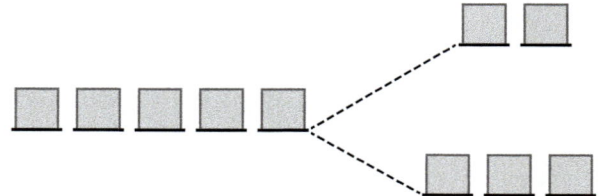

6 A solution containing hexaaquacobalt(II) ions is pink.

  a State the formula of the hexaaquacobalt(II) ion. *(1 mark)*

  b Give the coordination number of cobalt in the hexaaquacobalt(II) ion. *(1 mark)*

  c When concentrated hydrochloric acid is added to this solution, a blue solution containing $[CoCl_4]^{2-}$ ions is formed.

  Give the coordination number of cobalt in the $[CoCl_4]^{2-}$ ion and explain why this is different to your answer to part **b**. *(3 marks)*

7 In the reaction below, a ligand substitution reaction takes place between the water and NH_3 ligands:

$$[Ni(H_2O)_6]^{2+}(aq) + 6NH_3(aq) \rightarrow [Ni(NH_3)_6]^{2+}(aq) + 6H_2O(l)$$

  The first step in this reaction is given below:

$$[Ni(H_2O)_6]^{2+}(aq) + NH_3(aq) \rightarrow [Ni(NH_3)(H_2O)_5]^{2+}(aq) + H_2O(l)$$

  a Construct an equation to show the sixth step of the ligand substitution reaction. *(2 marks)*

  b Construct equilibrium constant expressions for the first and sixth steps. *(2 marks)*

  c The values for the stability constants for steps 1 and 6 are shown in the table below:

	Value of stability constant
K_1	630
K_6	1.07

  i The values of K_1 and K_6 are both greater than 1. Explain what this means. *(1 mark)*
  ii The value of K_1 is greater than the value of K_6. Explain what this means. *(1 mark)*

8 The formula below shows the structure of a complex.

  a Give the formula of the negatively charged ligands in the complex. *(1 mark)*

  b Give the formula of the neutral ligands in the compound. *(1 mark)*

  c This complex shows *cis–trans* isomerism.
  i State whether the isomer shown is the *cis*- or the *trans*-isomer. *(1 mark)*
  ii Draw the other isomer. *(1 mark)*

9 The formulae below represent different complexes:

  A $[Cr(H_2O)_5Cl]^{2+}$ **B** $[Cr(H_2O)_4Cl_2]^+$ **C** $[Pt(NH_3)_4]^{2+}$

  D $[Pt(NH_3)_2Cl_2]$ **E** $[Ni(NH_3)_6]^{2+}$ **F** $[Fe(SCN)(H_2O)_5]^{2+}$

  a Using the list of complexes above, identify the letters for:
  i two square planar complexes
  ii all octahedral complexes. *(2 marks)*

  b State the oxidation numbers of the metals in complexes A and B and F. *(3 marks)*

  c **i** Give the letters for the two complexes that show *cis–trans* isomerism. *(2 marks)*
  ii Draw the two isomers for each of the complexes. *(4 marks)*

  iii Explain why each of the other four complexes do not show *cis–trans* isomerism. *(2 marks)*

10 a Explain how the electron structures of transition elements differ from those of the elements across the main groups of the periodic table. *(1 mark)*

  b Other than variable oxidation states, give three characteristic features of transition metals or their compounds. *(3 marks)*

29 An introduction to A Level organic chemistry

29.1 Formulae, functional groups and the naming of organic compounds

Learning outcomes

On these pages you will learn to:

- identify functional groups
- interpret general, structural, displayed and skeletal formulae
- name aliphatic and aromatic molecules

The material in this chapter builds on the principles explained in Chapter 13. New functional groups are introduced here, and are discussed in detail in the following chapters.

Table 1 *Further classes of organic compounds*

Class of compound	Name of functional group	Structural formula of functional group	Displayed formula	Skeletal formula	Name
arene	arene	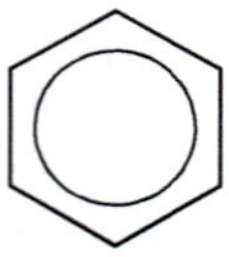	–	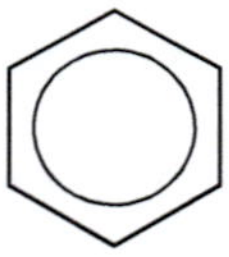	benzene
halogenoarene	halogen	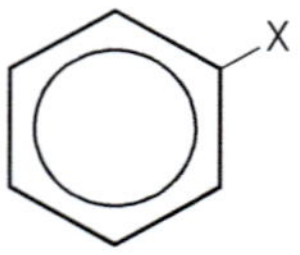	–	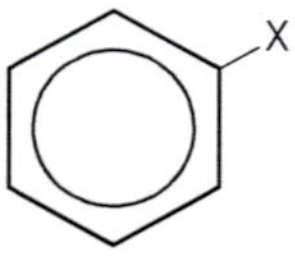	chlorobenzene (when X = Cl)
phenol	phenol	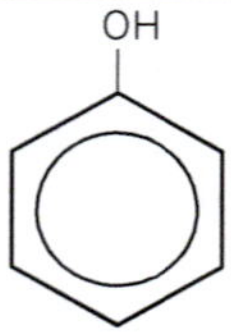	–	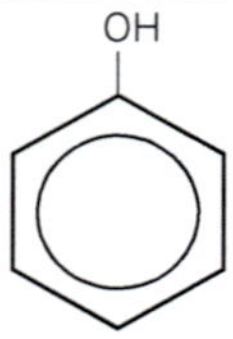	phenol
acyl chloride	acyl chloride	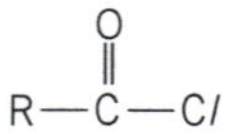 $R-\overset{\overset{O}{\|\|}}{C}-Cl$	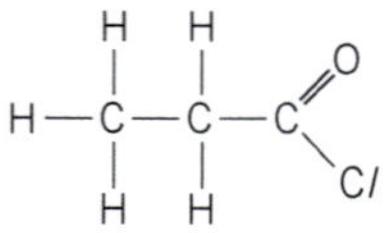	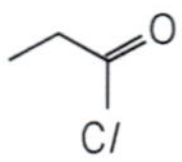	propanoyl chloride
amines (secondary and tertiary)	amine	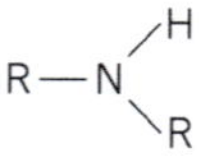 $R-N{\overset{H}{\underset{R}{}}}$	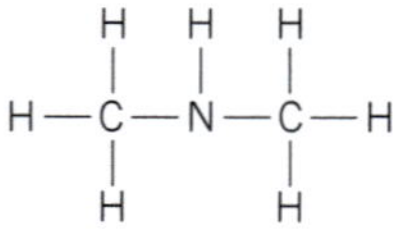	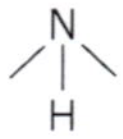	(naming of secondary and tertiary amines is not required)
amide (primary, secondary and tertiary)	amide	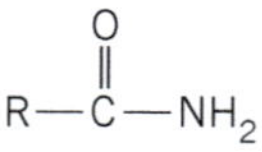 $R-\overset{\overset{O}{\|\|}}{C}-NH_2$	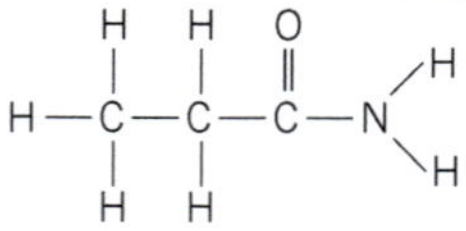	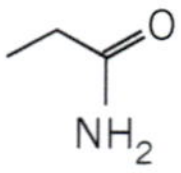	propanamide
amino acid	amine and carboxyl	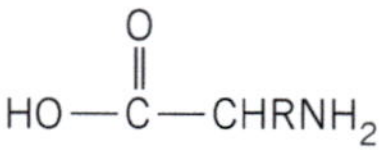 $HO-\overset{\overset{O}{\|\|}}{C}-CHRNH_2$	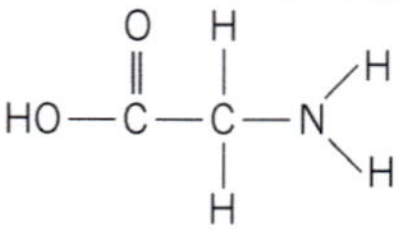	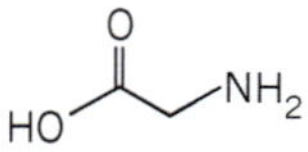	2-aminoethanoic acid

Exam tip

You will also need to be familiar with the classes of organic compounds in Table 1, Section 13.1.

Table 1 adds some extra functional groups to Table 1 of Section 13.1, in particular compounds based on the benzene ring. Benzene is a hydrocarbon of empirical formula CH and molecular formula C_6H_6. It is the 'parent' molecule of a group of organic compounds called **arenes** or **aromatic compounds**.

Although benzene is an unsaturated molecule, it is very stable and does not undergo addition reactions as easily as expected. It has a hexagonal (six-sided) ring structure with a special type of bonding.

Bonding and structure of benzene

The bonding and structure of benzene, C_6H_6, was a puzzle for a long time to organic chemists, because:

- in spite of being unsaturated, it does not readily undergo addition reactions
- all the carbon atoms are equivalent, which implies that all the carbon–carbon bonds are the same.

Benzene consists of a flat, regular hexagon of carbon atoms, each of which is sp^2 hybridised and bonded to a single hydrogen atom. The geometry of benzene is shown in Figure 1. Notice the difference between the flat benzene ring (Figure 2) and the puckered cyclohexane ring (Figure 3).

The C—C bond lengths in benzene are intermediate between those expected for a carbon–carbon single bond and a carbon–carbon double bond (Table 2). So, each bond is intermediate between a single and a double bond.

Table 2 Carbon–carbon bond lengths

Bond	Length / nm
C–C	0.154
C=C (in benzene)	0.140
C=C	0.134
C≡C	0.120

The symbol ⬡ is used to represent this. This is a skeletal formula, which does not show the carbon or hydrogen atoms. There is one carbon atom and one hydrogen atom at each point of the hexagon. The ring in the formula represents a delocalised system of electrons above and below the plane of the ring of carbon atoms. This is called the **aromatic system** and it makes benzene and its derivatives particularly stable. This is described in more detail in Section 29.3.

An arene can have other functional groups (substituents) replacing one or more of the hydrogen atoms in its structure.

Naming arenes

Substituted arenes are generally named as derivatives of benzene, so benzene forms the root of the name.

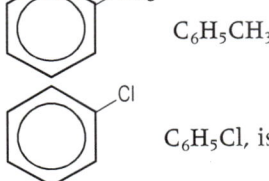

$C_6H_5CH_3$, is called methylbenzene.

C_6H_5Cl, is called chlorobenzene, and so on.

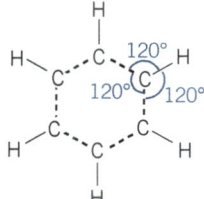

Figure 1 The geometry of benzene (the dashed lines show the shape and do not represent single bonds)

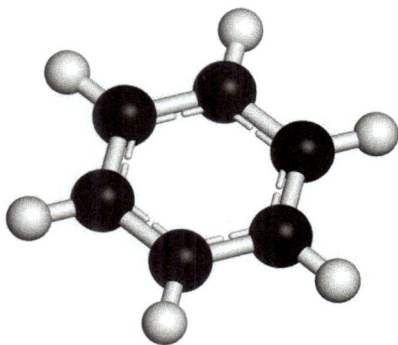

Figure 2 The flat benzene ring

Exam tip

The shorter the bond (between the same pair of atoms) the stronger it is.

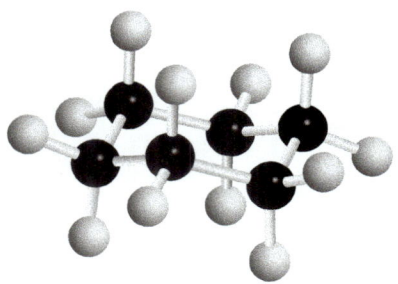

Figure 3 The puckered cyclohexane ring

Exam tip

A fully saturated chain hydrocarbon with six carbon atoms would have a molecular formula of C_6H_{14} and a saturated ring hydrocarbon would have the formula C_6H_{12}.

Summary test 29.1

1 Give the systematic name of the following compounds:

a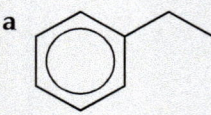

b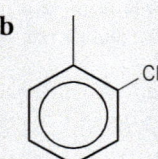

c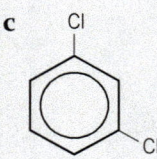

2 Sketch the skeletal formula of:
   a 1,3,5-tribromobenzene
   b 4-chloromethylbenzene
   c bromobenzene.
3 Explain why no compound is called 5-chloromethylbenzene.
4 Deduce how many molecules of hydrogen, H_2, would need to be added onto a benzene molecule to give a fully saturated product cyclohexane.
5 Give the systematic name of:

a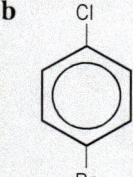

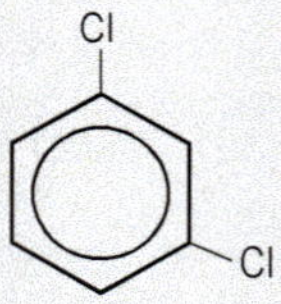

b

CI

Br

6 Sketch the structure of:
   a 1,4-dimethylbenzene
   b 1,2-dimethylbenzene

If there is more than one substituent, the ring is numbered:

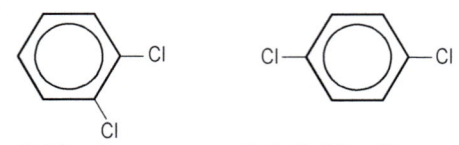

1,2-dichlorobenzene   1,4-dichlorobenzene

Examples
You can test yourself by covering the names or the structures.

ethylbenzene $C_6H_5C_2H_5$

nitrobenzene $C_6H_5NO_2$

1,2-dimethylbenzene $C_6H_4(CH_3)_2$

If a benzene ring is considered to be a substituent, the prefix *phenyl* is used, for example:

is usually called **phenylamine** rather than aminobenzene.

Extension

The stability of benzene
The enthalpy changes of hydrogenation of cyclohexene, benzene, and that predicted for the hypothetical compound 1,3,5-cyclohexatriene, are listed below.

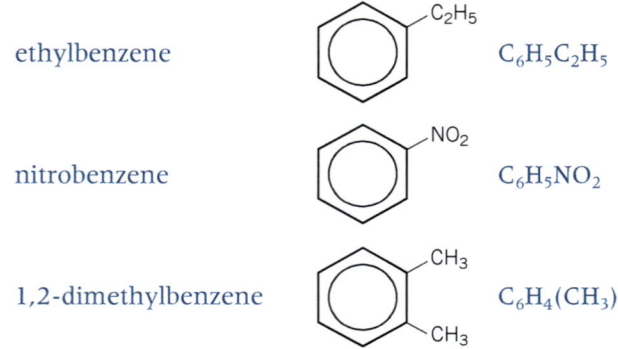

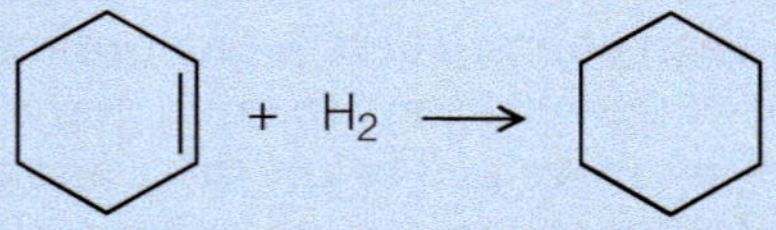

$+ H_2 \longrightarrow$

$\Delta H = -120$ kJ mol^{-1}

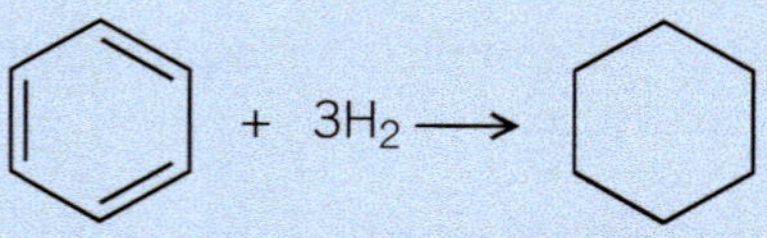

$+ 3H_2 \longrightarrow$

$\Delta H = -360$ kJ mol^{-1}

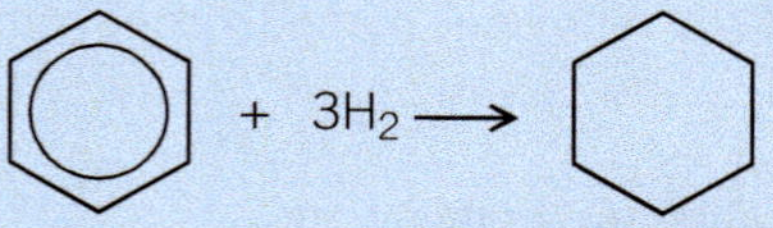

$+ 3H_2 \longrightarrow$

$\Delta H = -208$ kJ mol^{-1}

They show that benzene is 152 kJ mol^{-1} lower in energy than the hypothetical 1,3,5-cyclohexatriene. This is an indication of the extra stability associated with benzene's delocalised bonding system.

Characteristic organic reactions

In Chapter 13, we classified reagents as free radicals, nucleophiles and electrophiles. We classified types of reaction as addition, substitution and elimination.

Benzene and its derivatives usually undergo **electrophilic substitution** reactions for two reasons:

- Electrophiles are attracted to the ring of delocalised electrons.

- Substitution allows the stable aromatic ring system to stay intact while addition or elimination would not.

Mechanism of electrophilic substitution

The electrophile, represented by El$^+$, which has a positive charge, is attracted to the electron-rich delocalised aromatic system of benzene. A bond forms between one of the carbon atoms and the electrophile. But, to do this, the carbon must use electrons from the delocalised system. This destroys the stable aromatic system. To get back the stability of the aromatic system, the carbon loses an H$^+$ ion with the electrons in the C–H bond returning to the delocalised system. The overall effect of these reactions is the substitution of H$^+$ by El$^+$.

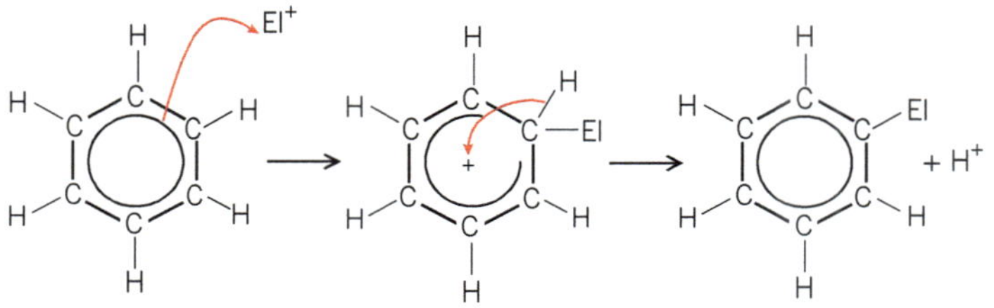

Mechanism of addition–elimination reactions

Another classification of organic reactions that is often used is **addition–elimination** reactions which could be regarded as a special case of substitution. One example is the reaction of acyl chlorides with a nucleophile, represented here by Nu. The overall reaction is:

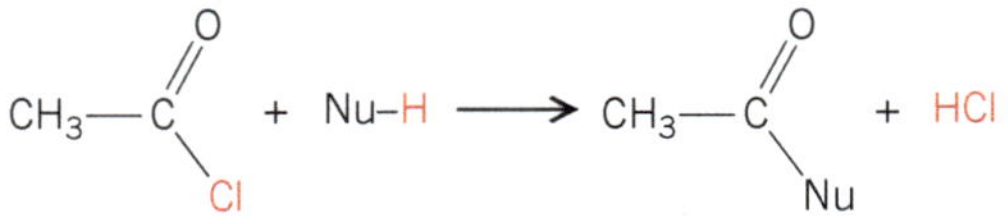

The first step is addition of the nucleophile:

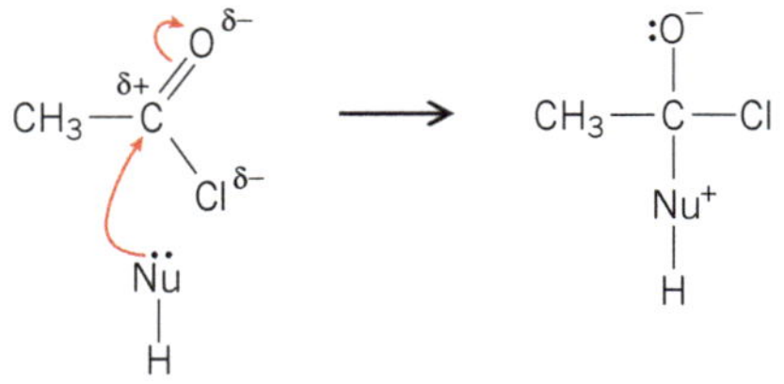

followed by the elimination:

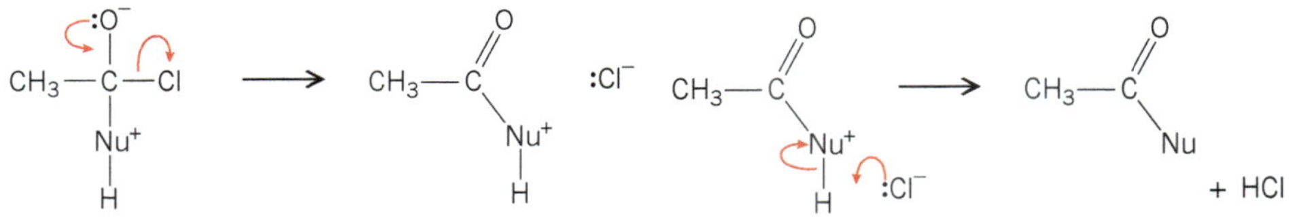

The overall result is substitution of the Cl atom by the nucleophile.

Learning outcomes

On these pages you will learn to:

- describe the mechanism for electrophilic substitution reactions
- describe the mechanism for addition-elimination reactions

Summary test 29.2

1 a Which of the following species is most likely to be attracted to the delocalised electron system in benzene?

Cl$^-$ H$^+$:NH$_3$ Br$^-$

b Identify each one as an electrophile, nucleophile or free radical.

357

Shapes of aromatic organic molecules; σ and π bonds

Learning outcomes

On these pages you will learn to:

- describe and explain the shape of benzene
- describe the bonding and sp² hybridisation in benzene

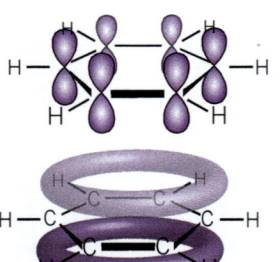

Figure 4 *Delocalisation of p-electrons in benzene to form areas of electron density above and below the ring*

In benzene, each carbon atom is hybridised sp². It forms three σ covalent bonds—one to a hydrogen atom and the other two to carbon atoms. The fourth electron of each carbon atom is in a p-orbital, and there are six of these—one on each carbon atom. The p-orbitals overlap and the electrons in them are delocalised. They form a region of electron density above and below the ring (Figure 4).

Overall, each carbon–carbon bond is intermediate between a single and a double bond. The delocalised system is very important in the chemistry of benzene and its derivatives. It makes benzene unusually stable. This is sometimes called **aromatic stability**.

Extension

The most important dream in history?

This is how Friedrich August von Kekulé's insight into a chemical mystery—the structure of benzene—has been described. Benzene, C_6H_6, had been discovered by Michael Faraday but its structure was a puzzle, as the proportion of carbon to hydrogen seemed to be too great for conventional theories. In 1865, the Belgian chemist Friedrich August von Kekulé published a paper in which he suggested that benzene's structure was based on a ring of carbon atoms with alternating double and single bonds.

His idea resulted from a dream of whirling snakes.

> "I turned my chair to the fire [after having worked on the problem for some time] and dozed. Again the atoms were gambolling before my eyes. This time the smaller groups kept modestly to the background. My mental eye, rendered more acute by repeated visions of this kind, could now distinguish larger structures, of manifold conformation—long rows, sometimes more closely fitted together—all twining and twisting in snakelike motion. But look! What was that? One of the snakes had seized hold of its own tail, and the form whirled mockingly before my eyes. As if by a flash of lighting I awoke."

However, even this insight left a number of problems:

- A cyclic triene should show addition reactions, which benzene rarely does.

$+ 3Br_2 \longrightarrow$

- Kekulé's structure should give rise to two isomeric di-substituted compounds as shown, using skeletal notation:

$+ Br_2 \longrightarrow$

or

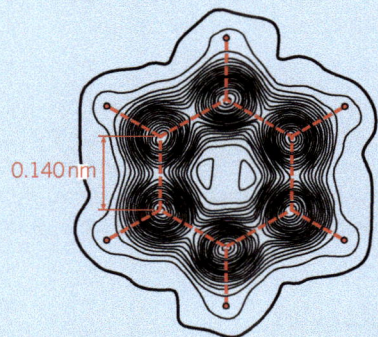

- The hexagon would not be symmetrical—double bonds are shorter than single bonds, Figure 5.

Figure 5 *A technique called X-ray diffraction shows a contour map of the electron density in an individual benzene molecule. This shows that the benzene molecule is a perfect hexagon and each carbon–carbon bond length is 0.140 nm*

Kekulé himself suggested a solution to the second dilemma by proposing that benzene consisted of structures in rapid equilibrium:

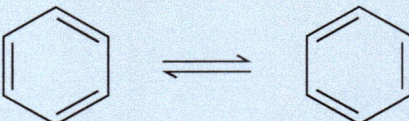

Later this rapid alternation of two structures evolved into the idea of resonance between two structures, both of which contribute to the actual structure. The actual structure was thought to be a hybrid (a sort of average) of the two. Such resonance **hybrids** were believed to be more stable than either of the separate structures.

Summary test 29.3

1. Give the angle between the three sp^2 hybrid orbitals.
2. State the empirical formula of benzene.
3. Deduce how many molecules of hydrogen, H_2, would need to be added onto a benzene molecule to give a fully saturated product cyclohexane.
4. Explain what is meant by delocalisation of electrons in the benzene ring.
5. Look at the two di-substituted compounds formed with the bromination of Kekulé's proposed structure of benzene. State which of the two hypothetical di-substituted compounds would have the shorter bond between the two carbon atoms bonded to the bromine atoms.

Exam tip

Benzene is more stable than the hypothetical molecule cyclohexa-1,3,5-triene because of delocalisation.

Optical isomerism is discussed in Section 13.4. It involves isomers that are non-superimposable mirror images of each other called **enantiomers**. In organic chemistry, optical isomerism occurs when one (or more than one) carbon atom has four different groups bonded to it. This carbon is called the chiral carbon or the chiral centre and is identified in formulae by an asterisk (*). It is perfectly possible for a molecule to have two or more chiral centres—the cholesterol molecule has eight!

Pairs of optical isomers have identical physical and chemical properties, except for their effect on polarised light (they are described as *optically active*). However, they may have different biological activities (as enzymes or as drugs, for example).

Many chemical reactions producing optical isomers form equal amounts of both enantiomers. This is called a **racemic mixture** and the enantiomers are sometimes called racemates. A racemic mixture would not affect the plane of polarisation of polarised light. Each enantiomer would produce the same angle of rotation but in opposite directions, one clockwise and one anticlockwise. So they would cancel each other out.

Biological activity and preparing drug molecules

Biological activity often involves compounds fitting into three dimensional shapes, rather like a hand fitting a glove. In the same way that your left hand will not fit properly into your right glove, one enantiomer may fit perfectly into the active site of an enzyme or a receptor in a cell while the other isomer will not, Figure 6. So pairs of enantiomers may have different biological activities.

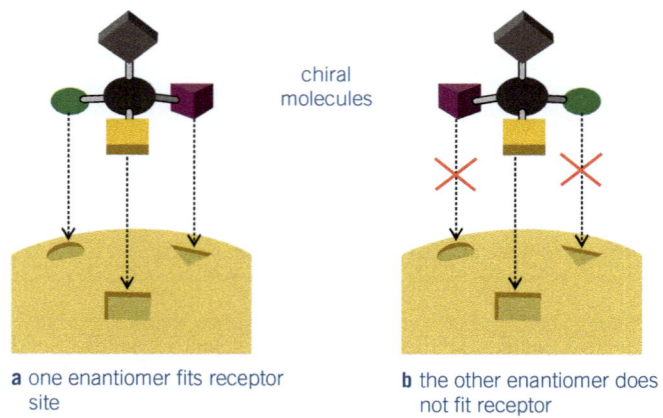

chiral molecules

a one enantiomer fits receptor site

b the other enantiomer does not fit receptor

Figure 6 *Only one enantiomer fits the receptor*

This leaves pharmaceutical chemists with a problem. They need to find a way of separating a pair of enantiomers with almost identical properties (melting point, boiling point, chemical reactions etc), or they need to find a reaction that will produce only one enantiomer. Both are difficult, but:

- Some catalysts (which are themselves chiral molecules) will catalyse a reaction to produce one enantiomer but not the other.
- Some stationary phases for chromatography, see Section 37.2, will separate enantiomers.

The first method is preferred. The second method involves producing a racemic mixture, separating it into two enantiomers and then discarding half of it. This is not very cost-effective.

Other medical examples are even more significant. For example, one enantiomer of ethambutol is an anti-tuberculosis drug while the other has been found to cause blindness. The pain-relief drug ibuprofen also exists as a pair of enantiomers, but only one is active as an anti-inflammatory painkiller.

Extension

The structure of ibuprofen

Ibuprofen is a popular remedy for mild pain and inflammation. The skeletal formula of ibuprofen in shown in Figure 7.

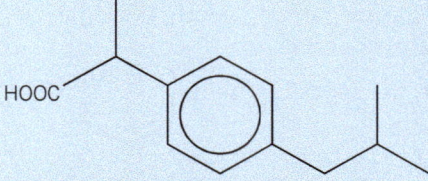

HOOC

Figure 7 *Skeletal formula of ibuprofen*

Optical activity of ibuprofen

Ibuprofen can exist as a pair of optical isomers that are mirror images of each other. These mirror images are non-superimposable. This mirror image property occurs in molecules that have a carbon atom to which four different groups are bonded. The two optical isomers of ibuprofen are identified by the prefixes R– and S+ (Figure 8).

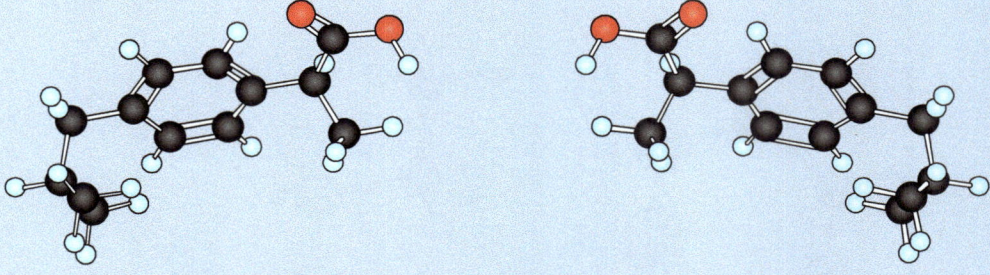

Figure 8 *S+ibuprofen (left) and R–ibuprofen (right) showing their mirror image relationship*

Mirror image isomers are identical in many of their properties, such as solubility, melting point, and boiling point. They can be distinguished by the fact that they rotate the plane of polarisation of polarised light in different directions—the (+)-isomer clockwise as the observer looks at the light, and the (–)-isomer anticlockwise. The symbols R and S refer to the 3D arrangement of the atoms in space. However, the two isomers do behave differently when they interact with other 'handed' molecules, such as prostaglandins in the body that are involved in the process of inflammation. It is the S+ form of ibuprofen which has the anti-inflammatory and pain-killing effect.

However, there is an enzyme in the body that converts the R– form into the S+ form. In fact 60% of the R– form is converted into the S+ form. In a typical dose of 400 mg of ibuprofen, 200 mg is S+ and 200 mg R–. Of this 200 mg of R–, 60% (i.e., 120 mg) is converted in the body to the active S+ form. This gives a total of 320 mg of active S+ form.

So there is little point synthesising the S+ form only. Ibuprofen is therefore sold as a racemic mixture (containing equal amounts of both optical isomers). However, a synthetic route is possible that produces a pure sample of just one of the isomers.

1 Identify the functional group in ibuprofen.

Answer: carboxylic acid

Summary test 29.4

1 Explain why distillation cannot be used to separate pairs of enantiomers.
2 Identify the chiral carbon atom on the formula of ibuprofen above.

There are exam-style questions to test your knowledge of the material in this chapter at the end of Chapter 31.

30.1 Arenes

Figure 1 *Arenes burn with a smoky flame*

Arenes are hydrocarbons based on benzene, C_6H_6 (see Section 29.1). Benzene is the simplest arene.

Arenes were first isolated from sweet-smelling oils, such as balsam, and this gave them the name aromatic compounds. Arenes are still called aromatic compounds, but this now refers to their structures rather than their aromas. Benzene and other arenes have characteristic properties.

As we saw in Section 29.1, benzene is given the special symbol:

Physical properties of arenes

Benzene is a colourless liquid at room temperature. It boils at 353 K and freezes at 279 K. Its boiling point is comparable with that of hexane (354 K) but its melting point is much higher than hexane's (178 K). This is because benzene's flat, hexagonal molecules (Figure 2) pack together very well in the solid state. They are therefore harder to separate and this must happen for the solid to melt.

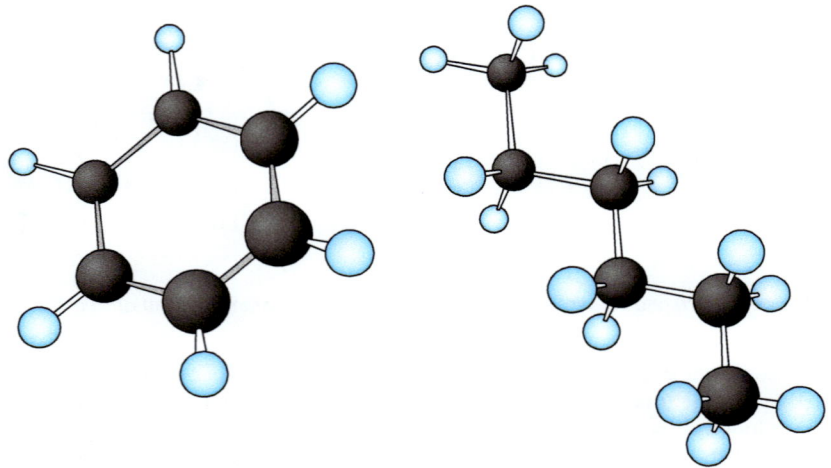

Figure 2 *Benzene molecules (left) can pack together better than hexane molecules (right), so benzene has a higher melting point than hexane*

Like other hydrocarbons that are non-polar, arenes do not mix with water but mix with other hydrocarbons and non-polar solvents.

Chemical reactivity

Benzene and other arenes are particularly stable considering they are unsaturated compounds. The delocalised system is very important in the chemistry of benzene and its derivatives. Most of the reactions of benzene are **electrophilic substitutions**.

Electrophilic substitution reactions

Two factors are important to the reactivity of aromatic compounds:

- The aromatic ring is an area of *high electron density* because of the delocalisation of the electrons. It is therefore attacked by electrophiles.
- The aromatic ring is very *stable* which means the ring almost always remains intact in the reactions of arenes.

So, the most typical reaction is an electrophilic substitution that leaves the delocalised aromatic system of electrons unchanged, rather than addition which would destroy the stability of the aromatic system.

The mechanism of electrophilic substitutions

This has been described in Section 29.2. The basic mechanism is shown below and this is the pattern followed by a variety of electrophiles.

The same overall process occurs in nitration and Friedel–Crafts acylation reactions.

The reaction with halogens

Halogenoarenes can be made by the substitution of a halogen for one of the hydrogen atoms on the benzene ring. Benzene will not react with halogens such as bromine and chlorine unless a catalyst is used, for example, aluminium bromide or aluminium chloride respectively.

The aluminium bromide reacts with the bromine to form $AlBr_4^-Br^+$. The Br^+ acts as the electrophile in the reaction with benzene, forming bromobenzene by substitution.

$$AlBr_3 + Br_2 \rightarrow AlBr_4^-Br^+$$

Then

and $AlBr_3$ is regenerated:

$$AlBr_4^- + H^+ \rightarrow AlBr_3 + HBr$$

Overall:

A similar substitution reaction takes place with chlorine using an aluminium chloride catalyst to form chlorobenzene.

Nitration

Nitration is the substitution of a NO_2 group for one of the hydrogen atoms on the benzene ring. The electrophile NO_2^+ is generated in the reaction mixture of concentrated nitric and concentrated sulfuric acids, and the reaction takes place at a temperature of 25–60°C:

$$H_2SO_4 + HNO_3 \rightarrow H_2NO_3^+ + HSO_4^-$$

Sulfuric acid is a stronger acid than nitric acid and donates a proton, H^+, to HNO_3.

$H_2NO_3^+$ then loses a molecule of water to give NO_2^+, which is called the **nitronium ion** or **nitryl cation**.

$$H_2NO_3^+ \rightarrow NO_2^+ + H_2O$$

The overall equation for the generation of the NO_2^+ electrophile is:

$$H_2SO_4 + HNO_3 \rightarrow NO_2^+ + HSO_4^- + H_2O$$

Exam tip

Electrophiles attack and form bonds with areas of high electron density.

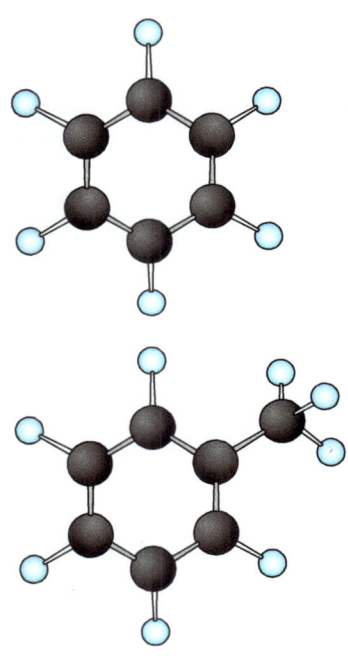

Figure 3 *3D representations of benzene and methylbenzene, both examples of arenes*

Exam tip

Here, nitric acid is acting as a base as it accepts a proton from sulfuric acid.

NO_2^+ is an electrophile and the following mechanism occurs:

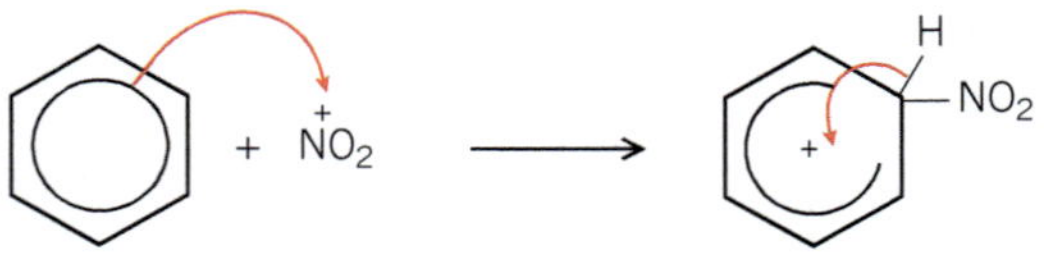

The overall product of the reaction of the nitronium ion, NO_2^+, with benzene is nitrobenzene:

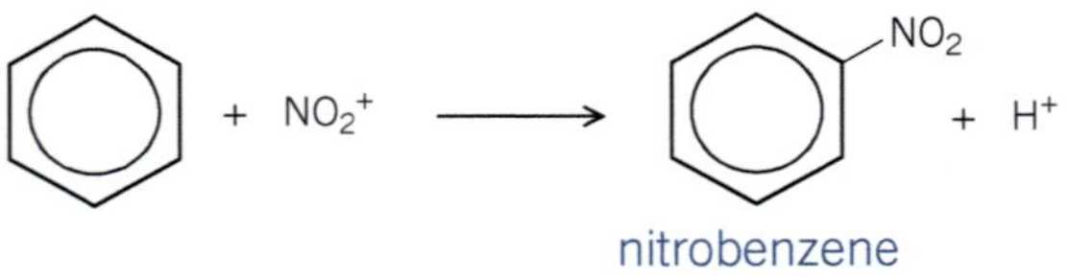

nitrobenzene

The H^+ then reacts with the HSO_4^- to regenerate H_2SO_4, making sulfuric acid a catalyst. The balanced equation is:

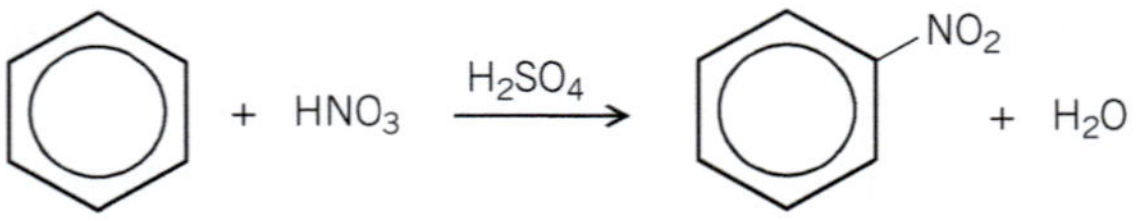

> **Exam tip**
>
> In organic chemistry, curly arrows are used to indicate the movement of a pair of electrons. They run from areas of high electron density to more positively charged areas.

Nitration is an important step in the production of explosives like trinitrotoluene (TNT). Nitration is the first step in making aromatic amines, and these in turn are used to make industrial dyes.

Friedel–Crafts reactions

Acylation
During acylation a hydrogen atom on an aromatic ring is substituted for a RCO group.

> **Exam tip**
>
> The systematic name of trinitrotoluene is 2,4,6-trinitromethylbenzene.

The method for doing this was discovered by Charles Friedel and James Crafts. The reactions use aluminium chloride as a catalyst. The aluminium atom in aluminium chloride has only six electrons in its outer shell and readily accepts a lone pair from the chlorine of an acyl chloride, to form $AlCl_4^-$. This leaves a positively charged electrophile, RCO^+.

$$RCOCl + AlCl_3 \rightarrow RCO^+ + AlCl_4^-$$

RCO^+ is a good electrophile that is attacked by the benzene ring to form substitution products. Its positive charge is localised on the carbon atom.

The aluminium chloride is a catalyst—it is reformed by reaction of the $AlCl_4^-$ ion with H^+ from the benzene ring:

$$AlCl_4^- + H^+ \rightarrow AlCl_3 + HCl$$

The mechanism for the reaction is:

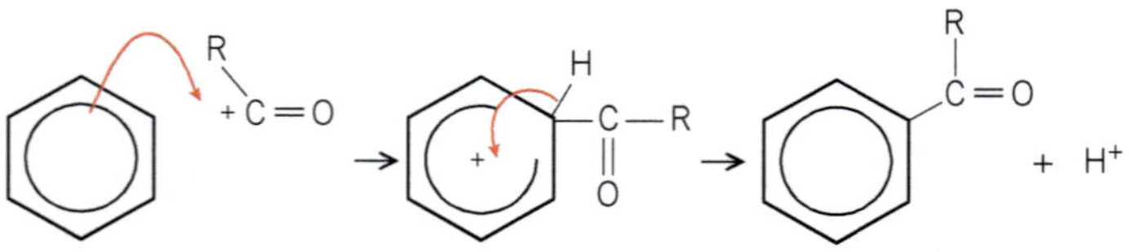

The products are acyl-substituted arenes. The overall reactions are:

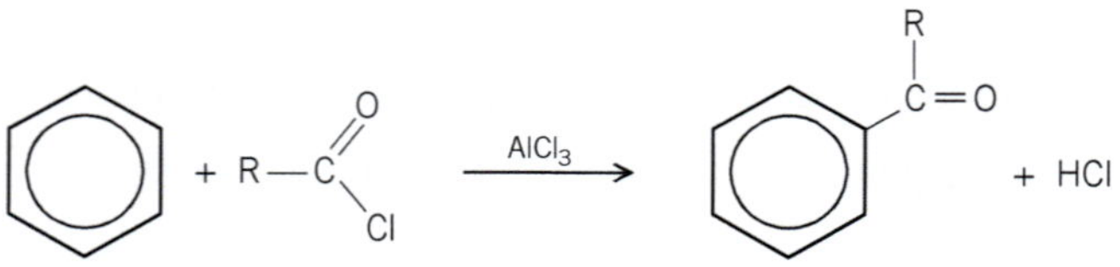

For example, ethanoyl chloride reacts with benzene to form the ketone phenyl ethanone:

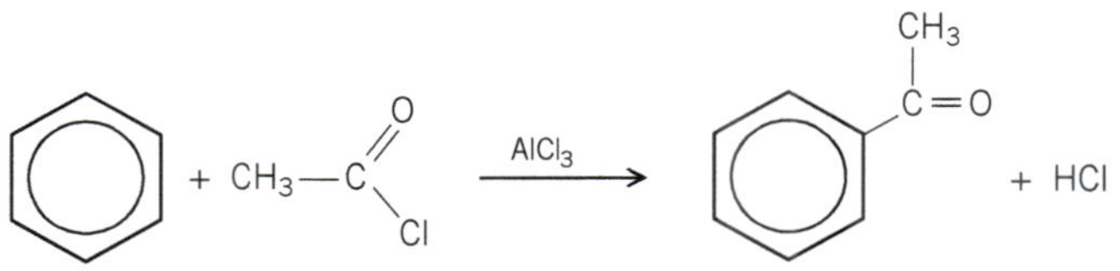

To carry out this reaction ethanoyl chloride is added to a mixture of benzene and aluminium chloride, and the mixture is refluxed gently until the reaction is completed.

Alkylation

During alkylation a alkyl group is substituted for a hydrogen on a benzene ring, in a reaction between a halogenoalkane and an arene.

As in acylation, aluminium chloride acts as a catalyst by attracting an electron from the halide and leaving a positively charged hydrocarbon group (Step 1).

For example, chloromethane will react with benzene to produce methylbenzene.

The mechanism of the reaction is:

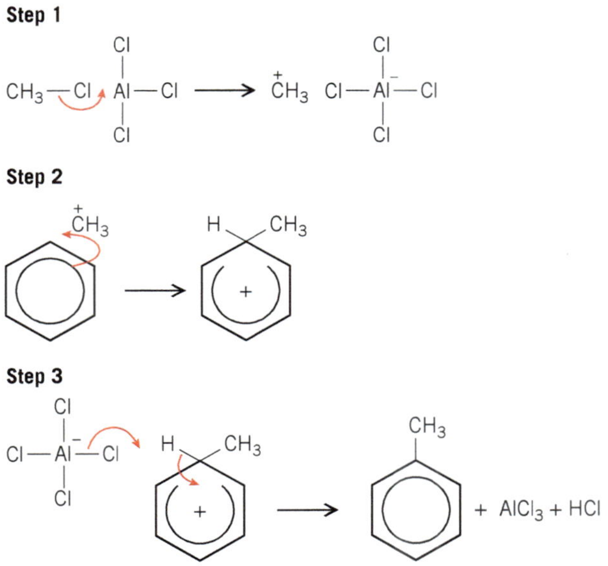

Step 1

Step 2

Step 3

Oxidation of arenes

Benzene itself is resistant to oxidation. However, carbon-based side chains can be oxidised while leaving the arene ring intact. For example methylbenzene reacts with hot alkaline potassium manganate(VII), $KMnO_4$, followed by addition of dilute acid, to give benzoic acid (benzenecarboxylic acid).

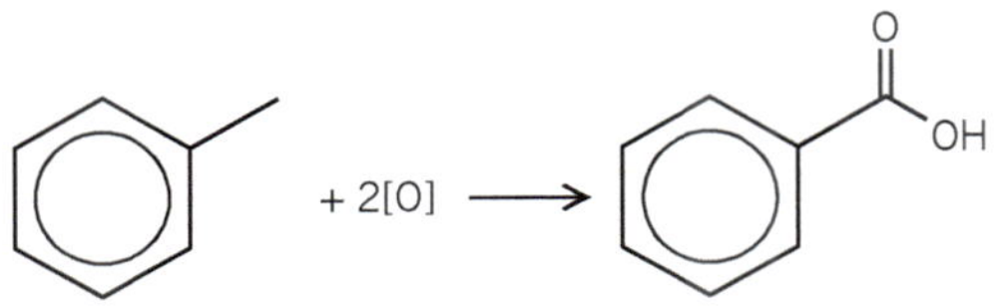

Longer carbon-based side chains will also be oxidised to give benzoic acid.

Exam tip

Acylation is a useful step in the synthesis of new substituted aromatic compounds.

Exam tip

Alkyl side chains of any length (including branched chains) are all susceptible to being oxidised to benzoic acid.

365

Further substitution of arenes

An atom or group of atoms already on a benzene ring will affect the further substitution reactions that take place. We can use the following rules to predict this.

1 R may *release electrons* onto the benzene ring. This will make it more susceptible to further electrophilic substitution reactions – these reactions will go faster and there may be more than one substituent.

Electron-releasing groups include $-CH_3$ and other alkyl groups, $-OH$, and $-NH_2$.

2 R may *withdraw electrons* from the benzene ring. This will make it less susceptible to further electrophilic substitution.

Electron withdrawing groups include $-NO_2$, $-COR$ and $-COOH$.

3 R will direct further substitution to particular positions on the ring:
- electron-releasing groups will direct further substitution to the 2, 4, and 6 positions
- electron-withdrawing groups direct further substitution to the 3 and 5 positions.

The exception is where R is a halogen—it will withdraw electrons but direct further substitution to the 2 and 4 positions.

These rules explain why phenol can easily be nitrated to 2,4,6-trinitrophenol because $-OH$ is electron-releasing.

So the nitration of methylbenzene would give a mixture of:

The nitration of benzenecarboxylic acid would give:

The nitration of chlorobenzene would give:

These are the main products formed. In practice small amounts of other isomers will be formed too.

Reaction of methylbenzene with chlorine

Methylbenzene is both an alkane and an arene. The reactions that can take place depend on the conditions used. In particular, the conditions govern whether the reaction takes place on the aromatic ring or on the alkane side-chain.

If chlorine gas is bubbled into methylbenzene in strong sunlight or under an ultraviolet lamp, a free radical substitution reaction of the methyl group (an alkane) occurs. This gives (chloromethyl)benzene (as well as small amounts of compounds with more substitution of the methyl group). The mechanism of this type of reaction is discussed in Section 14.1.

Here methylbenzene is reacting as an alkane.

In the absence of light and with a catalyst of $AlCl_3$, substitution takes place on the benzene ring at the 2- or 4-positions. This is because the methyl group is electron-releasing and directs further substitution to these positions.

Here methylbenzene is reacting as an arene.

Nucleophilic substitution of chlorobenzene

The typical reactions of arenes are electrophilic substitutions. However, even the electron-rich ring of chlorobenzene can undergo nucleophilic substitution, with a nucleophile such as OH^-. The electronegative chlorine atom can attract electrons away from the carbon to which it is bonded. This carbon becomes slightly $\delta+$ and therefore susceptible to attack by nucleophiles.

However, this reaction requires harsh conditions: 300°C and a pressure of 200 atmospheres. Compare this with the similar reaction of chloroalkanes, which takes place at a little above room temperature and atmospheric pressure.

Hydrogenation of benzene

Benzene is completely hydrogenated by heating to 150°C with hydrogen gas and a nickel/platinum catalyst to form cyclohexane.

There are exam-style questions to test your knowledge of the material in this chapter at the end of Chapter 31.

Summary test 30.1

1 Give equations for the steps in the formation of chlorobenzene from benzene using an aluminium chloride catalyst.

2 State which of the following is an electrophile:

R^+ $:NH_3$ NO_2 Cl^-

3 Give the names of the two isomers of 1,3-dinitrobenzene.

4 Explain why most of the reactions of benzene are substitutions rather than additions.

5 Give the equation for the reaction between propanoyl chloride and benzene. Identify the species that attacks the benzene ring.

6 Predict the likely products of the single nitration of ethylbenzene.

7 The nitration of methylbenzene can be done as a school practical exercise. Explain why there is no danger of the formation of the explosive 2,4,6-trinitromethylbenzene (trinitrotoluene, TNT).

8 a Predict the rate at which i benzenecarboxylic acid and ii phenol is nitrated compared with benzene

  b Explain your answers to part a.

31 Halogen compounds

31.1 Halogen compounds

Learning outcomes

On these pages you will learn to:

- describe the reactions used to produce halogenoarenes
- explain why halogenoarenes are less reactive than halogenoalkanes

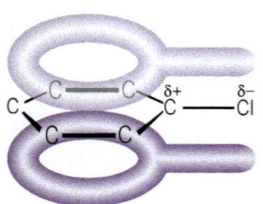

Figure 1 Bonding in chlorobenzene

> **Exam tip**
>
> The C–Cl bond in chlorobenzene is shorter than the C–Cl bond in chloroalkanes because of its double bond character.

This section looks at some of the halogen-containing derivatives of arenes. Remember from Chapters 29 and 30, that arenes are benzene-type rings with an aromatic system. In the halogenoarenes, the C–X bond is shorter and therefore stronger than in the halogenoalkanes. For example, the C–Cl bond in chloromethane is 0.177 nm and the C–Cl bond in chlorobenzene is 0.169 nm.

This is due to the overlap of p-orbitals on the Cl atom with the aromatic system in benzene. This provides extra π-type bonding, in addition to the C–Cl σ bond (see Figure 1).

This overlap also reduces the δ+ character of the carbon to which the halogen is bonded, so there are two effects:

1 a stronger bond
2 less positive charge on the carbon.

Both effects suggest that halogenoarenes will be less reactive than halogenoalkanes. This is what is found in practice.

For example, chlorobenzene will undergo nucleophilic substitution with aqueous sodium hydroxide only at 300°C and 200 atmospheres pressure.

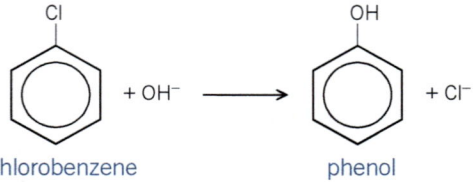

Compare this with chloroalkanes such as chloroethane, which will react with OH⁻ ions with gentle warming at room pressure, see Section 15.1.

Preparation of halogenoarenes

Halogenoarenes can be produced by an electrophilic substitution reaction of the parent arene with the appropriate halogen. For example:

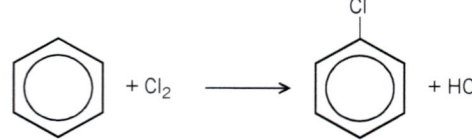

> **Exam tip**
>
> An alternative catalyst for this reaction is aluminium bromide.

This reaction takes place at room temperature but requires a catalyst of AlCl₃, which forms the electrophile Cl⁺ (as discussed in Section 30.1).

A similar reaction can be used to produce bromobenzene. In this case, the catalyst is AlBr₃ or FeBr₃ (made in the reaction vessel by reaction of some of the bromine with iron filings to make FeBr₃).

Methylbenzene is a little more reactive because the methyl group –CH$_3$ is a **ring-activating substituent**. This means that it releases electrons onto the benzene ring therefore making it more reactive to electrophiles (which have positive charges). The methyl group directs further substitution to the 2- or 4-positions in the ring, so the product is a mixture of 2-chloromethylbenzene and 4-chloromethylbenzene.

Again, similar reactions occur with bromine.

Figure 2 *Agent Orange was a mix of chlorine-subsituted arene compounds, used as a herbicide. It was used as a chemical weapon by the US army in the Vietnam War to devastating effect—3 million Vietnamese citizens suffered illnesses as a result of the 'spray runs' (pictured here), in addition to massive ecological disruption*

Summary test 31.1

1 Explain how the production of HCl in the reactions above indicates that a substitution reaction has occurred rather than an addition reaction.
2 Give a balanced equation for the formation of the electrophile Br$^+$ from iron and bromine.
3 Methylbenzene will react with bromine in the presence of UV light without a catalyst. Suggest what type of reaction is occurring (give the reagent type and reaction type). Give the formula of the main product that you would expect.

29–31 Exam-style questions

Launch additional digital resources for the chapter

1 Write structural formulae for the following benzene derivatives:

a 2,4-dinitrophenol *(1 mark)*

b 1,4-dichlorobenzene *(1 mark)*

c 4-nitrophenylamine *(1 mark)*

d 2-hydroxybenzoic acid *(1 mark)*

e 2-chlorophenylamine. *(1 mark)*

2 Consider the following compounds:

i
CH_3

ii

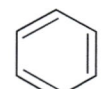

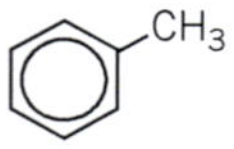

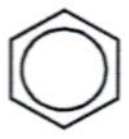

iii

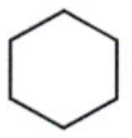

iv

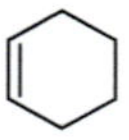

v $CH_3CH_2CH{=}CHCH_2CH_3$

Give the letter(s) for the substance(s) which:

a are aromatic hydrocarbons

b are cyclic compounds

c are unsaturated hydrocarbons

d have a planar ring in their molecule

e would decolourise aqueous bromine in the dark

f would evolve fumes of HBr when treated with bromine and iron filings in the dark

g would react with alkaline potassium manganate(VII) solution. *(7 marks)*

3 Consider the catalytic hydrogenation of cyclohexene:

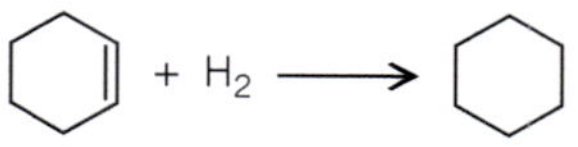

+ H_2

$\Delta H = -120\,\text{kJ mol}^{-1}$

a State a suitable catalyst for this reaction. *(1 mark)*

b Assuming that benzene has the Kekulé structure given below, predict a value for its enthalpy change of hydrogenation to cyclohexane, using the data above.

(1 mark)

c The experimental value for the enthalpy change of hydrogenation of benzene to cyclohexane is $-208\,\text{kJ mol}^{-1}$. Explain why this is different to your answer given to part **b**.

4 Give the displayed formulae for the major products for the following reactions:

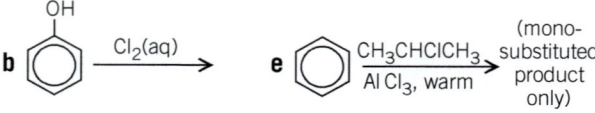

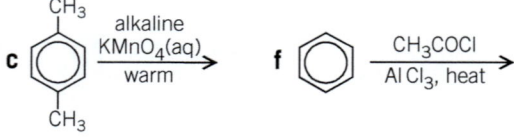

a
CH_3
H_2/Ni
300 °C

d
NO_2
conc. H_2SO_4
conc. HNO_3
120 °C
NO_2

b
OH
Cl_2(aq)

e
$CH_3CHClCH_3$
Al Cl_3, warm
(mono-substituted product only)

c
CH_3
alkaline
$KMnO_4$(aq)
warm
CH_3

f
CH_3COCl
Al Cl_3, heat

(6 marks)

5 Give the structural formulae for the main products that you would expect from the following substitution reactions. Assume monosubstitution occurs in each case.

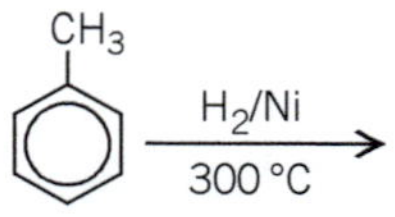

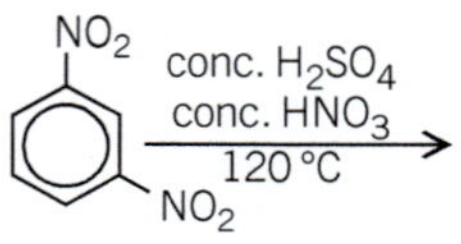

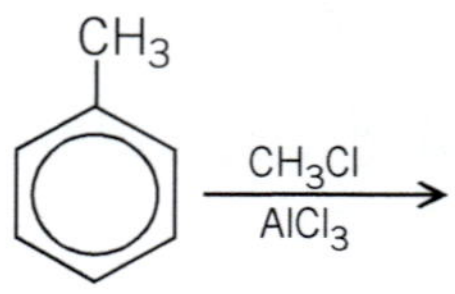

CH_3
CH_3Cl
$AlCl_3$

NO_2
Cl_2
$AlCl_3$

$CH_2CH_2CH_3$
conc. HNO_3
conc. H_2SO_4

(3 marks)

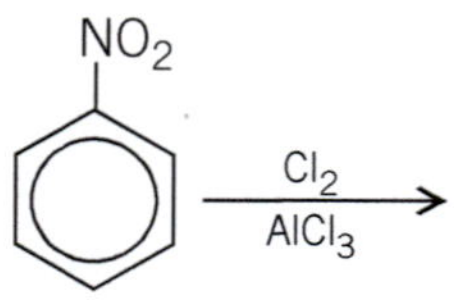

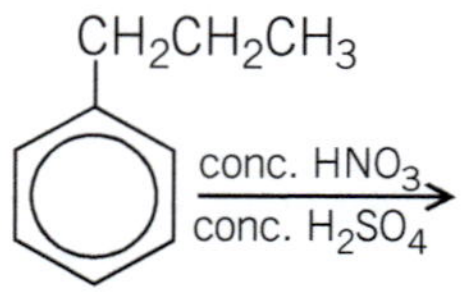

6 (1-methylethyl)benzene, commonly known as cumene, is an important intermediate in the manufacture of phenol and propanone (acetone). It has the following structure:

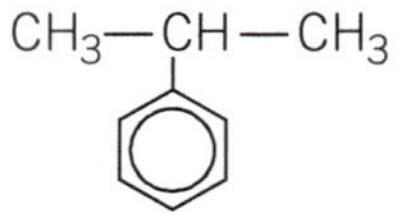

$$CH_3-CH-CH_3$$

(1-methylethyl)benzene is manufactured by a Friedel-Crafts type reaction between propene and benzene in the presence of an acid catalyst:

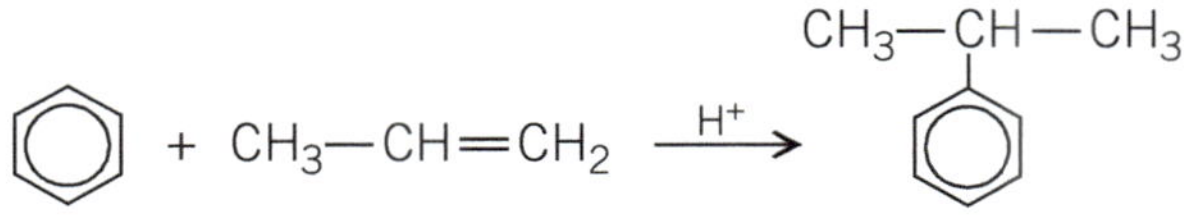

$$\bigcirc + CH_3-CH=CH_2 \xrightarrow{H^+} \bigcirc^{CH_3-CH-CH_3}$$

The reaction is believed to proceed via a carbocation intermediate.

a Give the structures of two possible carbocations formed by the attack on the H^+ on the double bond of propene. *(2 marks)*

b Using your answer to part **a**, predict which carbocation is more stable, explain your answer. *(2 marks)*

c Show how your answer to part **b** can attack and substitute the benzene ring. Explain why (1-methylethyl)benzene is virtually the only product of this reaction and hardly any propylbenzene is produced. *(2 marks)*

7 This question is about lactic acid, which has the IUPAC name 2-hydroxypropanoic acid.

a Draw the skeletal formula of lactic acid and label the chiral carbon atom. *(1 mark)*

b Draw 3D representations of both enantiomers of lactic acid. *(2 marks)*

c Describe how separate samples of enantiomers of lactic acid can be distinguished *(2 marks)*

8 Give the IUPAC name for the following benzene derivatives:

a

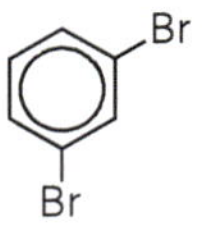

b

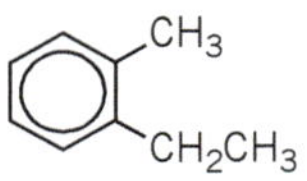

c

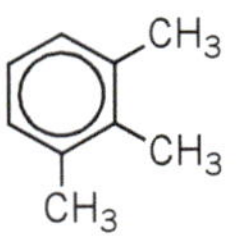

d

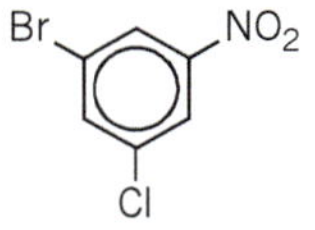

e

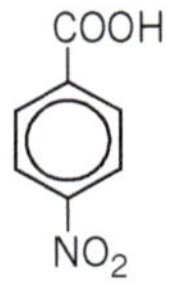

f

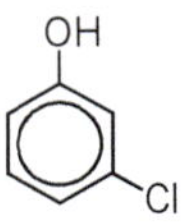

9 Ethylbenzene reacts with propanoyl chloride in the presence of aluminium chloride to produce a mixture of three isomers.

a Suggest the displayed formula of one of these isomers and name the type of isomerism present in the mixture of products. *(1 mark)*

b Give an equation to show the formation of the electrophile formed by the reaction of propanoyl chloride with aluminium chloride. *(2 marks)*

c Name and outline a mechanism for the reaction of the electrophile formed in part **b** with ethylbenzene to give the isomer in part **a**. *(5 marks)*

10 TNT is used in extracting oil and gas from shale formations. Also known as trinitrotoluene, TNT has the IUPAC name 2-methyl-1,3,5-trinitrobenzene.

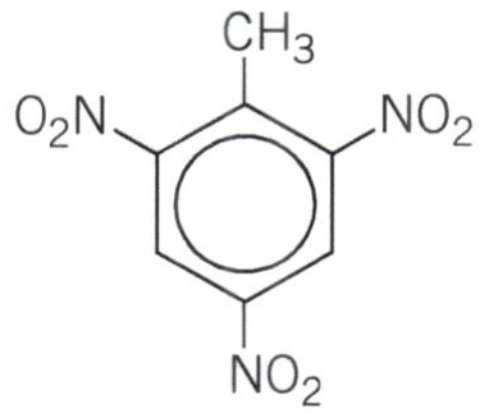

methyl-2,4,6-trinitrobenzene
(trinitrotoluene, TNT)

a TNT can be synthesised from the nitration of a precursor, **Q**.
  i Suggest the identity of **Q**. *(1 mark)*
  ii State the reagents required to synthesise TNT and give the identity of the electrophile intermediate formed. *(2 marks)*

b Nitration of **Q** also leads to the formation of 4-nitrobenzene.
  i Suggest a structure for 4-nitrobenzene. *(1 mark)*
  ii Name and give the mechanism for the synthesis of 4-nitrobenzene from precursor **Q** and the intermediate identified in part **aii**. *(4 marks)*

32.1 Alcohols

Formation of esters

Alcohols react readily with acyl chlorides at room temperature to form **esters** (see Section 33.2).

For example:

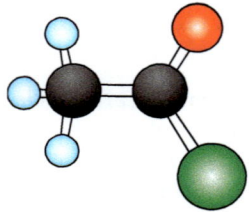

$$CH_3-C(=O)Cl \quad + \quad CH_3CH_2OH \longrightarrow CH_3-C(=O)O-CH_2CH_3 \quad + \quad HCl$$

ethanoyl chloride   ethanol   ethyl ethanoate

The reaction is very fast and exothermic. Steamy white clouds of hydrogen chloride gas are given off.

Acyl chlorides are also called acid chlorides. They are discussed in more detail in section 33.3.

Figure 1 *A 3D representation of ethanoyl chloride, an acyl chloride*

Summary test 32.1

1 Suggest an alternative to ethanoyl chloride for the formation of ethyl ethanoate. Explain why ethanoyl chloride is preferred.
2 State what you would react with ethanoyl chloride to produce methyl ethanoate. Give the structural formula of methyl ethanoate.

Phenol

Phenols are aromatic compounds in which one or more of the hydrogen atoms of the ring are replaced with an –OH group. Phenol itself is a pale pink crystalline solid.

Preparation of phenol

Phenylamine is reacted with nitrous acid, HNO_2 (produced by the reaction of sodium nitrite, $NaNO_2$, and dilute hydrochloric acid) at low temperature (below 10°C) to produce the diazonium salt benzenediazonium chloride.

On warming the aqueous benzenediazonium salt it is hydrolysed to phenol.

Coupling reactions

The benzenediazonium ion from the benzenediazonium chloride will react with phenol in a solution of sodium hydroxide. This is called a coupling reaction. The alkaline solution removes a H^+ ion from the OH group of the phenol to form a phenoxide ion. The phenoxide ion then couples with the benzenediazonium ion.

The bright yellow product is called 4-(phenylazo)phenol. This is the simplest of a family of compounds consisting of two aromatic rings joined by the –N=N– group, and called azo compounds. Because of the N=N double bond it exists as a pair of *cis–trans* isomers. Using different substituents on the aromatic rings produces a range of compounds with different colours. Many of them are used as dyes.

Reactions of the arene ring of phenol

Phenol, (also called hydroxybenzene) has both an aromatic ring and an OH group. However, the reactions of both groups are significantly altered when they are bonded together.

Extension

A test for phenol
Adding a small amount of phenol to a solution of iron(III) chloride produces a purple-coloured complex (in which the phenol is acting as a ligand). Substituted phenols also produce a range of colours.

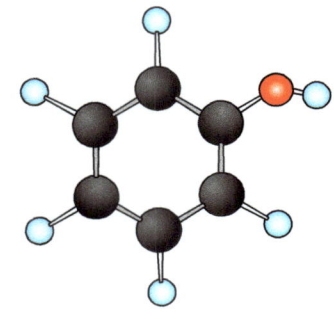

Figure 2 *A 3D representation of phenol*

Exam tip

You won't be expected to remember the name 4-(phenylazo)phenol!

Bromination

The arene ring of phenol undergoes the typical electrophilic substitution reactions of an aromatic compound. The OH group releases electrons onto the ring making it more reactive than benzene itself. The OH group directs substitution to the 2-, 4-, and 6-positions. So phenol will react with an aqueous solution of bromine to form 2,4,6-tribromophenol.

Compare this with the reaction of benzene itself, which requires liquid bromine and an iron catalyst and which only produces a single substitution to give bromobenzene.

Nitration

Nitration of phenol is also much easier than nitration of benzene. Nitration of phenol with dilute nitric acid at room temperature gives a mixture of 2-nitrophenol and 4-nitrophenol. Nitration of benzene requires a mixture of concentrated nitric and sulfuric acids. With this nitrating mixture, phenol produces 2,4,6-trinitrophenol.

2-nitrophenol   4-nitrophenol

2,4,6-trinitrophenol

Reactions of the OH group of phenol

The oxygen atom of the OH group has lone pairs of electrons in p-orbitals. One of these orbitals can overlap with the delocalised ring of electrons on the benzene ring, so there is less negative charge on the oxygen. This has the effect of strengthening the C–O bond (and giving it some of the character of a double bond, see Figure 3). This makes reactions such as substitution of the OH group more difficult than in alcohols such as ethanol, in which the OH is bonded to an alkyl group. It also reduces the δ+ character of the carbon bonded to the OH group, which makes it less likely to be attacked by nucleophiles. This, too, makes nucleophilic substitution reactions less likely.

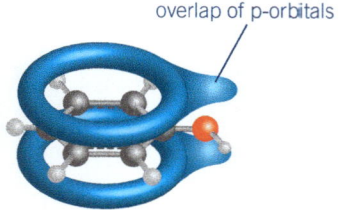

overlap of p-orbitals

Figure 3 Orbital overlap in phenol

Comparing the acidities of ethanol, water and phenol

Phenol is a weak acid. The following equilibrium is set up, forming the phenoxide ion and an H^+ ion.

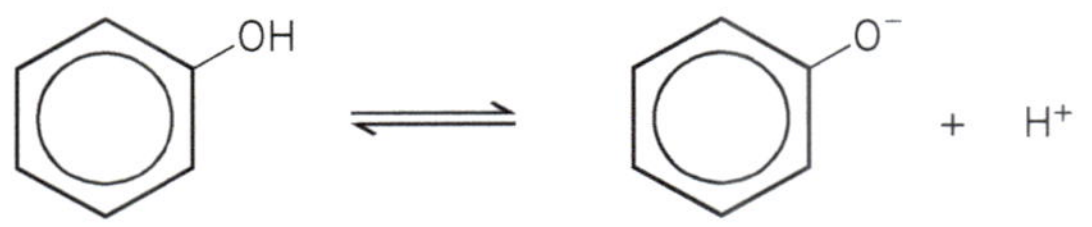

Phenol is a stronger acid than water, which is itself a stronger acid than ethanol.

The acidity of a molecule depends on how easily a hydrogen ion is lost. This depends on how strongly the hydrogen ion is held by the negative charge of the oxygen atom.

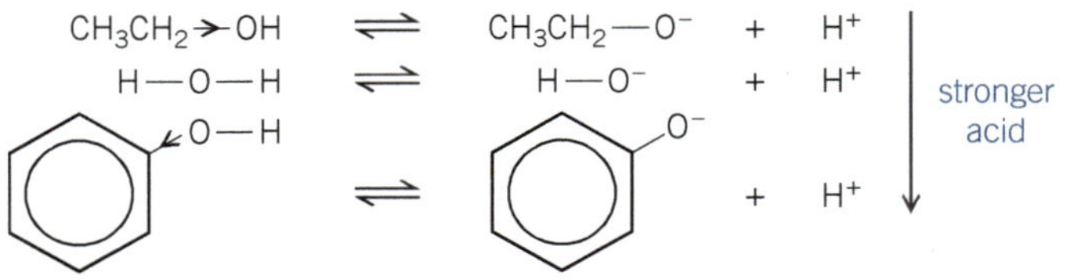

Figure 2 *The relative acid strengths of ethanol and phenol compared with water*

If we look at Figure 2, we can see that the ethyl group releases negative charge onto the oxygen atom via the inductive effect, see Section 15.1, making it more difficult to lose the H^+ ion. It is therefore a weaker acid than water.

In phenol, negative charge is drawn away from the oxygen atom, making it easier to lose an H^+ ion. It is therefore a stronger acid than water.

Acid–base reactions

Phenol will react with strong bases such as sodium hydroxide to produce sodium phenoxide.

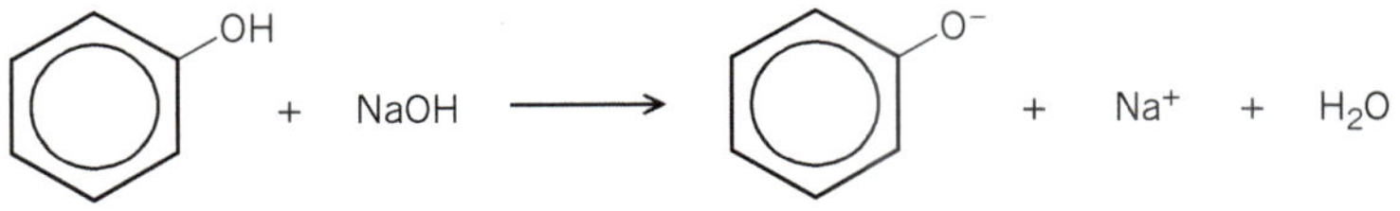

Phenol will also react with sodium to give sodium phenoxide and hydrogen.

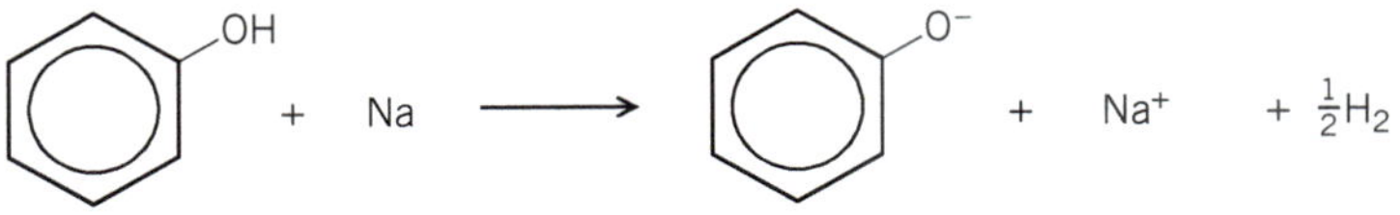

Naphthol

Naphthol is an example of an aromatic molecule with two joined rings and an OH group.

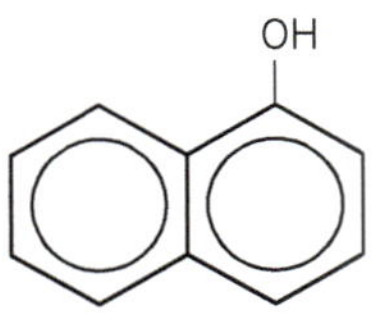

1-naphthol

1-naphthol has similar reactions to phenol.

The OH group is electron releasing, and directs electrophilic substitution onto the same ring as the OH group, mostly to the 4 position.

So, further electrophilic substitutions like nitration and bromination take place on the ring that has the OH.

Extension

In the phenoxide ion (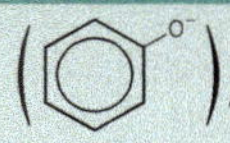), a lone pair of electrons on the oxygen atom is delocalised onto the benzene ring. Negative charge is therefore drawn away from the oxygen atom.

Exam tip

Alkyl groups are slightly electron releasing. This is the inductive effect.

Summary test 32.2

1. Give a balanced equation for the reaction of phenol with potassium.
2. Sketch the structure of 2,4,6-trinitrophenol.
3. Explain why it is necessary to use a locant when naming 1-naphthol, but not when naming phenol.
4. Give the structural formulae for the product of bromination of 1-naphthol.

There are exam-style questions to test your knowledge of the material in this chapter at the end of Chapter 33.

33 Carboxylic acids and derivatives

33.1 Carboxylic acids

The carboxylic acid functional group is

This chapter builds on the general properties of carboxylic acids discussed in Chapter 18. These include:

- naming carboxylic acids
- physical properties of carboxylic acids
- synthesis of carboxylic acids
- reactivity of carboxylic acids
- some reactions of carboxylic acids.

In particular it looks at the behaviour of the carboxylic acid group when attached to an aromatic ring, as in benzoic acid (also called benzenecarboxylic acid):

Learning outcomes

On these pages you will learn to:

- explain how benzoic acid can be produced
- describe the formation of acyl chlorides from carboxylic acids
- describe the further oxidation of some carboxylic acids
- explain the relative acidities of carboxylic acids, phenols, alcohols and chlorine-substituted carboxylic acids

The production of benzoic acid

Benzoic acid can be produced by the oxidation of the side chain of an alkyl-substituted benzene. Hot alkaline potassium manganate(VII) (potassium permanganate, $KMnO_4$) is used as the oxidising agent. The simplest alkylbenzene is methylbenzene, but longer side chains will also be oxidised to benzoic acid.

> **Exam tip**
>
> When using [O] to represent an oxidising agent, the equation should still be balanced.

As the reaction occurs in an alkaline solution, a benzoate salt is produced. Dilute acid is then added to convert this to benzoic acid.

The relative acidity of OH groups

Acids are substances that donate a proton (H^+ ion). The strength of an acid is measured by its pK_a value (see Section 25.1):

> **The smaller the pK_a value, the stronger the acid.**

pK_a values use a logarithmic scale, so a decrease of 1 pK_a unit means a tenfold increase in acid strength. Table 1 shows the differences in acid strength for organic compounds containing OH groups. How can we explain these values?

When a compound ROH acts as an acid, the following equilibrium is set up:

$$ROH(aq) \rightleftharpoons H^+(aq) + RO^-(aq)$$

The position of the equilibrium depends on the stability of the RO^- ion. The more stable the RO^- ion is, the further the equilibrium is to the right, the stronger is the acid ROH and the smaller is its pK_a value.

Table 1 *The strengths of some carboxylic acids compared with ethanol and phenol*

Compound	Formula	pK_a
ethanol	CH_3CH_2OH	16.0
phenol	C_6H_5OH	10.0
ethanoic acid	CH_3COOH	4.8
benzoic acid	C_6H_5COOH	4.2
chloroethanoic acid	$CH_2ClCOOH$	2.9
dichloroethanoic acid	$CHCl_2COOH$	1.3
trichloroethanoic acid	CCl_3COOH	0.7

The stability of the RO⁻ ion depends on how well the negative charge is spread over the ion. The more it is spread away from the oxygen atom, the easier it is to lose a H⁺ ion.

In ethanol, there is no tendency for the charge to be spread away from the oxygen atom. In phenol, the charge can be spread onto the benzene ring. This is why phenol is a stronger acid than ethanol.

We have seen in Section 18.1, that in carboxylic acids the charge is delocalised over the three atoms of the carboxylate group. This is why carboxylic acids are much stronger acids than ethanol. In benzoic acid, the negative charge can be further delocalised over the aromatic system as shown in Figure 1, making benzoic acid slightly stronger than alkyl carboxylic acids.

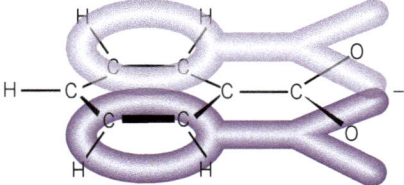

Figure 1 *Delocalisation of the negative charge in the benzoate ion*

In the chloroethanoate ion, the electronegative chlorine atom draws negative charge towards itself. This spreads out the negative charge further than in the ethanoate ion and makes chloroethanoic acid stronger than ethanoic acid. Further chlorine atoms increase this effect.

Acyl chlorides

Acyl chlorides, also called acid chlorides, have the general formula R—C⟨O⟩⟨Cl⟩ in which a chlorine atom has replaced the OH group of a carboxylic acid (R represents any alkyl group, CH_3, CH_3CH_2, etc). Like esters, they are acid derivatives.

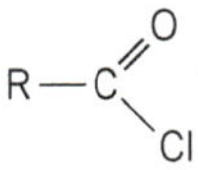

Formation of acyl chlorides
Acyl chlorides can be formed from carboxylic acids by any of the following three reactions with reagents that generate the nucleophile Cl:⁻

- with phosphorus trichloride:

$$3RCOOH + PCl_3 \rightarrow 3RCOCl + H_3PO_3$$

- with phosphorus pentachloride:

$$RCOOH + PCl_5 \rightarrow RCOCl + POCl_3 + HCl$$

- with thionyl chloride:

$$RCOOH + SOCl_2 \rightarrow RCOCl + SO_2 + HCl$$

All three reactions are nucleophilic substitution reactions. The nucleophile Cl⁻ attacks the $C^{\delta+}$ of the carboxylic acid, followed by loss of an OH⁻ group. The third reaction has the advantage that both the by-products are gases and so are easily separated from the acyl chloride.

The mechanism of these reactions can be represented using curly arrows to show the movement of electron pairs (see Figure 2).

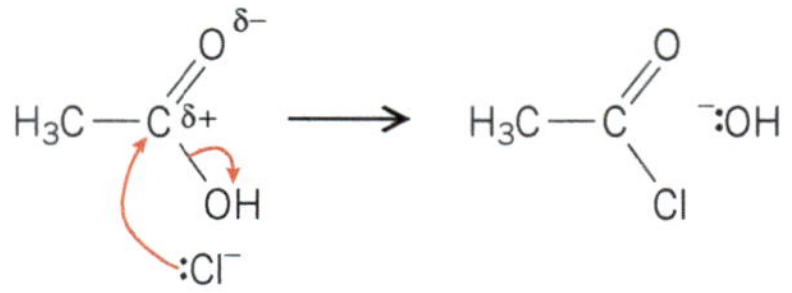

Figure 2 *The mechanism of nucleophilic substitution*

Oxidation of carboxylic acids

In general, carboxylic acids can only be oxidised under vigorous conditions and the products are the same as those of combustion—carbon dioxide and water.

Methanoic acid, HCOOH, is an exception. It can be oxidised (to carbon dioxide and water) by acidified potassium manganate(VII) (potassium permanganate, $KMnO_4$), or acidified potassium dichromate(VI) (potassium dichromate, $K_2Cr_2O_7$). However, it can even be oxidised by the gentle oxidising agents in either Fehling's solution or Tollens' reagent. In these cases the inorganic products are an red precipitate of copper(I) oxide and a silver mirror respectively. These are similar reactions to those used to distinguish aldehydes from ketones, see Section 17.1.

Ethanedioic acid (oxalic acid) 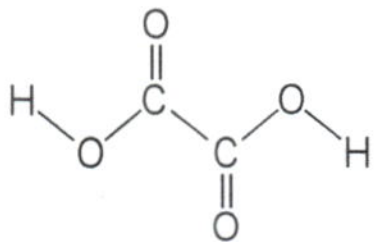 can be oxidised to carbon dioxide and water using a warm acidified solution of potassium manganate(VII) (potassium permanganate, $KMnO_4$).

Summary test 33.1

1 Carboxylic acids, being acidic, will react with the more reactive metals such as magnesium or zinc. Give three other reactions that are typical of acids.
2 Give balanced equations for the oxidation of **a** methanoic acid and **b** ethanedioic acid using [O] to represent the oxidising agent.
3 Suggest the order of acid strength of ethanoic acid, chloroethanoic acid, bromoethanoic acid and fluoroethanoic acid. Explain your answer.
4 Give the displayed formula for 3-chloropropanoic acid.
5 Explain why is it not necessary to call propanoic acid 1-propanoic acid.

33.2   Esters

Learning outcomes

On these pages you will learn to:

- explain how esters can be produced from the reaction of alcohols with acyl chlorides

Esters, general formula

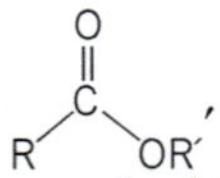

, are acid derivatives. They are named from the parent acid and the parent alcohol.

For example, $CH_3CH_2CH_2COOCH_3$ is derived from butanoic acid, $CH_3CH_2CH_2COOH$, and methanol, CH_3OH. It is called methyl butanoate.

Its isomer $HCOOCH_2CH_2CH_2CH_3$ is derived from methanoic acid, $HCOOH$, and butan-1-ol, $CH_3CH_2CH_2CH_2OH$. It is called butyl methanoate.

> **Exam tip**
>
> Take care with the names of esters. It is easy to get them the wrong way round. The part of the name relating to the acid comes last. Also, remember that the acid is named from the number of carbon atoms—including the carbon of the functional group.

Preparation of esters

Acyl chlorides (see Section 33.3 below) react with alcohols to produce esters. For example, the reaction with ethanoyl chloride and ethanol produces ethyl ethanoate.

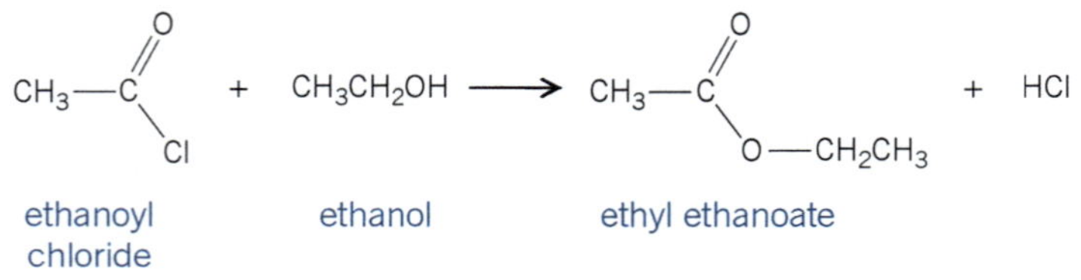

ethanoyl chloride   ethanol   ethyl ethanoate

The reaction is very fast and exothermic and steamy white clouds of hydrogen chloride gas are given off.

The reaction between ethanoyl chloride and phenol produces phenyl ethanoate.

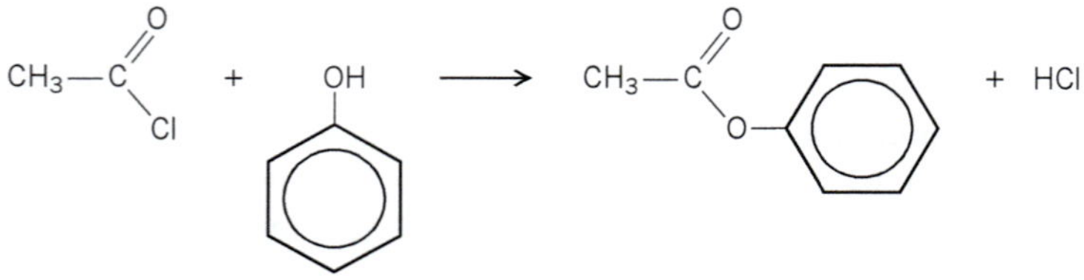

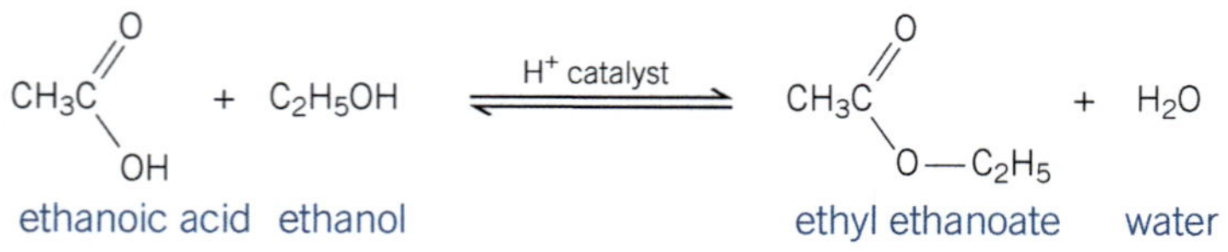

> **Exam tip**
>
> Hydrogen chloride gas (HCl) can be identified by the fact that it produces clouds of white smoke when mixed with ammonia gas.

This is a less vigorous reaction.

Carboxylic acids also react with alcohols to form esters. This reaction is speeded up by a strong acid catalyst. This is a reversible reaction and forms an equilibrium mixture of reactants and products. For example:

ethanoic acid   ethanol   ethyl ethanoate   water

Figure 3 *Esters are responsible for the flavours of many fruits*

Summary test 33.2

1 Give the name of the ester $CH_3CH_2COOCH_2CH_3$. Identify the acyl chloride and the alcohol you would react together to produce this compound.

2 Give the formula of ethyl ethanoate and an isomer of ethyl ethanoate that is also an ester. Give the formula of the isomer.

3 Identify the acyl chloride and the alcohol that would react together to give the ester methyl ethanoate.

4 Identify the acyl chloride and the alcohol that would react together to give the ester ethyl methanoate.

5 Methyl ethanoate and ethyl methanoate are a pair of isomers. Explain what this means.

6 Give the names of the following esters:

and

On these pages you will learn to:

- explain how acyl chlorides can be produced
- describe the addition-elimination reactions of acyl chlorides with water, alcohols, phenol, ammonia and amines
- describe the mechanisms for these addition–elimination reactions
- explain the relative ease of hydrolysis of acyl chlorides, halogenoalkanes and halogenoarenes

Exam tip

Acyl chlorides are also called acid chlorides.

Acyl chlorides (also called acid chlorides) are derivatives of carboxylic acids. The OH group has been replaced by a chlorine atom. For example, ethanoyl chloride:

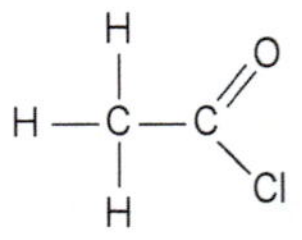

Acyl chlorides are prepared from carboxylic acids as described in Section 33.1. The reactions are:

- with phosphorus trichloride:

$$3RCOOH + PCl_3 \rightarrow 3RCOCl + H_3PO_3$$

- with phosphorus pentachloride:

$$RCOOH + PCl_5 \rightarrow RCOCl + POCl_3 + HCl$$

- with thionyl chloride:

$$RCOOH + SOCl_2 \rightarrow RCOCl + SO_2 + HCl$$

Reactions of acyl chlorides

The electronegative chlorine atom in acyl chlorides has the effect of drawing electrons away from the carbon atom of the carbonyl group. This increases its $\delta+$ character, as there are now two electronegative atoms bonded to this carbon.

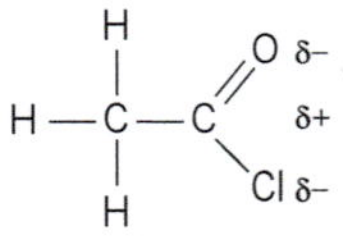

This makes this carbon atom more susceptible to nucleophilic attack than is ethanoic acid. So the important reactions of acyl chlorides are nucleophilic substitution reactions (also called addition–elimination reactions).

Acyl chlorides react with nucleophiles such as water, alcohols, phenols, ammonia and primary amines (see Section 19.1). These all have a lone pair of electrons that can attack and form a bond with the $C^{\delta+}$, followed by loss of a Cl^- ion as the leaving group. The overall reaction can be represented by:

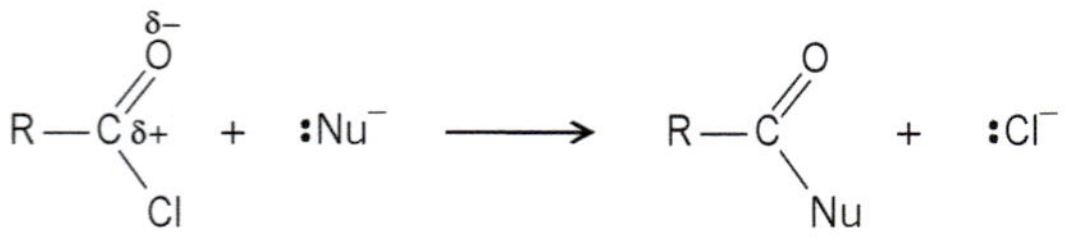

where $:Nu^-$ represents any negatively charged nucleophile.

The products are summarised in Table 2.

Table 2 *Compounds that react with acyl chlorides and the resulting products*

Reactant	Product
water	a carboxylic acid
alcohol	an ester
phenol	a phenyl ester
ammonia	an amide
amine	an N-substituted amide

Mechanisms of the reactions

The mechanisms of the reactions are shown below using curly arrows to show the movement of electron pairs. The mechanisms are essentially the same, so it is best to learn the general pattern rather than each one separately.

With a neutral nucleophile, a H^+ ion must be lost at some stage.

1 Ethanoyl chloride and water (called hydrolysis):

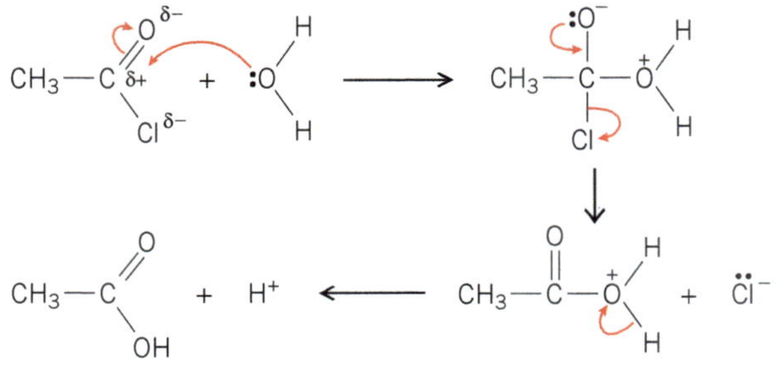

The overall equation may be written:

$$CH_3COCl + H_2O \rightarrow CH_3COOH + HCl$$

2 Ethanoyl chloride and ethanol:

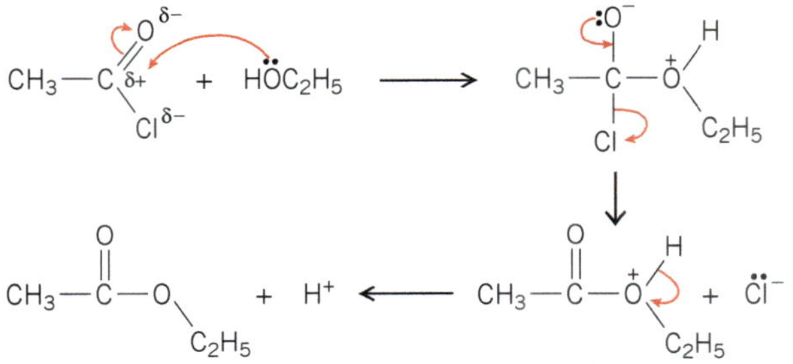

The overall equation may be written:

$$CH_3COCl + C_2H_5OH \rightarrow CH_3COOC_2H_5 + HCl$$

3 Ethanoyl chloride and ammonia (The H^+ ion that is lost then reacts with a second molecule of ammonia to form NH_4^+.)

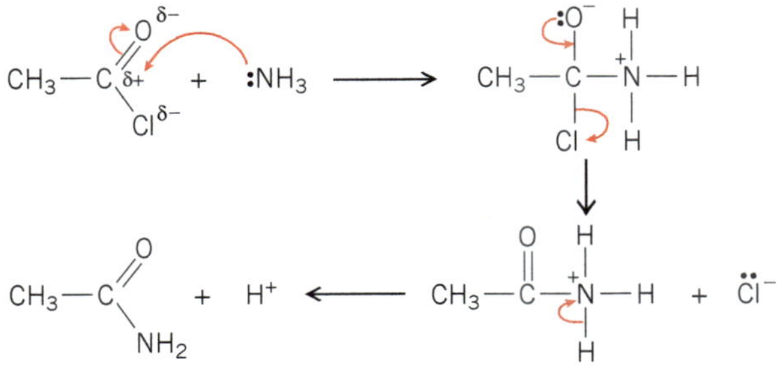

The overall equation may be written:

$$CH_3COCl + 2NH_3 \rightarrow CH_3CONH_2 + NH_4Cl$$

4 Ethanoyl chloride and methylamine:

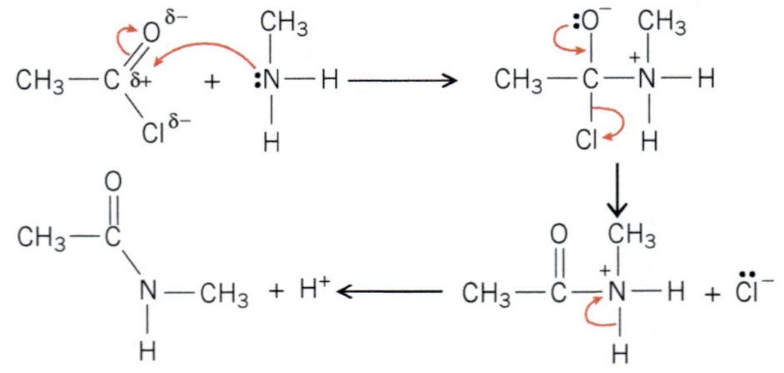

The overall equation may be written:

$$CH_3COCl + CH_3NH_2 \rightarrow CH_3CONHCH_3 + HCl$$

5 Ethanoyl chloride and dimethylamine:

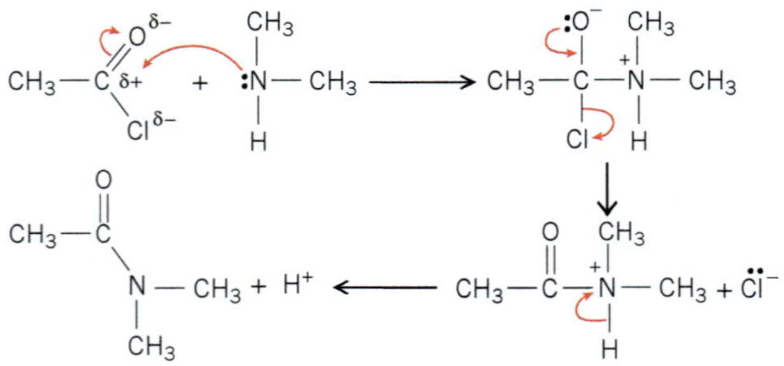

The overall equation may be written:

$$CH_3COCl + (CH_3)_2NH \rightarrow CH_3CON(CH_3)_2 + HCl$$

6 Ethanoyl chloride and phenol:

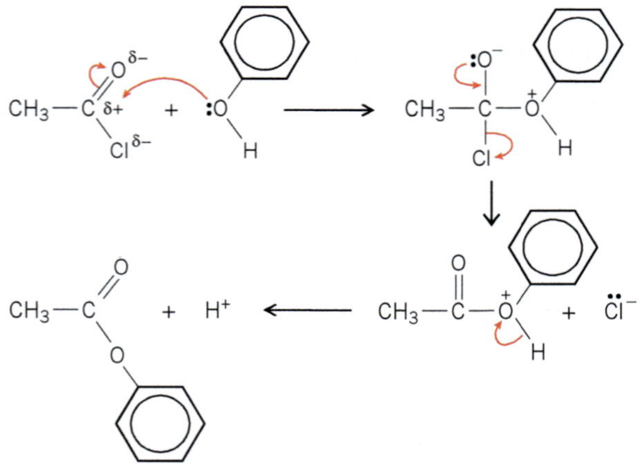

The overall equation may be written:

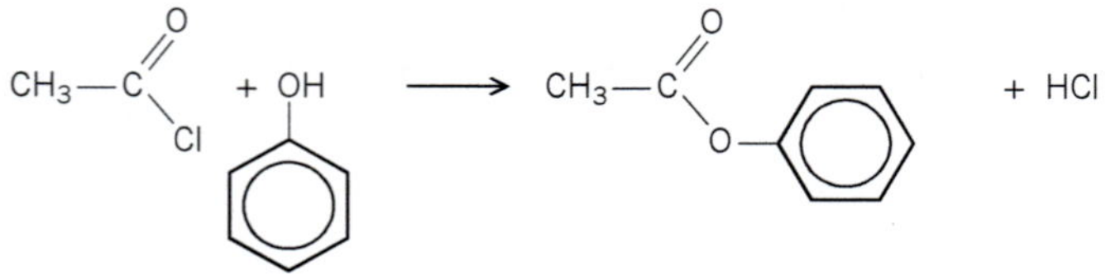

All the reactions 1–6 above take place at room temperature.

Hydrolysis

Hydrolysis is reaction with water, but it can also be used to describe the reaction with aqueous OH^- ions. There is a large difference in the reactivity of different classes of chlorine-substituted organic compounds, such as acyl chlorides, halogenoalkanes, and halogenoarenes.

Acyl chlorides

Ethanoyl chloride will react violently with water because of the strong δ^+ character of the carbonyl carbon which is bonded to two electronegative atoms, O and Cl. So even a neutral nucleophile such as water reacts rapidly.

$$CH_3COCl + H_2O \rightarrow CH_3COOH + HCl$$

Halogenoalkanes

Chloroethane will not react with water. The carbon to which the Cl atom is attached is less positive than that in ethanoyl chloride, so we need a better nucleophile. The OH^- ion will react because of its negative charge. Even so, warming is needed.

$$CH_3CH_2Cl + OH^- \rightarrow CH_3CH_2OH + Cl^-$$

Halogenoarenes

Chlorobenzene will not react with water. It will only react with sodium hydroxide at a high temperature and pressure. This is because the overlap of electrons from the lone pair of the chlorine atom and the delocalised system of the benzene ring strengthens the C–Cl bond and reduces the δ^+ character of the carbon atom.

$$C_6H_5Cl + OH^- \rightarrow C_6H_5OH + Cl^-$$

Summary test 33.3

1 Give the equation for the formation of propanamide from the reaction between ammonia and propanoyl chloride. Give the mechanism for this reaction.

2 The ionic compound sodium phenoxide has the ion:

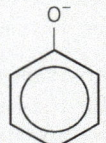

Give the mechanism for the reaction of this ion with ethanoyl chloride. Compare the rate of this reaction with the rate of the reaction of phenol with ethanoyl chloride.

3 State the reagents you would need to prepare the following from ethanoyl chloride:
   a methyl ethanoate, CH_3COOCH_3
   b N-propyl ethanamide $CH_3COONHCH_2CH_2CH_3$

4 Suggest how the reactivity of ethanoyl bromide compares with that of ethanoyl chloride. Explain your answer.

5 H_2O and OH^- are both nucleophiles. State the two features that a nucleophile must have.

(Launch additional digital resources for the chapter)

1 Ethanedioic acid, $(COOH)_2$ is oxidised by warming with potassium manganate(VII).

  a Write half equations for:

  i The oxidation of ethanedioic acid to form carbon dioxide and hydrogen ions.   *(1 mark)*

  ii The reduction of MnO_4^- ions in the presence of hydrogen ions to form Mn^{2+} ions.   *(1 mark)*

  b Use your answers to part **a** to give an overall equation for the oxidation of ethanedioic acid by potassium manganate(VII).   *(2 marks)*

2 Methanoic acid can be oxidised by Fehling's and Tollens' reagents.

  a Copy, complete and balance the equations below to show the reaction of methanoic acid with:

  i Fehling's solution:
$$HCOOH + 4OH^- + 2Cu^{2+} \rightarrow$$

  ii Tollens' solution:
$$HCOOH + 2O^{H-} + 2Ag(NH_3)^{2+} \rightarrow$$

    (4 marks)

  b Describe what you would observe for each reaction given in part **a**.   *(2 marks)*

3 Explain why:

  a Dichloroethanoic acid is a stronger acid than chloroethanoic acid.   *(3 marks)*

  b Benzoic acid is a stronger acid than ethanoic acid.   *(3 marks)*

  c Ethanedioic acid is a stronger acid for its first dissociation than for its second dissociation. *(3 marks)*

4 Phenylbenzoate is made from benzoyl chloride and phenol. Identify the reaction conditions and complete the equation below, giving the structure of phenylbenzoate and any other substances formed:

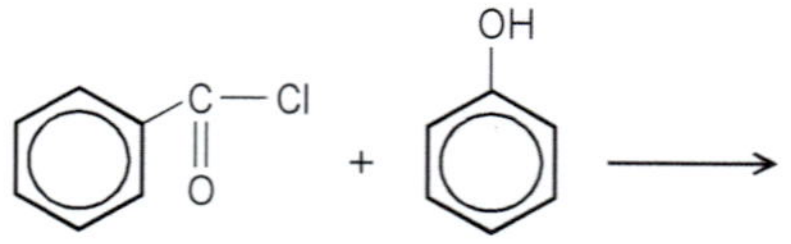

benzoyl chloride   phenol

    (3 marks)

5 Phenol, C_6H_5OH, was used as one of the earliest antiseptics and was once known as 'carbolic acid'.

  a Give the IUPAC name for phenol.   *(1 mark)*

  b Complete the equations below to show the products of the reaction between sodium hydroxide and phenol. Give the name of the organic product formed.

NaOH +

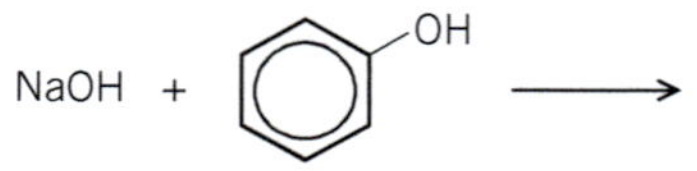

    (3 marks)

  c State the order, in increasing acidity of ethanol, water and phenol. Explain your answer.   *(4 marks)*

6 When aqueous bromine is added to a solution of phenol, the bromine is immediately decolourised and a white precipitate is formed. This reaction is shown below:

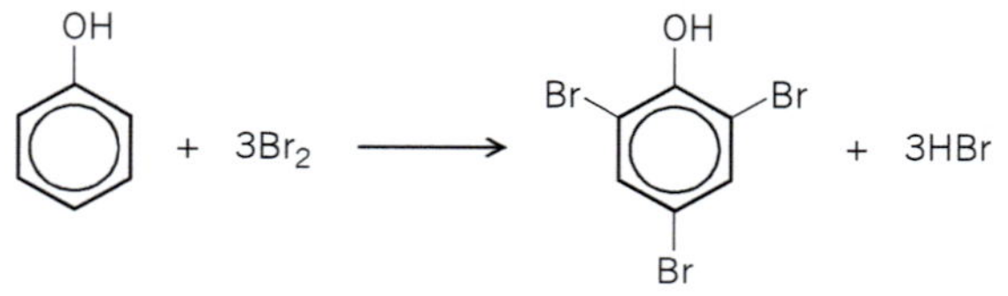

+ $3Br_2$ ⟶ + $3HBr$

  a Name the organic product formed in this reaction.   *(1 mark)*

  b State the type of reaction taking place.   *(1 mark)*

  c Explain why bromine adds to the benzene ring in these positions.   *(2 marks)*

7 Picric acid, 2,4,6-trinitrophenol, is a highly explosive substance produced by the reaction of phenol with a 'nitrating mixture'.

  a Draw the structure of picric acid.   *(1 mark)*

  b State the composition of the 'nitrating mixture'.   *(2 marks)*

  c Identify the products of the reaction of phenol with dilute nitric acid. Explain why these products are formed.   *(4 marks)*

8 Acyl chlorides are used in a variety of organic synthesis reactions:

   a Name the reagents that would be used to prepare:
     i propanoyl chloride
     ii butanoyl bromide.   *(2 marks)*

   b Name and give the structures of the organic products of the reactions between:
     i benzoyl chloride and ethanol
     ii propanoyl chloride and methylamine, CH_3NH_2
     iii ethanoyl chloride and phenol.   *(3 marks)*

9 Ethanoyl chloride reacts vigorously with alcohols at room temperature to produce esters.

   a Write an equation for the reaction of ethanoyl chloride with propanol. Name the organic product formed and explain why this reaction is not favoured for the industrial synthesis of esters.   *(4 marks)*

   b Give the names and structures of two reagents which can be used in the synthesis of phenyl benzoate, $C_6H_5COOC_6H_5$, when shaken with aqueous sodium hydroxide.   *(2 marks)*

10 Consider the following compounds:

   A $CH_3CHCOOH$
         |
         CH_3

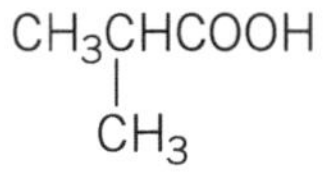

   B $CH_3(CH_2)_{16}COO^-Na^+$

   C $CH_3CH_2COOCH_3$

   D $HOOCCH_2CH_2COOH$

   a Identify which of the following compounds is:
     i an ester
     ii a dibasic acid
     iii a carboxylate salt.   *(3 marks)*

   b Give the IUPAC name for each compound. *(4 marks)*

   c Predict which compound would:
     i be almost insoluble in water, but would slowly dissolve when boiled with sodium hydroxide solution
     ii form a pleasant smelling liquid when warmed with ethanol and concentrated sulfuric acid.   *(2 marks)*

11 Predict the formulae of the products of the following reactions:

   a $CH_3CH_2CH_2OH(l) \xrightarrow{Cr_2O_7{}^{2-}(aq)/H^+(aq)}$

   b $CH_3CN(l) + HCl(aq) \xrightarrow{boil}$

   c $CH_3COO(CH_2)_4CH_3(l) + NaOH(aq) \xrightarrow{boil}$

   d $CH_3COOH(aq) + Ca(OH)_2 \longrightarrow$

   e $CH_3COCl(l) + CH_3CH_2NH_2(g) \longrightarrow$
                                         (5 marks)

Primary and secondary amines

Learning outcomes

On these pages you will learn to:

- describe how primary and secondary amines are produced
- explain the condensation reactions between ammonia/amines and acyl chlorides
- explain the behaviour of amines as bases

Exam tip

1°, 2°, and 3° are shorthand for primary, secondary, and tertiary, respectively.

Figure 1 *Phenylamine has almost the same density as water and is not soluble in it. Heat from a bulb at the base of the lava lamp changes the density enough for the phenylamine to float when hot and sink when cool*

This section is a continuation of Chapter 19, which introduced primary amines. **Amines** can be thought of as derivatives of ammonia in which one or more of the hydrogen atoms in the ammonia molecule have been replaced by alkyl or aryl groups to produce primary, secondary, and tertiary amines.

$$H-\overset{\cdot\cdot}{\underset{H}{N}}-H \qquad H-\overset{\cdot\cdot}{\underset{R}{N}}-H \qquad H-\overset{\cdot\cdot}{\underset{R}{N}}-R' \qquad R''-\overset{\cdot\cdot}{\underset{R}{N}}-R'$$

   ammonia   primary amine   secondary amine   tertiary amine

The terms **primary**, **secondary**, and **tertiary** are used for amines slightly differently from the way they are used with alcohols and halogenoalkanes (see Sections 15.1 and 16.1). In amines, 1°, 2°, and 3° refer to the number of substituents (R-groups) on the nitrogen atom. (In alcohols, 1°, 2°, and 3° refer to the number of substituents on the carbon atom bonded to the −OH group.)

How to name amines

Primary amines have the general formula RNH_2, where the R can be an alkyl or aryl group. Amines are named using the suffix *-amine*, for example:

CH_3-NH_2 is methylamine.

$C_2H_5-NH_2$ is ethylamine.

$C_6H_5-NH_2$, is phenylamine.

Secondary amines have the general formula RR′NH, for example:

$(CH_3)_2NH$, is dimethylamine.

Tertiary amines have the general formula RR′R″N, for example:

$(C_2H_5)_3N$, is triethylamine.

Different substituents are written in alphabetical order:

$CH_3(C_3H_7)NH$, is N-methylpropylamine.

The reactivity of amines

Amines have a lone pair of electrons and this is important in the way they react. The lone pair may be used to form a bond with:

- a H^+ ion, in which case we say the amine is acting as a base
- an electron-deficient carbon atom, in which case we say the amine is acting as a nucleophile.

The preparation of amines

Amines can be prepared by the reaction of halogenoalkanes with ammonia. Primary amines can be prepared by reduction of amides or nitriles.

Reactions of ammonia with halogenoalkanes

Primary aliphatic amines are produced when halogenoalkanes are heated with ammonia in ethanol, in a sealed vessel under pressure. There is nucleophilic substitution of the halogen by NH_2.

The reaction takes place in two stages: first a salt is formed, then the salt reacts with excess ammonia.

$$\overset{..}{N}H_3 + RX \rightarrow [RNH_3]^+ X^-$$
$$[RNH_3]^+ X^- + NH_3 \rightarrow R\overset{..}{N}H_2 + [NH_4]^+ X^-$$
$$\text{primary}$$
$$\text{amine}$$

However, the primary amine produced is also a nucleophile and this will react with the halogenoalkane to produce a secondary amine:

$$R\overset{..}{N}H_2 + RX \rightarrow [R_2NH_2]^+ X^-$$
$$[R_2NH_2]^+ X^- + NH_3 \rightarrow R_2\overset{..}{N}H + [NH_4]^+ X^-$$
$$\text{secondary}$$
$$\text{amine}$$

The secondary amine will react to give a tertiary amine:

$$R_2\overset{..}{N}H + RX \rightarrow [R_3NH]^+ X^-$$
$$[R_3NH]^+ X^- + NH_3 \rightarrow R_3\overset{..}{N} + [NH_4]^+ X^-$$
$$\text{tertiary}$$
$$\text{amine}$$

This in turn will react to a produce a quaternary ammonium salt:

$$R_3\overset{..}{N} + RX \rightarrow [R_4N]^+ X^-$$

So a mixture of primary, secondary, and tertiary amines and a quaternary ammonium salt is produced. This means that this is not a very efficient way of preparing an amine, although the products may be separated by fractional distillation. A large excess of ammonia gives a better yield of primary amine, and a large excess of halogenoalkane will favour a quaternary ammonium salt.

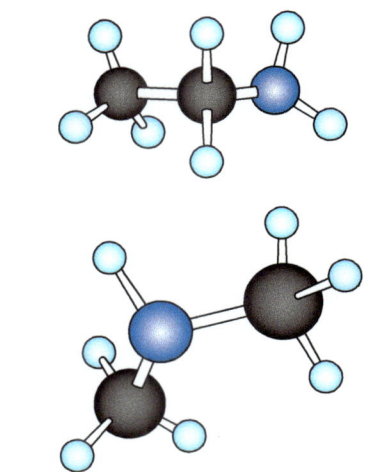

Figure 2 *3D representations of ethylamine (top), a primary amine, and dimethylamine (bottom), a secondary amine*

Exam tip

Halogenoalkanes are also called **haloalkanes**.

The mechanism of the reaction

For all the above reactions the mechanism is essentially the same:

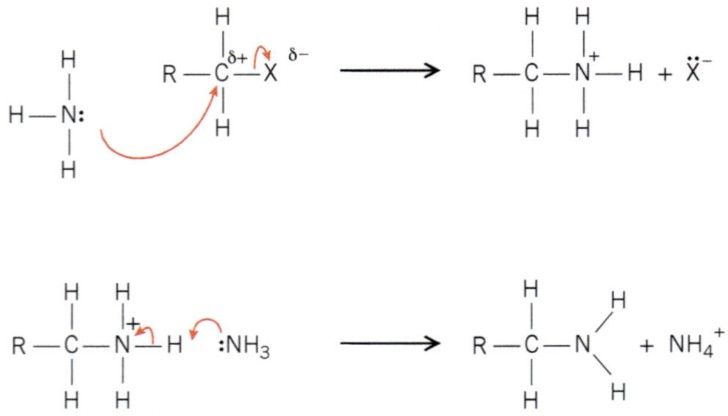

Initially ammonia acts as a nucleophile. In the second stage, it acts as a base.

Reduction of nitriles

Primary alkyl amines can be prepared more efficiently from halogenoalkanes in a two-step process:

Step 1   Halogenoalkanes react with the cyanide ion in aqueous ethanol. The cyanide ion replaces the halide ion by nucleophilic substitution to form a **nitrile**:

$$RBr + CN^- \rightarrow R{-}C{\equiv}N + Br^-$$

Step 2   Nitriles contain the functional group $-C{\equiv}N$. They can be reduced to primary amines, for example, with hydrogen using a nickel catalyst:

$$R{-}C{\equiv}N + 2H_2 \rightarrow R{-}CH_2NH_2$$

An alternative reducing agent is lithium tetrahydridoaluminate(III) (lithium aluminium hydride, $LiAlH_4$) in ether solution.

Reduction of nitriles gives a purer product than the reaction of a halogenoalkane and ammonia, because only the primary amine can be formed. The carbon chain of the product is one carbon atom longer than in the starting material.

Reduction of amides

Amides (see Section 34.3 below) can be reduced with lithium tetrahydridoaluminate(III) (lithium aluminium hydride, $LiAlH_4$). For example:

$$CH_3CONH_2 + 4[H] \rightarrow CH_3CH_2NH_2 + H_2O$$

Preparation of amides—condensation reactions

Ammonia or amines will react readily with acyl chlorides at room temperature to form **amides** or N-substituted amides, for example:

- ethanoyl chloride and ammonia form an amide:

$$CH_3COCl + 2NH_3 \rightarrow CH_3CONH_2 + NH_4Cl$$

- ethanoyl chloride and methylamine form a methyl substituted amide:

$$CH_3COCl + 2CH_3NH_2 \rightarrow CH_3CONHCH_3 + CH_3NH_3Cl$$

- ethanoyl chloride and dimethylamine form a dimethyl substituted amide:

$$CH_3COCl + (CH_3)_2NH \rightarrow CH_3CON(CH_3)_2 + (CH_3)_2NH_2Cl$$

Exam tip

The reaction of a halogenoalkane with cyanide ions is a way to increase the carbon chain length by one.

The mechanism of the first reaction takes place in three steps. It is an addition–elimination (or condensation) reaction:

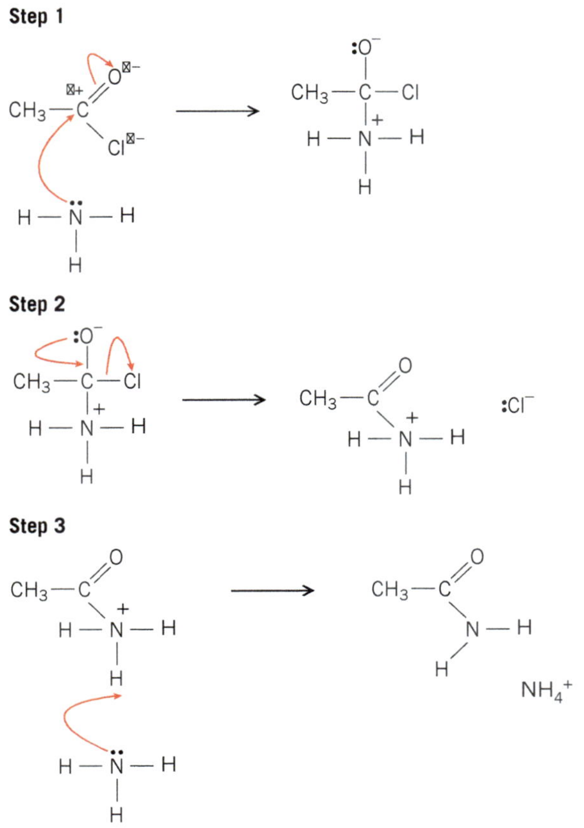

Step 1

Step 2

Step 3

You should be able to work out the mechanism of the other reactions using this one as a guide.

Amines as bases

Amines can accept a proton (an H^+ ion) so they are Brønsted–Lowry bases.

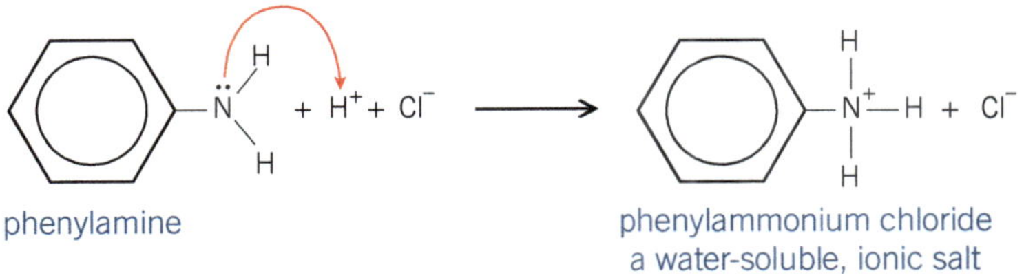

phenylamine

phenylammonium chloride
a water-soluble, ionic salt

Amines react with acids to form salts. For example, ethylamine, a soluble **alkylamine**, reacts with dilute hydrochloric acid:

$$C_2H_5NH_2 + H^+ + Cl^- \rightarrow C_2H_5NH_3^+ + Cl^-$$

ethylamine   ethylammonium chloride

The products are ionic compounds that will crystallise as the water evaporates. The salts of amines are sometimes named as hydrochlorides of the parent amine—for example, ethylamine hydrochloride.

Phenylamine, an **arylamine**, is relatively insoluble, but it will dissolve in excess hydrochloric acid because it forms the soluble ionic salt, sometimes called a hydrochloride, ie phenylamine hydrochloride.

$$\text{C}_6\text{H}_5\text{—NH}_2 + \text{H}^+ + \text{Cl}^- \longrightarrow \text{C}_6\text{H}_5\text{—NH}_3^+ + \text{Cl}^-$$

phenylamine

phenylammonium chloride
a water-soluble, ionic salt

If a strong base like sodium hydroxide is added, it removes the proton from the salt and regenerates the insoluble amine.

$$\text{C}_6\text{H}_5\text{—NH}_3^+ + \text{Cl}^- + \text{OH}^- \longrightarrow \text{C}_6\text{H}_5\text{—NH}_2 + \text{H}_2\text{O} + \text{Cl}^-$$

phenylamine

Summary test 34.1

1 Identify $\text{C}_2\text{H}_5\text{—}\overset{\displaystyle\text{H}}{\underset{\displaystyle}{\text{N}}}\text{—C}_3\text{H}_7$ as primary, secondary, or tertiary.

2 Give the name of the compound in question **1**.

3 Give the structural formula of trimethylamine.

4 Explain why nucleophilic substitution of a halogenoalkane is not a good method for preparing a primary amine.

5 **a** Give the equation for the reaction of chloroethane with an excess of ammonia. Give the reaction mechanism.

   b Give the other possible products of this reaction.

6 **a** Give the equation for dimethylamine reacting with hydrochloric acid.

   b Identify the product.

7 Phenylamine is not very soluble in water. It forms oily drops that float in the water. Predict what you would see if you:

   a add concentrated hydrochloric acid to a mixture of phenylamine and water.

   b then add sodium hydroxide solution to the resulting solution.

8 Suggest whether dimethylamine will be a weaker or stronger base than ethylamine. Explain your answer.

Extension

Solubility of drugs

A number of medicinal drugs are amines, for example, the nasal decongestant pseudoephedrine. Longer chain amines are relatively insoluble in water so when they are used in medicines they are often supplied as hydrochlorides to make them more soluble in the bloodstream.

Pseudoephedrine has the formula:

1 Is pseudoephedrine a primary, secondary, or tertiary amine?

2 Write the structural formula of pseudoephedrine hydrochloride.

3 Explain why pseudoephedrine hydrochloride is more soluble in water than pseudoephedrine.

4 Is pseudoephedrine hydrochloride likely to be a solid, liquid, or gas? Explain your answer.

5 As well as having an amine group, pseudoephedrine has two other functional groups. Name them.

6 Pseudoephedrine has two chiral centres in its molecule. Mark them with a * on your formula of pseudoephedrine hydrochloride. Hint: it may help if you draw in the hydrogen atoms that are not marked on the skeletal formula.

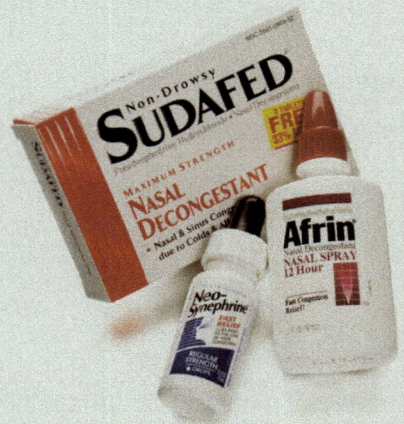

Figure 3 *Nasal decongestant sprays reduce swelling in the blood vessels inside your nose, helping to relieve breathing issues caused by colds or hayfever*

7 What problems might this chirality cause in pseudoephedrine's use as a drug?

Phenylamine is the simplest aryl amine. It is the starting point for making many other chemicals and is made in industry using benzene produced from crude oil.

Preparation of phenylamine

Phenylamine can be made from benzene in two steps.

Step 1   Benzene is reacted with a mixture of concentrated nitric and concentrated sulfuric acid. This produces nitrobenzene:

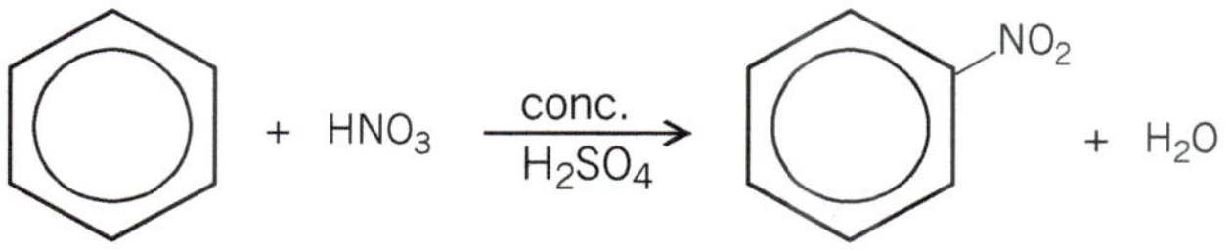

benzene   nitrobenzene

Step 2   Nitrobenzene is reduced to phenylamine, using tin and hot, concentrated hydrochloric acid as the reducing agent.

The tin and hydrochloric acid react to form hydrogen, which reduces the nitrobenzene by removing oxygen atoms of the NO_2 group and replacing them with hydrogen atoms.

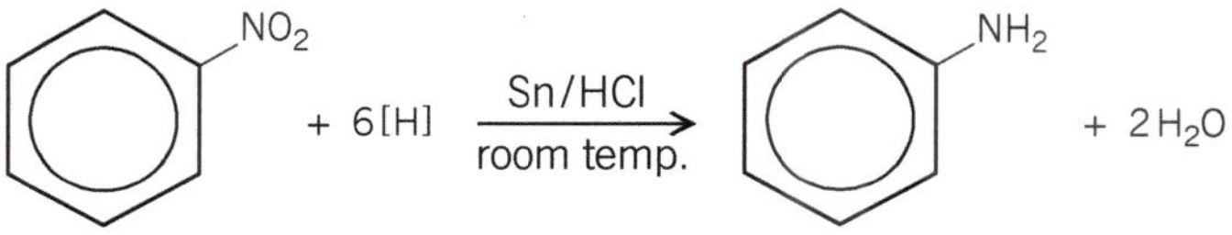

This could also be written:

$$C_6H_5NO_2 + 6[H] \rightarrow C_6H_5NH_2 + 2H_2O$$

Since the reaction is carried out in hydrochloric acid, the salt $C_6H_5NH_3^+Cl^-$ is formed and sodium hydroxide is then added to liberate the free amine:

$$C_6H_5NH_3^+Cl^- + NaOH \rightarrow C_6H_5NH_2 + H_2O + NaCl$$

Learning outcomes

On these pages you will learn to:

- describe how phenylamine is produced
- explain the reaction of phenylamine with bromine
- explain the reaction of phenylamine to produce a diazonium salt
- explain the behaviour of ammonia, ethylamine and phenylamine as bases
- describe the production and use of azo compounds

Exam tip

The formation of nitrobenzene is an electrophilic substitution reaction. The electrophile is NO_2^+ (see section 30.1).

Extension

Sulfa drugs

The story of the antibiotic penicillin is well known. It was the result of a chance observation of mould on a discarded Petri dish by Alexander Fleming. It was developed by Howard Flory and Ernst Chain (and a massive industrial effort) into a drug that saved thousands of lives in World War II and since. However, it was not the first anti-bacterial drug. Another class of drugs, the sulfanilamides, were already in use before penicillin. They may also have had an effect on the course of the war—by saving the life of British Prime Minister Winston Churchill.

Towards the end of the nineteenth century, it was noticed that some dyes used to stain bacteria to make them visible under the microscope could also kill them. These dyes were absorbed by the bacteria and not their surroundings, so they might be expected to kill the bacteria but not their host. Eventually the dye Prontosil Rubrum began to be used in medicine to fight bacterial infections.

By the early 1940s it was found that Prontosil was converted in the body into the compound sulfanilamide, the active ingredient:

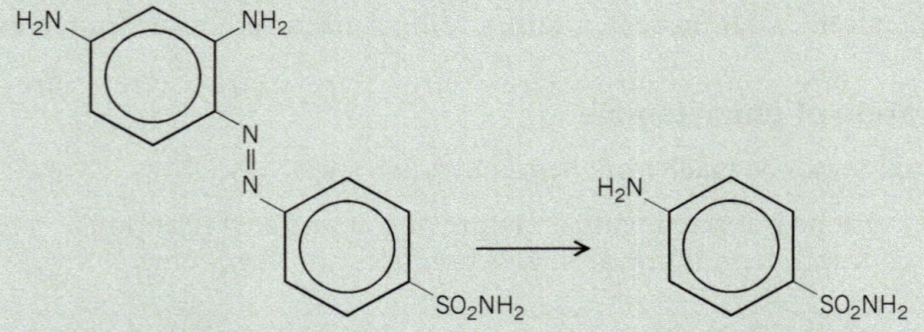

Prontosil Rubrum

sulfanilamide

The drug worked by preventing the bacteria from making folic acid, which they need to synthesise DNA and replicate. Bacteria make folic acid from a compound called *para*-aminobenzoic acid (PABA). The sulfanilamide molecule is a similar shape to PABA. The bacteria try to use sulfanilamide to make folic acid, but without success as it is the wrong molecule. Humans do not need to synthesise folic acid as they get it from their food. So sulfanilamide kills bacteria but is harmless to humans.

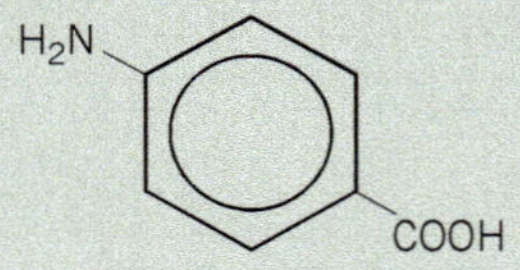

PABA

Since the 1940s, over 5000 variations on the sulfanilamide molecule have been synthesised by chemists. Their aim has been to find variations that are more effective, have fewer side effects, are absorbed at a different rate, and so on. This is one of the main methods for discovering new drugs. Chemists start with a molecule with a known benefit and make changes to enhance its activity or reduce any disadvantages. Nowadays, this process can be speeded up by the technique of combinatorial chemistry.

Although less common than they once were, sulfa drugs are still used today. The one that cured Winston Churchill's pneumonia in 1943 was sulfapyridine.

1 What is the systematic name of PABA?

Answer: 4-aminobenzenecarboxylic acid (or 4-aminobenzoic acid)

Exam tip

The **Brønsted–Lowry theory** describes bases as proton (H^+ ion) acceptors.

Comparing base strengths

The strength of a base depends on how readily it will accept a proton, H^+. Both ammonia and amines have a lone pair of electrons that attract a proton.

Alkyl groups *release* electrons away from the alkyl group and towards the nitrogen atom. This is called the **inductive effect** and is shown by an arrow in the direction of electron release (Figure 4).

$$R \rightarrow \ddot{N}H_2$$

Figure 4 *A primary amine. The arrow shows that R releases electrons. This is called the inductive effect*

The inductive effect of the alkyl group increases the electron density on the nitrogen atom and therefore makes it a better electron pair donor (ie more attractive to protons). So primary alkylamines are stronger bases than ammonia.

Secondary alkylamines have two inductive effects and are therefore stronger bases than primary alkylamines. However, tertiary alkylamines are not stronger bases than secondary ones because they are less soluble in water.

Aryl groups *withdraw* electrons from the nitrogen atom. This is because the lone pair of electrons overlaps with the delocalised system on the benzene ring, as shown for phenylamine.

[benzene ring structure with N bonded, showing lone pair, two H atoms]

The nitrogen is a weaker electron pair donor and therefore less attractive to protons, so arylamines are weaker bases than ammonia.

ethylamine > ammonia > phenylamine
strongest ⟶ weakest

Reactions of phenylamine

The $-NH_2$ group activates the benzene ring in the same way as the $-OH$ group of phenol does (see Section 32.2). The lone pair of electrons on the nitrogen atom interacts with the delocalised system on the benzene ring (see Figure 5).

This activates the ring as well as making the lone pair less available to accept an H^+ ion, so phenylamine is a very weak base—see Section 34.1.

Reaction with bromine
The activation of the benzene ring means that phenylamine will react with an aqueous solution of bromine (bromine water) at room temperature to substitute three bromine atoms. These go to the 2-, 4-, and 6-positions.

[reaction scheme: phenylamine + $3Br_2$ → 2,4,6-tribromophenylamine + $3HBr$]

2,4,6-tribromophenylamine

This is similar to the reaction of bromine with phenol. Remember that benzene itself will only react with *liquid* bromine in the presence of a catalyst (iron or aluminium bromide) to give a single substitution.

Formation of a diazonium salt
Phenylamine reacts with nitrous acid. Nitrous acid, HNO_2, is unstable and is formed as it is required by the reaction of sodium nitrite and dilute hydrochloric acid:

$$NaNO_2(aq) + HCl(aq) \rightarrow NaCl(aq) + HNO_2(aq)$$

The temperature is kept below 10°C (obtained by cooling the reaction vessel in ice) and, on adding phenylamine, a diazonium salt is formed:

[reaction scheme: phenylamine + HNO_2 + H^+ + Cl^- → diazonium salt (N≡N$^+$ Cl^-) + $2H_2O$]

On warming to room temperature with water, the diazonium salt decomposes to phenol and nitrogen.

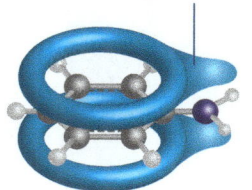

lone pair now delocalised with the ring electrons

Figure 5 *Orbitals overlap in phenylamine*

Exam tip

Sodium nitrite is also called sodium nitrate(III).

Figure 6 *Aromatic amines are used in the manufacture of dyestuffs*

Summary test 34.2

1 State what you would expect to *see* if phenylamine is reacted with a mixture of sodium nitrite and dilute hydrochloric acid at above 10°C.

2 Give the oxidation state of the nitrogen atom in nitrous acid.

3 Suggest the product of the reaction of phenylamine with chlorine. Identify and draw the structural formula of this product.

4 Draw the formulae of the *cis*- and *trans*-isomers of 4-(phenylazo)phenol.

Coupling reactions

The benzene diazonium salt can undergo a number of reactions. The most useful is a coupling reaction to produce **azo compounds**, which have the functional group R–N=N–R. Coupling reactions are very important for the dyestuffs industry as they can produce a range of compounds with intense colours. They are produced from the reaction of aromatic alcohols (or amines) with benzenediazonium ions.

Coupling with phenol

The benzenediazonium chloride is reacted with phenol in alkaline solution (this is the coupling reaction). The alkaline solution removes a H^+ ion from the OH group of the phenol to form a phenoxide ion, and this couples with the benzenediazonium ion.

The bright yellow product is called 4-(phenylazo)phenol. (You won't be expected to remember this name!) This is the simplest of a family of compounds consisting of two aromatic rings joined by the –N=N– group called diazo compounds. Because of the N=N double bond it exists as a pair of *cis*–*trans* isomers. Using different substituents on the aromatic rings produces a range of compounds with different colours.

Phenol can also be replaced in this reaction by naphthol. This is an aromatic alcohol in which two aromatic rings are joined (see Section 32.2).

1-naphthol

The product is

Coupling with phenylamine

A similar coupling reaction occurs with phenylamine

Coupling with substituted phenols or phenylamines can produce a range of intense colours used as dyestuffs. The product formed is yellow and is called 4-(phenylazo)aniline. Aniline is the non-systematic name of phenylamine.

Amides

Amides have the general formula $RCONH_2$.

R—C(=O)—N—H with H below (structure)

One or more of the hydrogen atoms of the NH_2 group may be replaced by alkyl or aryl groups to produce N-substituted amides.

The preparation of amides

Amines will react with acyl chlorides. These are nucleophilic substitution reactions (sometimes called addition–elimination reactions) and the products are N-substituted amides. The reactions take place readily at room temperature. (Ammonia will react with acyl chlorides under the same conditions, to produce unsubstituted amides.)

The mechanism is:

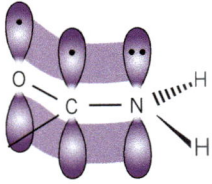

The amine adds on to the acid chloride and then HCl is eliminated.

This reaction is useful in forming polymers such as nylon, see Section 35.1.

Acid/base properties of amides

Ammonia and amines are basic because they have a lone pair of electrons on the nitrogen atom that can form a dative covalent bond with an H^+ ion (which has no electrons at all).

(structure showing $H-C-N$ with lone pair + $H^+ \longrightarrow H-C-N^+-H$)

The nitrogen atom of an amide also has a lone pair of electrons in a p-orbital. However, this can overlap with the π-orbital of the C=O to form a delocalised system. So the lone pair is no longer available to accept a H^+ ion and so amides are much weaker bases than amines.

Learning outcomes

On these pages you will learn to:

- describe how amides are produced
- describe the reactions of amides
- explain why amides are weaker bases than amines

Exam tip

Don't confuse *amides*, $RCONH_2$, with *amines*, RNH_2.

Exam tip

The reaction of an acyl chloride with ammonia produces an unsubstituted amide $RCONH_2$ (see Section 33.3).

Exam tip

Acyl chlorides and acid chlorides are alternative names for the same functional group: $COCl$.

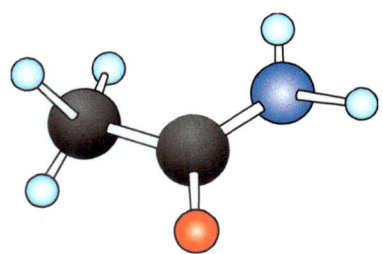

Figure 7 A 3D representation of ethanamide

Hydrolysis of amides

Amides can be hydrolysed in acidic or alkaline conditions to form a carboxylic acid and ammonia or an amine. In acid conditions:

$$CH_3CONH_2 + H_2O + HCl \rightarrow CH_3COOH + NH_4^+Cl^-$$

In alkaline conditions a salt is formed rather than the carboxylic acid itself:

$$CH_3CONH_2 + NaOH \rightarrow CH_3COONa + NH_3$$

Reduction of amides

Amides can be reduced by lithium tetrahydridoaluminate(III) (lithium aluminium hydride, $LiAlH_4$) in ether solution followed by addition of a dilute acid. An amine is formed. Using [H] to represent the reducing agent, a typical equation is:

$$CH_3CONH_2 + 4[H] \rightarrow CH_3CH_2NH_2 + H_2O$$

The oxygen of the amide functional group is lost in the molecule of water.

Figure 8 *Many important biological compounds contain amide groups, such as penicillin. The chemical structure of penicillin was determined by Dorothy Hodgkin, a British chemist, using a technique called X-ray crystallography. Her structure is shown above, with the two amide linkages circled in red*

Summary test 34.3

1 Give the names of the amide and the amine in the equation:
$$CH_3CONH_2 + 4[H] \rightarrow CH_3CH_2NH_2 + H_2O$$

2 State what type of amine you would get by reducing a singly N-substituted amide.

3 Explain why an amide is a very weak base.

Amino acids

Amino acids are the building blocks of proteins, which in turn are a vital component of all living systems.

Amino acids have two functional groups—a **carboxylic acid** and a primary amine. There are 20 important naturally occurring amino acids and they are all α-amino acids (also called 2-amino acids), which means that the amine group is on the carbon next to the $-CO_2H$ group (Figure 9).

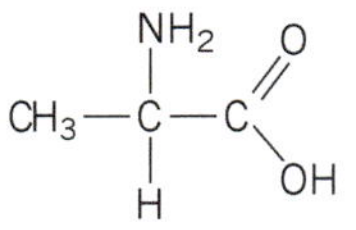

$$CH_3-\overset{\overset{\displaystyle NH_2}{|}}{\underset{\underset{\displaystyle H}{|}}{C}}-\overset{\overset{\displaystyle O}{\|}}{C}-OH$$

Figure 9 *2-Aminopropanoic acid, also called alanine, written in shorthand as $CH_3CH(NH_2)COOH$*

α-amino acids have the general formula:

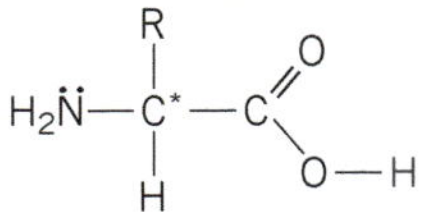

$$H_2\ddot{N}-\overset{\overset{\displaystyle R}{|}}{\underset{\underset{\displaystyle H}{|}}{C^*}}-\overset{\overset{\displaystyle O}{\|}}{C}-O-H$$

This structure has a carbon bonded to four different groups. The molecule is therefore **chiral**. Almost all naturally occurring amino acids exist as the (−) enantiomer.

Acid and base properties

Amino acids have both an acidic and a basic functional group.

* The carboxylic acid group has a tendency to lose a proton (act as an acid):

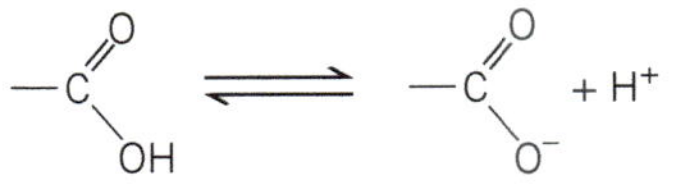

$$-\overset{\overset{\displaystyle O}{\|}}{C}-OH \rightleftharpoons -\overset{\overset{\displaystyle O}{\|}}{C}-O^- + H^+$$

* The amine group has a tendency to accept a proton (act as a base):

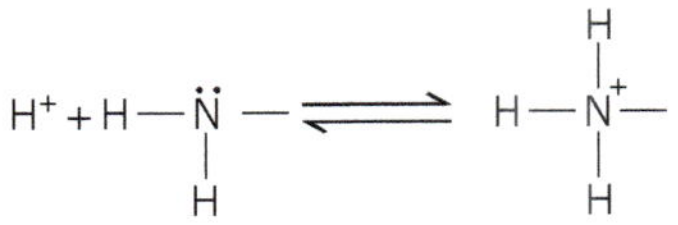

$$H^+ + H-\overset{\overset{\displaystyle H}{\ldots}}{\underset{\underset{\displaystyle H}{|}}{N}} \rightleftharpoons H-\overset{\overset{\displaystyle H}{|}}{\underset{\underset{\displaystyle H}{|}}{N^+}}-$$

Amino acids exist as **zwitterions**. Zwitterions have both a permanent positive charge and a permanent negative charge, although the compound is neutral overall (Figure 10).

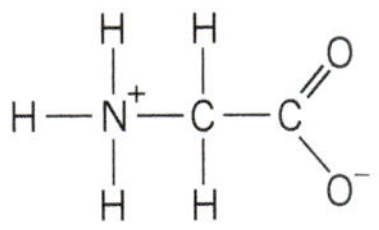

$$H-\overset{\overset{\displaystyle H}{|}}{\underset{\underset{\displaystyle H}{|}}{N^+}}-\overset{\overset{\displaystyle H}{|}}{\underset{\underset{\displaystyle H}{|}}{C}}-\overset{\overset{\displaystyle O}{\|}}{C}-O^-$$

Figure 10 *A zwitterion*

Because they are ionic, amino acids have high melting points and dissolve well in water but poorly in non-polar solvents. A typical amino acid is a white solid at room temperature and behaves very much like an ionic salt.

In strongly acidic conditions the lone pair of the H_2N-group accepts a proton (H^+ ion) to form the positive ion (Figure 12).

The amino group has gained a hydrogen ion—it is **protonated**.

In strongly alkaline solutions, the $-OH$ group loses a proton to form the negative ion (Figure 13).

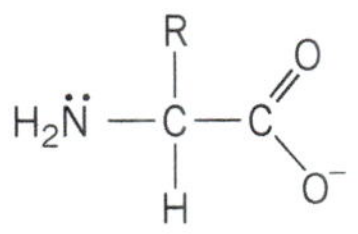

The carboxylic acid group has lost a hydrogen ion—it is **deprotonated**.

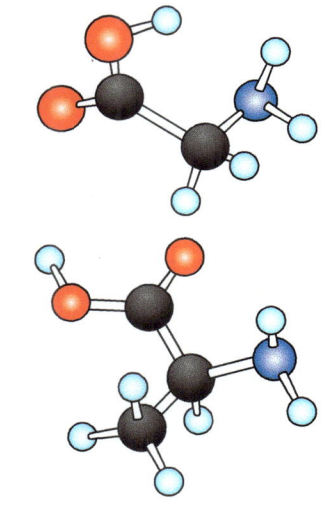

Figure 11 *3D representations of two amino acids, glycine (top) and alanine (bottom)*

$$H_2N^+-\overset{\overset{\displaystyle R}{|}}{\underset{\underset{\displaystyle H}{|}}{C}}-\overset{\overset{\displaystyle O}{\|}}{C}-O-H$$
(with $\overset{H}{|}$ on the N)

Figure 12 *A protonated amino acid*

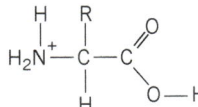

Figure 13 *A deprotonated amino acid*

Each amino acid has a pH at which it is neutral—this is called its **isoelectric point**.

Table 1 shows some naturally occurring amino acids. Each of these is usually referred to by its non-systematic name (the IUPAC names can be complex) and also by a three-letter abbreviation. This is useful when describing the sequences of amino acids in proteins.

Table 1 Names and structures of the 20 naturally occurring amino acids

Formula	Name and abbreviation	Formula	Name and abbreviation
H_2NCHCO_2H $\mid$ H	glycine (Gly)	H_2NCHCO_2H $\mid$ CHOH $\mid$ CH_3	threonine (Thr)
H_2NCHCO_2H $\mid$ CH_3	alanine (Ala)	H_2NCHCO_2H $\mid$ CH_2SH	cysteine (Cys)
H_2NCHCO_2H $\mid$ $CHCH_3$ $\mid$ CH_3	valine (Val)	H_2NCHCO_2H $\mid$ CH_2 $\mid$ $CONH_2$	asparagine (Asn)
H_2NCHCO_2H $\mid$ CH_2 $\mid$ $CH_3(CH_3)_2$	leucine (Leu)	H_2NCHCO_2H $\mid$ CH_2 $\mid$ CH_2CONH_2	glutamine (Gln)
H_2NCHCO_2H $\mid$ CHC_2H_5 $\mid$ CH_3	isoleucine (Ile)	H_2NCHCO_2H $\mid$ CH_2—⬡—OH	tyrosine (Tyr)
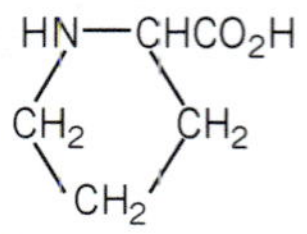 HN—$CHCO_2H$ CH_2 CH_2 CH_2	proline (Pro) (proline is a secondary amine)	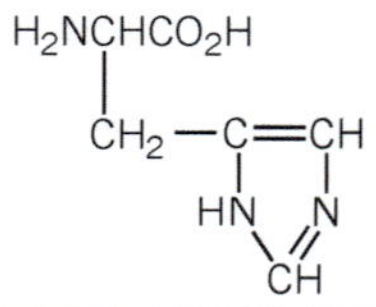 H_2NCHCO_2H $\mid$ CH_2—C=CH HN N CH	histidine (His)
 H_2NCHCO_2H $\mid$ CH_2 C CH NH	tryptophan (Try)	H_2NCHCO_2H $\mid$ $(CH_2)_3$ $\mid$ NH $\mid$ $NH=C—NH_2$	arginine (Arg)
H_2NCHCO_2H $\mid$ CH_2 $\mid$ CH_2SCH_3	methionine (Met)	H_2NCHCO_2H $\mid$ $(CH_2)_3$ $\mid$ CH_2NH_2	lysine (Lys)
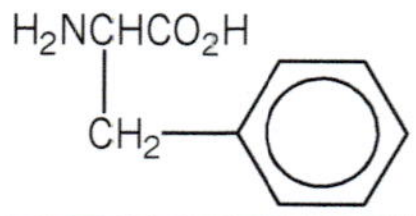 H_2NCHCO_2H $\mid$ CH_2—⬡	phenylalanine (Phe)	H_2NCHCO_2H $\mid$ CH_2CO_2H	aspartic acid (Asp)
H_2NCHCO_2H $\mid$ CH_2OH	serine (Ser)	H_2NCHCO_2H $\mid$ CH_2 $\mid$ CH_2CO_2H	glutamic acid (Glu)

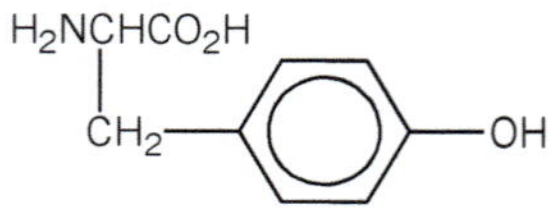

Peptides, polypeptides, and proteins

Amino acids link together to form **peptides**. Molecules containing up to about 50 amino acids are referred to as polypeptides. When there are more than 50 amino acids they are called proteins. Naturally occurring proteins are everywhere—enzymes, wool, hair, and muscles are all examples.

Amino acids and the peptide link

An amide has the functional group $-CONH_2$ or $-CONH_2$ or

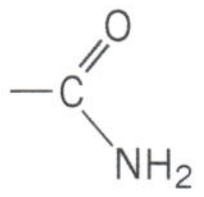

The amine group of one amino acid can react with the carboxylic acid group of another to form an **amide linkage** $-CONH-$.

This linkage is shown by shading in Figure 14.

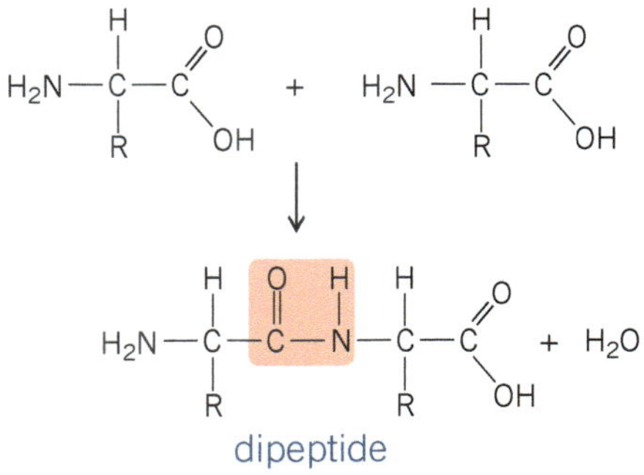

Compounds formed by the linkage of amino acids are called peptides, and the amide linkage is called a peptide linkage in this context. A peptide with two amino acids is called a **dipeptide**. The dipeptide still retains free $-NH_2$ and $-CO_2H$ groups and so can react further to give tri- and tetra-peptides, and so on (Figure 15).

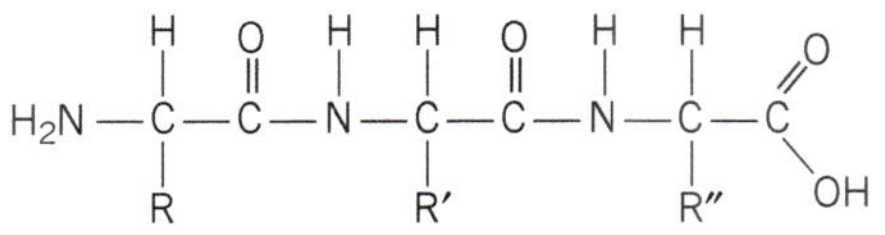

Figure 14 Formation of a dipeptide

Figure 15 A tripeptide. R, R', and R'' may be the same or different

A particular protein will have a fixed sequence of amino acids in its chain. This is called the **primary structure** of the protein and can be specified by writing the abbreviated names of the amino acids concerned in the correct order. For example, just one short sequence of the protein insulin (the hormone controlling sugar metabolism) runs:

-ala-glu-ala-leu-tyr-

Polypeptides and proteins are **condensation polymers** because a small molecule (in this case water) is eliminated as each link of the chain forms.

Hydrolysis

When a protein or a peptide is boiled with hydrochloric acid of concentration 6 mol dm^{-3} for about 24 hours, it breaks down completely to a mixture of all the amino acids that made up the original protein or peptide. All the peptide linkages are hydrolysed by the acid (Figure 16).

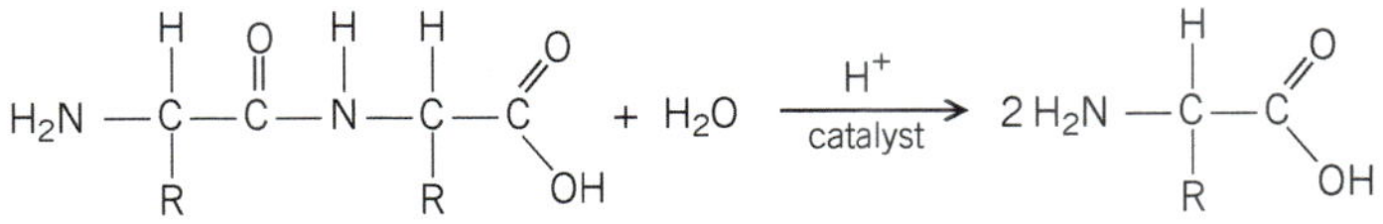

Figure 16 The hydrolysis of the peptide link

Electrophoresis

Finding the structure of proteins is an important analytical task. The first step is to hydrolyse the protein by refluxing with 6 mol dm^{-3} hydrochloric acid. This forms a mixture containing all the amino acids that make up the protein. These amino acids then need to be separated and identified.

Electrophoresis is a method of separating charged molecules, including amino acids. In acidic solutions (ones with low pH) amino acids accept a proton (an H^+ ion) and are therefore positively charged. In alkaline solutions, amino acids lose a proton to form a negative ion (see Figures 12 and 13 earlier in this section).

> **Exam tip**
>
> A polypeptide is a peptide with an unspecified number of amino acids.

> **Exam tip**
>
> **Hydrolysis** is a reaction with water (often boiling) that may be catalysed by an acid, an alkali, or an enzyme. As proteins can be hydrolysed, they are biodegradable.

In electrophoresis, a small volume of a mixture of amino acids is placed in a sample well on a gel or moist filter paper placed between two electrodes—see Figure 17.

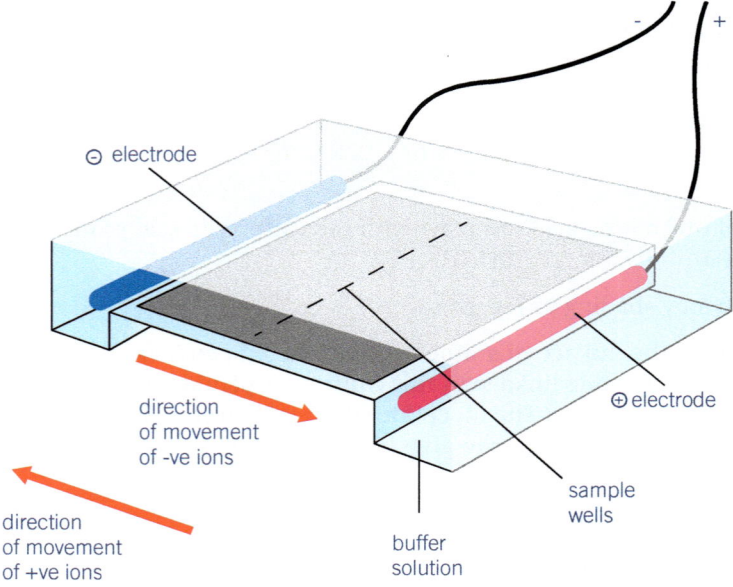

Figure 17 *Electrophoresis tank*

Negatively charged amino acids move towards the anode. Positively charged amino acids move towards the cathode. The rate of movement will depend on the size of the amino acid (larger amino acids move more slowly) and the properties of the R group, and the number of amino or carboxyl groups.

Amino acids are colourless. After separation the paper or gel is sprayed with a substance called ninhydrin. This reacts with the amino acids to form a purplish colour, so the amino acids can be located. Peptides and even proteins can be separated using this technique.

The actual amino acids may be identified by measuring the distance moved in a given time, and comparing this with the results for samples of known amino acids.

Worked example

Electrophoresis

The structure of alanine is:

H$_2$NCHCO$_2$H
|
CH$_3$

and the structure of phenylalanine is:

H$_2$NCHCO$_2$H
|
CH$_2$

1 Draw the structure of each amino acid in a strongly acidic solution (one of low pH) and in a strongly alkaline solution (one of high pH).

Solution to 1

Alanine in an acidic solution is protonated:

$H_3\overset{+}{N}CHCO_2H$
|
CH_3

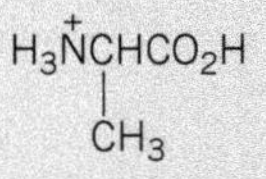

Phenylalanine in an acidic solution is protonated:

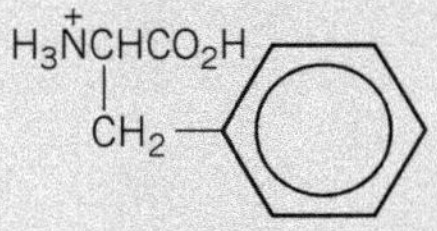

$H_3\overset{+}{N}CHCO_2H$
|
$CH_2-\bigcirc$

Alanine in an alkaline solution is deprotonated:

$H_2NCHCO_2^-$
|
CH_3

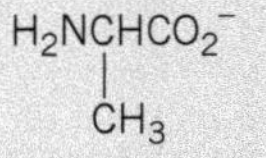

Phenylalanine in an alkaline solution is deprotonated:

$H_2NCHCO_2^-$
|
$CH_2-\bigcirc$

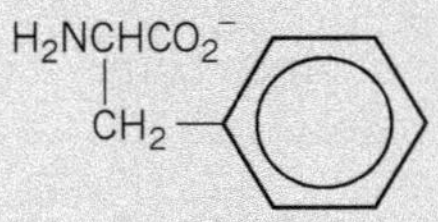

2 In which direction would each amino acid move in electrophoresis?

Solution to 2

In acid solution, both amino acids would move towards the cathode.

In alkaline solution, both amino acids would move towards the anode.

3 Predict which amino acid would move the faster in each case, and explain your predictions.

Solution to 3

Phenylalanine would be expected to move more slowly in both solutions because it has a bulky benzene ring to slow it down.

Summary test 34.4

1 The systematic name of glycine is 2-aminoethanoic acid:

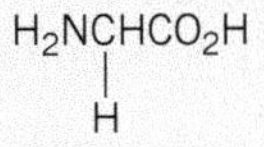

H_2NCHCO_2H
|
H

Give the systematic name of alanine.

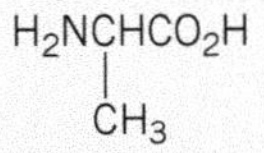

H_2NCHCO_2H
|
CH_3

2 Explain why alanine is chiral whereas glycine is not.

3 a State the functional groups in an amino acid.
   b State which group is acidic and which is basic.

4 Give the number of amide (peptide) linkages in a tripeptide.

5 State which form amino acid residues will exist in after a protein has been hydrolysed with 6 mol dm^{-3} hydrochloric acid. Sketch the structural formula of an alanine residue.

6 Sketch the formulae of the three amino acids that would be formed by the hydrolysis of the tripeptide shown below.

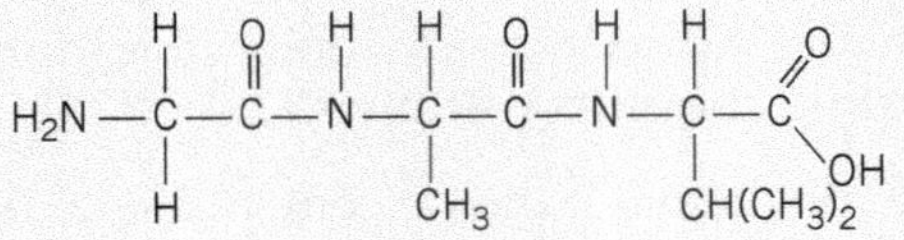

gly   ala   val

7 Explain why lysine, glutamic acid and valine can be separated by electrophoresis.

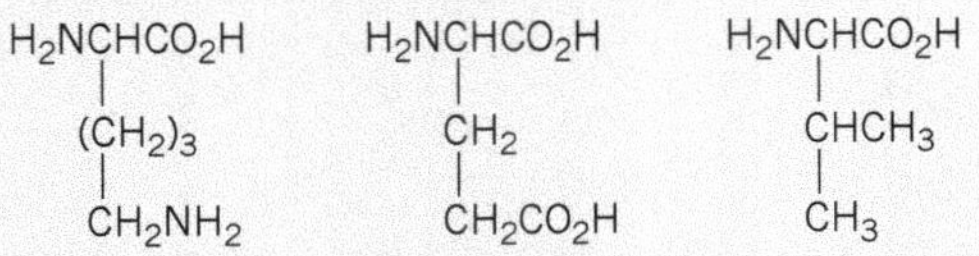

H_2NCHCO_2H H_2NCHCO_2H H_2NCHCO_2H
$(CH_2)_3$ CH_2 $CHCH_3$
CH_2NH_2 CH_2CO_2H CH_3
lysine   glutamic acid   valine

8 Suggest which way each amino acid is likely to move in an acidic buffer and justify your answer. (Hint: think about the side chains.)

> 📖 **Launch additional digital resources for the chapter**

1 Consider the following compounds:

A CH_3CHCH_3
      |
     CN

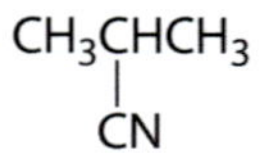

B $CH_3CH_2CH_2NH_2$

C O_2N ⬡ NO_2

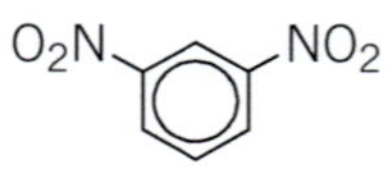

D $\begin{array}{c} CH_3 \\ \backslash \\ \quad NH \\ / \\ CH_3CH_2 \end{array}$

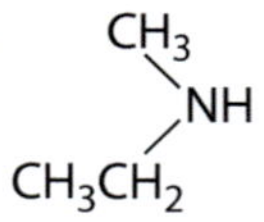

E $CH_3CH_2CONH_2$

F $(CH_3CH_2CH_2)_3N$

a Which is a primary amine? *(1 mark)*

b Which is a nitrile? *(1 mark)*

c Which is an amide? *(1 mark)*

d Which is a tertiary amine? *(1 mark)*

e Give the IUPAC name for each compound. *(6 marks)*

2 The diagram below summarises some methods for the synthesis of ethylamine:

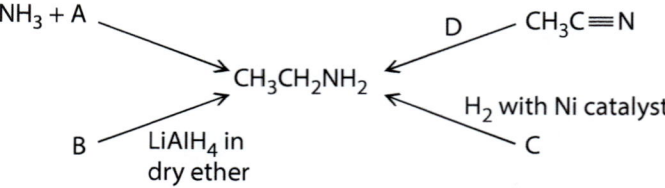

a Give the IUPAC names and displayed formulae of the starting reagents A, B and C. *(6 marks)*

b Identify the name of the reagent and state the conditions represented by reaction D. *(2 marks)*

3 This question is about three nitrogen compounds: ethylamine, phenylamine and ammonia.

State the order of increasing basicity of the three compounds. Give reasons for your answer. *(4 marks)*

4 Consider the following compounds:

A CH_3CHCH_3
      |
     CN

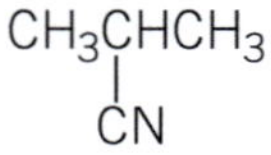

B $CH_3CH_2NH_2$

C $(CH_3CH_2)_2NH$

D $H_2N(CH_2)_5NH_2$

a Which would be among the products of the reaction of chloroethane with ammonia? *(1 mark)*

b Which could be converted to a diazonium compound by the reaction with nitrous acid below 10°C? *(1 mark)*

c Which could be made by reduction of ethanenitrile? *(1 mark)*

d Which would react with hydrochloric acid in the ratio one mole of the compound to two moles of hydrochloric acid? *(1 mark)*

e Which is the weakest base? *(1 mark)*

5 Salbutamol is a very effective treatment for asthma. It is the active ingredient in asthma inhalers.

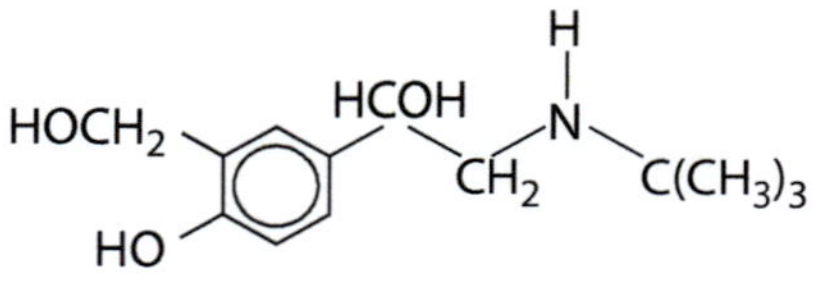

$$HOCH_2 \underset{HO}{\overset{}{\bigcirc}} \overset{HCOH}{\underset{CH_2}{}} \overset{\overset{H}{|}}{N} C(CH_3)_3$$

salbutamol

a State whether the amine group in salbutamol is primary, secondary or tertiary. *(1 mark)*

b Draw the structure of the product of the reaction of salbutamol with:
  i $HCl(aq)$
  ii CH_3COCl *(2 marks)*

6 This question is about amino acids.
  a The formula of alanine is shown below:

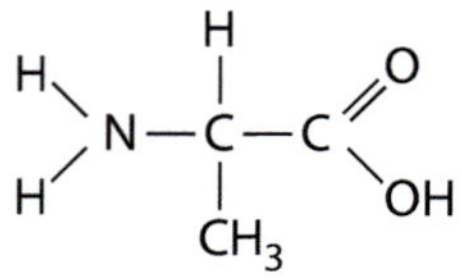

$$\begin{array}{c} H \\ \underset{H}{\overset{H}{\diagdown}} N - \overset{\overset{H}{|}}{\underset{\underset{CH_3}{|}}{C}} - C \overset{\diagup O}{\diagdown OH} \end{array}$$

  i Construct an equation to show the reaction of alanine with an acid, H^+. Include a displayed formula for the product(s). *(2 marks)*

  ii Construct an equation to show the reaction of alanine with an alkali, OH^-. Include a displayed formula for the product(s). *(2 marks)*

b In neutral solution, alanine forms a dipolar ion.

  i Draw a displayed formula to show the structure of the dipolar ion. *(1 mark)*

  ii Give the general name for this type of ion. *(1 mark)*

c Aspartic acid has the formula shown below. Explain why it is acidic overall.

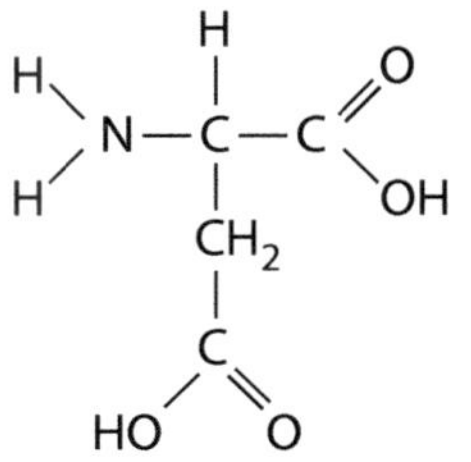

(1 mark)

d Lysine has the formula shown below. Describe whether lysine is acidic, basic or neutral overall? Explain your answer.

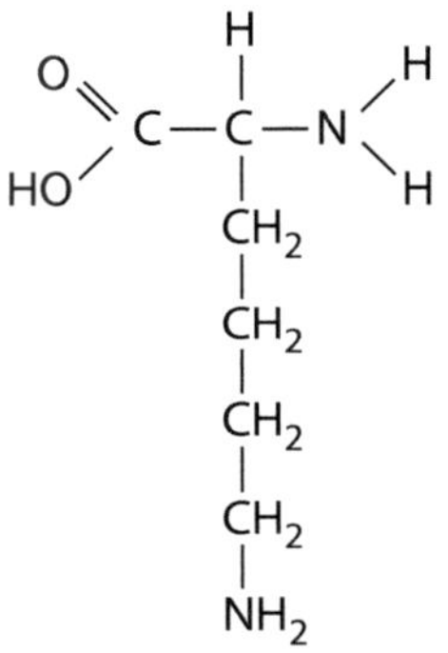

(2 marks)

7 Amino acids join together by forming peptide bonds.

a Draw the structural formulae of cysteine and methionine using the information below.

Amino acid	Abbreviated code	R group
cysteine	Cys	$-CH_2-SH$
methionine	Met	$-CH_2-CH_2-S-CH_3$

(2 marks)

b Write and annotate an equation to show how cysteine and methionine join together by the formation of a peptide bond. *(2 marks)*

c Is the product described in part **b** a dipeptide or tripeptide? Explain your answer. *(2 marks)*

d Give the name of the type of polymerisation that results in the formation of a peptide bond. *(1 mark)*

8 Complete the blank spaces for the following reaction sequences, giving the formulae of the products for each.

a

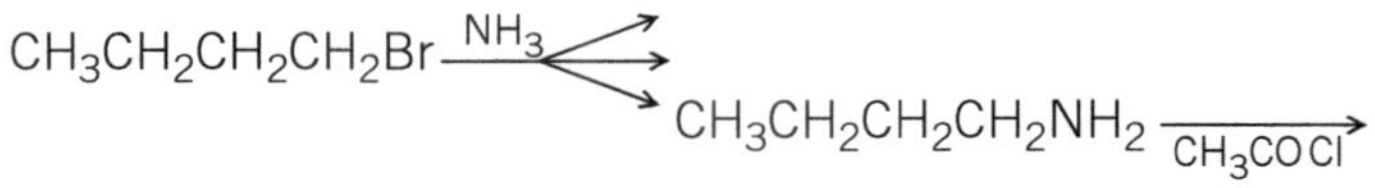

b

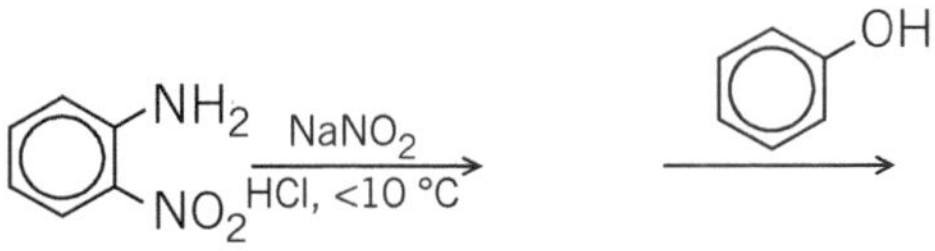

(5 marks)

9 Two isomeric products were formed from the reaction of compound **A**, C_7H_8, with concentrated nitric acid and concentrated sulfuric acid.

One of these, isomer **B**, has the molecular formula $C_7H_7O_2N$. **B** could be converted to **C**, C_7H_9N by reduction.

C reacts with dilute hydrochloric acid to form D, $C_7H_{10}NCl$. When cold sodium nitrite solution was added to a cold solution of **D** in hydrochloric acid, a solution of **E**, $C_7H_7N_2Cl$, was formed.

When a cold solution of **E** was added to a cold solution of phenol, a brightly coloured substance **F** was produced.

When the solution **E** was warmed alone, an unreactive gas was evolved and **G**, C_7H_8O, was formed.

Give possible structural formulae for **A** to **G**. *(7 marks)*

10 Give the displayed formulae for the products formed when:

a Phenylamine is dissolved in excess concentrated hydrochloric acid.

b Sodium nitrite solution is added to the cooled solution from part **a**.

c The product formed from part **b** is added to a fresh solution of phenylamine. *(3 marks)*

11 a Place the following compounds in order of increasing basicity. Give reasons for your answer.

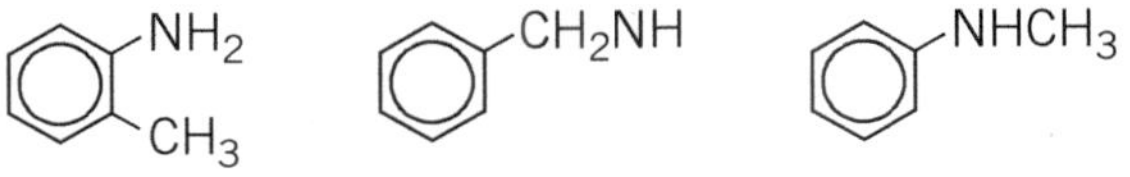

(4 marks)

b Place the following in order of increasing boiling point, giving reasons for your answer.

  i $CH_3CH_2CH_2NH_2$

  ii $CH_3CH_2CH_2CH_3$

  iii $CH_3CH_2CH_2OH$ *(4 marks)*

35.1 Condensation polymerisation

Learning outcomes

On these pages you will learn to:

- describe the formation of polyesters and polyamides
- deduce the repeat unit of a condensation polymer
- identify the monomers in a polymer

Figure 1 *There are examples of polyesters and polyamides (nylon) in the athletes' and fisher's clothing pictured here*

A condensation reaction occurs when two molecules react together and a small molecule, often water or hydrogen chloride, is eliminated. For example, esters are formed when carboxylic acids and alcohols react together. This is a **condensation reaction** and water, H_2O, is eliminated—hydrogen from the alcohol and an –OH group from the carboxylic acid:

$$R-C\underset{OH}{\overset{O}{\|}} + HO-R' \longrightarrow R-C\underset{O-R'}{\overset{O}{\|}} + H_2O$$

carboxylic acid　　alcohol　　　　　　　ester　　　　water

Condensation polymers are normally made from two different monomers, each of which has *two* functional groups. Both functional groups can react, forming long-chain polymers.

Polyesters, polyamides, and polypeptides are all examples of condensation polymers (Figure 1).

Polyesters

A *poly*ester has the ester linkage –COO– repeated over and over again.

To make a polyester, diols, which have two –OH groups, react with

- dicarboxylic acids, which have two carboxylic acid, –COOH, groups

or

- dioyl chlorides, which have two –COCl groups.

HO—A—OH　　　　dicarboxylic acid　　　dioyl chloride

diol

In each case, a small molecule is eliminated.

A and B represent unspecified organic groups, often $(CH_2)_n$. The functional groups on the ends of each molecule react to form a chain. For example, diols and dicarboxylic acids react together to give a polyester by eliminating molecules of water (Figure 2).

Figure 2 *Making a polyester*

The fibre PET (polyethylene terephthalate) is a polyester made from benzene-1,4-dicarboxylic acid and ethane-1,2-diol (Figure 3).

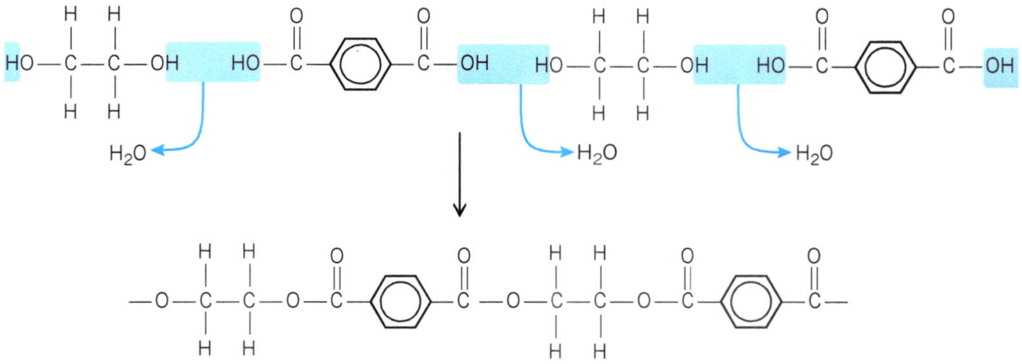

Figure 3 *PET is a polyester. Notice how the C–O is alternately to the left and to the right of the C=O*

Alternatively we can use one monomer with two different functional groups, such as 2-hydroxypropanoic acid. This has a carboxylic acid and an alcohol functional group on the same molecule (see Figure 4).

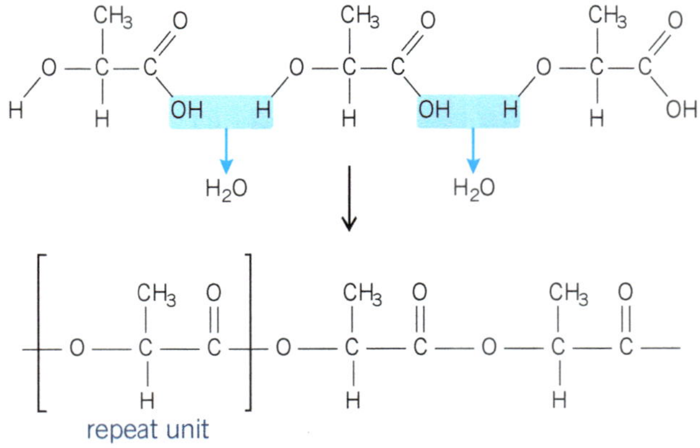

Figure 4 *Making a polyester from a hydroxycarboxylic acid monomer. Notice how the C–O group is always to the right of the C=O group*

Polyamides

An amide is formed when an amine and a carboxylic acid (or acyl chloride) react together and eliminate a small molecule as before:

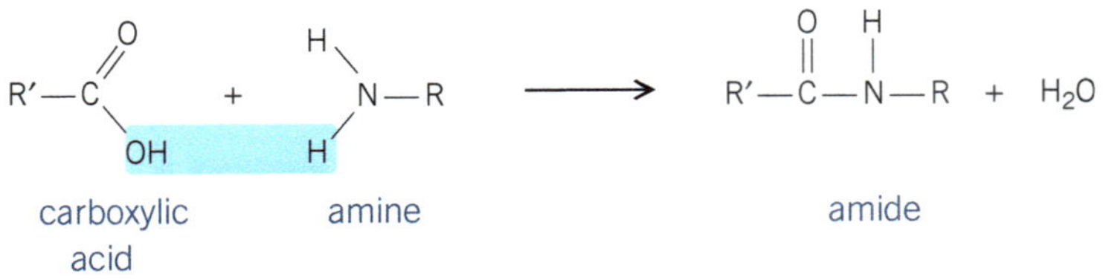

*Poly*amides have the amide linkage –CONH– repeated over and over again. To make polyamides from two different monomers, a diaminoalkane (which has two amine groups) reacts with:

• a dicarboxylic acid, which has two carboxylic acid groups (Figure 5 on the following page)

or

• a dioyl chloride, which has two acid chloride groups.

Notice how the NH group is alternately to the left and to the right of the C=O group.

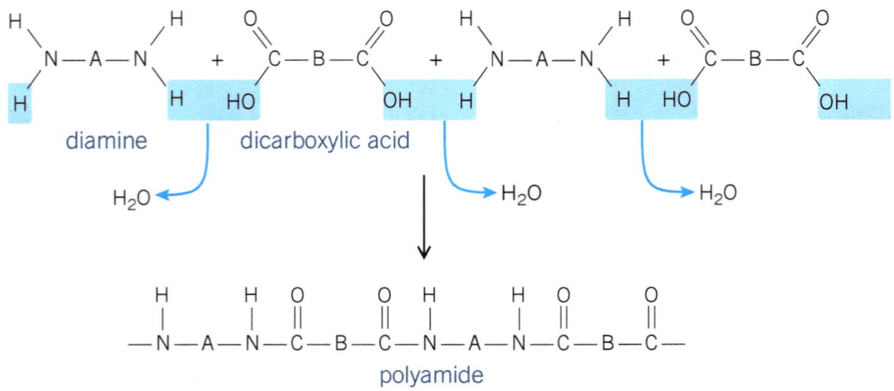

Figure 6 *When solutions of 1,6-diaminohexane and hexane-1,6-dioyl chloride meet, nylon-6,6 is formed at the interface. Nylon-6,6 was invented by Wallace Carothers in 1935, an American chemist*

Figure 8 *Formula 1 drivers' helmets need to be lightweight as the less weight it adds to a driver's head, the smaller the risk of whiplash injuries under the extreme G-forces experienced in accelerating and braking. Helmets worn by racing drivers contain Kevlar, which is five times stronger than steel, weight for weight*

Figure 5 *The general equation for making a polyamide, such as nylon-6,6 or Kevlar. Notice how N–H alternates to the left and right of C=O*

Both nylon and Kevlar are condensation polymers.

Nylon
Industrially, nylon-6,6 is made from 1,6-diaminohexane and hexane-1,6-dicarboxylic acid:

1,6-diaminohexane   hexane-1,6-dicarboxylic acid

nylon-6,6

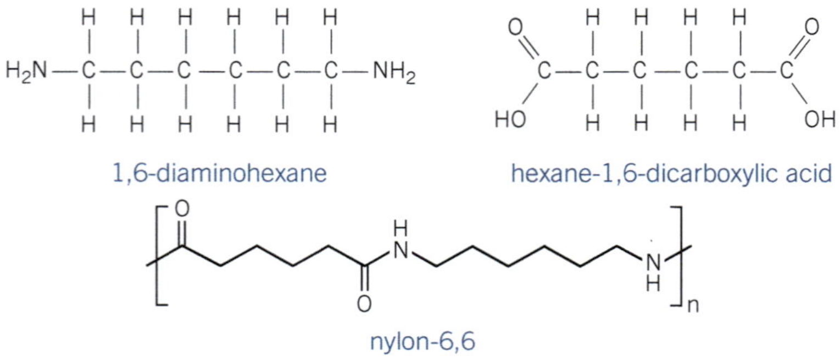

In the laboratory, the reaction goes faster if a dioyl chloride is used rather than the dicarboxylic acid, and in this case hydrogen chloride is eliminated. Nylon-6,10 is made from 1,6-diaminohexane and decane-1,10-dicarboxylic acid. Many other nylons are made each with slightly different properties (see Figure 6).

Kevlar
Kevlar is made from benzene-1,4-diamine and benzene-1,4-dicarboxylic acid (Figure 7).

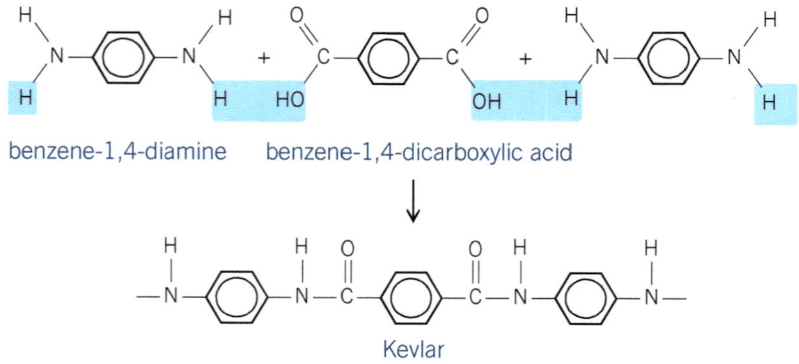

benzene-1,4-diamine   benzene-1,4-dicarboxylic acid

Kevlar

Figure 7 *Kevlar is a polyamide. Because the amide groups are linking rigid benzene rings, Kevlar has very different properties to nylon*

Kevlar's strength is due to the rigid chains and the ability of the flat aromatic rings to pack together held by strong intermolecular forces. The polymer was developed in the 1960s by American chemist Stephanie Kwolek, of the DuPont company. It is credited with saving some 3000 lives because of its use in bulletproof vests and anti-stab clothing as worn by the police. Kwolek remarked, "I don't think there's anything like saving someone's life to bring you satisfaction and happiness." You may even have Kevlar oven gloves at home!

Extension

The strength of Kevlar

When chains of Kevlar molecules pack together, they are held by hydrogen bonding between N–H groups on one chain and C=O groups on another:

N–H --- :O=C

Polypeptides

Polypeptides are also polyamides. They are formed from amino acids (that is aminocarboxylic acids). They may be made from one amino acid monomer, or from different amino acid monomers.

Each amino acid has both an amine group and a carboxylic acid group. To form a polyamide, the amine group of one amino acid can react with the carboxylic acid group of another amino acid. A molecule of water is eliminated and a condensation polymer can form:

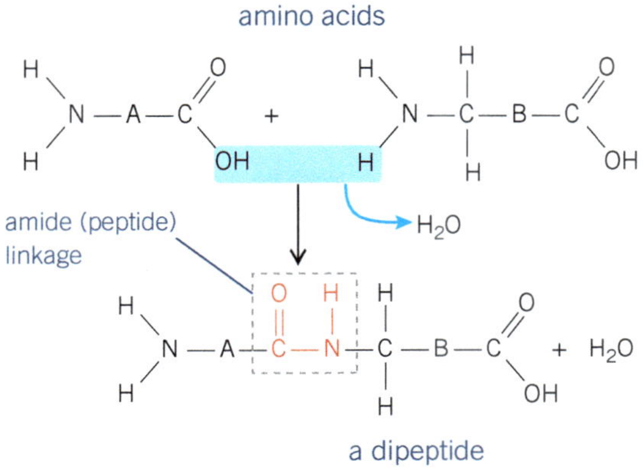

There is a difference between a polymer like nylon-6,6 and a polypeptide.

- In nylon-6,6, there are two monomers (one is a diamine, $H_2N–X–NH_2$, and one is a dicarboxylic acid, HOOC–Y–COOH). The different functional groups are on different monomers.
- In a polypeptide every monomer has one $–NH_2$ group and one –COOH group ($H_2N–X–COOH$). Both functional groups are on the same monomer.

A typical polypeptide, (in this case a tripeptide), shown in Figure 9, would have the following structure:

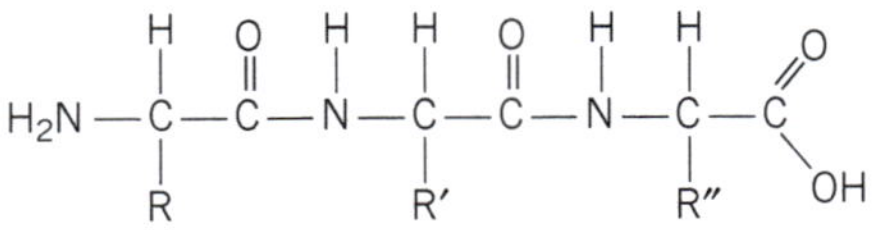

Figure 9 *A tripeptide. R, R′, and R″ may be the same or different. Notice how N–H is always to the right of C=O*

There are 20 naturally occurring varieties of amino acids.

Identifying the repeat unit of a condensation polymer

The repeat unit of a condensation polymer is found by starting at any point in the polymer and stopping when the same pattern of atoms begins again (Figure 10). This method works whether there are two different monomers or just a single type of monomer (as in addition polymerisation, see Chapter 20).

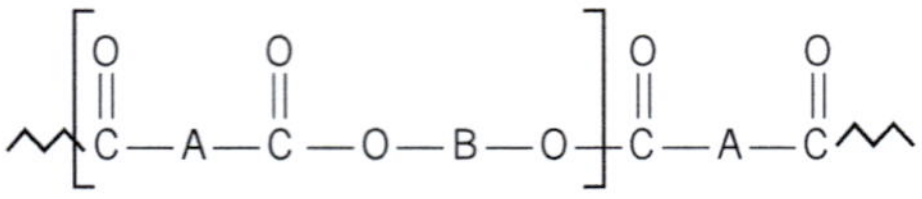

Figure 10 *The repeat unit is in brackets*

Identifying the monomer(s) of a condensation polymer

The best way to work out the monomer(s) in a condensation polymer is to try and recognise the links formed by familiar functional groups (Table 1).

Table 1 *Condensation polymers—the repeat unit is inside the bracket*

Monomer 1	Monomer 2	Polymer
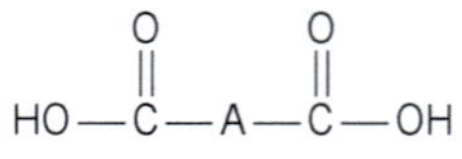 dicarboxylic acid	diol	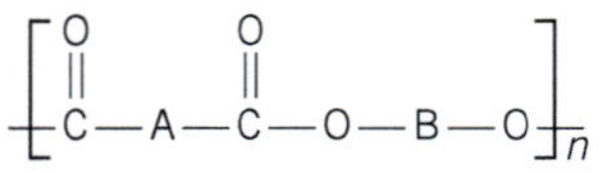
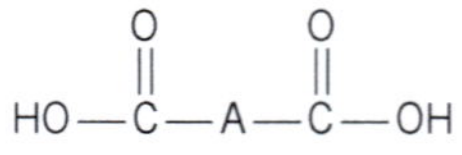 dicarboxylic acid	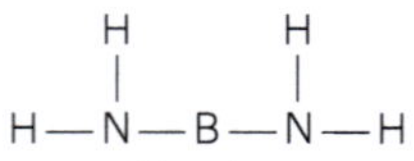 diamine	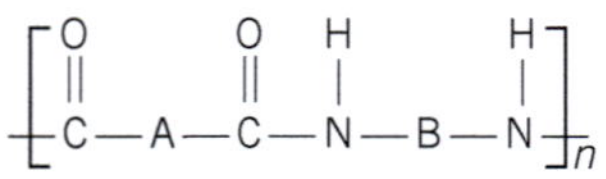
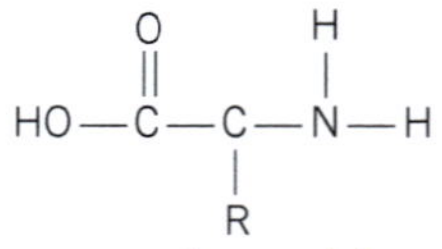 amino acid		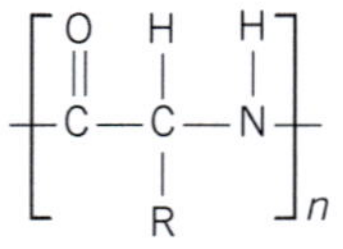

1 Start with the repeat unit.
2 Break the linkage (at the C–O for a polyester or C–N for a polyamide).
3 Add back the components of water for each ester or amide link.

For example:

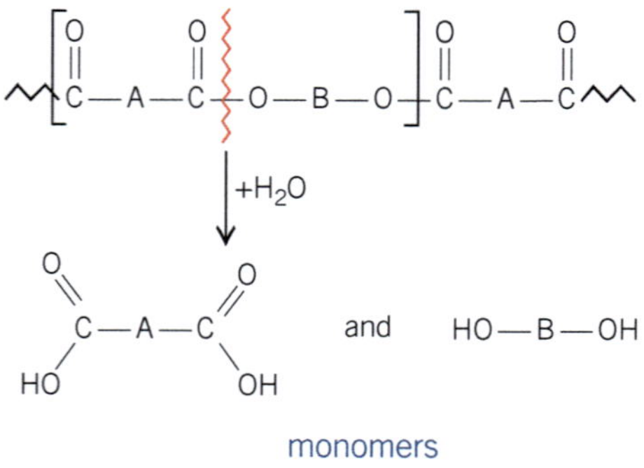

monomers

This is exactly the same process that occurs when condensation polymers are hydrolysed.

Exam tip

Monomers that have two acyl chloride functional groups may be called either **dioyl chlorides** or **dioyl dichlorides**.

Summary test 35.1

1 Sketch the elimination equation, as in Figure 2 earlier in this section, using a diol and a dioyl chloride to make a polyester (use A and B as the stems of the molecules).
2 There are a number of different types of nylon made from two monomers—a dicarboxylic acid and a diamine.
   a The one made from hexane-1,6-dicarboxylic acid and 1,6-diaminohexane is called nylon-6,6. Suggest where the numbers come from.
   b Nylon-6,10 is made from the same dicarboxylic acid as nylon-6,6. Give the name and formula of the other monomer.
3 Nylons are polyamides. Explain why proteins and peptides are also called *polyamides*.
4 PET is a polyester made from benzene-1,4-dicarboxylic acid and ethane-1,2-diol. Suggest another diol that would react with this acid to make a different polyester.
5 Give the names of the linkages in the two polymers below.
   a 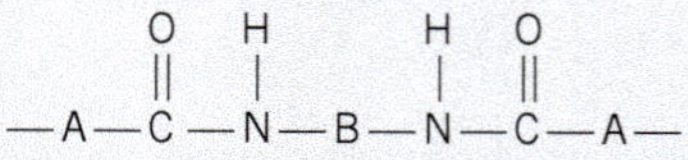
   b 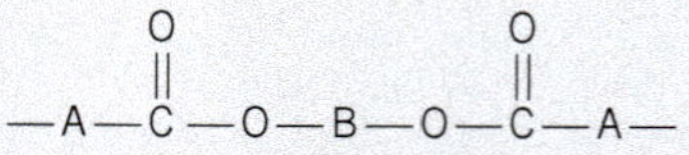
6 Give an equation for the hydrolysis of a polyester.

When you look at the structure of a monomer you can predict the polymer it will form.

Addition polymers

Addition polymerisation takes place between monomers that contain a carbon–carbon double bond—alk*enes* such as ethene or propene. The monomers add together with no other molecule formed.

Often all the monomers are identical. When each monomer has only one double bond, the resulting polymer is an alk*ane*. This can be confusing because the name of the polymer is derived from the name of the monomer. So ethene polymerises to form poly(ethene). The empirical formula of the polymer is exactly the same as the empirical formula of the monomer.

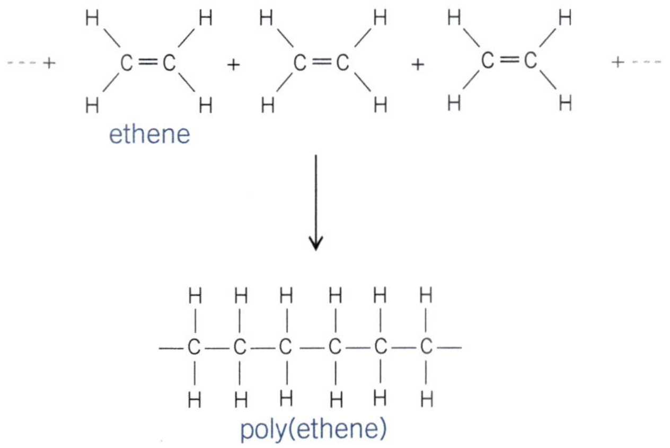

ethene

poly(ethene)

Condensation polymers

Condensation polymerisation reactions take place between monomers with two different functional groups. A small, stable molecule (often water) is ejected. This is why it is called condensation polymerisation. It produces the following polymers:

- **a polyester**: this is formed between a diol and a dicarboxylic acid (or dioyl chloride). The ester linkage –COO– is always present in the polymer and *the order of C=O and O alternates*.
- **a polyamide**: this is formed between a diamine and a dicarboxylic acid (or dioyl chloride). The amide linkage –CONH– is always present in the polymer and *the order of C=O and N–H alternates*.
- **a polypeptide**: this also a polyamide and is formed from amino acids with two different functional groups on either end—an amine and a carboxylic acid. The amide link –CONH– is always present *in the same order*.

The empirical formula of a condensation polymer is not the same as the empirical formula of either of its monomers. This is because:

- there may be two different monomers
- a molecule has been ejected.

Learning outcomes

On these pages you will learn to:

- predict the type of polymerisation for a monomer or pair of monomers
- deduce the type of polymerisation reaction that produces a given polymer

Worked example

Deducing the type of polymerisation from the polymer

What type of polymerisation has formed each of the polymers shown below?

a

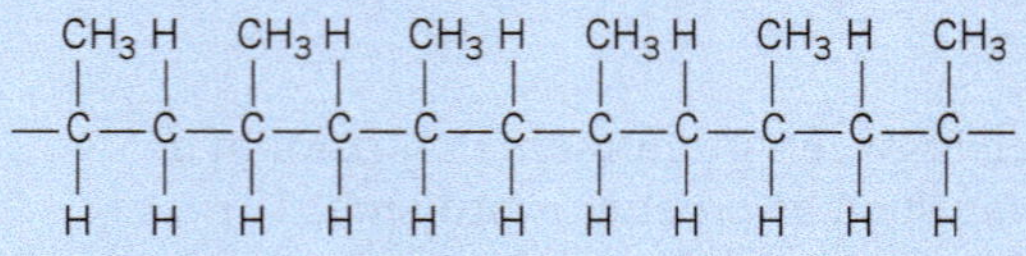

b

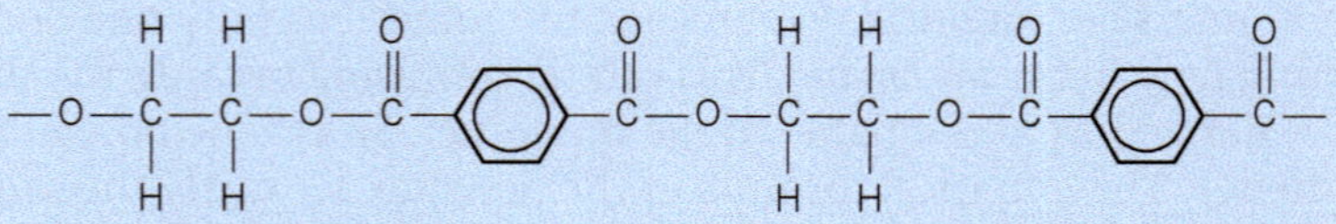

In each case work along the polymer chain from left to right looking for functional groups.

In polymer **a**, there are no functional groups. It is an alkane so it must be the product of an addition polymerisation of a monomer with a carbon-carbon double bond. There is a CH_3 group on every other carbon atom so the monomer was propene

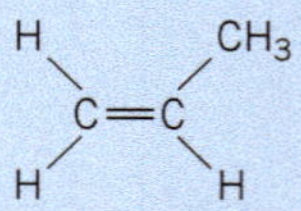

The polymer is called **poly(propene)**, common name polypropylene.

In polymer **b** we come across the group: $-O-\overset{\overset{O}{\|}}{C}-$

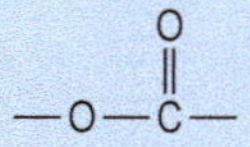

This is the **ester linkage**, so the polymer is a polyester made from the two monomers:

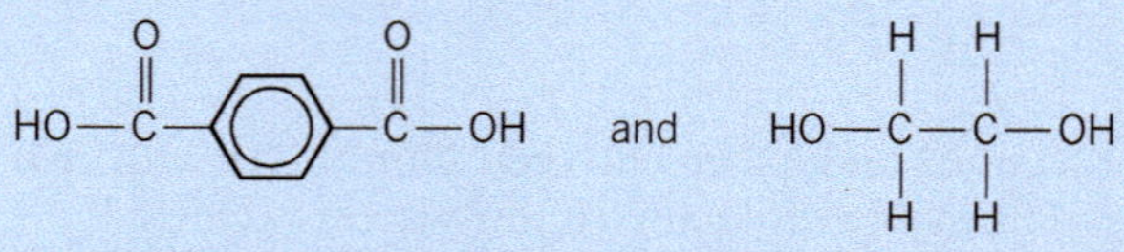

It is actually the polymer with the common name PET.

Exam tip

The systematic names of polymers are derived from the name of the monomer in brackets after the prefix poly. So poly(alkenes) are in fact alkanes.

Summary test 35.2

1 Look at the three sections of polymer chains, **A**, **B** and **C**. Identify **a** which was formed by addition polymerisation, **b** which was formed from two different monomers and **c** which was formed from a single monomer with two different functional groups.

A

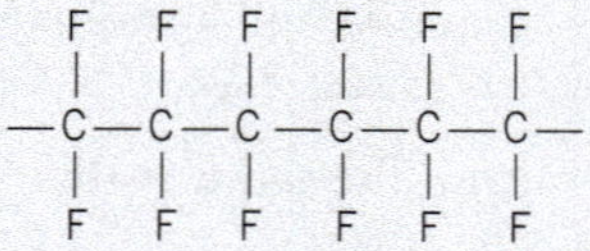

B

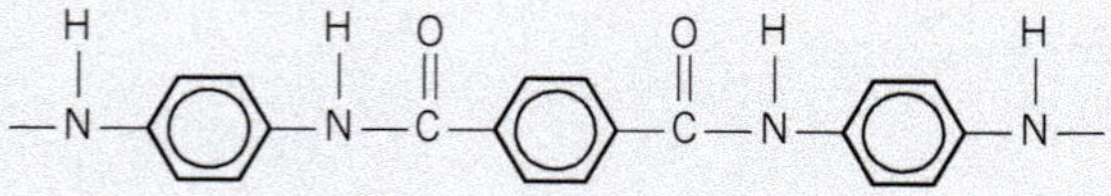

C

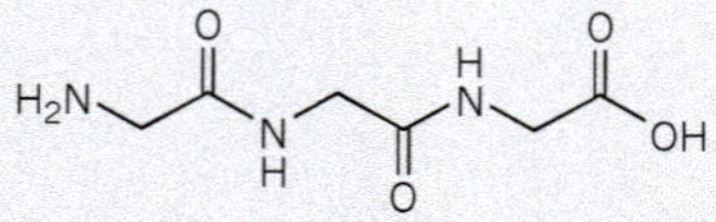

Degradable polymers

Addition polymers such as poly(ethene) and poly(propene) are alkanes. They have strong covalent bonds and little polarity and so are inert to most chemical reagents.

When poly(alkenes) were first introduced after World War II they were prized for their resistance to degradation. Nowadays this is a serious problem, as seen by plastic litter in the world's oceans. Poly(alkenes) can be burnt to carbon dioxide and water, to create energy for district heating schemes for example. However, this releases the greenhouse gas carbon dioxide into the atmosphere.

Learning outcomes

On these pages you will learn to:

- recognise that poly(alkenes) are difficult to biodegrade
- describe how some polymers can be degraded by light
- describe how polyesters and polyamides are biodegradable by hydrolysis

![Undecomposed poly(ethene) and poly(propene) cause problems for wildlife](image)

Figure 11 *Undecomposed poly(ethene) and poly(propene) cause problems for wildlife*

One way in which these polymers may be broken down or degraded is by the action of ultraviolet light from sunlight. Quanta of UV light have sufficient energy to break the carbon–carbon bonds in these polymers. This produces free radicals which start the breakdown of the polymer. To prevent this type of degradation, some polymers have stabilisers added. These chemicals absorb the UV light instead of it being absorbed by the polymers.

Polyesters and polyamides can be degraded by the action of water. Their ester and amide bonds are susceptible to hydrolysis in acidic or alkaline conditions. The hydrolysis of the amide bonds in nylon is shown below.

Throughout the polymer, the N—C bond is broken.

The amide bonds in proteins can be hydrolysed in the same way.

Summary test 35.3

1 Give an equation (in the style of the one for nylon on the left) for the hydrolysis of a polyester.
2 State what type of chemical species is likely to be formed when a carbon–carbon bond is broken by UV light. Explain your answer.

Launch additional digital resources for the chapter

1 Answer the questions that follow for each of the following monomers or pairs of monomers:

a 1,1-dichloroethene,

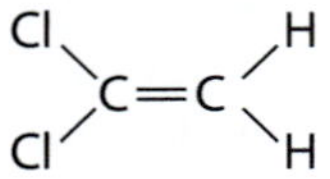

b $H_2N(CH_2)_5NH_2$ and $HOOC(CH_2)_5COOH$

c 3-aminobenzoic acid, $HOOC$—⟨O⟩—NH_2

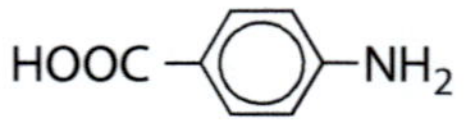

d ethene

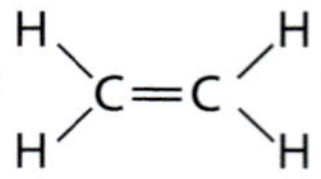

and propene

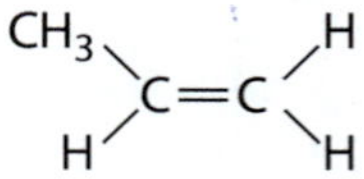

For each of parts **a–d**:

  i State whether you would expect an addition polymer or a condensation polymer. *(4 marks)*

  ii Draw the structure of a section of the polymer chain showing three monomers or pairs of monomers. *(8 marks)*

  iii Give the structure of the repeating units for each polymer. *(6 marks)*

2 Every year, huge amounts of polymer waste, including polyalkenes such as poly(ethene) and poly(propene), must be disposed of.

a Poly(ethene) is non-biodegradable and so cannot be decomposed by bacteria. Explain why. *(2 marks)*

b Low density poly(ethene) and poly(propene) are degraded by the action of ultraviolet radiation in sunlight. This is called photodegradation.

  i Suggest one problem caused by photodegradation of a poly(propene) rope. *(1 mark)*

  ii Suggest why photodegradation is not likely to become an important technique in the disposal of polymer waste. *(1 mark)*

3 Short sections of different polymer chains are shown below:

a $\cdots$—O—CH_2—CH_2—O—C—CH_2—CH_2—C— $\cdots$

                                                  ║   ║
                                                  O   O

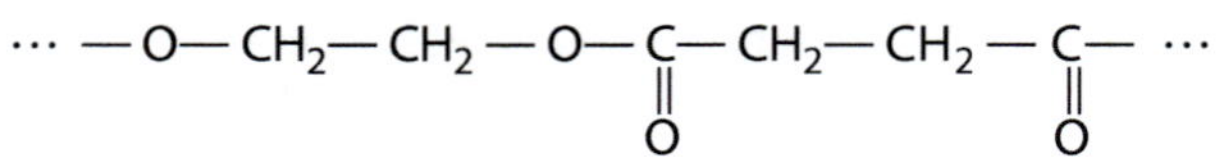

b $\ldots$—CF_2—CF_2—CF_2—CF_2—CF_2—CF_2—$\ldots$

c $\ldots$—HN—$(CH_2)_9$—$NHCO$—$(CH_2)_7$—$CO\ldots$

For each of polymers **a–c**:

  i State whether they are examples of addition or condensation polymers. *(3 marks)*

  ii Give the structure of each monomer from which each polymer is made. *(3 marks)*

4 A section of the polymer chain in poly(ethenol) is shown below:

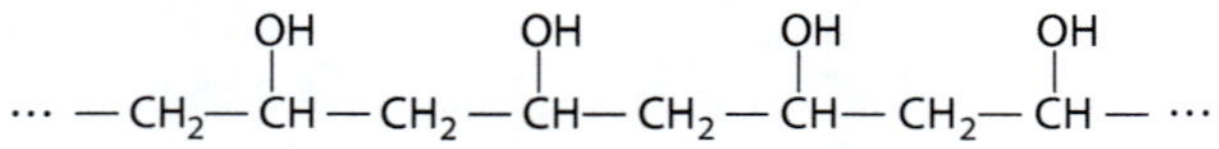

a State whether poly(ethenol) is an addition polymer or a condensation polymer. *(1 mark)*

b Pure poly(ethenol) is insoluble in water. However, a soluble form of the polymer can be made by replacing some of the –OH groups with CH_3COO^- groups.

  i Draw a section of the poly(ethenol) chain in which half the –OH groups have been replaced by CH_3COO^- groups. *(2 marks)*

  ii Explain why replacing some of the –OH groups helps to make the polymer soluble in water. *(2 marks)*

5 Kevlar is a polymer material that is used to make bulletproof vests. It is formed from a polymerisation reaction between two monomers given below:

H_2N-⟨benzene ring⟩$-NH_2$ $ClOC-$⟨benzene ring⟩$-COCl$

1,4-diaminobenzene   1,4-benzenedicarbonyl chloride

a State the type of polymerisation that occurs when Kevlar is made. *(1 mark)*

b Draw the repeating unit for Kevlar. *(2 marks)*

c Predict whether Kevlar would be biodegradable, giving reasons for your answer. *(2 marks)*

6 Nylon-6,10 is used as an engineering plastic. The repeating unit in nylon-6,10 is shown below:

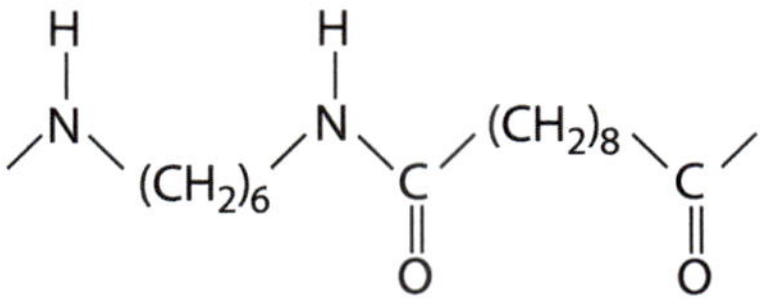

⟨structure of nylon-6,10 repeating unit: N–H, $(CH_2)_6$, N–H, $C=O$, $(CH_2)_8$, $C=O$⟩

a Draw the structures of the monomers of nylon-6,10. *(2 marks)*

b Use the structure of nylon-6,10 given above to predict the structure of nylon-6,4. *(2 marks)*

7 Polyaniline is a conductive polymer used in microelectronics. The structure below shows part of a polyaniline chain:

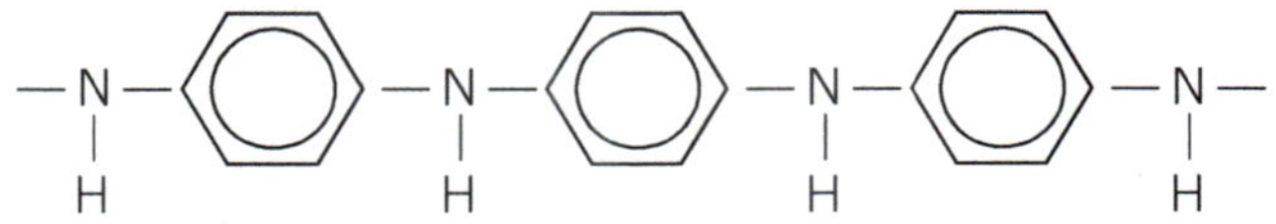

⟨structure: $-N(H)-$⟨benzene ring⟩$-N(H)-$⟨benzene ring⟩$-N(H)-$⟨benzene ring⟩$-N(H)-$⟩

a State the IUPAC name for polyaniline. *(1 mark)*

b Draw the structure of the repeat unit of the polymer. *(3 marks)*

c Explain why polyaniline is an electrical conductor. *(2 marks)*

d Suggest one property of polyaniline, other than its electrical conductivity, which would make it suitable for use in electrically conductive fabrics. *(1 mark)*

8 A tripeptide is formed from the reaction of two amino acids, valine and alanine:

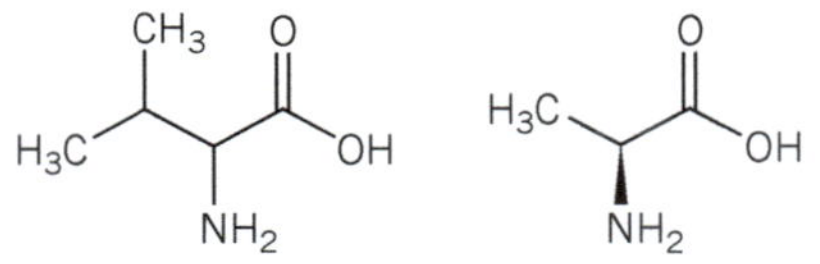

⟨structure of valine: CH_3, H_3C, NH_2, $COOH$⟩   ⟨structure of alanine: H_3C, NH_2, $COOH$⟩

valine (Val)   alanine (Ala)

a Identify the asymmetric carbon atom in alanine. *(1 mark)*

b Draw the zwitterion of valine. *(1 mark)*

c A tripeptide with the sequence Val-Ala-Ala was formed. Draw the structure of this tripeptide and name the type of polymerisation involved. *(4 marks)*

36.1 Organic synthesis

The principles of devising a synthesis of an organic compound from a given starting material are discussed in Chapter 21. This chapter builds on this by adding extra classes of compounds and functional groups including arenes.

Alkyl compounds

The reactions discussed in this section are those of alkyl compounds ie ones derived from alkanes.

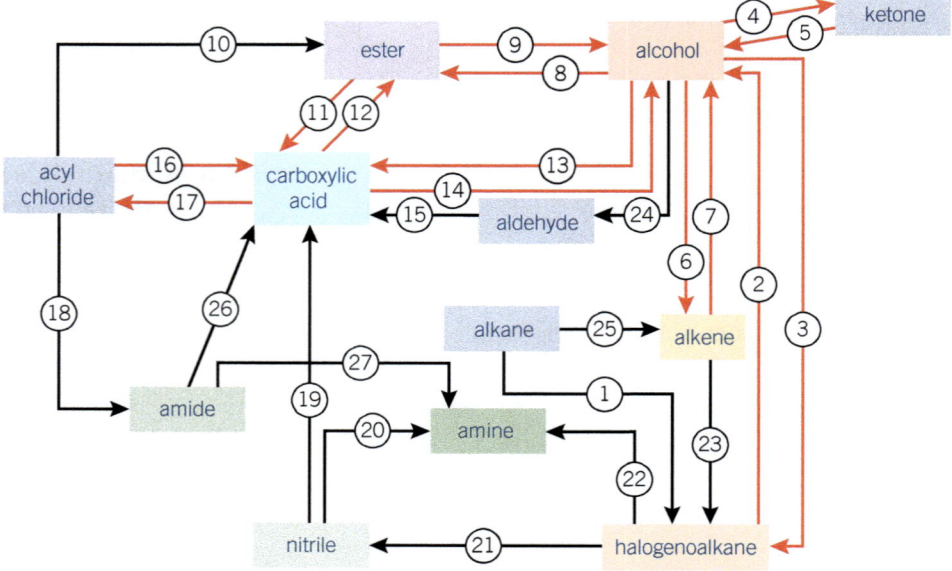

Figure 1 *Interrelationships between functional groups. You can use this chart to revise your knowledge of organic reactions*

Figure 1 is shows some of the alkyl compounds with the functional groups you have met. It shows the relationships between them. You will need to know the reagents that convert one functional group into another, along with reaction conditions such as temperature, pressure, catalysts, solvents, and excess of reagents if appropriate. Examples are summarised below; further details can be found in the appropriate chapter(s).

To plan the synthesis of a target compound from a given starting material, locate the two on Figure 1 and follow the arrows from starting material to product. You can find equations and brief notes on reaction conditions below. There may be alternative methods and the one chosen in practice will depend on factors such as:

- yield of reaction
- ease of separating the desired product from the reaction mixture
- by-products
- cost and availability of reagent
- hazards involved with reactants or procedure.

The procedure may vary with the chain length of the starting material and product. For example, many short chain organic compounds are gases.

The examples below use mainly short chain compounds of the relevant functional group.

1 Alkane to halogenoalkane (Chapter 14)

For example, propane to bromopropane:

$$CH_3CH_2CH_3 + Br_2 \rightarrow CH_3CH_2CH_2Br + HBr$$

Free radical substitution by a halogen using UV light to initiate the reaction.

This is not a particularly useful synthetic reaction because it produces a mixture of products including isomers and multi-substituted compounds.

However, it is one of the few reactions undergone by alkanes.

2 Halogenoalkane to alcohol (Chapter 15)

For example, 2-bromobutane to butan-2-ol:

$$CH_3CHBrCH_2CH_3 + NaOH \rightarrow CH_3CH(OH)CH_2CH_3 + NaBr$$

Nucleophilic substitution by hydroxide ion in aqueous solution. The nucleophile is $^-$:OH.

This reaction occurs very slowly at room temperature. To speed up the reaction it is necessary to warm the mixture. Halogenoalkanes do not mix with water, so ethanol is used as a solvent in which the halogenoalkane and the aqueous sodium (or potassium) hydroxide both mix.

Fluoroalkanes do not react at all whilst iodoalkanes react rapidly.

3 Alcohol to halogenoalkane (Chapter 16)

For example, ethanol to halogenoethane:

The reactions used depend on which halogen is being substituted

Typical equations are:

a $CH_3CH_2OH + HBr \rightarrow CH_3CH_2Br + H_2O$

   (The HBr is generated in the reaction flask from KBr and concentrated sulfuric acid.)

b $3CH_3CH_2OH + PCl_3 \rightarrow 3CH_3CH_2Cl + H_3PO_3$

c $CH_3CH_2OH + PCl_5 \rightarrow CH_3CH_2Cl + POCl_3 + HCl$

d $2P + 3Br_2 \rightarrow 2PBr_3$

   followed by:
$$3CH_3CH_2OH + PBr_3 \rightarrow 3CH_3CH_2Br + H_3PO_3$$
   Iodine can be used instead of bromine.

e $CH_3CH_2OH + SOCl_2 \rightarrow CH_3CH_2Cl + SO_2 + HCl$

4 Secondary alcohol to ketone (Chapter 16)

For example, propan-2-ol to propanone:

$$CH_3CH(OH)CH_3 + [O] \rightarrow CH_3COCH_3 + H_2O$$

Secondary alcohols are oxidised to ketones by heating with acidified potassium dichromate(VI), $K_2Cr_2O_7$ or potassium manganate(VII), $KMnO_4$.

Unlike primary alcohols, secondary alcohols cannot easily be oxidised further so the concentration of oxidising agent is not critical. The product can be distilled from the reaction mixture.

Exam tip

Alkanes are a readily available starting material as they are obtained by fractional distillation of crude oil.

Exam tip

This substitution reaction takes place in aqueous solution with some ethanol as a co-solvent. Do not confuse it with the elimination reaction that takes place in ethanol alone.

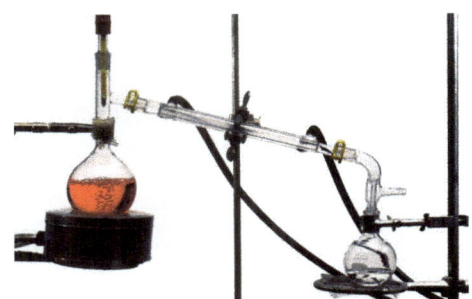

Figure 2 This kind of experimental apparatus is used to oxidise secondary alcohols to ketones. The alcohol is heated in potassium dichromate(VI) (left) and the resulting ketone is distilled into the round-bottomed flask on the right

5 Ketone to secondary alcohol (Chapter 17)

For example, propanone to propan-2-ol:

$$CH_3COCH_3 + 2[H] \rightarrow CH_3CH(OH)CH_3$$

The reducing agent is lithium tetrahydridoaluminate(III) (lithium aluminium hydride, $LiAlH_4$) in ether solution followed by addition of dilute acid; or sodium tetrahydridoborate(III) (sodium borohydride, $NaBH_4$) in aqueous solution.

6 Alcohol to alkene (Chapter 16)

For example, ethanol to ethene:

$$CH_3CH_2OH \rightarrow CH_2{=}CH_2 + H_2O$$

This dehydration can be done in two ways:

- by passing the heated alcohol vapour over a solid catalyst such as aluminium oxide
- by heating the alcohol with a dehydrating agent such as concentrated sulfuric acid or phosphoric(V) acid.

Phosphoric(V) acid is preferred as a dehydrating agent because sulfuric acid is also an oxidising agent and will oxidise some of the alcohol to carbon dioxide and water.

7 Alkene to alcohol (Chapter 14)

For example, ethene to ethanol:

$$CH_2{=}CH_2 + H_2O \rightarrow CH_3CH_2OH$$

Industrially ethene is reacted with steam with a catalyst of phosphoric acid, H_3PO_4.

8 Alcohol to ester (Chapters 16 and 33)

For example, ethanol to ethyl ethanoate:

$$CH_3CH_2OH + CH_3COCl \rightarrow CH_3COOCH_2CH_3 + HCl$$

Alcohols will react readily with acyl chlorides

Or

$$CH_3CH_2OH + CH_3COOH \rightleftharpoons CH_3COOCH_2CH_3 + H_2O$$

Alcohols will react with carboxylic acids. Concentrated sulfuric acid is used as a catalyst but the reaction does not go to completion.

9 Ester to alcohol (Chapter 33)

For example, ethyl ethanoate to ethanol:

$$CH_3COOCH_2CH_3 + H_2O \rightarrow CH_3CH_2OH + CH_3COOH$$

This is the reverse of reaction 8 above.

10 Acyl chloride to ester (Chapter 16)

For example, ethanoyl chloride to ethyl ethanoate:

$$CH_3COCl + CH_3CH_2OH \rightarrow CH_3COOCH_2CH_3 + HCl$$

Acyl chlorides (acid chlorides) will react readily with alcohols.

11 Ester to carboxylic acid

For example, ethyl ethanoate to ethanoic acid:

$$CH_3COOCH_2CH_3 + H_2O \rightarrow CH_3COOH + CH_3CH_2OH$$

This hydrolysis uses dilute sulfuric acid or sodium hydroxide (in which case the sodium salt of the carboxylic acid is formed and is followed by acidification to form the acid itself).

12 Carboxylic acid to ester (Chapter 18)

For example, ethanoic acid to ethyl ethanoate:

$$CH_3COOH + CH_3CH_2OH \rightleftharpoons CH_3COOCH_2CH_3 + H_2O$$

This reversible reaction needs a strong acid catalyst and produces an equilibrium mixture. It can be driven to the right by using concentrated sulfuric acid as the catalyst. Concentrated sulfuric acid is a powerful dehydrating agent and removes water thus moving the equilibrium to the right and favouring the production of the ester.

13 Alcohol to carboxylic acid (Chapter 16)

For example, ethanol to ethanoic acid:

$$CH_3CH_2OH + 2[O] \rightarrow CH_3COOH + H_2O$$

This oxidation takes place under more vigorous conditions than the oxidation to an aldehyde. Excess potassium dichromate(VI) is used under reflux with concentrated sulfuric acid to make sure that all the ethanol is oxidised to the carboxylic acid rather than the aldehyde. The acid can be removed from the reaction mixture by distillation.

14 Carboxylic acid to alcohol (Chapter 18)

For example, ethanoic acid to ethanol:

$$CH_3COOH + 4[H] \rightarrow CH_3CH_2OH + H_2O$$

The reducing agent is lithium tetrahydridoaluminate(III) (lithium aluminium hydride, $LiAlH_4$) in ether solution followed by addition of dilute acid.

15 Aldehyde to carboxylic acid (Chapter 17)

For example, ethanal to ethanoic acid:

$$CH_3CHO + [O] \rightarrow CH_3COOH$$

This oxidation takes place with acidified potassium dichromate(VI) under reflux followed by distillation to separate the acid produced.

16 Acyl chloride to carboxylic acid (Chapter 33)

For example, ethanoyl chloride to ethanoic acid:

$$CH_3COCl + H_2O \rightarrow CH_3COOH + HCl$$

Acyl chlorides react readily with water at room temperature.

17 Carboxylic acid to acyl chloride (Chapter 33)

For example, ethanoic acid to ethanoyl chloride:

Three reagents will bring about this conversion:

a Phosphorus pentachloride:

$$CH_3COOH + PCl_5 \rightarrow CH_3COCl + POCl_3 + HCl$$

This reaction occurs readily at room temperature.

b Phosphorus trichloride:

$$3CH_3COOH + PCl_3 \rightarrow 3CH_3COCl + H_3PO_3$$

This also takes place at room temperature but less vigorously than the reaction with phosphorus pentachloride.

c Thionyl chloride:

$$CH_3COOH + SOCl_2 \rightarrow CH_3COCl + SO_2 + HCl$$

This also takes place at room temperature. It is preferred, since both by-products are gases and are easily removed from the reaction mixture.

18 Acyl chloride to amide (Chapter 33)

For example, ethanoyl chloride to ethanamide:

$$CH_3COCl + NH_3 \rightarrow CH_3CONH_2 + HCl$$

The reaction takes place at room temperature.

19 Nitrile to carboxylic acid (Chapter 18)

For example, ethanenitrile to ethanoic acid:

$$CH_3C \equiv N + 2H_2O + HCl \rightarrow CH_3COOH + NH_4Cl$$

The nitrile is refluxed with dilute hydrochloric acid.

20 Nitrile to amine (Chapter 34)

For example, ethanenitrile to ethylamine:

$$CH_3C \equiv N + 4[H] \rightarrow CH_3CH_2NH_2$$

Three reducing agents can be used:

- lithium tetrahydridoaluminate(III) (lithium aluminium hydride, $LiAlH_4$) in ether solution followed by addition of dilute acid
- sodium tetrahydridoborate(III) (sodium borohydride, $NaBH_4$) in aqueous solution
- hydrogen with a nickel catalyst.

21 Halogenoalkane to nitrile (Chapter 15)

For example, bromoethane to propanenitrile:

$$CH_3CH_2Br + KC \equiv N \rightarrow CH_3CH_2C \equiv N + KBr$$

The halogenoalkane is refluxed with a solution of potassium cyanide in ethanol.

This reaction increases the carbon chain length by one.

22 Halogenoalkane to amine (Chapter 15)

For example, bromoethane to ethylamine:

$$CH_3CH_2Br + NH_3 \rightarrow CH_3CH_2NH_2 + HBr$$

The amine is heated under pressure in ethanol solution in a sealed tube (to stop the ammonia gas escaping). The HBr reacts with excess ammonia to form ammonium bromide.

An excess of halogenoalkane gives mostly primary amine, but secondary and tertiary amines will also form along with quaternary ammonium salts.

Exam tip

This reaction is not useful in synthesis because of the mixture of products.

23 Alkene to halogenoalkane (Chapter 14)

For example, ethene to bromoethane:

$$CH_2=CH_2 + HBr \rightarrow CH_3CH_2Br$$

Other hydrogen halides will also react.

With longer chain alkenes, such as propene, two isomeric products are possible—1-bromopropane and 2-bromopropane. Markovnikov's rule applies and the main product is the one where the hydrogen atom bonds to the carbon atom which already has the most hydrogen atoms, ie 2-bromopropane.

Halogens, eg Br_2, will also react to give dihalogenoalkanes, for example:

$$CH_3CH=CH_2 + Br_2 \rightarrow CH_3CHBrCH_2Br$$

The two halogen atoms end up on adjacent carbon atoms.

24 Alcohol to aldehyde (Chapter 16)

For example, ethanol to ethanal:

$$CH_3CH_2OH + [O] \rightarrow CH_3CHO + H_2O$$

Gentle oxidation with excess alcohol, dilute acid and potassium dichromate is carried out. The aldehyde is distilled off as it is formed to prevent further oxidation to carboxylic acid.

25 Alkane to alkene (Chapter 14)

For example, hexane to ethene:

$$CH_3CH_2CH_2CH_2CH_2CH_3 \rightarrow CH_2=CH_2 + CH_3CH_2CH_2CH_3$$

This is called cracking and is done by heating with a suitable catalyst. A variety of different products is formed as the alkane chain may break anywhere.

Exam tip

The equations above are written using structural formulae. Practise writing them using displayed formulae and skeletal formulae.

26 Amide to carboxylic acid (Chapter 34)

For example, ethanamide to ethanoic acid:

$$CH_3CONH_2 + H_2O \rightarrow CH_3COOH + NH_3$$

This can be done in acid conditions or alkaline conditions (in which case the salt of the carboxylic acid is formed).

27 Amide to primary amine (Chapter 34)

For example, ethanamide to ethylamine:

$$CH_3CONH_2 + 4[H] \rightarrow CH_2CH_2NH_2 + H_2O$$

The reducing agent is lithium tetrahydridoaluminate(III) (lithium aluminium hydride, $LiAlH_4$) in ether, followed by the addition of dilute acid.

Synthetic routes

In Section 21.1, we devised a two-step route to convert 1-bromopropane into propanoic acid, and a three-step route to make propylamine from ethene. We will look again at these routes more closely.

Propan-1-ol into propanoic acid

Step 1 $CH_3CH_2CH_2Br$ —— heat with NaOH(aq) ——> $CH_3CH_2CH_2OH$
                1-bromopropane   propan-1-ol

Step 2 $CH_3CH_2CH_2OH$ —— heat with $H^+/K_2Cr_2O_7$(aq) ——> $CH_3CH_2CO_2H$
                propan-1-ol   propanoic acid

Step 1 may be problematic. Two products are possible from the reaction of a halogenoalkane with sodium hydroxide:

- substitution of the halogen atom by the OH group to give propan-1-ol, or
- elimination of water to give propene.

To produce the largest percentage of the required substitution product, we must tailor the conditions. A warm (rather than hot) temperature and aqueous sodium hydroxide (rather than an ethanol solution of sodium hydroxide) are the conditions that favour substitution, so we must choose these.

Step 2 can also give two possible products. The oxidation could stop at propanal or go further to propanoic acid. To ensure that we get propanoic acid, we reflux the mixture so that the aldehyde will drip back into the reaction mixture and react further. We also use excess potassium dichromate so that there is more than enough oxidising agent to convert all the alcohol into the carboxylic acid. Finally ethanoic acid is distilled from the reaction mixture.

Ethene to propylamine

Step 1 $CH_2=CH_2$ —— HBr ——> CH_3CH_2Br
                ethene   bromoethane

Step 2 CH_3CH_2Br —— KCN in ethanol ——> $CH_3CH_2C≡N$
                bromoethane   propanenitrile

Step 3 $CH_3CH_2C≡N$ —— Ni/H_2 ——> $CH_3CH_2CH_2NH_2$
                propanenitrile   propylamine

Step 1 The hydrogen bromide is generated in the reaction vessel by the reaction of potassium bromide with concentrated sulfuric acid. Ethene is a gas and can be bubbled through the reaction mixture.

Step 2 Sodium cyanide is dissolved in ethanol to avoid substitution by OH^- ions instead of CN^- that would take place if an aqueous solution were used.

Step 3 The nickel catalyst is a solid and so is easily separated from the liquid product.

Aromatic compounds (arenes)

Aromatic compounds (those based on benzene rings) have their own characteristic reactions. These are summarised in Figure 3.

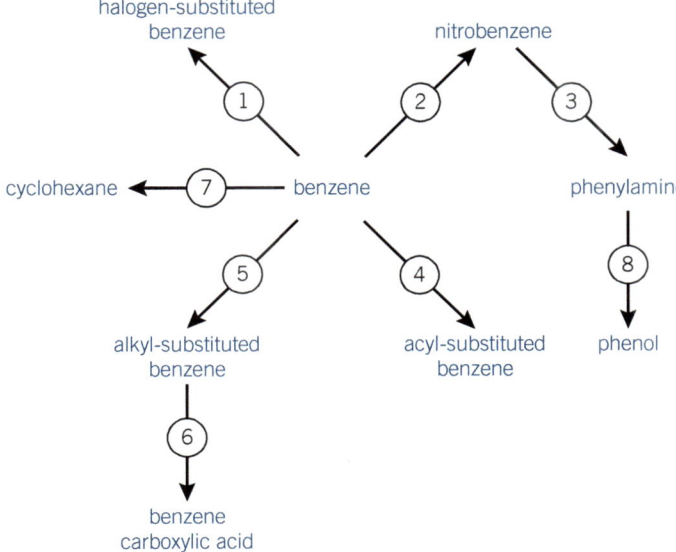

Figure 3 *Some important reactions of arenes*

You will need to know the reagents that convert one functional group into another along with reaction conditions such as temperature, pressure, catalysts, solvents, and excess of reagents if appropriate. Examples are summarised below; further details can be found in the appropriate chapter(s). These reactions are those of the benzene ring. Substituted arenes will also undergo reactions of substituents attached to alkyl side chains, although these may be modified to some extent by any interaction with the arene ring.

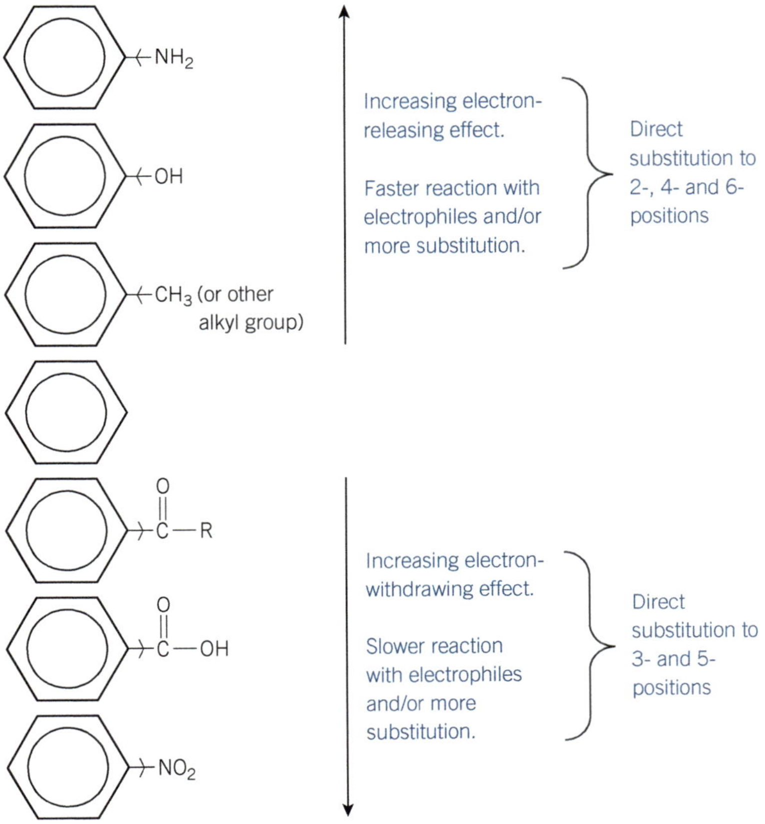

Exam tip

In all of the reactions in Figure 3 (except **7**) the delocalised aromatic system remains intact.

NH$_2$

OH

CH$_3$ (or other alkyl group)

Increasing electron-releasing effect.

Faster reaction with electrophiles and/or more substitution.

Direct substitution to 2-, 4- and 6- positions

C—R

C—OH

NO$_2$

Increasing electron-withdrawing effect.

Slower reaction with electrophiles and/or more substitution.

Direct substitution to 3- and 5- positions

Figure 4 *The effect of substituents on the benzene ring*

Figure 4 summarises the effect of various substituents on the activity of the benzene ring towards electrophilic substitution. **Activating substituents** produce more substitution as well as making the reactions faster. They direct further substitution to the 2-, 4-, and 6-positions. **Deactivating substituents** direct further substitution to the 3- and 5-positions. The exception is halogens which deactivate the ring but direct to the 2- and 4-positions.

1 Benzene to halogen-substituted benzene (Chapter 30)
For example, benzene to bromobenzene:

$$\text{benzene} + Br_2 \longrightarrow \text{bromobenzene(Br)} + HBr$$

This requires liquid bromine and a catalyst of aluminium bromide ($AlBr_3$).

2 Nitration (Chapter 30)
For example, benzene to nitrobenzene:

$$\text{benzene} + HNO_3 \longrightarrow \text{nitrobenzene} + H_2O$$

A mixture of concentrated nitric and sulfuric acids is used.

3 Nitrobenzene to phenylamine (Chapter 34)

$$\text{nitrobenzene} + 6[H] \longrightarrow \text{phenylamine}(NH_2) + 2H_2O$$

The reducing agent is tin and concentrated hydrochloric acid.

 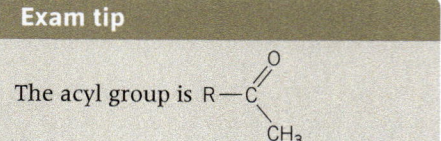
4 Benzene to acyl-substituted benzene (Chapter 30)
For example, benzene to phenylethanone:

$$\text{benzene} + CH_3COCl \longrightarrow \text{phenylethanone}(COCH_3) + HCl$$

A catalyst of aluminium chloride ($AlCl_3$) is used.

5 Benzene to alkyl-substituted benzene (Chapter 30)
For example, benzene to methylbenzene:

$$\text{benzene} + CH_3Cl \longrightarrow \text{methylbenzene}(CH_3) + HCl$$

A catalyst of aluminium chloride ($AlCl_3$) is used.

6 Alkyl-substituted benzenes to benzenecarboxylic acids (Chapter 30)
For example, methylbenzene to benzoic acid:

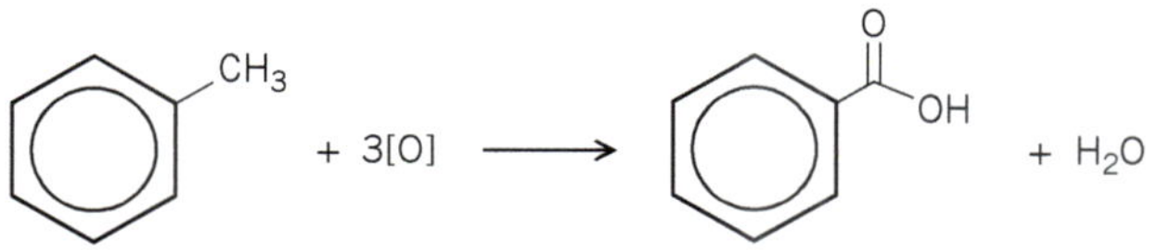

The oxidising agent is hot alkaline potassium manganate(VII) (potassium permanganate, $KMnO_4$) followed by acidification to produce benzoic acid itself. Longer alkyl side chains will also be oxidised to benzoic acid.

7 Benzene to cyclohexane (Chapter 30)

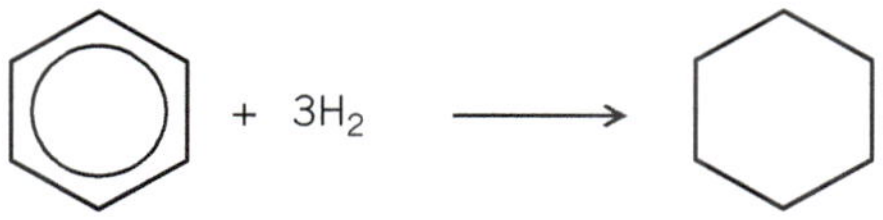

The reducing agent is hydrogen with a nickel/platinum catalyst.

Exam tip

Remember that cyclohexane has two hydrogen atoms on each carbon atom.

8 Phenylamine to phenol (Chapter 32)

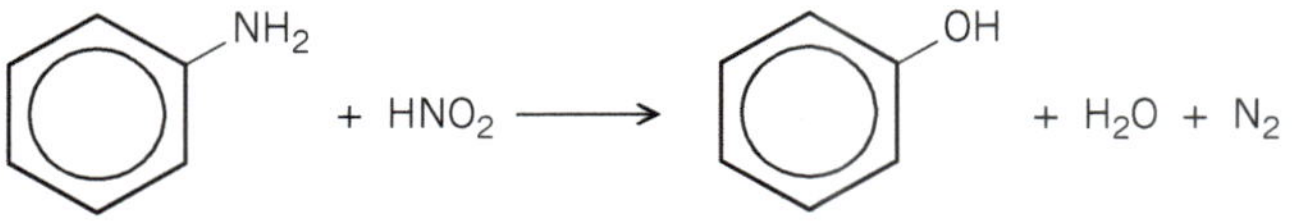

The nitrous acid is made in the reaction vessel from sodium nitrite ($NaNO_2$) and hydrochloric acid (HCl). The initial reaction is carried out at low temperature and the resulting diazonium salt is allowed to warm up.

Summary test 36.1

1 Explain how you could change propan-1-ol into propan-2-ol.

2 a Identify the functional groups (**i–v**) ringed in the molecule below, which is an artificial sweetener:

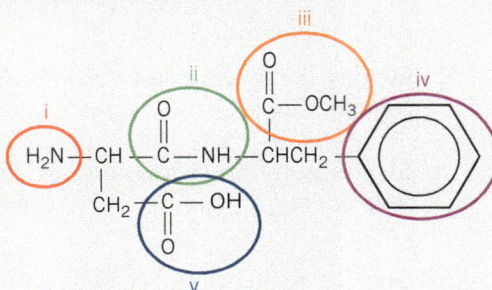

b For each functional group give two characteristic reactions that it will undergo.

3 Explain how you would convert in three steps:

a propan-1-ol to propanone

b ethane to ethane-1,2-dioyl chloride.

4 For each of the steps below, give the name for the type of reaction taking place and the reagents required.

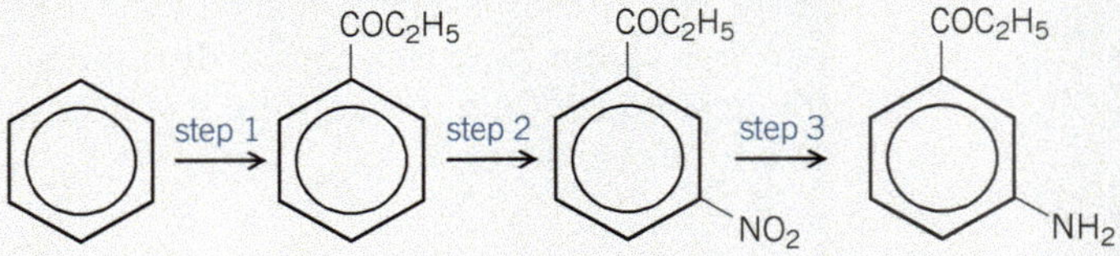

> Launch additional digital resources for the chapter

1 Describe a two-step synthesis for each conversion below. For each step, name the reagent required and the conditions of the reaction. Identify the organic compound involved in each of the conversions.

a ethene to ethyl butanoate

b phenol to phenylamine

c propan-2-ol to propane. *(9 marks)*

2 Cinnamaldehyde is an organic compound that is present in the bark of cinnamon trees. Its formula is shown below:

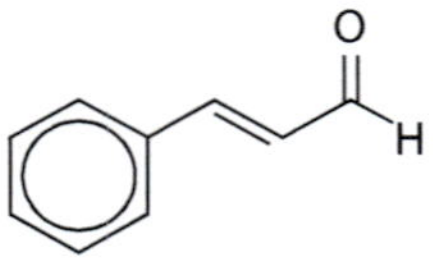

a Name the two functional groups in cinnamaldehyde. *(2 marks)*

b A sample of cinnamaldehyde is added to bromine water.
   i Predict which functional group will react. *(1 mark)*
   ii Predict the colour change that would occur in this reaction. *(1 mark)*

c A separate sample of cinnamaldehyde is added to Tollens' reagent, which contains Ag^+ complexed in alkaline solution.
   i State what, if anything, would be observed. *(1 mark)*
   ii Draw the structural formula of the organic product of the reaction. *(1 mark)*

d Cinnamyl alcohol smells of hyacinth flowers. It is used in perfumes and deodorants. The compound occurs naturally in only small amounts, so it is synthesised industrially from cinnamaldehyde.
   i The IUPAC name of cinnamyl alcohol is 3-phenylprop-2-en-1-ol. Draw its structural formula. *(1 mark)*
   ii Suggest a reagent and give the conditions needed to convert cinnamaldehyde to cinnamyl alcohol. *(2 marks)*

3 Chloromethane can be converted to ethyl methanoate by a three-step synthesis. The three steps are outlined below:

Step 1 Chloromethane is heated with dilute sodium hydroxide solution to form compound **A**.

Step 2 Compound **A** is oxidised to compound **B**.

Step 3 Compound **B** reacts with reagent **C** to make ethyl methanoate

a Give the structural formula of ethyl methanoate *(1 mark)*

b Name compound **A**. *(1 mark)*

c State the mechanism of the reaction that converts chloromethane to compound **A**. *(3 marks)*

d Name compound **B** and give its formula. *(2 marks)*

e Name the reagent and conditions required to convert compound **A** to compound **B**. *(2 marks)*

f Name reagent **C** and give its formula. *(2 marks)*

g Describe the conditions required for the reaction of compound **B** with reagent **C** to make ethyl methanoate. *(1 mark)*

4 Ethene can be converted into propanoic acid by a three-step synthesis. The steps are partly shown below:

Step 1 Ethene → bromoethane

Step 2 Heat bromoethane with KCN in dilute sulfuric acid to make **X**.

Step 3 Convert **X** to propanoic acid.

a Name the reagent required to convert ethene to bromoethane. *(1 mark)*

b Name compound **X**. *(1 mark)*

c Identify the reagent and conditions required to convert **X** to propanoic acid. *(2 marks)*

5 a Give reagents and conditions for the conversion of propene to propanone in a two-step synthesis. *(3 marks)*

b Give the reagents and conditions for the three-step synthesis of butanoic acid from propan-1-ol. *(5 marks)*

c Give the structure of a by-product formed from the first stage of the reaction described in part **a**), explain how this is formed and why it is only present in small amounts. *(3 marks)*

d Explain how the substance given in your answer to part **c** reacts in the second stage of the reaction of part **a**. Explain how this can be distinguished from the desired product, propanone using simple test-tube reactions. *(3 marks)*

6 The reaction scheme below shows a series of reactions involving ethanol. Copy and complete the diagram below to show the structures of compounds **A** and **C** and identify the conditions required for Reaction **2**. Name Compound **B**.

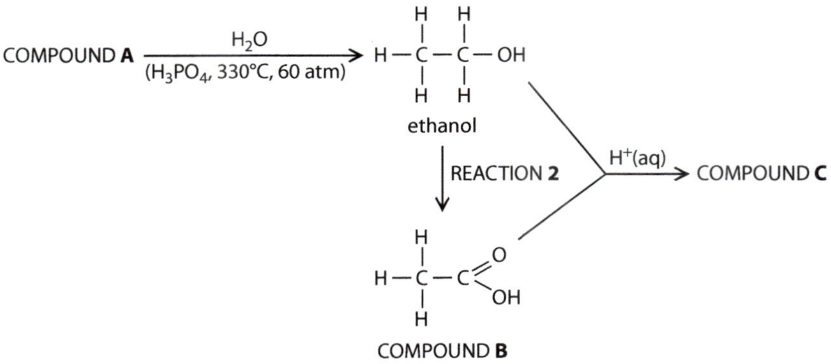

(4 marks)

7 a Outline a three-step synthesis for the formation of the N-substituted amide $C_5H_{11}NO$, using propene as the starting material.

For each reaction, give the reagents, conditions and structure of the product, naming the type of mechanism involved. *(12 marks)*

b Give the mechanism for the first step in this synthesis. Name the product formed. *(5 marks)*

8 Salicylic acid is a naturally occurring compound found in willow tree bark. For many years, salicylic acid was used as a natural painkiller. Salicylic acid was then used to synthesise acetylsalicylic acid, also known as aspirin. Aspirin was found to be more effective at treating pain. The structures of both molecules are given below:

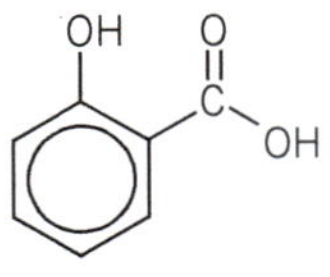

salicylic acid 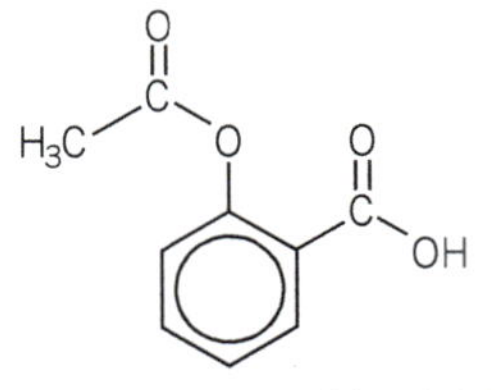   acetylsalicylic acid (aspirin)

a Suggest a suitable reagent and conditions for the synthesis of aspirin from salicylic acid. *(2 marks)*

b Using your answer to part **a**, draw the mechanism for this reaction and name this type of reaction. *(4 marks)*

9 Propanal reacts with reagent **Q** to produce a racemic mixture of compound **R**. Compound **R** undergoes an elimination reaction to produce compound **S** which exists as a pair of geometrical isomers.

a Define the term *racemic mixture* and predict the structures of the two enantiomers of compound R. *(3 marks)*

b Identify reagent Q and name the mechanism by which compound R is formed. *(2 marks)*

c Suggest suitable reagents and conditions for the formation of compound S and give a mechanism for this reaction. *(4 marks)*

d i Suggest structures for each geometrical isomer of compound S.

ii Predict, if at all, which of the structures in part **d i** would be the major product of this reaction. *(3 marks)*

Thin-layer chromatography

You will be familiar with paper chromatography, which is often used to separate the mixture of dyes in, for example, felt-tip pens.

Chromatography describes a whole family of separation techniques. They all depend on the principle that a mixture can be separated by a liquid or gas (the mobile phase) moving through a solid (the stationary phase) and by carrying the components of the mixture at different rates.

- The moving or mobile phase carries the soluble components of the mixture with it. The more soluble that the component is in the mobile phase, the faster it moves. The solvent in the moving phase is sometimes called the **eluent**. The solvent may be polar in character like water or ethanol, or it may be non-polar (for example, a hydrocarbon).
- The stationary phase will hold back the components in the mixture that are attracted to it. The more affinity that a component in the mixture being separated has for the stationary phase, the slower it will move with the solvent.

So, if suitable moving and stationary phases are chosen, a mixture of similar substances can be separated completely. This is because every component of the mixture has a unique balance between its affinity for the stationary phase and its solubility in the mobile phase. In fact, chromatography is often the only way that very similar components of a mixture can be separated.

TLC

TLC is an abbreviation for **thin-layer chromatography**, which is a development of paper chromatography. The filter paper is replaced by a glass, metal, or plastic sheet coated with a thin layer of silica gel (silicon dioxide, SiO_2) or alumina (aluminium oxide, Al_2O_3) which acts as the stationary phase. These are often called plates. Plastic- and metal-backed sheets can be cut to size with scissors.

TLC has several advantages over paper chromatography:

- it runs faster
- smaller amounts of mixtures can be separated
- the spots usually spread out less
- the plates are more robust than paper.

When the chromatogram has run, the position of colourless spots may have to be located by shining ultraviolet light on the plate, or chemically by spraying the plate with a locating agent which reacts with the components of the mixture to give coloured compounds.

Example

Thin-layer chromatography is often used to separate compounds that are very similar to each other, such as amino acids.

1 A small spot containing the mixture of amino acids to be separated is placed on a **baseline** about 1 cm up the plate. The plate is then placed in a tank containing a suitable solvent to a depth of about 0.5 cm. The baseline must be above the initial level of the solvent. The solvent (or mixture of solvents) is called the mobile phase (or eluent).

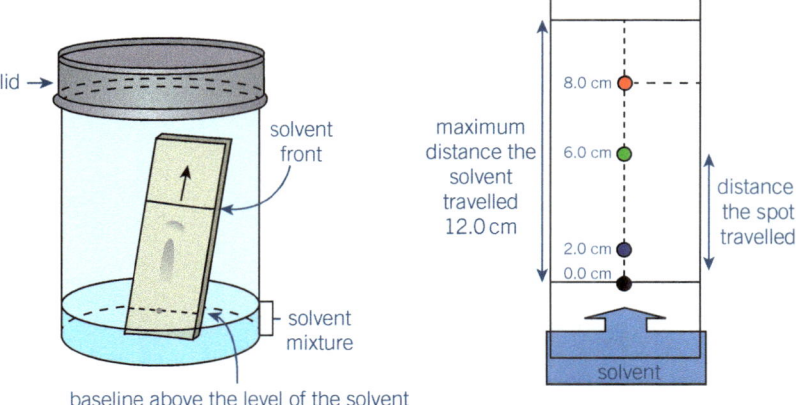

Figure 1 A thin-layer chromatography experiment and the chromatogram that results

2 A lid is placed on the tank so that the inside of it is saturated with solvent vapour and the solvent is allowed to rise up the plate (Figure 1). As it does so, it carries the amino acids with it. Each amino acid lags behind the **solvent front** to an extent that depends on its affinity for the solvent compared with its affinity for the stationary phase. This depends on the intermolecular forces that act between the amino acid and the solvent—the stronger they are, the closer the amino acid is to the solvent front.

3 When the solvent has almost reached the top of the plate, the plate is removed from the tank and the position to which the solvent front has moved is marked. Amino acids are colourless, so the positions they have reached have to be made visible. This is done by spraying the plate with a developing agent such as ninhydrin, which reacts with amino acids to form a purple compound, or by shining ultraviolet light on the plate. If the solvent is suitable, the amino acids will be completely separated.

R_f **values** are then calculated for each amino acid spot:

$$R_f = \frac{\text{distance moved by the spot}}{\text{distance moved by the solvent}}$$

For example:

The red spot has moved 8.0 cm; the solvent has moved 12 cm. So the R_f value for the red spot in Figure 1 is $\dfrac{8.0\,\text{cm}}{12.0\,\text{cm}} = 0.67$

This allows each amino acid in the mixture to be identified by comparing the R_f value of each spot with the values obtained by known pure amino acids run in the *same* solvent mixture.

The R_f values can be used to help identify each component, because for a given solvent a component will have a particular R_f value.

Column chromatography

Column chromatography uses a powder, such as silica, aluminium oxide, or a resin, as the stationary phase. This is packed into a narrow tube—the column—and a solvent (the eluent) is added at the top (Figure 2). As the eluent runs down the column, the components of the mixture move at different rates and can be collected separately in flasks at the bottom. More than one eluent may be used to get a better separation. This method has the advantage that fairly large amounts can be separated and collected. For example, a mixture of amino acids can be separated into its pure components by this method.

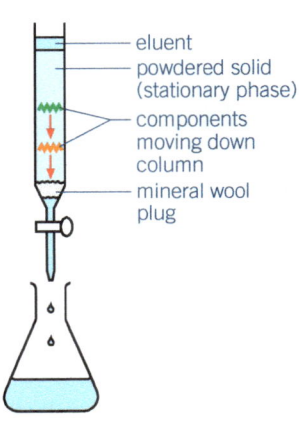

Figure 2 Column chromatography

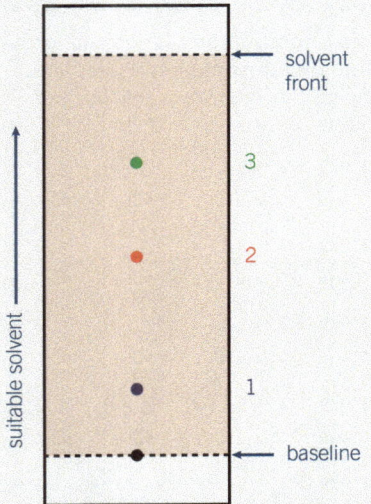

Learning outcomes

On these pages you will learn to:

- explain the terms stationary phase; mobile phase; retention time
- interpret gas/liquid chromatograms

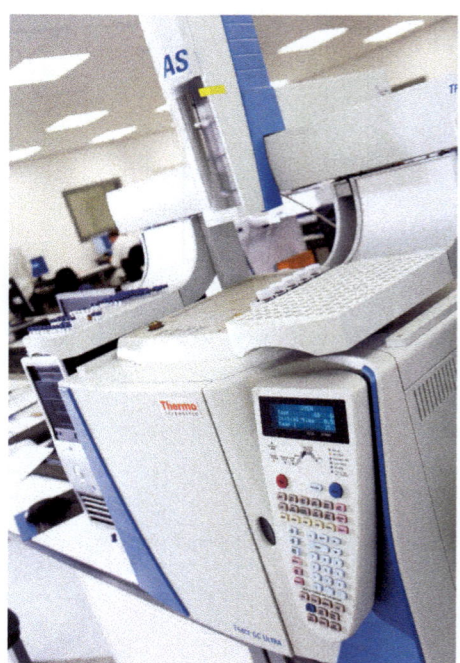

Figure 5 A gas chromatography instrument

Gas/liquid chromatography (GC) is one of the most important modern analytical techniques. The basic apparatus is shown in Figure 3.

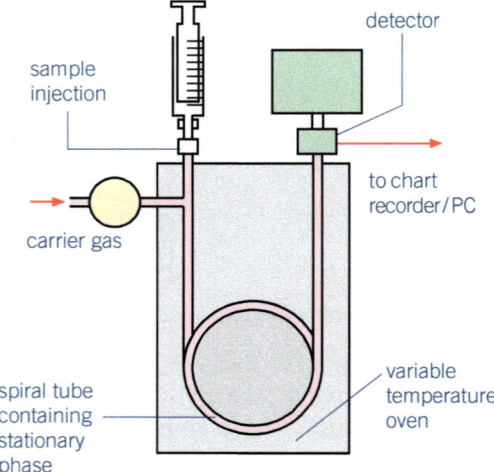

Figure 3 Gas/liquid chromatography (GC)

The stationary phase is a non-polar liquid with a high boiling point. It is held in position by a powder which is packed in or coated onto the inside of a long capillary tube. The tube is up to 100 m long and less than 0.5 mm in diameter. It is coiled up and placed in an oven whose temperature can be varied.

The mobile phase is usually an unreactive gas, such as nitrogen or helium. After injection, the sample is carried along by the gas. The mixture separates as some of the components move along with the gas and some are retained by the stationary phase, each to a different extent. This means that the components leave the column at different times after injection – they have different **retention times**.

Various types of detectors are used, including ones that measure the thermal conductivity of the emerging gas. The results may be presented on a graph (Figure 4). The area under each peak is proportional to the amount of that component. So, the percentage of each component is the area under its peak divided by the total area under all the peaks. Modern instruments will calculate this automatically.

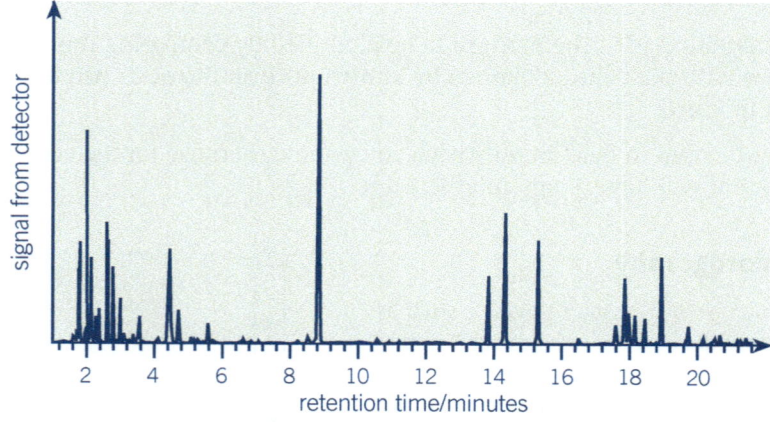

Figure 4 Typical GC trace—each peak represents a different component

In some instruments the components are fed directly into a mass spectrometer, infrared spectrometer, or NMR spectrometer, see Section 37.3, for identification. Today the whole process is automated and computer controlled (Figure 5).

As an analytical method for separating mixtures, GC is extremely sensitive. It can separate minute traces of substances in foodstuffs, and even link crude oil pollution found on beaches with its tanker of origin by comparing oil samples. Perhaps its best-known use is for testing athletes' blood or urine for banned substances.

The identification of a component is done by matching its retention time with that of a known substance under the same conditions. This is then confirmed by comparing the mass spectra of the two substances.

The retention time of a compound depends on its interaction with the mobile phase compared with its interaction with the stationary phase. A polar compound will move faster (ie have a shorter retention time) if the mobile phase is polar and the stationary phase is non-polar.

Worked example

The gas chromatogram on the right shows the separation of some pollutants found in a sample of air.

1 Which pollutant came off the column first?
2 Which pollutant had the longest retention time?
3 Which pollutant was present in the smallest amount?
4 Which pollutant was present in the largest amount?
5 Which two pollutants appear to be present in approximately the same quantities?
6 Why can the quantities not be compared exactly?

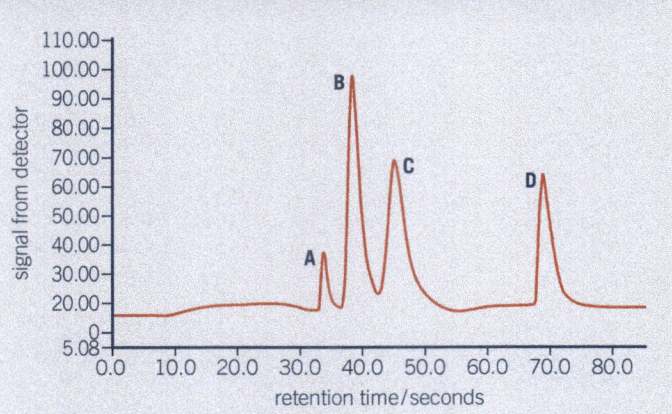

Solutions

1 A   2 D   3 A
4 B   5 C and D
6 The quantities are related to the area under the peaks (which is hard to estimate by eye), not the peak height.

Extension

HPLC stands for high pressure liquid chromatography or high performance liquid chromatography. Both names are appropriate. Here the mixture to be separated is forced through a column containing the stationary phase by a solvent driven by a high pressure pump. It is similar to column chromatography except that the pump drives the solvent (the eluent) rather than gravity. A variety of materials can be used as the stationary phase including chiral ones that can separate optical isomers. A variety of detection methods can be used, for example, the absorption of ultraviolet light.

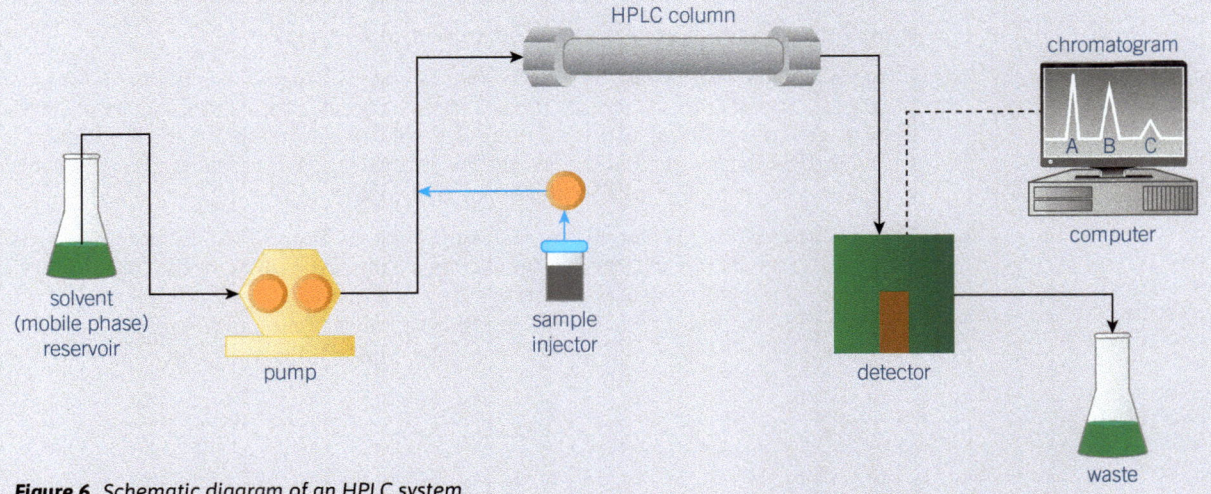

Figure 6 *Schematic diagram of an HPLC system*

Summary test 37.2

1 Explain why GC is so important in forensic detective work. Give a possible example not in the text.
2 From the chromatogram in Figure 6 above, identify from A, B, and C:

a the most abundant component in the mixture
b the one with the greatest affinity for the solid phase
c the one with the greatest affinity for the mobile phase
d the one with the greatest retention time.

Learning outcomes

On these pages you will learn to:

- analyse and interpret ^{13}C NMR spectra of simple molecules
- predict the number of peaks in a ^{13}C NMR spectrum of a given molecule

Nuclear magnetic resonance spectroscopy (NMR) is used particularly in organic chemistry. It is a powerful technique that can help us find the structures of even quite complex molecules. It has also revolutionised our ability to 'see' inside our body without using X-rays.

A magnetic field is applied to a sample, which is surrounded by a source of radio waves and a radio receiver. This generates an energy change in the nuclei of atoms in the sample that can be detected. Electromagnetic energy is emitted, which can then be interpreted by a computer.

Extension

A brief theory of NMR

Although you will only be examined on *interpreting* NMR spectra, this background reading may help you to understand how NMR works, although in some respects it is an oversimplification.

Many nuclei with odd mass numbers, such as ^{1}H, ^{13}C, ^{15}N, ^{19}F, and ^{31}P, have the property of spin (as do electrons). This gives them a magnetic field like that of a bar magnet.

If bar magnets are placed in an external magnetic field, they will line up parallel to the field (Figure 7a).

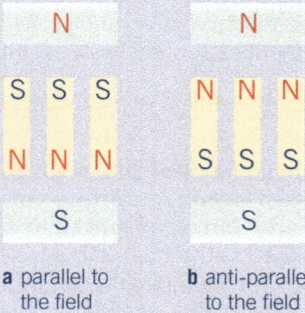

a parallel to the field   **b** anti-parallel to the field

Figure 7 *The two possible orientations of bar magnets in a magnetic field*

It is also possible that the bar magnets could line up anti-parallel to the field, as in Figure 7b. However this orientation has a higher energy, as the bar magnets have to be forced into position against the repulsion of the external magnetic field. The stronger the external magnetic field and the stronger the bar magnets, the larger the energy gap between the parallel and anti-parallel states.

Something similar applies to nuclei with spin, such as ^{1}H and ^{13}C. There will be some of the nuclei in each energy state but more of them will be in the lower (parallel) one. If electromagnetic energy just equal in energy to the difference between the two positions (ΔE in Figure 8) is supplied, some nuclei will flip between the parallel and anti-parallel positions. This is called resonance.

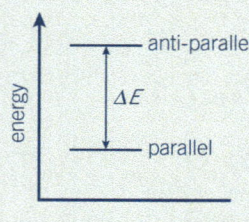

Figure 8 *Energy level diagram of the two orientations of bar magnets in a magnetic field*

The energy required to cause this is in the radio region of the electromagnetic spectrum. It is supplied by a radio frequency source, and the resonances are detected by a radio receiver (Figure 9). The frequency of the radio waves required to cause flipping for a particular magnetic field is called the resonant frequency of that atomic nucleus. A higher frequency corresponds to a larger energy gap between the two states. If the magnetic field is kept constant and the radio frequency gradually increased, different atomic nuclei will come into resonance at different frequencies depending on the strength of their atomic magnets.

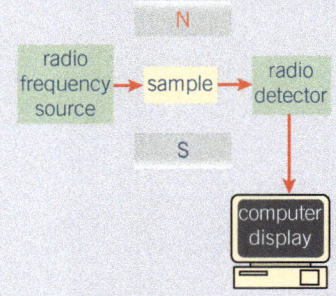

Figure 9 *Schematic diagram of an NMR spectrometer*

In fact, modern instruments use pulses of radio waves of a range of frequencies all at once, and analyse the response by a computer technique called Fourier transformation. However they use the same principle of finding the frequencies at which different nuclei resonate.

1 What do you notice about the relative atomic masses of the nuclei with spin?

Carbon-13, ^{13}C, NMR

NMR is most often used with organic compounds. Although carbon-12, ^{12}C, has no nuclear spin carbon-13, ^{13}C, does have one. Although only 1% of carbon atoms are carbon-13, modern instruments are sensitive enough to obtain a carbon-13 spectrum.

Not all the carbon-13 atoms in a molecule resonate at exactly the same magnetic field strength. Carbon atoms in different functional groups feel the magnetic field differently. This is because all nuclei are **shielded** from the external magnetic field by the electrons that surround them. Nuclei with more electrons around them are better shielded. The greater the electron density around a carbon-13 atom, the smaller the magnetic field felt by the nucleus and the lower the frequency at which it resonates. The NMR instrument produces a graph of energy absorbed (from the radio signal) vertically against a quantity called **chemical shift** (which is related to the resonant frequency) horizontally.

The chemical shift

Chemical shift δ is measured in units called parts per million (ppm) from a defined zero related to a compound called **tetramethylsilane, TMS** (see Section 37.4). Chemical shift is related to the difference in frequency between the resonating nucleus and that of TMS. In ^{13}C NMR values of δ range from 0 to around 200 ppm.

The main point about ^{13}C NMR is that carbon atoms in different environments will give different chemical shift values. Figure 11 overleaf shows the ^{13}C NMR spectrum of ethanol. It has two peaks, one for each carbon, because the carbon atoms are in different environments—one is further from the oxygen atom than the other. The oxygen atom, being electronegative, draws electrons away from the carbon atom to which it is directly bonded.

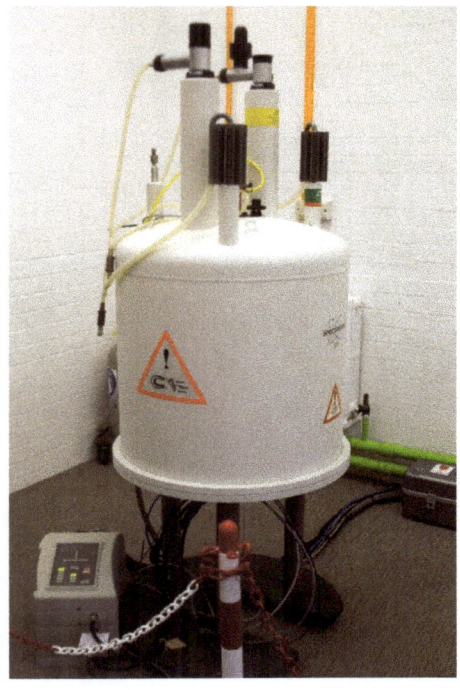

Figure 10 *Modern NMR instruments use electromagnets with superconducting coils to produce the strong magnetic fields required. The large white tank holds a jacket of liquid nitrogen surrounding an inner jacket of liquid helium which cools the magnet coils to 4 K*

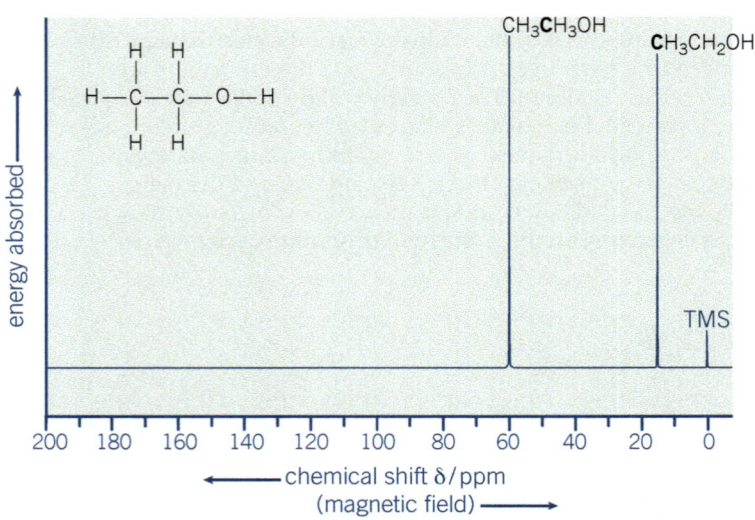

Figure 11 *Carbon-13 NMR spectrum of ethanol*

Table 1 *^{13}C chemical shift values*

Type of carbon	δ / ppm
$-\overset{\vert}{\underset{\vert}{C}}-\overset{\vert}{\underset{\vert}{C}}-$	0–50
$R-\overset{\vert}{\underset{\vert}{C}}-Cl$ or Br	30–60
$R-\overset{\vert}{\underset{\underset{O}{\parallel}}{C}}-\overset{\vert}{\underset{\vert}{C}}-$	30–65
$R-\overset{\vert}{\underset{\vert}{C}}-\overset{\vert}{C}=C\overset{/}{\underset{\backslash}{}}$	25–50
$-\overset{\vert}{\underset{\vert}{C}}-O-$ alcohols, ethers, or esters	50–70
$\overset{\backslash}{\underset{/}{C}}=C\overset{/}{\underset{\backslash}{}}$	110–160
$R-C\equiv N$	100–125
benzene ring	110–160
$R-\overset{}{\underset{\underset{O}{\parallel}}{C}}-$ esters or acids	160–185
$R-\overset{}{\underset{\underset{O}{\parallel}}{C}}-$ aldehydes or ketones	190–220

Table 1 shows values of ^{13}C chemical shifts for carbon atoms in a variety of environments. The carbon atom at δ = 60 ppm in the ethanol spectrum is the carbon bonded to the oxygen (CH_3**C**H_2OH), whilst that at δ = 15 ppm is the other carbon (**C**H_3CH_2OH).

This is because the electronegative oxygen atom draws electrons away from the carbon bonded to it CH_3**C**H_2OH. It is deshielded and feels a greater magnetic field and so resonates at a higher frequency and therefore has a *greater* δ value than the other carbon. The other carbon **C**H_3CH_2OH is surrounded by more electrons and therefore shielded and has a *smaller* δ value.

More examples of ^{13}C NMR spectra

Figures 12 and 13 show the ^{13}C NMR spectra of the isomers propanone, CH_3COCH_3, and propanal, CH_3CH_2CHO. In propanone, there are just two different environments for the carbon atoms—the two CH_3 groups and the C=O. The spectrum shows two peaks:

- At δ = 205 ppm due to the C=O.
- At δ = 30 ppm due to the CH_3 groups.

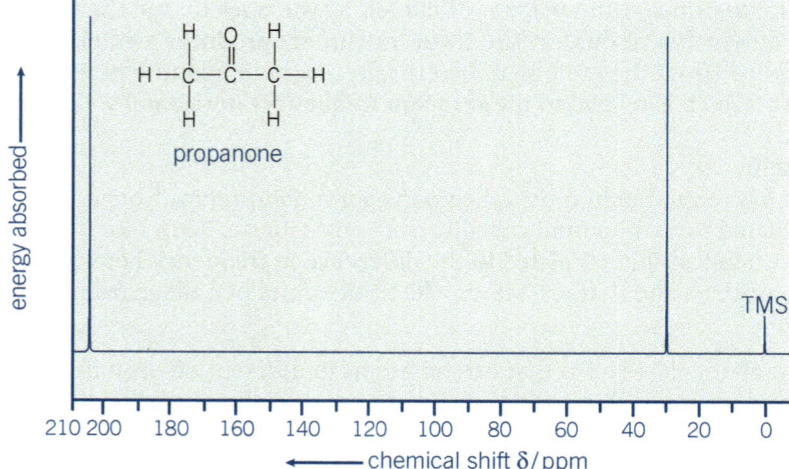

Figure 12 *^{13}C NMR spectrum of propanone*

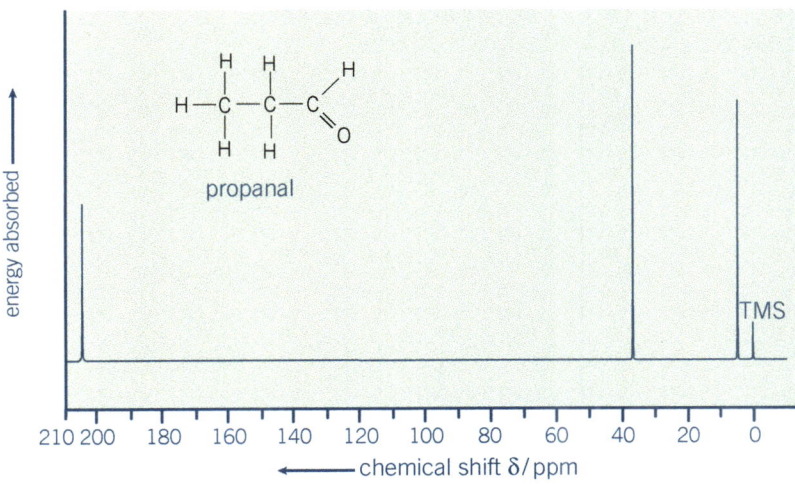

Figure 13 ^{13}C NMR spectrum of propanal

Propanal has three different carbon environments and so shows three peaks:

- The CH_3 group at $\delta = 5$ ppm.
- The CH_2 at $\delta = 37$.
- The CHO group at $\delta = 205$ ppm.

Worked example

Figure 14 shows the ^{13}C NMR spectrum of a compound of relative molecular mass 44. What can we deduce about it?

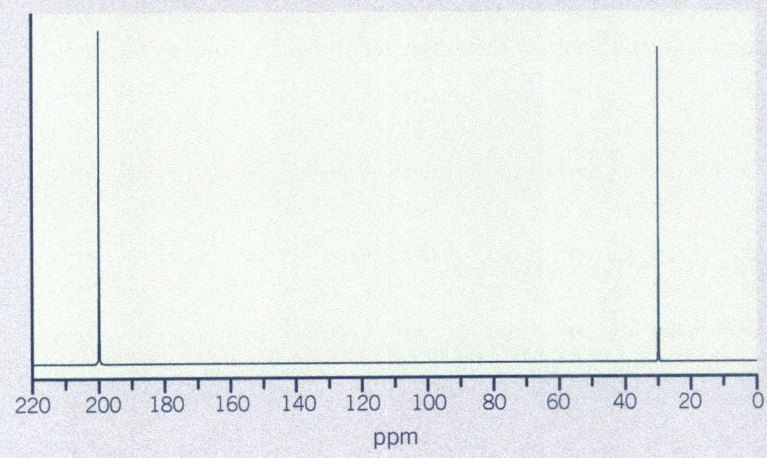

Figure 14

Firstly there are two peaks and therefore two environments for carbon atoms.

Looking at Table 1:

- The peak as 200 ppm is consistent with the C atom in aldehydes or ketones, ie the C atom of a carbonyl group, C=O.
- The peak at 30 ppm is consistent with a carbon atom next to a carbonyl group.

^{13}C NMR does not tell us about *how many* of each carbon atom there are, so the compound could, for example be ethanal CH_3CHO or propanone, CH_3COCH_3.

However, if the M_r is 44, it must be ethanal, rather than propanone, which has $M_r = 58$.

Notice how similar the spectrum is to that of propanone (on the previous page) and how it may be necessary to have some extra information such as the M_r to be help indentifying the compound.

> **Exam tip**
>
> The heights of the peaks in ^{13}C NMR spectra are not significant. It is their δ values that are important in interpreting spectra.

> **Exam tip**
>
> It will help to draw the displayed or structural formula of the two isomers.

Figures 15 and 16 show the ^{13}C NMR spectra of the isomers 1,2-dihydroxybenzene and 1,4-dihydroxybenzene.

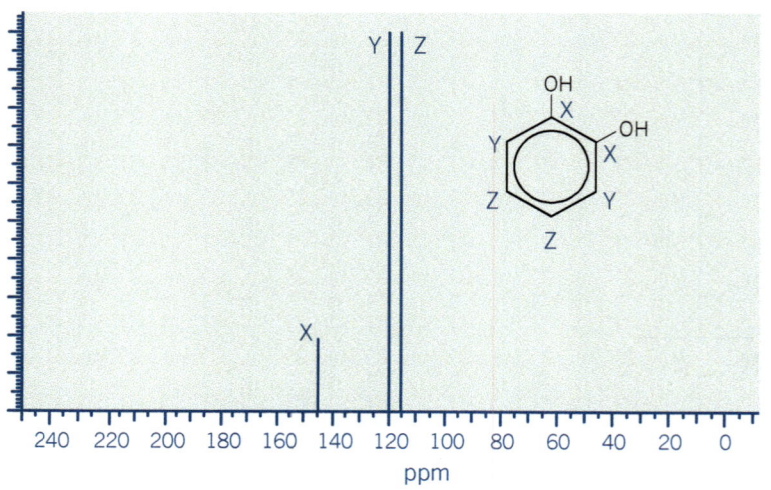

Figure 15 ^{13}C NMR spectrum of 1,2-dihydroxybenzene

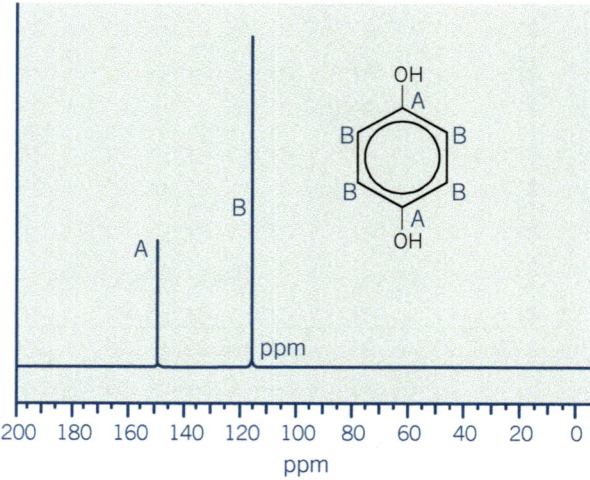

Figure 16 ^{13}C NMR spectrum of 1,4-dihydroxybenzene

Figure 15 shows three peaks because there are three different environments of carbon atoms in the molecule, depending on how far away they are from the OH groups.

Figure 16 has two peaks because there are only two environments for the carbon atoms—those bonded to the OH groups and those that are not.

Summary test 37.3

1 The ^{13}C NMR spectrum of ethanol is discussed above and has two peaks. Methoxymethane is an isomer of ethanol:

H—C—O—C—H

methoxymethane

  a State how many peaks you would expect to find in its ^{13}C NMR spectrum?
  b Explain your answer.
2 The ^{13}C NMR spectra of propan-1-ol and propan-2-ol are given below.
  State which is which and explain your answer.

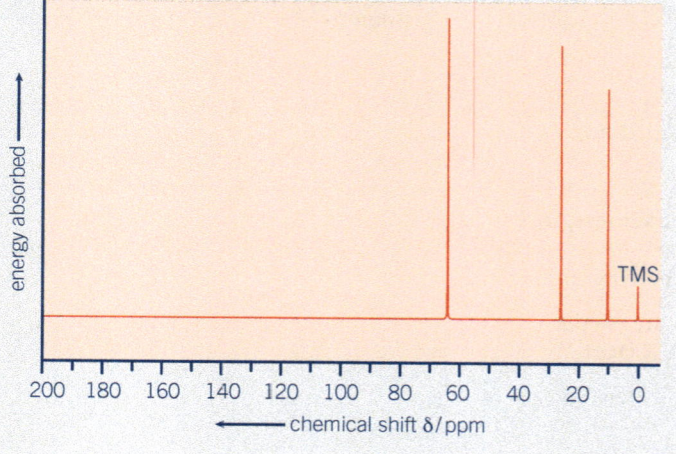

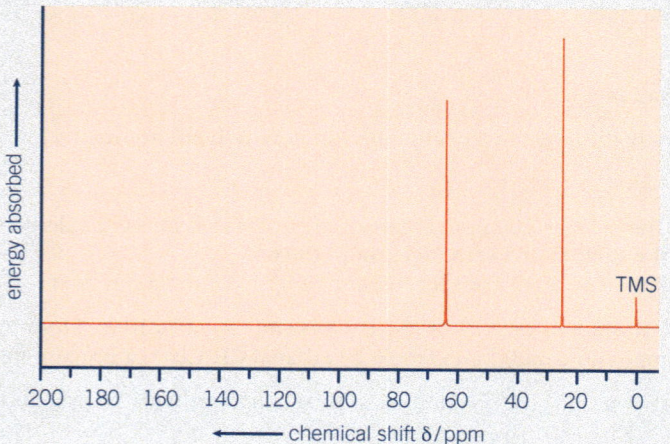

Proton (^{1}H) NMR spectroscopy

In **proton NMR**, it is the ^{1}H nucleus that is being examined. Nearly all hydrogen atoms are ^{1}H so it is easier to get an NMR spectrum for ^{1}H than for ^{13}C.

Here it is hydrogen atoms attached to different functional groups that feel the magnetic field differently, because all nuclei are shielded from the external magnetic field by the electrons that surround them. Nuclei with more electrons around them are better shielded. The greater the electron density around a hydrogen atom, the smaller the chemical shift δ. The values of chemical shift in proton NMR are smaller than those for ^{13}C NMR—most are between 0 and 10 ppm.

If all the hydrogen nuclei in an organic compound are in identical environments, you get only one chemical shift value. For example, all the hydrogen atoms in methane, CH_4, are in the same environment and have the same chemical shift.

But in a molecule like methanol there are hydrogen atoms in two different environments—the three on the carbon atom, and the one on the oxygen atom. The NMR spectrum will show the two environments (Figure 17).

Learning outcomes

On these pages you will learn to:

- analyse and interpret ^{1}H NMR spectra of simple molecules
- deduce chemical shifts and splitting patterns of molecules
- described how TMS, D_2O, and $CDCl_3$ are used in ^{1}H NMR

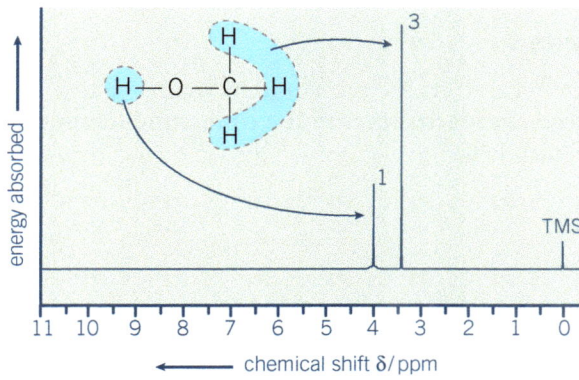

Figure 17 *The NMR spectrum of methanol—the peak areas are in the ratio 1 : 3*

In general, the further away a hydrogen atom is from an electronegative atom (such as oxygen) the smaller its chemical shift. In ethanol, CH_3CH_2OH, there are three values of δ.

In ^{1}H NMR the areas under the peaks (shown here by the numbers next to them) are proportional to the number of hydrogen atoms of each type—in this case three and one.

The integration trace

In proton NMR spectra, the area of each peak is related to the number of hydrogen atoms producing it. So, in the spectrum of methanol, CH_3OH, the CH_3 peak is three times the area of the OH peak. This can be difficult to evaluate by eye, so the instrument produces a line called the **integration trace**, shown in red in Figure 18 overleaf. The relative heights of the steps of this trace give the relative number of each type of hydrogen—3 : 1 in this case.

The chemical shift value at which the peak representing each type of proton appears tells you about its environment—the type of functional group of which it is a part.

Table 2 *Chemical shift values for ^{1}H NMR*

Type of proton	δ / ppm
ROH	0.5–6.0
Ar–OH	4.5–7.0
RCH$_3$, R$_2$CH$_2$, R$_3$CH	0.9–1.7
RNH$_2$	1.0–5.0
Ar–NH$_2$	3.0–6.0
CH$_3$–Ar, RCH$_2$–Ar, R$_2$CH–Ar	2.3–3.0
R—C—C— ∥ ∣ O H	2.2–3.0
R—O—C— ∣ H	3.2–4.0
RCH$_2$Cl or Br	3.2–4.0
R—C—O—C— ∥ ∣ O H	3.2–4.0
R H \ / C=C / \	4.5–6.0
H–Ar	6.0–9.0
R—C=O ∣ H	9.3–10.5
R—C=O ∣ O—H	9.0–13.0
R—C=O ∣ N—H ∣ R'	5.0–12.0

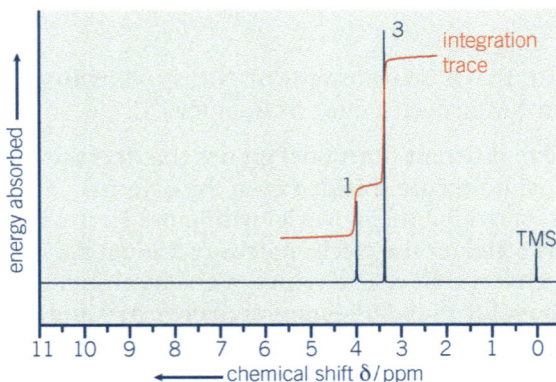

Figure 18 *The NMR spectrum of methanol showing the integration trace in red*

Chemical shift values

Hydrogen atom(s) in any functional group have a particular chemical shift value (Table 2).

Tetramethylsilane

The δ values of chemical shifts are measured by reference to a standard – the chemical shift of the hydrogen atoms in the compound tetramethylsilane, Si(CH$_4$)$_4$, TMS (Figure 19).

Figure 19 *Tetramethylsilane (TMS)—all 12 hydrogen atoms are in exactly the same environment, so they produce a single ^{1}H NMR signal*

The chemical shift value of these hydrogen atoms is zero by definition. A little TMS, which is a liquid, may be added to samples before their NMR spectra are run. TMS gives a peak at a δ value of exactly zero ppm to calibrate the spectrum (although modern techniques do not require this). All the spectra in this book show a TMS peak at $\delta = 0$.

Other reasons for using TMS are that it is inert, non-toxic, and easy to remove from the sample.

> **Exam tip**
>
> For simplicity, the integration trace has been omitted from NMR spectra in this book, and the relative number of hydrogen atoms that each peak represents is given.

Proton exchange

Proton exchange is a useful technique for identifying hydrogen atoms bonded to highly electronegative atoms such as oxygen or nitrogen. If a sample containing an –OH, –NH or –NH$_2$ group is dissolved in deuterium oxide, D$_2$O, these ^{1}H NMR peaks disappear from the spectrum. Deuterium oxide consists of water molecules in which the hydrogen atoms have been replaced by deuterium, ^{2}H—hydrogen isotopes with a neutron and a proton in the nucleus.

This is because the hydrogen atoms of the –OH, –NH or –NH$_2$ groups exchange with deuterium atoms in D$_2$O. For example:

$$CH_3CH_2OH + D_2O \rightleftharpoons CH_3CH_2OD + HOD$$

Deuterium does not show up on ^{1}H NMR spectra, which is why the peak disappears.

Interpreting proton, ^{1}H, NMR spectra

If you are presented with a spectrum of an organic compound, such as in Figure 20, you can find out a lot about its structure.

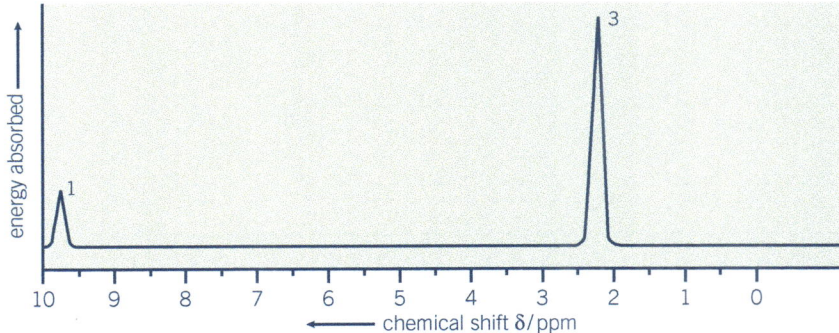

Figure 20 *The ^{1}H NMR spectrum of an organic compound*

The chemical shift values in Table 2 tell you that the single hydrogen at δ 9.7 is the hydrogen from a –CHO (aldehyde) group and the three hydrogens at δ 2.2 are those of a –COCH$_3$ group. (This peak could also be caused by –COCH$_2$R, but since there are three hydrogens it must be –COCH$_3$.)

So the compound is likely to be ethanal, CH$_3$CHO (Figure 21).

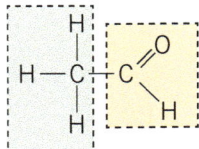

Figure 21 *The two groups that make up ethanal*

Spin–spin coupling
If you zoom in on most ^{1}H NMR peaks, they are split into particular patterns—this is called **spin–spin coupling** (also called spin–spin splitting). It happens because the applied magnetic field felt by any hydrogen atom is affected by the magnetic field of the hydrogen atoms on the neighbouring carbon atoms. This spin–spin splitting gives information about the neighbouring hydrogen atoms, which can be very helpful when working out structure.

Figure 22 shows the spin–spin splitting patterns.

The *n* + 1 rule
If there is one hydrogen atom on an adjacent carbon, this will split the NMR signal of a particular hydrogen into two peaks each of the same height.

If there are two hydrogen atoms on an adjacent carbon, this will split the NMR signal of a particular hydrogen into three peaks with the height ratio 1 : 2 : 1.

Three adjacent hydrogen atoms will split the NMR signal of a particular hydrogen into four peaks with the height ratio 1 : 3 : 3 : 1.

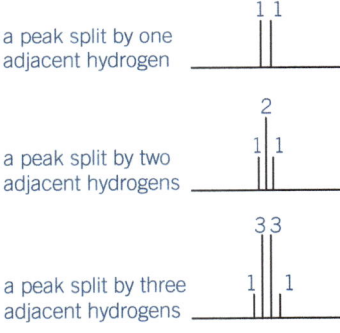

a peak split by one adjacent hydrogen

a peak split by two adjacent hydrogens

a peak split by three adjacent hydrogens

Figure 22 *NMR splitting patterns*

Exam tip

- Peaks with no splitting are often called singlets. Those with two are called doublets, with three, triplets, and with four, quartets.

- Spin–spin coupling is not usually seen in ^{13}C NMR spectra due to the low abundance of ^{13}C. This is one reason why ^{13}C spectra are simpler than ^{1}H spectra.

This is called the $n + 1$ rule:

> n **hydrogens on an adjacent carbon atom will split a peak into $n + 1$ smaller peaks. These groups of peaks are called multiplets.**

Some examples of interpreting ^{1}H NMR spectra

Ethanal
If you zoom in on the peaks in the spectrum of ethanal shown in Figure 20, you will see spin–spin splitting (Figure 23).

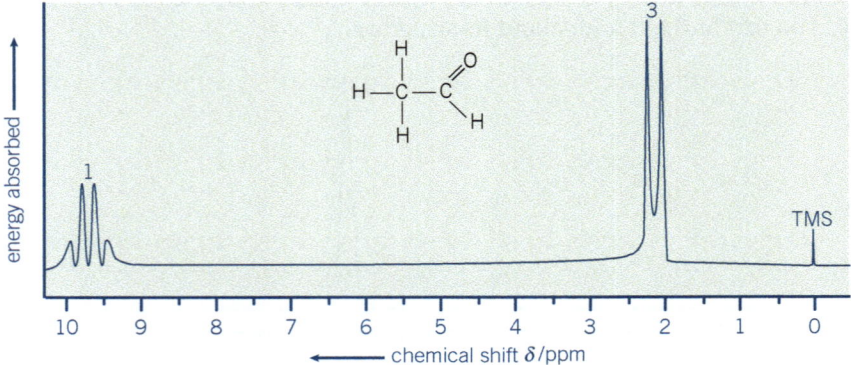

Figure 23 *The NMR spectrum of ethanal, CH_3CHO, at high resolution*

There are two types of hydrogen environments:

- A single peak of δ 9.7. This is the hydrogen of a –CHO group. This peak is split into four (height ratios 1 : 3 : 3 : 1) by the three hydrogens of the adjacent –CH_3 group.
- The peak with δ 2.2 is caused by three hydrogens of a –CH_3 group. This peak is split into two (height ratios 1 : 1) by the one hydrogen of the adjacent –CHO group.

Propanoic acid
Figure 24 shows the NMR spectrum of propanoic acid.

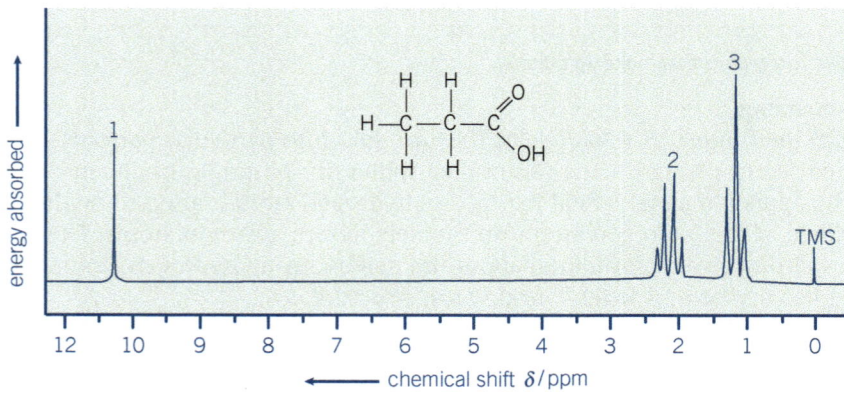

Figure 24 *The NMR spectrum of propanoic acid, CH_3CH_2COOH*

It is useful to make a table (Table 3) of the chemical shift of the peaks and what group they could correspond to by reference to Table 2 above.

From the chemical shift value alone, the peak at 2.4 could be caused by either –$COCH_2R$ or –$COCH_3$. However the fact that there are just two hydrogens means that it must correspond to –$COCH_2R$.

Table 3 *Chemical shift of the peaks of the ^{1}H NMR spectrum of propanoic acid, and what groups they could correspond to*

Chemical shift δ	Type of hydrogen	Number of hydrogens
11.7	–COOH	1
2.4	–$COCH_2R$ or –$COCH_3$	2
1.1	RCH_3	3

Looking at the spin–spin splitting:

- The peak at 11.7 is not split. This is because the adjacent carbon has no hydrogens bonded to it, –COOH.
- The peak at 2.4 is split into four. This indicates that the adjacent carbon has three hydrogens bonded to it. So, the R in –COCH$_2$R must be –CH$_3$.
- The peak at 1.1 is split into three. This indicates that the adjacent carbon has two hydrogens bonded to it. So, the R in RCH$_3$ must be –CH$_2$.

So, if you put these groups together you make propanoic acid:

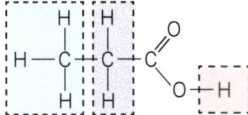

Propanone

The NMR spectrum of propanone (Figure 25) has just one peak. This means that all the hydrogen atoms in the molecule are in identical environments. The chemical shift value of 2.1 indicates that this corresponds to –COCH$_3$ or –COCH$_2$R.

> **Exam tip**
>
> Spin–spin coupling is not seen when equivalent hydrogens are on adjacent carbon atoms. For example, HOCH$_2$CH$_2$OH has only two single peaks in its NMR spectrum.

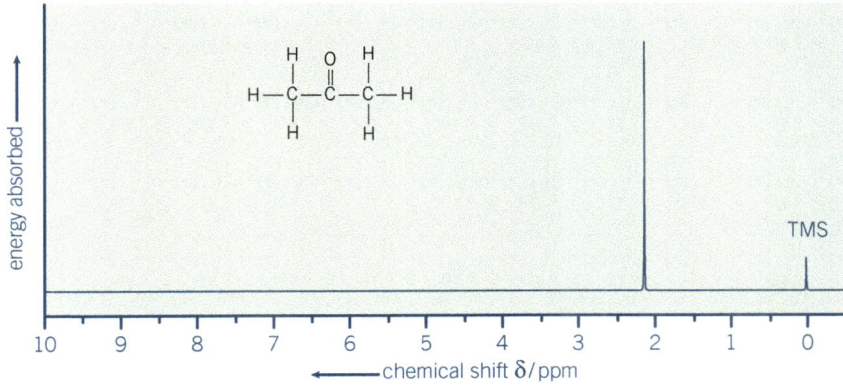

Figure 25 *The NMR spectrum of propanone*

> **Exam tip**
>
> There are six hydrogen atoms in a molecule of propanone, all of which are in an identical environment. However, you cannot tell the number from the spectrum because the peak areas in NMR spectra are only relative. This means that you can tell that in methanol, for example, the area under the –CH$_3$ peak is three times that of the –OH peak. But you cannot tell the absolute number of each type of hydrogen (the absolute numbers could be two and six, for example).

Solvents for ^{1}H NMR

NMR spectra are normally run in solution. The solvent must not contain any hydrogen atoms, otherwise the signal from the hydrogen atoms in the solution would swamp the signals from hydrogen atoms in the sample, because there are vastly more of them.

One solvent commonly used is tetrachloromethane, CCl$_4$, which has no hydrogen atoms. Other solvents contain deuterium, an isotope of hydrogen. Deuterium does not produce an NMR signal in the same range as hydrogen, though it has the same chemical properties. Some examples of deuterium-based solvents are deuterotrichloromethane, CDCl$_3$, deuterium oxide, D$_2$O, and perdeuterobenzene, C$_6$D$_6$.

Extension

Magnetic resonance imaging (MRI)

NMR can be used to investigate the human body—this was first realised by Felix Bloch, who found he got a strong signal by placing his finger in an NMR spectrometer. This signal was coming from protons in the water molecules that make up a large proportion of the human body. Water in different parts of the body (eg, normal cells and cancer cells) gives slightly different NMR signals. MRI scanning of parts of the body, to help diagnose medical conditions, is now routine. The patient passes through a scanner where the magnetic field varies across the body. This, along with sophisticated computer processing of the NMR signal, allows a three-dimensional image of the body to be built up. The technique is harmless as, unlike X-rays, neither the radio waves nor the magnetic field can damage cells. However, the name 'magnetic resonance imaging' is used rather than 'nuclear magnetic resonance' because of the association of the word 'nuclear' with radioactivity in the mind of the public.

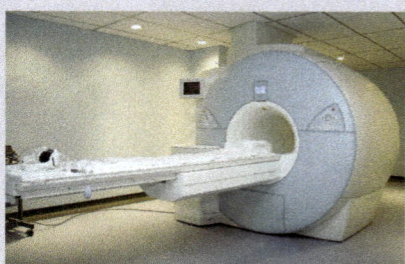

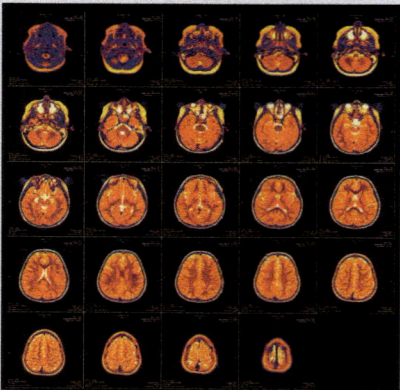

Figure 27 MRI scanner and a scan of a female child's brain obtained by using this technique

Predicting NMR spectra

Chemists making new compounds may predict the spectrum of a compound they are making. They then compare their prediction with that of the compound they actually produce, to check that their reaction has gone as intended.

Ethyl ethanoate

There are three sets of hydrogen atoms in different environments. The values of chemical shift are predicted using Table 2 above.

You can predict the spectrum shown in Figure 26 by dividing up the molecule as shown:

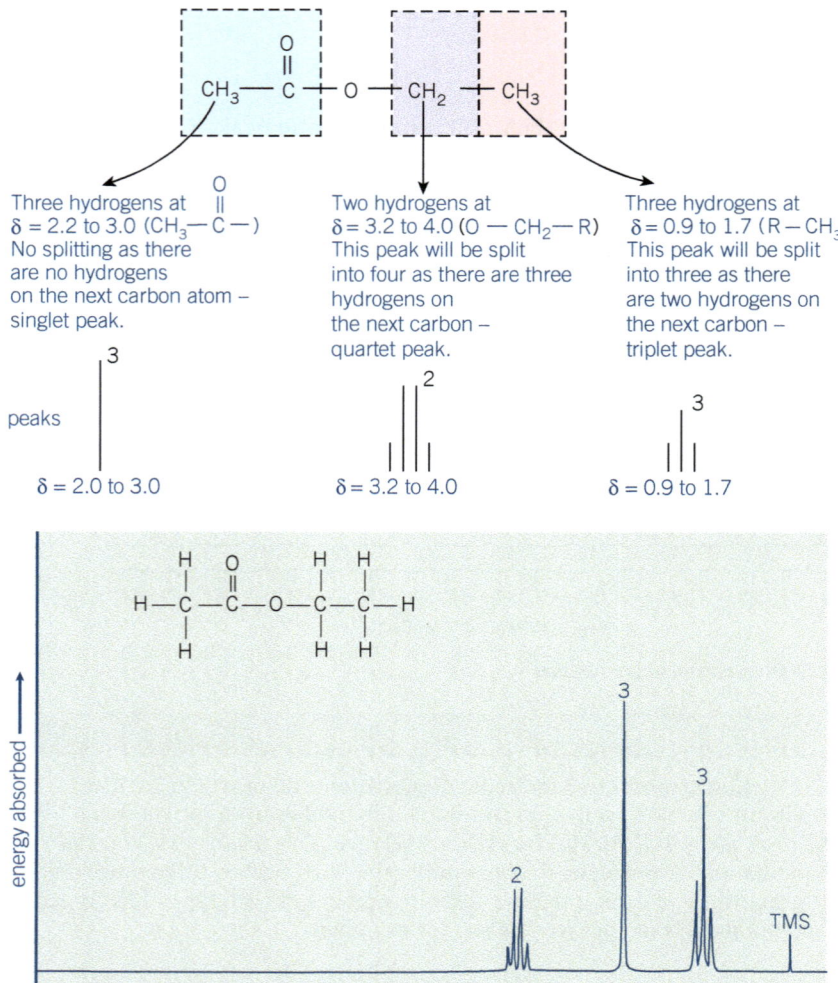

Figure 26 The NMR spectrum of ethyl ethanoate

Summary test 37.4

1 This question is about the isomers propan-1-ol and propan-2-ol.

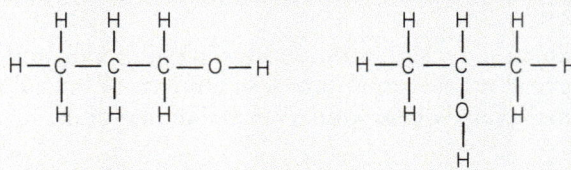

   a State what is meant by the term *isomer*.

   b Draw the structures of propan-1-ol and propan-2-ol and mark each of the hydrogen atoms A, B, and so on, to show which are in different environments.

   c Give the number of different environments for the hydrogen atoms in:

     i propan-1-ol

     ii propan-2-ol.

   d Give the number of hydrogen atoms in each of the different environments, A, B, and so on, for:

     i propan-1-ol

     ii propan-2-ol.

2 The ^{1}H NMR spectra shown are those of ethanol and of methoxymethane:

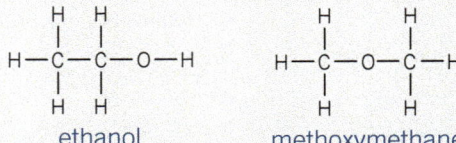

ethanol   methoxymethane

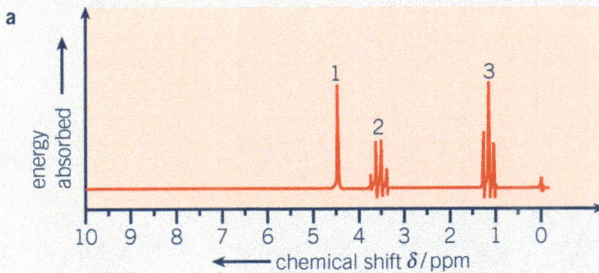

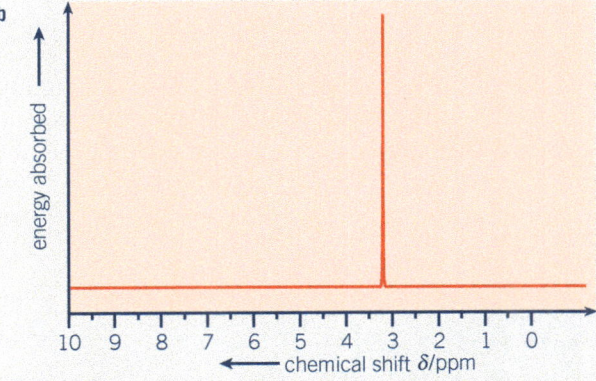

   a Deduce which spectrum represents which compound.

   b State what type of hydrogen each peak represents.

   c Give the numbers of each type of hydrogen.

   d In spectrum **a**, the peak at $\delta = 4.5$ ppm disappears if the sample is dissolved in D_2O. Explain what this suggests about this peak.

3 Predict the ^{1}H NMR spectrum of methyl ethanoate, CH_3COOCH_3, using the same procedure as for ethyl ethanoate above.

(🗇 **Launch additional digital resources for the chapter**)

1 Thin-layer chromatography was used to investigate four different food dyes, A, B, C and D. The results are shown below:

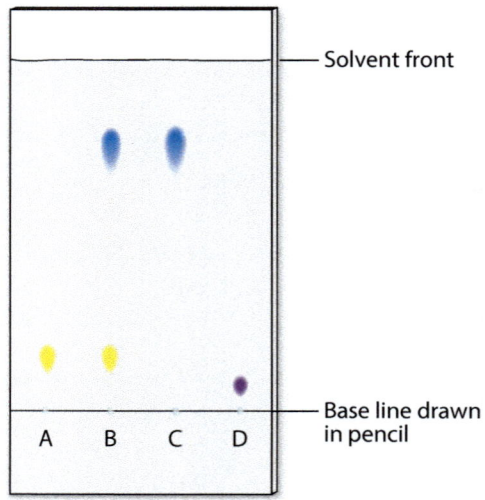

— Solvent front

— Base line drawn in pencil

A   B   C   D

a Describe what information the chromatogram gives for food dye B. *(1 mark)*

b Explain what conclusion can be made about the solubility of food dye D. *(2 marks)*

c Explain why it is important that the base line is drawn in pencil. *(1 mark)*

d Calculate the R_f values for dyes A and C. *(2 marks)*

2 The chromatogram below was collected from a sample of natural gas:

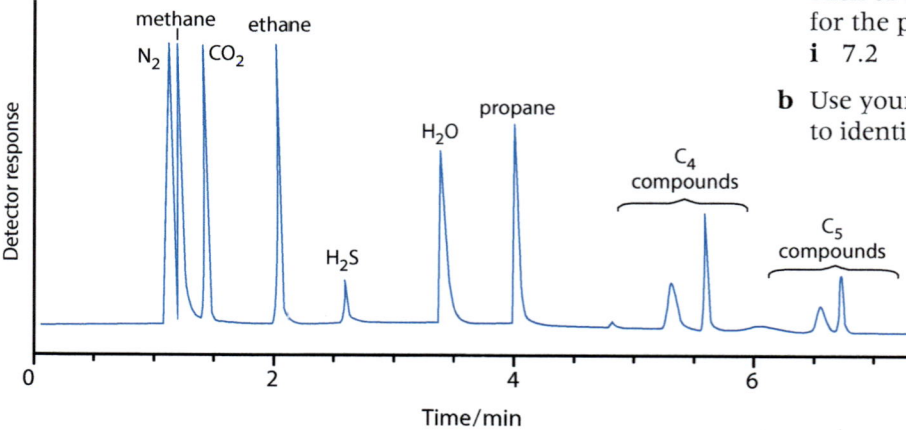

a Estimate the retention times of:
  i CO_2 **ii** H_2S *(2 marks)*

b Explain why some compounds like methane and N_2 have short retention times, so elute from the column more quickly than compounds like propane. *(2 marks)*

c Use the peak heights to estimate the ratio of methane to propane in this sample of natural gas. *(2 marks)*

d The chromatogram shows two peaks assigned to C_4 alkanes. Draw structural formulae for each peak. *(2 marks)*

3 The diagram below shows a low resolution ^1H NMR spectrum of 1-phenylbutan-2-one.

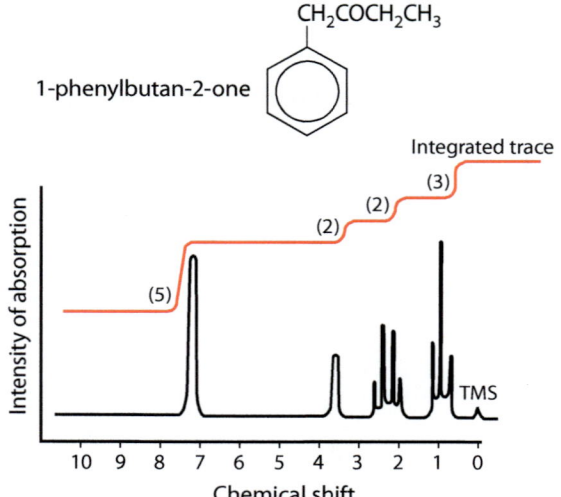

$CH_2COCH_2CH_3$

1-phenylbutan-2-one

Integrated trace

a Use the table of ^1H NMR chemical shift data at the back of the book to identify the protons responsible for the peaks at:
  i 7.2   **ii** 2.3   **iii** 0.9   *(3 marks)*

b Use your answer to part a and a process of elimination to identify the protons responsible for the peak at 3.6. *(1 mark)*

c Explain in detail the relative areas under the peaks in the NMR spectrum. *(2 marks)*

d The NMR spectrum above shows a peak for TMS.
  i Write down what TMS stands for. *(1 mark)*
  ii Explain the purpose of TMS in NMR spectroscopy. *(2 marks)*

4 This question is about ¹H NMR spectroscopy.

a Glycine is an amino acid. Its formula is H_2NCH_2COOH.
  i Draw the displayed formula of glycine. *(1 mark)*
  ii Predict the number of peaks, and their relative heights, in the ¹H NMR spectrum of glycine. Explain your answer. *(2 marks)*
  iii Predict the number of peaks that would be expected from the ¹H NMR spectrum of a solution of glycine in D_2O. Give reasons for your answer. *(2 marks)*

b Valine is another amino acid. The ¹H NMR spectrum of valine (dissolved in H_2O) is shown below.

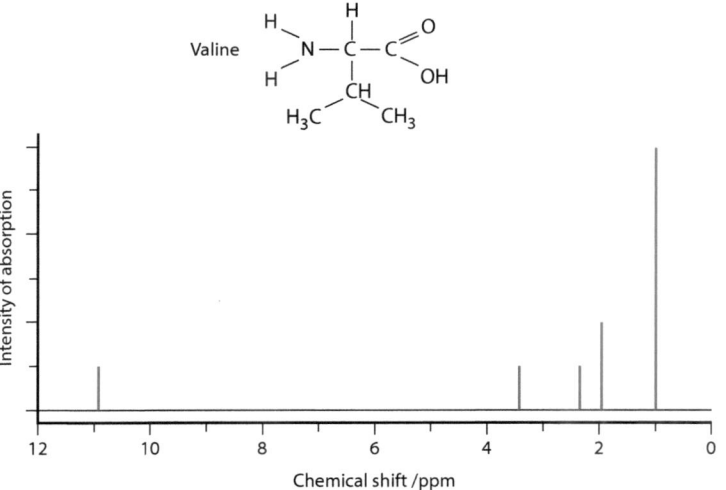

  i Use the data from the table of values at the back of the book to identify the protons responsible for the peaks on the spectrum at 11.0 and 0.9. *(2 marks)*

ii Use the structural formula and relative peak heights of the peaks on the spectrum, to suggest which of the three remaining peaks is produced by the protons attached to the nitrogen atom in valine. *(1 mark)*

iii Identify which two peaks on the spectrum would not be observed if valine is dissolved in D_2O instead of H_2O? Explain your answer. *(3 marks)*

5 a Identify the number of separate peaks in the ¹³C NMR spectra of:
  i methane
  ii ethane
  iii propane
  iv butane
  v 1,2-dichlorobenzene
  vi 1,3-dichlorobenzene. *(6 marks)*

b The diagram below shows the ¹³C NMR spectrum for an unknown substance X, whose molecular formula is C_4H_9Br.

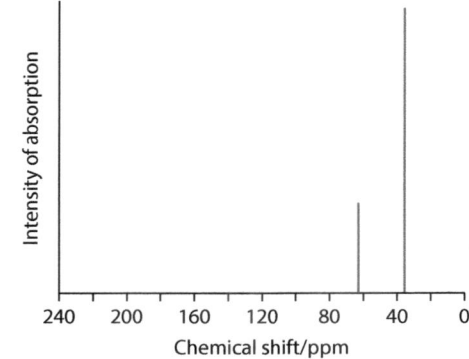

  i Draw possible isomeric structures for **X**. *(4 marks)*
  ii Use the spectrum to deduce which isomer is correct for structure X. *(1 mark)*

Glossary

A

Acid Brønsted–Lowry: a proton (H^+ ion) donor.

Acid derivative An organic compound, related to a carboxylic acid, of formula RCOZ, where Z = –Cl, $-NH_2$, –NHR or –NRR′.

Acid dissociation constant, K_a The equilibrium constant for the dissociation of a weak acid.

Activated complex An intermediate species that is formed and exists briefly at the maximum energy point in the reaction profile of a chemical reaction. Also called the transition state.

Activation energy The minimum energy required for a collision to be effective; the energy (enthalpy) difference between the reactants and the activated complex or transition state.

Addition polymer A polymer formed from a monomer (or monomers) with a carbon-carbon double bond.

Addition reaction A reaction in which two molecules join together to make a bigger molecule.

Addition–elimination A reaction in which the addition of a nucleophile is followed by elimination of another atom or group. The overall result is a substitution.

Adsorption The formation of weak temporary bonds that hold a gaseous reactant onto the surface of a catalyst.

Alcohol An organic compound with the general formula ROH.

Aldehyde An organic compound with the general formula RCHO.

Alkaline earth metals The metals in Group 2 of the Periodic Table.

Alkane A hydrocarbon with C–C and C–H single bonds only, with the general formula C_nH_{2n+2} for a chain and C_nH_{2n} for a ring.

Alkene A hydrocarbon with one or more C=C double bonds, with the general formula C_nH_{2n} for a chain and C_nH_{2n-2} for a ring with one double bond.

Alkylamine An amine where one or more hydrogen atoms of ammonia are replaced by an alkyl group.

Allotropes Pure elements which can exist in different physical forms in which their atoms are arranged differently. For example, diamond, graphite and buckminsterfullerene are allotropes of carbon.

Amide linkage A carbonyl carbon–nitrogen bond, represented as RCONR.

Amine An organic compound formed by replacing one or more hydrogen atoms of the ammonia molecule by organic groups.

Anhydrous A compound whose water of crystallisation has been removed.

Anode The positive electrode in an electrochemical cell.

Arenes Hydrocarbons that contain a benzene ring or a similar delocalised system of electrons. Also called aromatic compounds.

Aromatic compounds Hydrocarbons that contain a benzene ring or a similar delocalised system of electrons. Also called arenes.

Aromatic stability The increased stability of the benzene ring and other arenes given by their delocalised electrons.

Aromatic system The delocalised electron system above and below the plane of the ring of carbons in benzene and other arenes.

Arylamine An amine in which at least one of the hydrogen atoms in ammonia has been replaced by an aryl group.

Asymmetric carbon atom A carbon atom that is attached to four different types of atoms or groups of atoms. Also called a chiral centre.

Atomic number The number of protons in the nucleus of an atom; the same as the proton number.

Atomic orbital A region of space around an atomic nucleus where there is a high probability of finding an electron.

Atomic radius Half the distance between the centres of a pair of atoms.

Avogadro constant The number of particles in a mole of substance. Also called the Avogadro number. Its value is 6.02×10^{23}.

Avogadro's law Equal volumes of all gases contain the same number of particles (molecules or separate atoms) at the same temperature and pressure.

Azo compounds Compounds with the functional group R–N=N–R′.

B

Base Brønsted–Lowry: a proton (H^+ ion) acceptor.

Baseline A pencil line drawn on a chromatography plate to show the starting position of the mixture being separated.

Bidentate Describes a ligand that can form two coordinate (dative) bonds with a transition metal ion.

Bond energy (bond enthalpy) The energy required to break one mole of a particular covalent bond in the gaseous state.

Bond length The distance between the nuclei of two covalently bonded atoms.

Bonding A general term used to describe the forces that hold atoms together.

Born–Haber cycle A thermochemical cycle that includes all the enthalpy changes involved in the formation of an ionic compound.

Brittle Substances that shatter easily when hit.

Buffer A solution that resists change of pH when small amounts of acid or base are added or on dilution.

C

Calorimeter An instrument for measuring the heat changes that accompany chemical reactions.

Carbocation An organic ion in which one of the carbon atoms has a positive charge.

Carbon-13 (^{13}C) NMR Nuclear magnetic resonance spectroscopy used to identify the carbon atoms in different environments in a molecule.

Carboxylic acid An organic compound with the general formula R–COOH that contains a carboxyl group (–COOH).

Catalyst A substance that alters the rate of a chemical reaction but is not used up in the reaction.

Catalytic cracking The breaking, with the aid of a catalyst, of long-chain alkane molecules (obtained from crude oil) into shorter chain hydrocarbons (some of which are alkenes).

Cathode The negative electrode in an electrochemical cell.

Chain isomerism Isomers with their side-chain at different positions on the main hydrocarbon chain.

Charge density The amount of electric charge per unit volume. Charge density decreases as the size of an ion increases.

Chelates Complex ions with polydentate ligands.

Chelation The process by which a polydentate ligand replaces monodentate ligands in forming coordinate (dative) bonds to a transition metal ion.

Chemical feedstock The starting materials in an industrial chemical process.

Chemical shift, δ The difference in position between a hydrogen (or carbon) atom and a reference compound (usually tetramethylsilane) in an NMR spectrum. It is measured in parts per million (ppm).

Chiral centre An atom to which four different atoms or groups are bonded. The presence of such an atom causes the parent molecule to exist as a pair of non-identical mirror images.

Chiral This means 'handed'. A chiral molecule exists in two mirror image forms that are not identical.

Chromatography A family of techniques used to separate mixtures. It includes paper chromatography, thin-layer chromatography, column chromatography and gas/liquid chromatography. The components of the mixture are carried at different rates by a mobile phase through a stationary phase.

Closed system A reaction system where the reactants and products cannot escape.

Column chromatography A chromatography technique where the stationary phase is a solid packed into a narrow tube (the column). The liquid mobile phase is added at the top and runs down the column.

Common ion effect The effect on an equilibrium when a common ion (an ion that is already contained in the solution) is added to a solution. The common ion effect causes precipitation.

Complex ion A transition metal ion surrounded by ligands bonded to it by coordinate (dative) bonds.

Condensation polymer A polymer formed from monomers with two functional groups, which react together with the elimination of a small molecule.

Condensation reaction A reaction in which two molecules join together with the loss of a small molecule (often water).

Conjugate acid In Brønsted–Lowry theory, a conjugate acid is a chemical species formed by adding a H$^+$ ion to a base.

Conjugate acid–base pair In Brønsted–Lowry theory, a conjugate acid–base pair are the Brønsted-Lowry acid and base forms of a chemical species. They differ by one proton (H$^+$ ion). For example NH$_4^+$ and NH$_3$.

Conjugate base In Brønsted-Lowry theory, a conjugate base is a chemical species formed by the removal of a H$^+$ ion from an acid.

Coordinate bond A covalent bond in which both the electrons in the bond come from one of the atoms forming the bond. (Also called a dative covalent bond.)

Coordination number The number of ligand molecules bonded to a metal ion.

Covalent bonding A chemical bond formed by the electrostatic attraction between the nuclei of two atoms and a shared pair of electrons.

Cracking The breakdown of a large alkane into shorter chain alkanes and alkenes.

Cyclic Describes a compound where some of the atoms are connected to form a ring.

D

Dative covalent bonding Covalent bonding in which both the electrons in the bond come from one of the atoms in the bond. (Also called coordinate bonding.)

Degenerate orbitals A set of orbitals at exactly the same energy level.

Delocalisation Describes the process by which electrons are spread over several atoms and help bond them together.

Deprotonated Describes an atom, molecule or ion which has lost a proton (an H$^+$ ion).

Desorption The breaking of weak temporary bonds to enable the escape of a gaseous product from the surface of a catalyst.

Dimer A chemical structure formed when two monomers join together.

Dipeptide A peptide with two amino acids.

Dipole–dipole forces An intermolecular force that results from the attraction between molecules with permanent dipoles.

Displacement reaction A chemical reaction in which one atom or group of atoms replaces another in a compound, for example, $Zn + CuO \rightarrow ZnO + Cu$

Displayed formula The formula of a compound drawn out so that each atom and each bond is shown.

Disproportionation Describes a redox reaction in which the oxidation number of some atoms of a particular element increases and that of other atoms of the same element decreases.

Dynamic equilibrium A situation in which the composition of a reaction mixture does not change because both forward and backward reactions are proceeding at the same rate.

E

Effective collisions Collisions between reactant particles that lead to a reaction.

Electrochemical series A list of ions in order of the voltage they produce in an electrochemical cell.

Electrode The positive and negative terminals by which the current enters and leaves an electrochemical cell.

Electrolysis The process by which ionic compounds are decomposed by passing electricity through them.

Electrolyte The ionic liquid in an electrochemical cell.

Electron A subatomic particle with a negative charge.

Electron density The probability of electrons being found in a particular volume of space.

Electron pair repulsion theory A theory which explains the shapes of simple molecules by assuming that pairs of electrons around a central atom repel each other and thus take up positions as far away as possible from each other in space.

Electronegativity The power of an atom to attract electrons towards itself.

Electrophile An electron-deficient atom, ion or molecule that takes part in an organic reaction by attacking areas of high electron density in another reactant.

Electrophilic addition An organic reaction in which a molecule with an area of high electron density is attacked by an electron-deficient atom, ion or molecule (an electrophile). It results in a larger molecule.

Electrophilic substitution A substitution reaction where the first step is attack by an electrophile on an area of high electron density in another molecule. It is followed by loss of an atom, ion, or group of atoms.

Electropositive A term used to describe atoms with low electronegativity e.g. metals.

Electrostatic forces The forces of attraction and repulsion between electrically charged particles.

Elimination reaction A reaction in which an atom or group of atoms is removed from a reactant.

Eluent The solvent used as the mobile phase in chromatography.

Empirical formula The simplest whole number ratio of atoms of each element in a compound.

Enantiomer One of a pair of non-identical mirror image isomers.

End point The point in a titration when the volume of reactant added just causes the colour of the indicator to change.

Endothermic reactions A reaction in which heat energy is taken in as the reactants change to products—the temperature therefore drops.

Energy cycle A sequence of chemical reactions (with their enthalpy changes) that convert a reactant into a product. The total enthalpy change of the sequence of reactions will be the same as that for the conversion of the reactant to the product directly (or by any other route). It is also called a thermochemical cycle.

Enthalpy change A measure of heat energy given out or taken in when a chemical or physical change occurs at constant pressure.

Enthalpy change of first ionisation, ΔH_{i1} The enthalpy change when 1 mole of gaseous atoms forms 1 mole of gaseous ions with a single positive charge.

Enthalpy change of hydration, ΔH_{hyd} The enthalpy change when 1 mole of gaseous ions dissolves in water to give a very dilute solution.

Enthalpy change of solution, ΔH_{sol} The enthalpy change when 1 mole of a substance dissolves in water to give a very dilute solution.

Enthalpy change of atomisation, ΔH_{at} The enthalpy change when 1 mole of an element in its standard state is atomised to produce 1 mol of gaseous atoms.

Enthalpy level diagrams (energy level diagrams) Diagrams in which the enthalpies (energies) of the reactants and products of a chemical reaction are plotted on a vertical scale to show their relative levels.

Entropy, S A numerical measure related to the number of possible arrangements of the particles and their energy in a given system.

Equilibrium constant, K_c for a reaction $a\text{A} + b\text{B} + c\text{C} \rightleftharpoons x\text{X} + y\text{Y} + z\text{Z}$

$$K_c = \frac{[\text{X}]_{eqm}{}^x\,[\text{Y}]_{eqm}{}^y\,[\text{Z}]_{eqm}{}^z}{[\text{A}]_{eqm}{}^a\,[\text{B}]_{eqm}{}^b\,[\text{C}]_{eqm}{}^c}$$

Equilibrium constant, K_p for a reaction $a\text{A(g)} + b\text{B(g)} \rightleftharpoons y\text{Y(g)} + z\text{Z(g)}$

$$K_p = \frac{p^y\text{Y(g)}_{eqm}\,p^z\text{Z(g)}_{eqm}}{p^a\text{A(g)}_{eqm}\,p^b\text{B(g)}_{eqm}}$$

Equilibrium mixture The mixture of reactants and products formed when a reversible reaction is allowed to proceed in a closed container until no further change occurs. The forward and backward reactions are still proceeding but at the same rate.

Equivalence point The point in a titration at which the reaction is just complete.

Ester An organic compound with the general formula RCOOR'.

Excess A reactant that is left over after a chemical reaction finishes.

Excited state A chemical species in which one or more electrons have moved from the lowest energy state to a higher one.

Exothermic reactions A reaction in which heat energy is given out as the reactants change to products – the temperature therefore rises.

F

Faraday constant, F The amount of electric charge needed to produce 1 mole of atoms of product from singly charged ions during electrolysis.

Fingerprint region The area of an infrared spectrum below about 1500 cm^{-1}. It is caused by complex vibrations of the whole molecule and is characteristic of a particular molecule.

First electron affinity, EA The energy released when 1 mole of gaseous atoms each acquire an electron to form 1 mole of gaseous ions with a single negative charge.

First-order A reaction is first order with respect to a certain species if the power to which its concentration is raised in the rate equation is 1.

Fraction A mixture of hydrocarbons collected over a particular range of boiling points during the fractional distillation of crude oil.

Fractional distillation A distillation process used to separate a mixture containing chemicals with boiling points close to each other. Used to separate the mixture of compounds in crude oil.

Fragmentation The process where ions from the sample molecule break up as they pass through a mass spectrometer.

Free radical A chemical species with one or more unpaired electrons—usually highly reactive.

Free radical substitution A reaction in which a free radical species replaces a hydrogen atom.

Frequency of collisions The rate of collisions between particles of reactants.

Functional group An atom or group of atoms in an organic molecule which is responsible for the characteristic reactions of that molecule.

Functional group isomerism Isomers with the same molecular formula but different functional groups.

G

Gas/liquid chromatography A type of chromatography where the stationary phase is a liquid with a high boiling point, held in position by a powder packed in (or coated inside) a long capillary tube. The mobile phase is an unreactive gas.

Geometric isomerism Isomers with their substituents either on opposite sides or on the same side of a carbon-carbon double bond (also called *cis–trans* isomerism).

Giant structure A structure where the atoms form an extended geometrical arrangement.

Gibbs free energy change, ΔG A measure of whether a reaction is thermodynamically feasible at a particular temperature. If ΔG is negative, the reaction is feasible. If ΔG is positive is the reaction is not feasible.

Ground state The lowest energy electron arrangement of an atom.

Group A vertical column of elements in the Periodic Table. The elements have similar properties because they have the same outer electron arrangement.

H

Half-cell An electrode made by placing a piece of metal in a solution of its own ions. Two half-cells are connected to make an electrochemical cell.

Half equation An equation for a redox reaction which considers just one of the species involved and shows explicitly the electrons transferred to or from it.

Half reaction An equation for a redox reaction which considers just one of the species involved and shows explicitly the electrons transferred to or from it. Also called a half equation.

Half-life The half-life of a reaction, $t_{1/2}$, is the time taken for the concentration of a reactant to fall from any value to half that value.

Halogenoalkane An alkane in which one or more hydrogen atoms have been replaced by halogen atoms.

Henderson–Hasselbalch equation A formula used to find the pH in calculations involving buffers;

$$pH = pKa - \log\left(\frac{[HA]}{[A^-]}\right)$$

Heterogeneous catalysts A catalyst that is in a different phase (sold, liquid or gas) to the reactants.

Heterolytic fission Bond breaking between two atoms of different elements, where both electrons go to the more electronegative atom.

Hexadentate A ligand that can form six coordinate bonds with a transition metal ion.

Homogeneous catalysts A catalyst that is in the same phase (solid, liquid or gas) as the reactants.

Homologous series A set of organic compounds with the same functional group. The compounds differ in the length of their hydrocarbon chains by CH_2.

447

Homolytic fission Bond breaking between two atoms of the same element, where one electron goes to each atom to form two free radicals.

HPLC High pressure liquid chromatography (or high performance liquid chromatography). It is similar to column chromatography but the solvent is forced though the column by a pump.

Hybrid orbitals Orbitals containing outer electrons that mix together and are used in bonding.

Hybridisation The formation of hybrid orbitals.

Hydrated A compound that contains water of crystallisation.

Hydration A reaction in which water is added.

Hydrocarbon A compound made up of carbon and hydrogen atoms only.

Hydrogen bonding A type of intermolecular force in which a hydrogen atom ($H^{\delta+}$) interacts with a more electronegative atom with a $\delta-$ charge.

Hydrolysis A reaction of a compound with water.

I

Ideal gas An imaginary gas with zero particle volume and no intermolecular forces of attraction. It obeys the ideal gas equation.

Incomplete combustion A combustion reaction in which there is insufficient oxygen for all the carbon in the fuel to burn to carbon dioxide. Carbon monoxide and/or carbon (soot) are formed.

Initial rate The instantaneous rate at the start of a reaction, when $t = 0$.

Initiation The first step in a chain reaction, where the free radicals are formed.

Integration trace A technique used in proton nuclear magnetic resonance (^{1}H-NMR) spectroscopy. It accurately measures the relative areas of the peaks and shows them as a line on the spectrum.

Intermediate A short-lived species formed during one of the steps in a reaction mechanism.

Intermolecular forces Weak forces between molecules (the three types of intermolecular forces are London dispersion forces, dipole–dipole forces and hydrogen bonding).

Ionic bonding The electrostatic attraction between oppositely charged ions (positively charged cations and negatively charged anions).

Ionic product of water, K_w The equilibrium constant for the dissociation of water; $K_w = [H^+(aq)]\ [OH^-(aq)] = 1.0 \times 10^{-14}\ mol^2\,dm^{-6}$ (at 298 K)

Ionisation energy The energy required to remove a mole of electrons from a mole of isolated gaseous atoms or ions.

Isoelectric point The pH of at which an amino acid is neutral.

Isomer One of two (or more) compounds with the same molecular formula but different arrangement of atoms in space.

Isotopes Atoms with the same number of protons but different numbers of neutrons.

K

Ketone An organic compound with the general formula RR'CO in which there is a C=O double bond.

Kinetics The branch of chemistry dealing with reaction rates.

L

Lattice A regular three-dimensional arrangement of atoms, ions or molecules.

Lattice formation energy, ΔH_{latt} The energy released when gaseous ions come together to form an ionic solid.

Leaving group In an organic substitution reaction, the leaving group is an atom or group of atoms that is ejected from the starting material, normally taking with it an electron pair and forming a negative ion.

Ligand A species that contains a lone pair of electrons that forms a coordinate (dative) covalent bond to a central metal atom or ion.

Limiting reagent A reactant that is completely used up in a chemical reaction. Once this reactant is used up, the reaction stops.

Locant A number used to indicate the position of branching or the position of a functional group on a hydrocarbon chain.

Lone pair A pair of electrons in the outer shell of an atom that is not involved in bonding.

M

Mass number The total number of protons and neutrons in the nucleus of an atom. Also called the nucleon number.

Maxwell–Boltzmann distribution The distribution of energies (and therefore speeds) of the molecules in a gas or liquid.

Mean bond energy The average value of the bond dissociation energy for a given type of bond taken from a range of different compounds.

Metallic bonding The electrostatic attraction between positive metal ions and delocalised electrons.

Mobile phase The solvent used in chromatography. The mobile phase carries the mixture being separated through the chromatography apparatus.

Mole A quantity of a substance that contains the Avogadro number (6.02×10^{23}) of particles (eg atoms, molecules or ions).

Mole fraction The amount of one component of a mixture (in moles) divided by the total number of moles of all the components.

Molecular formula A formula that tells us the actual numbers of atoms of each different element that make up a molecule of a compound.

Molecular ion peak In mass spectrometry this peak is from a molecule of the sample which has been ionised but which has not broken up during its journey through the instrument.

Molecular structure A structure that consists of separate molecules.

Monodentate A ligand with a single lone pair of electrons that forms a coordinate (dative) bond with a transition metal ion.

Monomer A small molecule that combines with many other monomers to form a polymer.

N

Nernst equation An equation used to calculate how electrode potential changes when the concentration of one or more species is changed:

$$E = E^\circ + \left(\frac{0.059}{z}\right) \log \frac{[\text{oxidised species}]}{[\text{reduced species}]}$$

Neutron A subatomic particle with no charge found in the nucleus of an atom.

Nitrile An organic compound with the general formula $RC{\equiv}N$.

Nitronium ion (nitryl cation) The ion NO_2^+.

Non-degenerate orbitals A set of orbitals that have been split into groups with slightly different energy levels.

Non-effective collisions Collisions between reactant particles that do not lead to a reaction.

Nuclear magnetic resonance spectroscopy (NMR) An analytical technique used identify the hydrogen (^{1}H NMR) or carbon (^{13}C NMR) atoms in different environments in a molecule.

Nucleon number The total number of protons plus neutrons in the nucleus of an atom; also called the mass number.

Nucleons Protons and neutrons—the sub-atomic particles found in the nuclei of atoms.

Nucleophile A negative ion or molecule that is able to donate a pair of electrons and takes part in an organic reaction by attacking an electron-deficient area in another reactant.

Nucleophilic addition An organic reaction in which a reagent with a negative charge or partially negatively charged area (a nucleophile) adds to a molecule with an area of low electron density.

Nucleophilic substitution An organic reaction in which a molecule with a partially positively charged carbon atom is attacked by a reagent with a negative charge or partially negatively charged area (a nucleophile). It results in the replacement of one of the groups or atoms on the original molecule by the nucleophile.

Nucleus The tiny, positively charged centre of at atom composed of protons and neutrons.

O

Optical activity The property of rotating the plane of polarisation of plane polarised light.

Optical isomerism Refers to pairs of molecules that are non-identical mirror images.

Order of reaction In the rate equation, rate = $k[A]^a[B]^b$, this is the power to which the concentration of each species is raised. For species A the order of reaction is a, and for species B the order of reaction is b.

Overall order of the reaction In the rate equation, this is the sum of the powers to which the concentrations of all the species involved in the reaction are raised. If rate = $k[A]^a[B]^b$, the overall order of the reaction is a + b.

Oxidation A reaction in which an atom or group of atoms loses electrons.

Oxidation number (oxidation state) The number of electrons lost or gained by an atom in a compound compared to the uncombined atom. It forms the basis of a way of keeping track of redox (electron transfer) reactions. Also called oxidation state.

Oxidising agent A reagent that oxidises (removes electrons from) another species.

Oxonium ion The ion H_3O^+ formed when water accepts a proton. It is also called the hydronium or hydroxonium ion.

P

Partial pressure The contribution made by each gas in a mixture of gases to the overall pressure.

Partition coefficient The ratio of the concentrations of a solute in two imiscible liquids when it is in equilibrium across the interface between them.

Peptides Compounds formed by the linkage of amino acids.

Percentage yield In a chemical reaction this is the actual amount of product produced divided by the theoretical amount (predicted from the chemical equation) expressed as a percentage.

Period A horizontal row of elements in the Periodic Table. There are trends in the properties of the elements as we cross a period.

Periodicity The regular recurrence of the properties of elements when they are arranged in atomic number order as in the Periodic Table.

pH A scale for measuring acidity and alkalinity. pH = $-\log_{10}[H^+]$ in a solution.

Planar A flat structure where all the atoms are in the same plane.

Plasticisers Small molecules added to polymers to make them more flexible.

Polar Describes a molecule in which the charge is not symmetrically distributed so that one area is slightly positively charged and another slightly negatively charged.

Polarimeter A scientific instrument used to measure the angle of rotation when polarised light is passed through an optically active substance.

Polarised light Light waves where the vibrations occur in a single plane.

Polaroid A filter used to produce polarised light.

Polydentate A ligand that can form more than one coordinate (dative) bond with a transition metal ion.

Polymer A chemical compound in which small molecules are bonded together to form long chains.

Positional isomerism Isomers with their functional group at a different position on the hydrocarbon chain.

Positive inductive effect Describes the tendency of some atoms or groups of atoms to release electrons *via* a covalent bond.

Potential difference The 'push' that causes charges to move in an electrical conductor. Also called voltage.

Primary carbocation A positively charged carbon atom with one alkyl group.

Primary halogenoalkane A halogenoalkane where the halogen is at the end of the hydrocarbon chain.

Primary structure The fixed sequence of amino acids in the chain of a protein.

Principal quantum number The number given to an electron shell.

Propagation The second step in a chain reaction, in which free radicals react to form a molecule of product and another free radical.

Proton (^{1}H) NMR Nuclear magnetic resonance spectroscopy used to identify the hydrogen atoms in different environments in a molecule.

Proton A subatomic particle with a positive charge found in the nucleus of an atom.

Proton exchange A technique used in proton nuclear magnetic resonance (^{1}H-NMR) spectroscopy to identify hydrogen atoms bonded to highly electronegative atoms (such as oxygen or nitrogen). The sample is dissolved in deuterium oxide (D_2O) and exchange takes place between hydrogen and deuterium atoms, so the peak disappears from the spectrum.

Proton number The number of protons in the nucleus of an atom; also called the atomic number.

Protonated Describes an atom, molecule or ion to which a proton (an H^+ ion) has been added.

R

Racemic mixture A mixture of equal amounts of two optical isomers of a chiral compound. It is optically inactive. Also called a racemate.

Rate constant The constant of proportionality in the rate equation, e.g. k in the expression rate = $k[A]^a[B]^b$.

Rate equation A mathematical expression showing how the rate of a chemical reaction depends on the concentrations of various chemical species involved. It is also called the rate expression, e.g. rate = $k[A]^a[B]^b$

Rate of the reaction The change in concentration of reactants (or products) with time.

Rate-determining step The slowest step in the reaction mechanism. It governs the rate of the overall reaction.

Reaction mechanism The series of simple steps that lead from reactants to products in a chemical reaction.

Reaction pathway The conversion of reactants into products as shown on an enthalpy level diagram.

Real gas An actual gas. Real gases do not obey the ideal gas equation exactly.

Redox reaction Short for reduction–oxidation reaction, it describes reactions in which electrons are transferred from one species to another.

Reducing agent A reagent that reduces (adds electrons to) another species.

Reduction A reaction in which an atom or group of atoms gain electrons.

Relative atomic mass, A_r $A_r = \dfrac{\text{average mass of an atom}}{\frac{1}{12\text{th}} \text{ mass of 1 atom of } ^{12}C}$

Relative formula mass, M_r $M_r = \dfrac{\text{average mass of an entity}}{\frac{1}{12\text{th}} \text{ mass of 1 atom of } ^{12}C}$

Relative isotopic mass The relative mass of a particular isotope measured on the carbon-12 scale.

Relative molecular mass, M_r $M_r = \dfrac{\text{average mass of a molecule}}{\frac{1}{12\text{th}} \text{ mass of 1 atom of } ^{12}C}$

Retention time The time taken for a component in a mixture to pass through a chromatography column (or tube).

Reversible reaction A chemical reaction where the products can react together to produce the original reactants.

R_f values In chromatography, $R_f = \dfrac{\text{distance moved by the spot}}{\text{distance moved by the solvent}}$

Ring-activating substituent A group attached to a benzene ring that releases electrons onto the ring, making it more reactive to electrophiles and directing further substitution to the 2-, 4- and 6-positions.

Ring-deactivating substituents A group attached to a benzene ring that withdraws electrons from the ring, making it less reactive to electrophiles and directing further substitution to the 3- or 5-positions.

S

Salt bridge A method to connect the half-cells in an electrochemical cell without using a metal wire. Often made using a filter paper soaked in a solution of a salt.

Saturated hydrocarbon A compound containing only hydrogen and carbon with only C–C and C–H single bonds, ie one to which no more hydrogen can be added.

Second-order A reaction is second order with respect to a certain species if the power to which its concentration is raised in the rate equation is 2. A reaction is second order overall if the sum of the powers to which the concentrations of all the species involved in the reaction are raised is 2.

Secondary carbocation A positively charged carbon atom with two alkyl groups.

Secondary halogenoalkane A halogenoalkane where the halogen is in the body of the hydrocarbon chain.

Shielding Inner shell electrons acting as a 'shield' and decreasing the attraction between an outer shell electron and the nucleus.

Skeletal formula A way of writing the formula of an organic compound in which the carbon atoms are not drawn at all. Straight lines represent carbon–carbon bonds and carbon atoms are assumed to be where the bonds meet.

S_N1 A mechanism for nucleophilic substitution where the rate determining step involves one species.

S_N2 A mechanism for nucleophilic substitution where the rate determining step involves two species.

Solubility product The equilibrium constant used for a sparingly soluble solid in a saturated solution at a given temperature.

Solvent extraction A method used to remove a substance from a solution or mixture by dissolving it in another immiscible solvent in which it is more soluble.

Solvent front The furthest point reached by the solvent in paper or thin-layer chromatography.

Specific heat capacity, c The amount of heat needed to raise the temperature of 1 g of substance by 1 K.

Spectator ions Ions that are unchanged during a chemical reaction, that is, they take no part in the reaction.

Spin–spin coupling A term used in proton nuclear magnetic resonance (^{1}H-NMR) spectroscopy. The applied magnetic field felt by any hydrogen is affected by the magnetic fields of the hydrogens on the neighbouring carbon atoms. This splits ^{1}H-NMR peaks into patterns. Also called spin-spin splitting.

Stability constant, K_{stab} The equilibrium constant for the formation of a complex ion.

Standard cell potential, E°_{cell} The voltage obtained by connecting two standard electrodes together.

Standard electrode potential, E° The measured voltage of a half-cell when connected to a standard hydrogen electrode. It is also called the standard reduction potential.

Standard hydrogen electrode An electrode consisting of platinum metal dipped into a solution of 1 mol dm^{-3} H$^+$ ions through which hydrogen gas is bubbled (at 298 K and 100 kPa). It is used as a reference electrode to compare the tendency of different metals to release electrons.

Standard molar enthalpy of combustion, ΔH_c° The enthalpy change when 1 mole of a substance is completely burned in oxygen with all reactants and products in their standard states (298 K and 100 kPa).

Standard molar enthalpy of formation, ΔH_f° The enthalpy change when 1 mole of substance is formed from its elements with all reactants and products in their standard states (298 K and 100 kPa).

Standard molar enthalpy of neutralisation, ΔH_{neut}° The enthalpy change when solutions of an acid and an alkali react together to produce one mole of water.

Standard reduction potential The measured voltage of a half-cell when connected to a standard hydrogen electrode. It is also called the standard electrode potential.

State symbols Letters, in brackets, which can be added to the formulae in equations to say what state the reactants and products are in—(s) means solid, (l) means liquid, (g) means gas and (aq) means aqueous solution (dissolved in water).

Stationary phase A porous solid used in chromatography to separate mixtures. The more affinity that a component of the mixture has for the stationary phase, the slower it will move with the solvent.

Stereoisomerism Isomers with the same molecular formula and the same structure, but a different position of atoms in space.

Stoichiometry Describes the simple whole number ratios in which chemical species react.

Strong acid An acid that is fully dissociated into ions in solution.

Strong nuclear force The force that holds protons and neutrons together within the nucleus of the atom.

Structural formula A way of writing the formula of an organic compound in which bonds are not shown but each carbon atom is written separately with the atoms or groups of atoms attached to it.

Structural isomer Isomers with the same molecular formula but a different structure.

Structure A general term used to describe the geometrical arrangement of atoms.

Substitution reaction A reaction in which an atom or group of atoms is replaced by another atom or group of atoms.

Synthesis The making of new molecules via one or more chemical reactions.

T

Target molecule The molecule resulting from a series of organic reactions in a chemical synthesis.

Termination The final step in a chain reaction, when the free radicals are removed.

Tertiary carbocation A positively charged carbon atom with three alkyl groups.

Tertiary halogenoalkane A halogenoalkane where the halogen is at a branch in the hydrocarbon chain.

Tetradentate A ligand that can form four coordinate (dative) bonds with a transition metal ion.

Tetramethylsilane, TMS A reference compound used to measure chemical shift in nuclear magnetic resonance (NMR) spectroscopy.

Thermal cracking A process in which hydrocarbons in crude oil are heated to a high temperature to break bonds and produce shorter chain products (some of which are alkenes).

Thermochemical cycle A sequence of chemical reactions (with their enthalpy changes) that convert a reactant into a product. The total enthalpy change of the sequence of reactions will be the same as that for the conversion of the reactant to the product directly (or by any other route). It is also called an energy cycle.

Thermodynamics The branch of chemistry dealing with enthalpy and entropy changes.

Thin-layer chromatography, TLC A chromatography technique where the stationary phase is a glass, metal or plastic sheet coated with a thin layer of a porous solid (usually silica gel or aluminium oxide).

Third-order A reaction is third order with respect to a certain species if the power to which its concentration is raised in the rate equation is 3. A reaction is third order overall if the sum of the powers to which the concentrations of all the species involved in the reaction are raised is 3.

Transition state See **Activated complex**

U

Unified atomic mass unit The standard unit for measuring atomic masses, defined as $\frac{1}{12}$ of the mass of an atom of carbon-12; it is sometimes called the Dalton.

Unsaturated hydrocarbon A compound containing a carbon–carbon double or triple bond, i.e. one to which hydrogen can be added.

V

van der Waals forces Weak, short-range electrostatic attractive forces between uncharged molecules, arising from the interaction of permanent or transient electric dipole moments.

Voltage The 'push' that causes charges to move in an electrical conductor. Also called potential difference.

W

Water of crystallisation Water that is part of the crystal structure of a compound.

Weak acid An acid that is only partially dissociated into ions in solution.

Y

Yield $\text{Yield} = \dfrac{\text{the mass of a product actually obtained}}{\substack{\text{the theoretical mass of product predicted} \\ \text{by the equation}}}$

Yield is usually given as a percentage.

Z

Zero-order A reaction is zero order with respect to a certain species if it is not involved in the rate equation.

Zwitterion A compound with both a permanent positive charge and a permanent negative charge.

Data section

Values of constants

Constant	Value
molar gas constant, R	8.31 J K^{-1} mol^{-1}
Faraday constant, F	9.65 × 10^4 C mol^{-1}
Avogadro constant, L	6.02 × 10^{23} mol^{-1}
electronic charge, e	−1.60 × 10^{-19} C
molar volume of gas, V_m	22.4 dm^3 mol^{-1} (at s.t.p.)
	24.0 dm^3 mol^{-1} (at room conditions)
ionic product of water, K_w	1.00 × 10^{-14} mol^2 dm^{-6} (at room temperature)
specific heat capacity of water, c	4.18 kJ kg^{-1} K^{-1}

Bond energies in diatomic molecules

The following values are exact measurements.

Bond	Energy / kJ mol^{-1}	Bond	Energy / kJ mol^{-1}
H–H	436	Br–Br	193
N≡N	944	I–I	151
O=O	496	H–F	562
P≡P	485	H–Cl	431
S=S	425	H–Br	366
F–F	158	H–I	299
Cl–Cl	242	C≡O	1077

Bond energies in polyatomic molecules

The following values are averaged values from various compounds.

Bond	Energy / kJ mol^{-1}	Bond	Energy / kJ mol^{-1}
C–C	350	O–O	150
C=C	610	O–H	460
C≡C	840	Si–Si	225
C=C (in benzene)	520	Si–H	320
C–H	410	Si–Cl	360
C–Cl	340	Si–O (in SiO_2(s))	460
C–Br	280	Si=O (in SiO_2(g))	640
C–I	240	P–P	200
C–N	305	P–H	320
C=N	610	P–Cl	330
C≡N	890	P–O	340
C–O	360	P=O	540
C=O	740	S–S	265
C=O (in CO_2)	805	S–H	340

Continued from page 453

Bond	Energy / kJ mol^{-1}	Bond	Energy / kJ mol^{-1}
N–N	160	S–Cl	250
N=N	410	S–O	360
N–H	390	S=O	500
N–Cl	310		

Ionisation energies of elements in kJ mol^{-1}

Element	Proton number	First	Second	Third	Fourth
H	1	1310	–	–	–
He	2	2370	5250	–	–
Li	3	519	7300	11800	–
Be	4	900	1760	14800	21000
B	5	799	2420	3660	25000
C	6	1090	2350	4610	6220
N	7	1400	2860	4590	7480
O	8	1310	3390	5320	7450
F	9	1680	3370	6040	8410
Ne	10	2080	3950	6150	9290
Na	11	494	4560	6940	9540
Mg	12	736	1450	7740	10500
Al	13	577	1820	2740	11600
Si	14	786	1580	3230	4360
P	15	1060	1900	2920	4960
S	16	1000	2260	3390	4540
Cl	17	1260	2300	3850	5150
Ar	18	1520	2660	3950	5770
K	19	418	3070	4600	5860
Ca	20	590	1150	4940	6480
Sc	21	632	1240	2390	7110
Ti	22	661	1310	2720	4170
V	23	648	1370	2870	4600
Cr	24	653	1590	2990	4770
Mn	25	716	1510	3250	5190
Fe	26	762	1560	2960	5400
Co	27	757	1640	3230	5100
Ni	28	736	1750	3390	5400
Cu	29	745	1960	3350	5690
Zn	30	908	1730	3828	5980
Ga	31	577	1980	2960	6190
Br	35	1140	2080	3460	4850
Rb	37	403	2632	3900	5080
Sr	38	548	1060	4120	5440

Continued from page 454

Element	Proton number	First	Second	Third	Fourth
Ag	47	731	2074	3361	5000
I	53	1010	1840	3000	4030
Cs	55	376	2420	3300	4400
Ba	56	502	966	3390	4700

Standard electrode potentials at 298 K

Reduction half equation	$E^{\ominus}$ / V	Reduction half equation	$E^{\ominus}$ / V
$Ag^+ + e^- \rightleftharpoons Ag$	+0.80	$MnO_2 + 4H^+ + 2e^- \rightleftharpoons Mn^{2+} + 2H_2O$	+1.23
$Al^{3+} + 3e^- \rightleftharpoons Al$	−1.66	$MnO_4^- + e^- \rightleftharpoons MnO_4^{2-}$	+0.56
$Ba^{2+} + 2e^- \rightleftharpoons Ba$	−2.90	$MnO_4^- + 4H^+ + 3e^- \rightleftharpoons MnO_2 + 2H_2O$	+1.67
$Br_2 + 2e^- \rightleftharpoons 2Br^-$	+1.07	$MnO_4^- + 8H^+ + 5e- \rightleftharpoons Mn2+ + 4H_2O$	+1.52
$Ca^{2+} + 2e^- \rightleftharpoons Ca$	−2.87	$NO_3^- + 2H^+ + e^- \rightleftharpoons NO_2 + H_2O$	+0.81
$Cl_2 + 2e^- \rightleftharpoons 2Cl^-$	+1.36	$NO_3^- + 3H^+ + 2e^- \rightleftharpoons HNO_2 + H_2O$	+0.94
$2HOCl + 2H^+ + 2e^- \rightleftharpoons Cl_2 + 2H_2O$	+1.64	$NO_3^- + 10H^+ + 8e^- \rightleftharpoons NH_4^+ + 3H_2O$	+0.87
$ClO^- + 2H_2O + 2e^- \rightleftharpoons Cl^- + 2OH^-$	+0.89	$Na^+ + e^- \rightleftharpoons Na$	−2.71
$Co^{2+} + 2e^- \rightleftharpoons Co$	−0.28	$Ni^{2+} + 2e^- \rightleftharpoons Ni$	−0.25
$Co^{3+} + e^- \rightleftharpoons Co^{2+}$	+1.82	$[Ni(NH_3)_6]^{2+} + 2e^- \rightleftharpoons Ni + 6NH_3$	−0.51
$[Co(NH_3)_6]^{2+} + 2e^- \rightleftharpoons Co + 6NH_3$	−0.43	$H_2O_2 + 2H^+ + 2e^- \rightleftharpoons 2H_2O$	+1.77
$Cr^{2+} + 2e^- \rightleftharpoons Cr$	−0.91	$HO^{2-} + H_2O + 2e^- \rightleftharpoons 3OH^-$	+0.88
$Cr^{3+} + 3e^- \rightleftharpoons Cr$	−0.74	$O_2 + 4H^+ + 4e^- \rightleftharpoons 2H_2O$	+1.23
$Cr^{3+} + e^- \rightleftharpoons Cr^{2+}$	−0.41	$O_2 + 2H_2O + 4e^- \rightleftharpoons 4OH^-$	+0.40
$Cr_2O7^{2-} + 14H^+ + 6e^- \rightleftharpoons 2Cr^{3+} + 7H_2O$	+1.33	$O_2 + 2H^+ + 2e^- \rightleftharpoons 2H_2O_2$	+0.68
$Cu+ + e^- \rightleftharpoons Cu$	+0.52	$O_2 + H_2O + 2e^- \rightleftharpoons HO_2- + OH^-$	−0.08
$Cu^{2+} + 2e^- \rightleftharpoons Cu$	+0.34	$Pb^{2+} + 2e^- \rightleftharpoons Pb$	−0.13
$Cu^{2+} + e^- \rightleftharpoons Cu^+$	+0.15	$Pb^{4+} + 2e^- \rightleftharpoons Pb^{2+}$	+1.69
$[Cu(NH_3)_4]^{2+} + 2e^- \rightleftharpoons Cu + 4NH_3$	−0.05	$PbO_2 + 4H^+ + 2e^- \rightleftharpoons Pb^{2+} + 2H_2O$	+1.47
$F_2 + 2e^- \rightleftharpoons 2F^-$	+2.87	$SO_4^{2-} + 4H^+ + 2e^- \rightleftharpoons SO_2 + 2H_2O$	+0.17
$Fe^{2+} + 2e^- \rightleftharpoons Fe$	−0.44	$S_2O_8^{2-} + 2e^- \rightleftharpoons 2SO_4^{2-}$	+2.01
$Fe^{3+} + 3e^- \rightleftharpoons Fe$	−0.04	$S_4O_6^{2-} + 2e^- \rightleftharpoons 2S_2O_3^{2-}$	+0.09
$Fe^{3+} + e^- \rightleftharpoons Fe^{2+}$	+0.77	$Sn^{2+} + 2e^- \rightleftharpoons Sn$	−0.14
$[Fe(CN)_6]^{3-} + e^- \rightleftharpoons [Fe(CN)_6]^{4-}$	+0.36	$Sn^{4+} + 2e^- \rightleftharpoons Sn^{2+}$	+0.15
$Fe(OH)_3 + e^- \rightleftharpoons Fe(OH)_2 + OH^-$	−0.56	$V^{2+} + 2e^- \rightleftharpoons V$	−1.20
$2H+ + 2e- \rightleftharpoons H2$	0.00	$V^{3+} + e^- \rightleftharpoons V^{2+}$	−0.26
$2H_2O + 2e^- \rightleftharpoons H_2 + 2OH^-$	−0.83	$VO^{2+} + 2H^+ + e^- \rightleftharpoons V^{3+} + H_2O$	+0.34
$I_2 + 2e^- \rightleftharpoons 2I^-$	+0.54	$VO_2^+ + 2H^+ + e^- \rightleftharpoons VO^{2+} + H_2O$	+1.00
$K^+ + e^- \rightleftharpoons K$	−2.92	$VO_3^- + 4H^+ + e^- \rightleftharpoons VO^{2+} + 2H_2O$	+1.00
$Li^+ + e^- \rightleftharpoons Li$	−3.04	$Zn^{2+} + 2e^- \rightleftharpoons Zn$	−0.76
$Mg^{2+} + 2e^- \rightleftharpoons Mg$	−2.38	$VO_3^- + 4H^+ + e^- \rightleftharpoons VO^{2+} + 2H_2O$	+1.00
$Mn^{2+} + 2e^- \rightleftharpoons Mn$	−1.18	$Zn^{2+} + 2e^- \rightleftharpoons Zn$	−0.76
$Mn^{3+} + e^- \rightleftharpoons Mn^{2+}$	+1.49	$Zn^{2+} + 2e^- \rightleftharpoons Zn$	−0.76

Chemical shift values for ^{13}C NMR (left) and ^1H NMR (right)

Table 1 ^{13}C chemical shift values

Type of carbon	δ / ppm
	0–50
R—C—Cl or Br	30–60
R—C—C— (C=O)	30–65
R—C—C=C	25–50
—C—O— alcohols, ethers, or esters	50–70
C=C	110–160
R—C≡N	100–125
	110–160
R—C— esters or acids (C=O)	160–185
R—C— aldehydes or ketones (C=O)	190–220

Table 2 Chemical shift values for ^1H NMR

Type of proton	δ / ppm
RO**H**	0.5–6.0
Ar–O**H**	4.5–7.0
RC**H**₃, R₂C**H**₂, R₃C**H**	0.9–1.7
RN**H**₂	1.0–5.0
Ar–N**H**₂	3.0–6.0
C**H**₃–Ar, RC**H**₂–Ar, R₂C**H**–Ar	2.3–3.0
R—C(=O)—C—**H**	2.2–3.0
R—O—C—**H**	3.2–4.0
RC**H**₂Cl or Br	3.2–4.0
R—C(=O)—O—C—**H**	3.2–4.0
R C=C **H**	4.5–6.0
H–Ar	6.0–9.0
R—C(=O)—**H**	9.3–10.5
R—C(=O)—O—**H**	9.0–13.0
R—C(=O)—N—**H** (R')	5.0–12.0

Values of electronegativity for elements H to Br

H																	
2.1																	
Li	Be												B	C	N	O	F
1.0	1.6												2.0	2.5	3.0	3.5	4.0
Na	Mg												Al	Si	P	S	Cl
0.9	1.3												1.5	1.9	2.2	2.6	3.0
K	Ca	Sc	Ti	V	Cr	Mn	Fe	Co	Ni	Cu	Zn	Ga	Ge	As	Se	Br
0.8	1.0	1.4	1.5	1.6	1.7	1.5	1.8	1.9	1.9	1.9	1.6	1.8	2.0	2.2	2.6	2.6

Characteristic infrared absorption frequencies in organic molecules

Bond	Functional group	Wavenumber / cm^{-1}
C–O	hydroxy, ester	1040–1300
C=C	aromatic compound, alkene	1500–1680
C=O	amide	1640–1690
	carbonyl, carboxyl	1670–1740
	ester	1710–1750
C≡N	nitrile	2200–2250
C–H	alkane	2850–2950
N–H	amine, amide	3300–3500
O–H	carboxyl	2500–3000
	hydroxy	3200–3600

Index